Plant Biology

Series Editor
Nina Strömgren Allen
Wake Forest University, Department of Biology, Winston-Salem, North Carolina

Titles in the Series

Vol. 1: Phloem Transport
James Cronshaw, William J. Lucas, and Robert T. Gianquinta, *Editors*

Vol. 2: Regulation of Chloroplast Differentiation
George Akoyunoglou and Horst Senger, *Editors*

Vol. 3: Plant Tissue and Cell Culture / C.E. Green D.A. Sommers, W.P. Hackett, and D.D. Biesboaer, *Editors*

Vol. 4: Tomato Biotechnology
Donald J. Nevins and Richard A. Jones, *Editors*

Vol. 5: Plant Cold Hardiness
Paul H. Li, *Editor*

Vol. 6: Second Messengers in Plant Growth and Development
Wendy F. Boss and D. James Morré, *Editors*

Vol. 7: Algae as Experimental Systems / Annette W. Coleman, Linda J. Goff, and Janet R. Stein-Taylor, *Editors*

Vol. 8: Photosynthesis
Winslow R. Briggs, *Editor*

Photosynthesis

Photosynthesis

Proceedings of the C.S. French Symposium on Photosynthesis Held in Stanford, California, July 17–23, 1988

Editor

Winslow R. Briggs
Department of Plant Biology
Carnegie Institution of Washington
Stanford, California

Alan R. Liss, Inc., New York

Address all Inquiries to the Publisher
Alan R. Liss, Inc., 41 East 11th Street, New York, NY 10003

Printed in the United States of America

The publication of this volume was facilitated by the authors and editors who submitted the text in a form suitable for direct reproduction without subsequent editing or proofreading by the publisher.

Library of Congress Cataloging-in-Publication Data

C.S. French Symposium (1988 : Stanford, Calif.)
Photosynthesis : proceedings of the C.S. French Symposium held in Stanford, California, July 17-23, 1988 / editor, Winslow R. Briggs.
p. cm. -- (Plant biology ; v. 8)
Includes bibliographies and index.
ISBN 0-8451-1807-2
1. Photosynthesis--Congresses. I. French, Charles Stacy, 1907- . II. Briggs, Winslow R. III. Title. IV. Series.
QK882.C24 1988
581.1'3342--dc20 89-8332
CIP

Contents

Contributors

Claude Alban, Laboratoire de Physiologie Cellulaire Végétale, UA CNRS n576, Département de Recherche Fondamentale, Centre d'Etudes Nucléaires et Université Joseph Fourier, 85 X, F-38041 Grenoble, France **[331]**

J. Amesz*, Department of Biophysics, Huygens Laboratory, University of Leiden, 2300 RA Leiden, The Netherlands **[123]**

Jan M. Anderson*, CSIRO, Division of Plant Industry, Box 1600, Canberra ACT 2601, Australia **[161]**

Rina Ariel, Life Science Institute, The Hebrew University of Jerusalem, 91904 Jerusalem, Israel **[243]**

John Biggins*, Section of Biochemistry, Brown University, Providence, RI 02912 **[363]**

Olle Björkman*, Department of Plant Biology, Carnegie Institution of Washington, Stanford, CA 94305 **[45]**

Maryse A. Block, Laboratoire de Physiologie Cellulare Végétale, UA CNRS n576, Département de Recherche Fondamentale, Centre d'Etudes Nucléaires et Université Joseph Fourier, 85 X, F-38041 Grenoble, France **[331]**

Winslow R. Briggs*, Department of Plant Biology, Carnegie Institution of Washington, Stanford, CA 94305-1297 **[xv,xvii,xix]**

Doug Bruce, Department of Biology, Brock University, St. Catherines, Ontario L2S 3A1, Canada **[363]**

B.B. Buchanan*, Division of Molecular Plant Biology, University of California at Berkeley, Berkeley, CA 94720 **[425]**

Gisela G. Chiang, Department of Biological Sciences, Purdue University, West Lafayette, IN 47907 **[437]**

Nam-Hai Chua, Laboratory of Plant Molecular Biology, The Rockefeller University, New York, NY 10021-6399 **[257]**

N.A. Crawford, Division of Molecular Plant Biology, University of California at Berkeley, Berkeley, CA 94720 **[425]**

Franz-Christian Czygan, Institut für Botanik und Pharmazeutische Biologie der Universität Würzburg, 8700 Würzburg, Federal Republic of Germany **[375]**

The numbers in brackets are the opening page numbers of the contributors' articles.
*Denotes a speaker at the symposium.

Barbara Demmig-Adams*, Institut für Botanik und Pharmazeutische Biologie der Universität Würzburg, 8700 Würzburg, Federal Republic of Germany; present address: Department of Environmental, Population, and Organismic Biology, University of Colorado, Boulder, CO 80309-0334 **[375]**

Xing-wang Deng, Department of Botany, University of California at Berkeley, Berkeley, CA 94720 **[269]**

Richard A. Dilley*, Department of Biological Sciences, Purdue University, West Lafayette, IN 47907 **[437]**

Roland Douce, Laboratoire de Physiologie Cellulare Végétale, UA CNRS n576, Département de Recherche Fondamentale, Centre d'Etudes Nucléaires et Université Joseph Fourier, 85 X, F-38041 Grenoble, France **[331]**

Susan E. Douglas, Department of Botany, University of Georgia, Athens, GA 30602; present address: Atlantic Research Laboratory, Halifax, Nova Scotia B3H 3Z1, Canada **[311]**

M. Droux, Division of Molecular Plant Biology, University of California at Berkeley, Berkeley, CA 94720 **[425]**

John R. Evans, Plant Environmental Biology Group, Research School of Biological Sciences, Australian National University, Box 475, Canberra, ACT 2601 Australia **[183]**

C.B. Field, Department of Plant Biology, Carnegie Institution of Washington, Stanford, CA 94305 **[19]**

Devorah Friedberg, Life Science Institute, The Hebrew University of Jerusalem, 91904 Jerusalem, Israel **[243]**

Patricia E. Gamble, Department of Biochemistry and Biophysics, Texas A&M University, College Station, TX 77843-2128 **[285]**

Anthony A. Gatenby*, Central Research and Development Department, Experimental Station, E.I. DuPont Nemours and Company, Wilmington, DE 19880-0402 **[299]**

Pamela B. Gibbs, Section of Biochemistry, Brown University, Providence, RI 02912 **[363]**

Pierre Goloubinoff, Central Research and Development Department, Experimental Station, E.I. DuPont de Nemours and Company, Wilmington, DE 19880-0402 **[299]**

Govindjee*, Departments of Physiology and Biophysics, and Plant Biology, University of Illinois at Urbana-Champaign, Urbana, IL 61801 **[1,71]**

Wilhelm Gruissem*, Department of Botany, University of California at Berkeley, Berkeley, CA 94720 **[269]**

Ulrich Heber*, Institute of Botany and Pharmaceutical Botany, University of Würzburg, D-87 Würzburg, Federal Republic of Germany **[359]**

Peter Horton*, Department of Biochemistry and Research Institute for Photosynthesis, University of Sheffield, Sheffield, S10 2TN, United Kingdom **[393]**

Jacques Joyard*, Laboratoire de Physiologie Cellulaire Végétale, UA CNRS n576, Département de Recherche Fondamentale, Centre d'Etudes Nucléaires et Université Joseph Fourier, 85 X, F-38041 Grenoble, France **[331]**

Aaron Kaplan*, Life Science Institute, The Hebrew University of Jerusalem, 91904 Jerusalem, Israel; present address: Department of Botany, The Hebrew University of Jerusalem, 91904 Jerusalem, Israel **[243]**

Kenneth Keegstra*, Department of Botany, University of Wisconsin at Madison, Madison, WI 53706 **[347]**

Robert R. Klein, Department of Biochemistry and Biophysics, Texas A&M University, College Station, TX 77843-2128 **[285]**

N.S. Kosower, Division of Molecular Plant Biology, University of California at Berkeley, Berkeley, CA 94720 **[425]**

Almuth Krüger, Institut für Botanik und Pharmazeutische Biologie der Universität Würzburg, 8700 Würzburg, Federal Republic of Germany **[375]**

Cris Kuhlemeier*, Laboratory of Plant Molecular Biology, The Rockefeller University, New York, NY 10021-6399 **[257]**

Richard C. Leegood*, Research Institute for Photosynthesis and Department of Plant Sciences, University of Sheffield, Sheffield, S10 2TN, United Kingdom **[457]**

George H. Lorimer, Central Research and Development Department, Experimental Station, E.I. DuPont de Nemours and Company, Wilmington, DE 19880-0402 **[299]**

Thomas H. Lubben, Central Research and Development Department, Experimental Station, E.I. DuPont de Nemours and Company, Wilmington, DE 19880-0402 **[299]**

Mitsue Miyao, National Institute for Basic Biology, Myodaiji, Okazaki 444, Japan **[59]**

H.A. Mooney*, Department of Biological Sciences, Stanford University, Stanford, CA 94305 **[19]**

Ishita Mukerji, Department of Chemistry and Chemical Biodynamics Division, Lawrence Berkeley Laboratory, University of California at Berkeley, Berkeley, CA 94720 **[105]**

John E. Mullet*, Department of Biochemistry and Biophysics, Texas A&M University, College Station, TX 77843-2128 **[285]**

Norio Murata*, National Institute for Basic Biology, Myodaiji, Okazaki 444, Japan **[59]**

John M. Norman*, Department of Soil Science, University of Wisconsin at Madison, Madison, WI 53706 **[227]**

Charles Barry Osmond*, Botany Department, Duke University, Durham, NC 27706 **[5]**

William W. Parson*, Department of Biochemistry, University of Washington, Seattle, WA 98195 **[135]**

Robert W. Pearcy*, Department of Botany, University of California at Davis, Davis, CA 95616 **[407]**

John Pierce*, Agricultural Products Department, Experimental Station, E.I. DuPont de Nemours and Company, Inc., Wilmington, DE 19880-0402 **[149]**

F. Gerald Plumley, Department of Botany, University of Georgia, Athens, GA 30602 **[311]**

Wayne Polley, USDA-ARS, Temple, TX 76502 **[227]**

Leonora Reinhold, Life Science Institute, The Hebrew University of Jerusalem, 91904 Jerusalem, Israel **[243]**

Kenneth Sauer*, Department of Chemistry and Chemical Biodynamics Division, Lawrence Berkeley Laboratory, University of California at Berkeley, Berkeley, CA 94720 **[105]**

Gregory W. Schmidt*, Department of Botany, University of Georgia, Athens, GA 30602 **[311]**

Rakefet Schwarz, Life Science Institute, The Hebrew University of Jerusalem, 91904 Jerusalem, Israel **[243]**

Jeffrey R. Seemann*, Department of Biochemistry, University of Nevada, Reno, NV 89557 **[183]**

Julie Seijffers, Life Science Institute, The Hebrew University of Jerusalem, 91904 Jerusalem, Israel **[243]**

Alison Branagan Switzer, Department of Botany, University of Georgia, Athens, GA 30602 **[311]**

Ichiro Terashima*, Plant Environmental Biology Group, Research School of Biological Sciences, Australian National University, Box 475, Canberra, ACT 2601, Australia; present address: Department of Botany, Faculty of Science, University of Tokyo, Hongo, Tokyo 113, Japan **[207]**

William W. Thomson, Department of Botany and Plant Sciences, University of California, Riverside, CA 92521 **[161]**

John Tonkyn, Department of Botany, University of California at Berkeley, Berkeley, CA 94720 **[269]**

M.R. Wasielewski, Chemistry Division, Argonne National Laboratory, Argonne, IL 60439 **[71]**

Klaus Winter, Institut für Botanik und Pharmazeutische Biologie der Universität Würzburg, 8700 Würzburg, Federal Republic of Germany **[375]**

Ian E. Woodrow*, Bureau of Rural Resources, Department of Primary Industries and Energy, Canberra, ACT 2600, Australia **[475]**

PREFACE

A group of former fellows of the Department of Plant Biology of the Carnegie Institution of Washington were holding a reunion of sorts at a local establishment during the course of the VIIth International Congress on Photosynthesis being held at Brown University, Providence, Rhode Island, during the summer of 1986, when it occurred to one of them that some time in the not too distant future, Charles Stacy French, who had been director of the Department from 1947 to 1973, would be celebrating his 80th birthday. The suggestion was made that perhaps the Department of Plant Biology might be interested in sponsoring a symposium to honor this event, perhaps to be held in the summer of 1988. This suggestion immediately found its way onto a paper, and in due course, the present Director received in the mail a slightly crumpled letter suggesting such a symposium. The signatories were in order William Hagar, Aaron Kaplan, Norio Murata, John Coleman, PraSanna Mohanty, Ulrich Schreiber, Murray Badger, John Downton, Graham Farquhar, Mervyn Ludlow, Barbara Demmig, and Barry Osmond. The Director promptly took up the matter with the other Carnegie Faculty members - Jeanette Brown, Joseph Berry, Olle Björkman, Christopher Field, David Fork, and Arthur Grossman, and the suggestion was enthusiastically adopted. The notion was to present a very broad integrated program to bring together highly creative people whose interests and expertise in photosynthesis ranged from the most biophysical to the most global, and to provide the speakers and other participants the best possible opportunity for cross fertilization. We almost immediately found that if we put into the program all of the people we could identify as potential speakers, we would have a program that would last at least five weeks. We obviously had to make some difficult choices! The list was narrowed down, and we were immensely gratified that virtually all of the speakers we invited responded with enormous enthusiasm - including a small handful who had other commitments that preventing them from coming. The volume that follows is the final physical consequence of the meeting. We feel that both the meeting and the volume go a long way toward meeting our objectives, but the reader will obviously have to judge for him- or herself.

We respectfully dedicate this book to Charles Stacy French on the occasion of his 80th birthday, December 13, 1987.

Winslow R. Briggs
March 8th, 1989

ACKNOWLEDGEMENTS

We acknowledge with thanks support from Sandoz Crop Protection (Zoecon Research Institute), Monsanto Company, E. I. duPont de Nemours & Company, and Merck and Company. We are also grateful to the United States Department of Agriculture Competitive Research Grants Office (grant 88-37234-3492) and the Department of Energy (grant DE-FG05-88ER13953) for partial funding. Without this support the meeting simply could not have taken place.

We also thank Ann McKillop, Lynn Collatz, and Ann Briggs for their generous help during the meeting. Very special thanks are due to Mary Smith for the making all of the administrative arrangements, arranging the non-scientific parts of the meeting, and keeping the meeting running smoothly. Finally we owe an enormous amount to Loretta Tayabas who converted everybody's manuscript from their own word-processing language (have you any idea how many there are out there?) into MS Word, taking care of innumerable formatting and make-up problems, and giving us the final camera-ready copy on which this book is based.

INTRODUCTION

Winslow R. Briggs

Department of Plant Biology, Carnegie Institution of Washington, 290 Panama St., Stanford, CA 94305

This symposium, dedicated to Charles Stacy French on the occasion of his 80th birthday, was designed with one over-riding scientific objective - to bring together a group of active researchers representing a wide range of approaches to an understanding of photosynthesis, put them together, and hope that the chemistry would be right to encourage a broadly ranging interdisciplinary discussion and exchange of information.

Any given year may produce more than one conference on photosynthesis, and attendance is frequently international. Indeed there is an International Congress on Photosynthesis every three years. However, the smaller meetings tend to be highly focused - for example on carbon metabolism, on molecular biology of regulation of chloroplast proteins, on Photosystem II, or on photoinhibition. Even at international congresses, at which presumably all of the various diverse subdisciplines are represented, people gravitate of necessity to sessions close to their own particular interests. Unless the precise structure of a photosynthetic reaction center as revealed by X-ray crystallography is offered as a plenary lecture, a physiological ecologist would probably not be able to hear it, feeling compelled to keep up-to-date in his or her own special area.

It was the hope of the faculty of the Department of Plant Biology of the Carnegie Institution of Washington in organizing this meeting that these deficiencies could be remedied. The basic hypothesis was that if you provided the proper ingredients - good speakers doing first rate work - and arranged things so that a molecular biologist, for example, would both learn from and educate a modeler on the one hand and a biophysicist on the other, some transfer of viewpoint might take place. Thus the objectives in organizing the sequence of presentations were partially sociological. The organization of the flow of the sessions was slightly Machiavellian: each day was as mixture of topics representing several different biological levels of organization. The reasoning was that on any given half day, everybody would find something both within and without their scope of interest, and would hence be more likely to hear a talk outside their own special field than if the topics were less mixed. We weren't at all certain that this plan would work, but felt that it was well worth attempting.

In thinking and planning in this fashion, we almost certainly made extraordinary demands of our speakers and participants. They responded splendidly: attendance was extremely high from beginning to end, most speakers made special effort to reach outside of their immediate disciplines in their presentation, and the quality and quantity of interaction both within and outside of the meeting session exceeded our most optimistic expectations.

The arrangement of the articles that follow is the consequence of trying to meet scientific rather than sociological objectives. Two chairpersons provided us with short introductory essays - one by Govindjee that is a delightful vignette about Stacy French, and the other a thoughtful preamble by Ulrich Heber to one of the sessions on regulation. Both of these essays clearly belong in this volume. The first three research articles take three quite different broad perspectives: integrative approaches from molecule to biosphere; extrapolation of studies of plants or small plant communities to a global scale; and investigations of ways in which plants respond to stress. The next group considers structure and function from the level of the reaction center to the level of the whole canopy. A paper on oxygen evolution is followed by two contributions considering early events in Photosystem II and Photosystem I, respectively. Consideration of energy capture and storage by photosynthetic bacteria precedes a paper on the structure of bacterial reaction centers. A detailed structure-function study of Rubisco is then followed by a series of papers on organization: at the level of the thylakoid membrane, at the level of nitrogen partitioning, at the level of the whole leaf, and at the level of the entire canopy.

A paper on the CO_2 concentrating mechanism of cyanobacteria is followed by a series of papers on regulation - at the level of the nuclear gene, at the level of the chloroplast gene, at the level of translation, at the level of enzyme assembly, at the level of the assembly of multi-peptide chlorophyll-protein complexes, at the level of membrane biogenesis, and at the level of whole chloroplast biogenesis. These papers are followed by a series that discusses regulation in other senses - the distribution of excitation energy as a function of light spectral quality; regulation of the dissipation of excess light energy; regulatory cross talk between electron transport and carbon flow; regulation of CO_2 assimilation in the face of sudden dramatic changes in photon fluence rate; regulation of the activation state of chloroplast enzymes; regulation of local proton gradients in thylakoids; and regulation by phosphate of carbon flow to sucrose. The volume concludes with detailed mathematical analysis of limitation to CO_2 fixation that includes biochemical, physiological, and environmental parameters.

It is our hope that these contributions will highlight the continuity of efforts to understand photosynthesis from electron to ecosystem, and will illustrate the great importance of studying the partial processes of photosynthesis through several levels of organization.

Photosynthesis, pages 1–3

MY ASSOCIATION WITH STACY FRENCH

Govindjee

Departments of Physiology and Biophysics, and Plant Biology, University of Illinois at Urbana-Champaign, 289 Morrill Hall, 505 South Goodwin Avenue, Urbana, IL 61801

An exciting series of articles on the form and function of the photosynthetic apparatus is presented in this book. This is an area in which Stacy French has made enormous contributions (French, 1979). In particular, Stacy's pioneering work (see *e.g.* French, 1971) on the spectral forms of chlorophyll *a* and their functions continues to be at the cornerstone of discussions in photosynthesis.

I shall fail in my job as chairman of the session on the "Form and Function of the Photosynthetic Apparatus" if I did not include the remarks I made, about my association with Stacy, at the time of the conference.

I had heard about Stacy when I was a student of Robert Emerson and Eugene Rabinowitch from 1956-1960 (first two years with Emerson until his death, and then with Rabinowitch). During that time, I had heard about the French press, of course, but I thought it was one of those American things like the French toast! Emerson, however, educated me on its origin, *i.e.*, its creation/invention by Stacy French. Rabinowitch, on the other hand, instructed me on Stacy's work on fluorescence and energy transfer (Franck *et al.*, 1941; French and Young, 1952). I was already aware of Stacy's work on the Hill reaction (French *et al.*, 1946), but it was Rabinowitch who told me that it was French who had named the chloroplast reaction, discovered by Robin Hill, as the Hill reaction (see French, 1979).

In 1960, I published (Govindjee and Rabinowitch, 1960) action spectra of the Emerson Enhancement Effect in the green alga *Chlorella* and the diatom *Navicula* showing peaks in the chlorophyll *a* (Chl *a*) region at 670nm, suggesting that some Chl *a* belonged to the short-wave (accessory or auxiliary pigment) system (the PSII of today). Later, I saw papers from Stacy's laboratory showing a similar peak (see *e.g.* French *et al.*, 1960). Being inexperienced in the ways of scientific life, I thought, wrongly, of course, that this was the end of my scientific career since my work was not recognized and cited by an authority in the field. I thought that the award of my Ph.D. was threatened since the University of Illinois requires "original" work for Ph.D. I discussed this with Rabinowitch, my thesis advisor. He was quite cool about it and said why don't you write a letter to French. I wrote, perhaps, a very stupid letter to Stacy; but, I remember receiving a very kind and wonderful reply. I believe it said something about independent observations in science. I was relieved and decided that I must work with this wonderful person sometimes in the future.

When Rajni (my wife) and I were at the Carnegie Institution of Washington at Stanford in the summer of 1963, we learned from Stacy how doing experiments can be really fun, and it was fun. We enjoyed the use of the automatic action spectrophotometer he had constructed; Stacy also helped us do the experiments on oxygen evolution in the red alga *Porphyridium* in alternating light flashes (Govindjee and Govindjee, 1965). His input in this paper was greater than mine, and he should have been a coauthor of this paper. It was indeed easy and fun to work with Stacy. He never looked over our shoulders when we were doing experiments, and, I have followed this rule in my life with my own graduate students.

On the lighter side, I learned during the 1963 visit that the Carnegie Inst. of Wash. (as it was listed in the telephone directory) was sometimes mistaken by local residents to be a laundry service! A few years ago, I received an interesting letter from H.I. Virgin of Sweden. He described an anecdote about the discovery of a new fluorescence band which appeared and disappeared depending upon Stacy's presence or absence in the room. Its origin was ultimately traced to Stacy's fat cigar. We need not worry about such emission bands as Stacy no longer smokes cigars.

Finally, it is association with people like Stacy that makes science so much fun. Happy 80th Birthday Stacy.

REFERENCES

Franck J, French CS, Puck TT (1941). Fluorescence of chlorophyll and photosynthesis. J Phys Chem 45:1268-1300.

French CS (1971). The distribution and action in photosynthesis of several forms of chlorophyll. Proc Natl Acad Sci USA 68: 2893-2897.

French CS (1979). Fifty years of Photosynthesis. Annu Rev Plant Physiol 30:1-26.

French CS and Young VMK (1952). The fluorescence of red algae and the transfer of energy from phycoerythrin to phycocyanin and chlorophyll. J Gen Physiol 35:873-890.

French CS, Holt AS, Powell RD, Anson ML (1946). The evolution of oxygen from illuminated suspensions of frozen, dried and homogenized chloroplasts. Science 103:505-506.

French CS, Myers J, McLeod GC (1960). Automatic recording of photosynthesis action spectra used to measure the Emerson Enhancement. Symp Comp Biol 1:361-365.

Govindjee, Govindjee R (1965). Two different manifestations of enhancement in the photosynthesis of *Porphyridium cruentum* in flashing monochromatic light. Photochem Photobiol 4:401-415.

Govindjee and Rabinowitch E (1960). Two distinct forms of chlorophyll *a in vivo* with distinct photochemical functions. Science 132:355-356.

Photosynthesis, pages 5–17

PHOTOSYNTHESIS FROM THE MOLECULE TO THE BIOSPHERE: A CHALLENGE FOR INTEGRATION

Charles Barry Osmond

Botany Department, Duke University, Durham, NC 27706 USA

INTRODUCTION

Photosynthesis is taken for granted, and as such research in photosynthesis rarely attracts the support or the recognition commensurate with its role in sustaining all other biospheric activities. Yet 1988 may come to be regarded as a turning point in the popular appreciation of autotrophy. On the one hand a Nobel Prize in Chemistry has been awarded to Diesenhoffer *et al.* (1985) for structural elucidation of the photosynthetic reaction center of *Rhodopseudomonas viridis*. On the other, gathering consensus as to inexorable global climatic change, mediated in part by increasing atmospheric CO_2 concentration, with implications for photosynthetic performance throughout the biosphere, is slowly dawning on everyman. In these circumstances it seems particularly appropriate to celebrate the very special contributions of Stacy French (and of the Department of Plant Biology under his leadership) to photosynthetic research; contributions which extend from the molecule to the biosphere.

It is difficult to identify a more discreet source of creative ideas, of key observations and integration in modern photosynthetic research. As such it is presumptuous for an outsider to comment on the continuing need for integration in photosynthetic research when it is done so well at Carnegie. However, as much of which is presented in this paper derives from collaborations with Carnegie staff members, it may reasonably reflect the source. The Carnegie-Canberra axis (Fig. 1), which was built up following the visits of Anderson to Carnegie and Björkman to Canberra, led to a well-documented pulse of contributions in photosynthetic research, ranging from thylakoid membrane architecture to ecophysiology. It has been my immense good fortune, personally and professionally, to have participated in a little of this action, and to have had a hand in deflecting some of the impact back to the desert, from whence the Department of Plant Biology came.

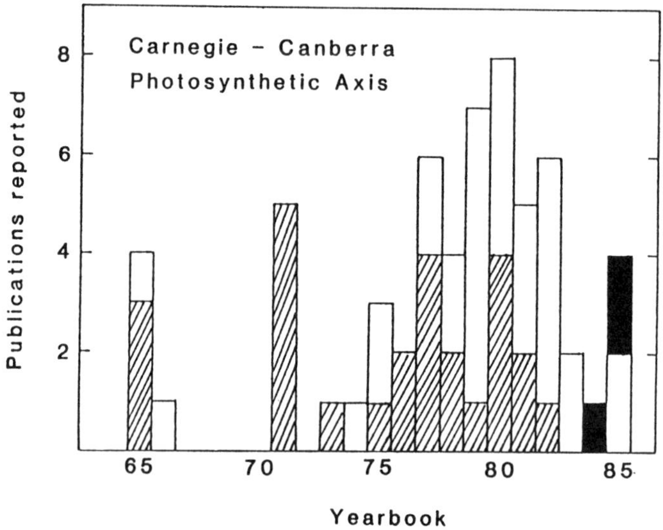

Figure 1. Annual number of research papers authored jointly by staff of the Department of Plant Biology, in collaboration with staff and postdoctoral fellows from the Division of the Plant Industry, CSIRO and the Department of Environmental Biology Research School of Biological Sciences ANU, as reported in Yearbook of the Carnegie Institution of Washington 1965-85: open bars, other journals; hatched bars, Carnegie Yearbook; black bars, collaborations with Desert Research Institute, Reno.

In this reflective context it seems natural to seek the roots which sustain such flowering, and to ask if it is likely to be more than ephemeral. In my view the roots lie in a philosophy which seeks to "encourage academic excellence without danger from the almost inevitable degradation that follows from the spirit of too much competition" (French, 1979). I believe that Stacy French endorses an atmosphere for research in which serendipity has a role and in which the spirit of the quest itself is paramount. Our keepers of today question whether the public purse can afford such attitudes. They do not question the cost of the bureaucratization of science; nor do they cost the waste of competition. Clearly, the measure of independence enjoyed by the Carnegie Institution, and Stacy French's philosophy of collaboration in research across national and cultural boundaries, rather than in local competitions, have provided unique opportunities for integration of photosynthetic research.

There are parallels between this philosophy of the practice of research and the framework we use to pose questions. Thus, just as it may be more profitable to promote photosynthetic research by engaging researchers from widely differing human environments in a noncompetitive gameplay, extension of this philosophy at Carnegie to the study of plants from contrasting environments has uncovered fascinating differences in photosynthetic processes. However, we should not loose sight of the fact, that these plants, like the researchers, are rarely in competition with each other. It may be unwise to assume the processes elucidated are cornerstones of competitive interactions for plant production or survival. Surely plant success in a stochastic environment has as much to do with opportunity, and the art of compromise, as does the success of homeostatic heterotrophs in the creative pursuit of understanding. Perhaps then, the circumstances in which one has been privileged to do research lie behind my question of whether application of strict philosophy of fitness, based on fine tuning of photosynthetic processes to the environment, and worked out through the competition and the economics of resource capture, is the best framework for integration. There is overwhelming evidence that photosynthetic processes often make the difference between really interesting plants, and the mere vegetable. There is much less evidence that fine tuning of photosynthesis makes the difference between life and death to a plant, and perhaps this is why photosynthesis really deserves to be taken for granted.

A FRAMEWORK FOR INTEGRATION

Integration in photosynthetic research, like that in any other science, begins with observations of the state of things over a wide range of scale. The range of scale is most conveniently expressed in terms of powers of ten, of orders of magnitude (Morrison *et al.*, 1982). According to de Wit (1970) individuals tend to develop an observational expertise defined by about 3 orders of magnitude in scale. One needs different tools to describe the state of things at different scales, but occasionally the numbers obtained at vastly different scales imply an underlying importance. For example, Anderson and Osmond (1987), compared the gradient in light transmission through a 55 m deep rainforest (Yoda, 1974) with that through a 400 μm leaf of *Camellia* (Terashima and Saeki, 1983). Though differing in scale, approximately 138,000 x, these analyses promote analogies between the photosynthetic properties of leaves of seedlings on the rainforest floor, and properties of chloroplasts in the lowermost mesophyll cells of *Camellia* leaves (Fig. 2). These analogies also seem to be born out by other state descriptions (Fig. 3), based on the relationships between chlorophyll *a/b* ratio and the ultrastructure of membranes in chloroplasts through a spinach leaf (Terashima and Inoue, 1985) or in chloroplasts from leaves of plants in a range of habitats from sun to shade (Aro *et al.*, 1986). Whether this amounts to vertical integration or lateral thinking is a matter of opinion. However, observers

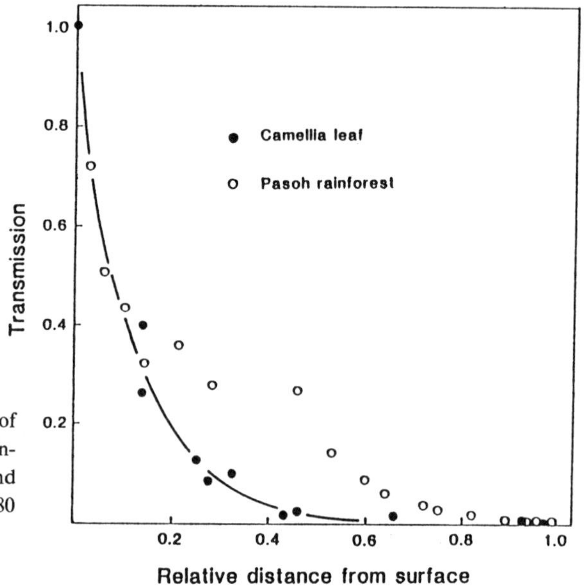

Figure 2. Transmission of light through a tropical rainforest (400-700 nm) and through a *Camellia* leaf (680 nm).

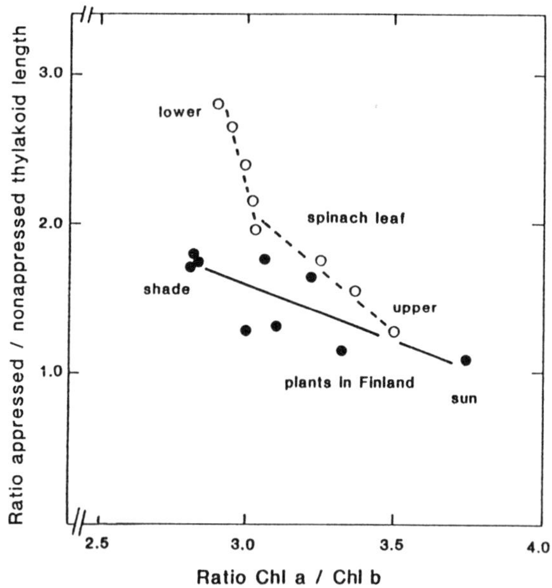

Figure 3. Correlation between chloroplast ultrastructure and chlorophyll *a/b* ratio in leaves of plants from sun and shade habitats, and in chloroplasts through a spinach leaf.

are wont to seek the functional significance of such relationships and strive to observe the key processes (the changes in state with time) that belie these states. The relaxation time of processes, that is the half time for a state to return to its original condition after perturbation, seems to be naturally linked with the scale of things by some unexplained nexus between seconds and microns (Fig. 4). The staircase of natural subdisciplines in photosynthetic research which emerges owes much to Kamen (1964), who may have been the first to describe the continuum of photosynthetic processes defined by their relaxation times, and to de Wit (1970) who used relaxation times to distinguish levels of integration in productivity modeling. The nexus between size and relaxation time was developed during protracted exchanges between Osmond, Björkman, and Anderson (1980), in the heyday of the Carnegie-Canberra axis. It's present 2-dimensional structure is the result of two further iterations

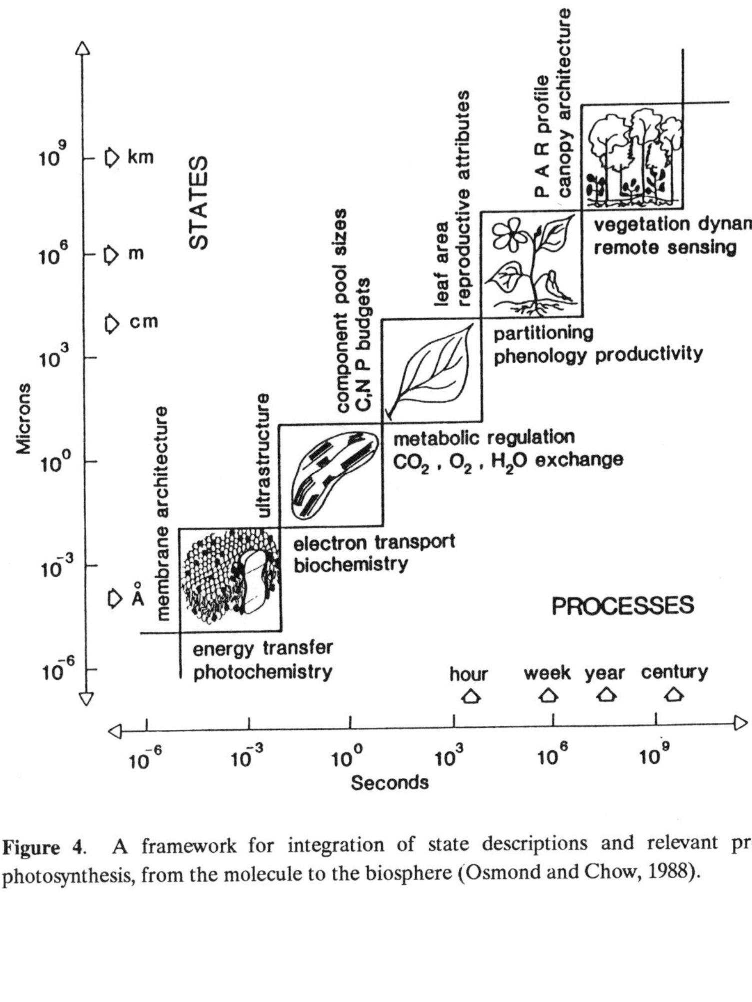

Figure 4. A framework for integration of state descriptions and relevant processes in photosynthesis, from the molecule to the biosphere (Osmond and Chow, 1988).

(Osmond, 1987; Osmond and Chow, 1988) prompted by a scintillating afternoon with Joe Berry, between connections at San Francisco airport. The simple truisms of this framework are widely encountered in the prospectus for interdisciplinary research, from landscape ecology to earth system science.

The potential merits of this framework for integration have been discussed elsewhere (Osmond *et al.*, 1980; Osmond, 1987). It probably can do more than give individual researchers a better picture of where they most commonly operate in the scheme of things. For some it will provide a clear reason to confine oneself to a particular patch defined by three or more orders of ten. As de Wit (1970) suggested, it may conserve computing power during integration of the daily grind of carbon acquisition and water loss in productivity modeling. It might help define the scales of stochastic and potentially limiting environmental variables in relationship to key physiological processes. In the integration process one must hope to discover economical insights (as opposed to the insights of economics) which help dispense with the less important. Usually one seeks to identify small scale processes which serve as indicators or intergrators of larger scale processes (and which allow some relaxation time for the investigator!). For example, applications of chlorophyll fluorescence analyses as indicators of photosynthesis under stress, and of carbon isotope analyses as integrators of CO_2 and H_2O exchange, show considerable promise.

INTEGRATING PHOTOSYNTHETIC RESPONSES TO LIGHT

There has been enormous progress in the molecular through to biospheric integration of the way the photosynthetic apparatus responds to light. At the molecular level we have witnessed many methodological, and conceptual breakthroughs in description of the state of things, which suggest a wide range of dynamic and purposeful changes. (Anderson and Anderson, 1988) These enthusiasms need to be tempered by the limited evidence for dynamic responses *in vivo*. One is reminded for example, that something as well established as granal stacking, which can be so dynamic *in vitro*, is not very dynamic *in vivo* (Terashima *et al.*, 1986) and presents a difficult case to argue functionally, in terms of light capture (Evans, 1986). Phenomena such as LHCII phosphorylation (Bennett *et al.*, 1980), which seem so significant *in vitro*, may have scarcely detectable consequences when put to test *in vivo* (Demmig *et al.*, 1987; Horton and Hague, 1988). Nevertheless, the quest to improve our understanding of processes is driven by the limits of our imagination when applied to state descriptions at scales some orders of magnitude lower than those we can investigate *in vivo*.

There is now reasonable agreement as to the changes in states and processes of the photosynthetic apparatus which accompany adjustment (acclimation?) to modest changes in growth irradiance in mere vegetables such as spinach and radish (Anderson and Osmond, 1987). However wild plants remain full of surprises. The archetypical shade plant, *Alocasia macrorrhiza*, has shown an unexpected capacity for adjustment over a greater range of growth irradiance (Chow *et al.*, 1988). Moreover, it seems to apply different rules for the negotiation of membrane and stromal protein nitrogen in the process (Seeman *et al.*, 1987). Admittedly, data from both laboratories are distinguished by the scatter of points, but this may reflect more our inexperience with cultivation of the plants than anything else. Although *Alocasia* obviously luxuriates in the low light of the rainforest floor, it is also found cohabitating rainforest margins with *Panicum maximum*, an archetypical, light-demanding C_4 pathway plant. Our conventional wisdoms as to the fine tuning of the photosynthetic apparatus to light, and its relationship to plant distribution, seem to need significant qualification in these and several other instances (PE Kriedeman and WA Thompson unpublished). One can imagine, for example, that the capacity to adjust respiratory costs in the shade, and the capacity to dissipate excess excitation, or to repair damage resulting from excess excitation, represent important and different compromises for *Alocasia* at the extremes of its distribution with respect to irradiance.

Things seem to be much worse at sea, however. It is difficult to categorize the many perplexing patterns of photosynthetic adjustment to irradiance in micro-algae (Richardson *et al.*, 1983). The significance of some of the most intriguing molecular insights has been questioned by ecophysiological observation. Saffo (1987) summarized present misgivings as the role of chromatic adaptation in marine macrophyte zonation with a quotation from Steinbeck and Ricketts in "The Sea of Cortez":

"There is one great difficulty with a good hypothesis. When it is completely rounded, the corners smooth and the content cohesive and coherent, it is likely to become a work of art... One hates to disturb it. Even if subsequent information should shoot a hole in it, one hates to tear it down because it once was beautiful and whole ..."

So far as understanding the ways plants deal with excess excitation is concerned, we may be now approaching a beautiful hypothesis, but the approach has been anything but smooth. Photoinhibition is an anathema to all but the most pessimistic researcher in photosynthesis. Yet understanding photoinhibition is an important approach to defining the intrinsic limits to photosynthetic efficiency. Tracking hypotheses and evidence in this field over the last decade is testimony to the dictum (French, 1967) that:

"In one sense no scientific problem has ever been completely solved, and in another sense, in terms of the state of knowledge at the time and the type of solution that would then have had meaning, most scientific questions have found their answer again and again - but each time within the limits of contemporary knowledge and vocabulary."

In recent times the vocabulary used to describe photoinhibitory damage revealed spirited disagreement as to the involvement of events on the oxidizing or reducing side of the PSII reaction center (Kyle *et al.*, 1984; Cleland and Critchley, 1985). It is fascinating to observe how rapidly the state description of Diesenhoffer *et al.* (1985) led to reconciliation and compromise in this contest (Arntz and Trebst, 1986). Much remains to be resolved however, and our recent attempts (Chow *et al.*, 1988) to integrate changes in key states and processes throughout the lower three black boxes of Fig. 4 are summarized in Fig. 5. This figure describes the onset of processes of photoprotection and photoinhibitory damage in leaves of a shade grown (120 μmol photons m^{-2} s^{-1}), cold tolerant vegetable (spinach), exposed to excess excitation (1500 μmol photons m^{-2} s^{-1}) in air at low temperature (10°C), and their reversal in air at 25°C in weak light (5 μmol photons m^{-2} s^{-1}). Explanations of the larger scale processes have been sought in changes in the state of the thylakoid membrane.

Under the conditions of these and similar experiments (Ögren and Öquist, 1984; Björkman, 1987) it seems that the efficiency of primary photochemistry of PSII declines rapidly during excess excitation as indicated by changes in F_v/F_m and fluorescence yield (Fig. 5c, d). This change is due to both photoinhibitory damage of the reaction centers, indicated by an increase in F_o (Fig. 5d), and a decrease in concentration of functional PSII centers (Fig. 5b), and to an increase in photoprotection due to nonphotochemical quenching (qNP=qE of Schreiber and Bilger, 1987); to an increase in excitation dissipation as heat (Fig. 5e). In this experiment the capacity for photoprotection was exceeded, because photoinhibitory quenching (qI of Schreiber and Bilger, 1987) was soon observed. The recovery kinetics are not clearly defined. Moreover, nonphotochemical quenching relaxed quickly (Fig. 5e) with little effect on fluorescence yield (Figs. 5c, d) indicating the persistence of photoinhibitory damage. Photoinhibitory quenching (qI) relaxed more slowly than nonphotochemical quenching (cf. Figs. 5 e, f) and did not return to the initial level. In this it resembled the extent of recovery of F_v/F_m (Figs. 5c, d) and of functional PSII reaction centers (Fig. 5b).

In this experiment the odds were stacked against changes in membrane properties which were dependent on biochemical, as opposed to photochemical processes. The low Q_{10} of photochemistry means that

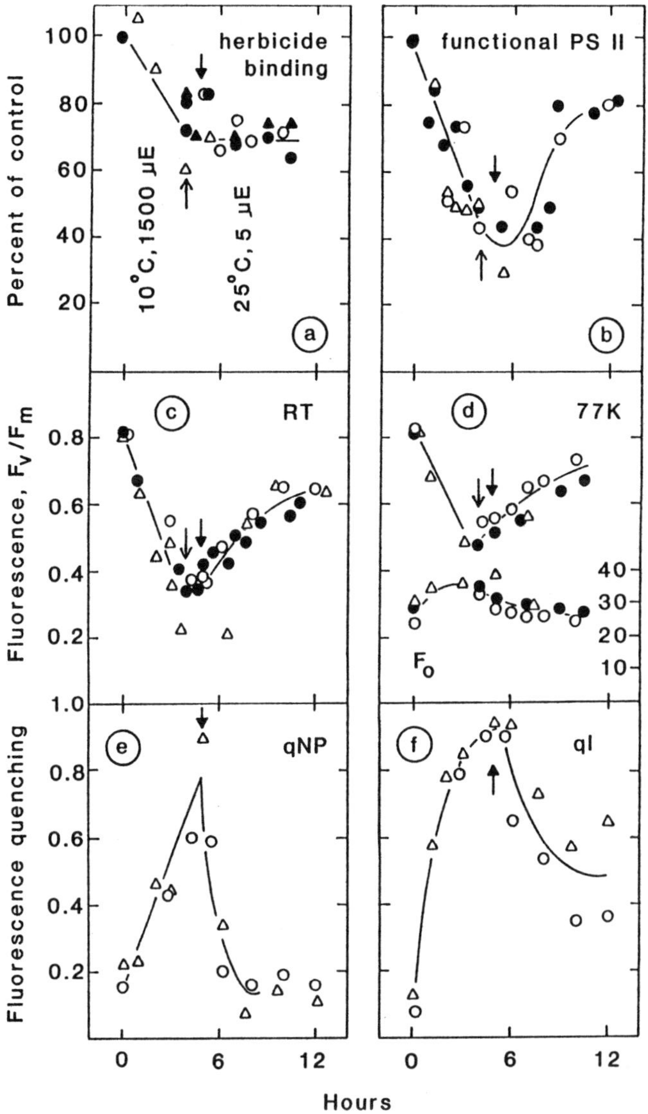

Figure 5. Characterization of photoinhibition in shade- grown spinach in air at 10°C, and recovery at 25°C in air. Symbols refer to different experiments: 4 Sept 87 (Δ); 14 Sept 87 (o) photoinhibited 5h (↓); and 19 Oct 87 (•) photoinhibited 4h (↓) and herbicide binding, measured with atrazine (•) and DCMU (▲).

such processes would be relatively insensitive to the change from 25° to 10°C. However the high Q_{10} of biochemistry means that membrane protein turnover, presumably dependent on proteases and protein synthesis, would be seriously impaired at 10°C (Greer *et al.*, 1986). Thus it is not surprising that the relative decrease in D-1 protein concentration did not track the more extensive loss of PSII reaction centers during photoinhibition (cf. Fig. 5a, b). Similar results, which dissociate changes in PSII reaction center function from D-1 protein turnover during photoinhibition, have been obtained with thylakoids (Arntz and Trebst, 1986;Chow *et al.*, 1988; Virgin *et al.*, 1988).

Integration of photoinhibitory processes, from the molecule to the biosphere now seems within grasp, thanks to research which stems largely from the Department of Plant Biology (Demmig and Björkman, 1987; Demmig *et al.*, 1987; Weiss and Berry, 1987). Whichever way we look at it photoprotection and photoinhibitory damage seem to be necessary compromises, or essential inefficiencies, in the highly dangerous business of excitation transformation in photosynthesis. That relatively few shade plants have been pressed to engineer themselves for light avoidance (Powles and Björkman, 1982), or that relatively few sun plants avoid light when stressed (Ludlow and Björkman, 1984) suggests that these compromises are adequate. Like the compromises of carboxylation and oxygenation in the dark reactions of photosynthesis (Andrews and Lorimer, 1987), the compromises of photoprotection and photoinhibition deserve nomination among the "evolutionary spandrels" of autotrophy (Gould and Lewontin, 1979). As we seek to integrate photosynthesis from the molecule to the biosphere, these approximations to a good thing seem to multiply at each step until some of them, most of the time, seem capable of being taken for granted.

The real challenge is to discover what matters and when. Some ecologists among us may conclude that photosynthesis matters so far as it affects reproduction, a view not much different from that of a crop physiologist employed by a seed company. If we consider for a moment the ecology of research in photosynthesis, then it has to be conceded again that it is the flowering of talent that matters. Clearly, the flowering of talent which can be traced to the influence of Stacy French is unlikely to be ephemeral. There will be good years in some habitats, and bad years in others, but widespread dissemination of the spirit ensures that flowering will continue, even when the older men have been obliged to sit in the shade (Bush, 1967). In some habitats, at some times, it becomes exceedingly difficult to sustain this most complex and creative biological activity, and well nigh impossible to impress our keepers of the need to do so. In these circumstances the French model at Carnegie becomes increasingly significant. It has been transplanted widely and I hope it may take root again in the near South, in the near future.

ACKNOWLEDGMENTS

This has become a welcome opportunity for reflection on what continues to be a daunting task, and I thank the Department of Plant Biology for sustaining the challenge. I am also grateful to Fred Chow, Ichiro Terashima, Huang, Lin-Ke and many others in the former Department of Environmental Biology, ANU, who have put some numbers among the speculations.

REFERENCES

Anderson JM, Anderson B (1988). The dynamic photosynthetic membrane and regulation of solar energy conversion. TIBS 13:351-355.

Anderson JM, Osmond CB (1987). Shade-sun responses: compromises between acclimation and photoinhibition. In Kyle DJ, Osmond CB, Arntzen CJ (eds): "Photoinhibition" Amsterdam, Elsevier, pp 1-38.

Andrews TJ, Lorimer GH (1987). Rubisco: structure, mechanisms and prospects for improvement. In Hatch MD, Boardman HK (eds): "The Biochemistry of Plants" Vol. 10 New York, Academic Press. pp 131-218.

Arntz B, Trebst A (1986). On the role of the QB protein in photoinhibition. FEBS Letts 194:43-49.

Aro E-M, Rintamäki E, Korhonen P, Mäenpää P (1986). Relationship between chloroplast structure and O_2 evolution rate of leaf discs in plants from different biotopes in South Finland. Plant Cell Environ 9:87-94.

Bennett J, Steinback KE, Arntzen CJ (1980). Chloroplast phosphoproteins: regulation of excitation energy transfer by photophosphorylation of thylakoid membrane polypeptides. Proc Natl Acad Sci USA 77:5253-5257.

Björkman O (1987). Low-temperature chlorophyll fluorescence in leaves and its relationship to photon yield of photosynthesis. In Kyle DJ, Osmond CB, Arntzen CJ (eds): "Photoinhibition". Amsterdam, Elsevier, pp 123-144.

Bush V (1967). The builders. In Haskins CP (ed). "The Search for Understanding" Washington, Carnegie Institution, pp 3-5.

Chow WS, Osmond CB, Huang L-K (1988). Photosystem II function and herbicide binding sites during photoinhibition of spinach chloroplasts *in vivo* and *in vitro.* Photosynth Res (in press).

Chow WS, Qian L-P, Goodchild DJ, Anderson JM (1988). Photosynthetic acclimation of *Alocasia macrorrhiza* (L.). G. Don to growth irradiance: structure function and composition of chloroplasts. Aust J Plant Physiol 15:107-122.

Cleland RE, Critchley C (1985). Studies on the mechanism of photoinhibition in higher plants II. Inactivation by high light of photosystem II reaction center function in isolated spinach thylakoids and O_2 evolving particles. Photobiochem Photobiophys 10:83-92.

Demmig B, Björkman O (1987). Comparison of the effect of excessive light on chlorophyll fluorescence and photon yield of O_2 evolution in leaves of higher plants. Planta 171:171-184.

Demmig B, Cleland RE, Björkman O (1987). Photoinhibition, 77K fluorescence quenching and phosphorylation of the light-harvesting chlorophyll-protein complex of photosystem II in soybean leaves. Planta 172:378-385.

Demmig B, Winter K, Kruger A, Czygan F-C (1987). Photoinhibition and zeaxanthin formation in intact leaves. A possible role of the xanthophyll cycle in the dissipation of excess light energy. Plant Physiol. 84:218-224.

Diesenhofer J, Eff O, Miki K, Huber R, Michel M (1985). Structure of the protein subunits in the photosynthetic reaction centre of *Rhodopseudomonas viridis* at 3A° resolution. Nature 318:618-624.

deWit CT (1970). Dynamic concepts in biology. In "Prediction and Measurement of Photosynthetic Productivity" Wageningen, Center Agric Publ Documentation, pp 17-23.

Evans JR (1986). A quantitative analysis of light distribution between the two photosystems, considering variation in both the relative amounts of the chlorophyll protein complexes and the spectral quality of light. Photobiochem Photobiophys 10:137-147.

French CS (1979). Fifty years of photosynthesis. Annu Rev Plant Physiol 30:1-26.

French, CS (1967). Photosynthesis. In Haskins CP (ed). "The Search for Understanding". Washington, Carnegie Institution, pp 155-178.

Gould SJ, Lewontin RC (1979). The spandrels of San Marco and the Panglossian paradigm: a critique of the adaptionist programme. Proc R Soc Lond B 205:581-598.

Horton P, Hague A (1988). Studies on the induction of chlorophyll fluorescence in isolated barley protoplasts, IV. Resolution of non-photochemical quenching. Biochim Biophys Acta 932:107-115.

Kamen MD (1964). "Primary Processes in Photosynthesis." New York, Academic Press.

Kyle DJ, Ohad I, Arntzen CJ (1984). Membrane protein damage and repair: selective loss of quinone-protein function in chloroplast membranes. Proc Natl Acad Sci USA 81:4070-4074.

Ludlow MM, Björkman O (1984). Paraheliotropic leaf movement in Siratro as a protective mechansim against drought-induced damage to primary photosynthetic reactions: damage by excessive light and heat. Planta 161:508-518.

Morrison P, Morrison P, Eames C, Eames R (1982). "Powers of Ten" New York, Freeman-Scientific American Books.

Ögren E, Öquist G (1984). Photoinhibition of photosynthesis in *Lemna gibba* as induced by the interaction between light and temperature III. Chlorophyll fluorescence at 77K. Physiol Plant 62:193-200.

Osmond CB, (1987). An overview of plant ecophysiology In "Proceedings of International Conference on Tropical Plant Ecophysiology" Bogor, Biotrop Spec Publ 31, pp 1-10.

Osmond CB, Björkman O, Anderson DJ (1980). "Physiological Processes in Plant Ecology" Berlin, Springer-Verlag.

Osmond CB, Chow W S (1988). Ecology of photosynthesis in the sun and shade: summary and prognostications. Aust J Plant Physiol 15:1-9.

Powles SB, Björkman O (1982). Leaf movement in the shade species *Oxalis oregana.* II. Role in protection against injury by intense light. Carnegie Inst Wash Yearbk 81:63-66.

Richardson K, Bendall J, Raven JA (1983). Adaptation of unicellular algae to irradiance: an analysis of strategies. New Phytol 93:157-191.

Saffo MB (1987). New light on seaweeds. Bioscience 37:654-664.

Schreiber U, Bilger W (1987). Rapid assessment of stress effects on plant leaves by chlorophyll fluorescence measurements. In Tenhunen JD, Catarino FM, Lange OL, Oechel WC (eds): "Plant Response to Stress-Functional Analysis in Mediterranean Ecosystems." Berlin, Springer-Verlag pp 27-53.

Seemann JR, Sharkey TD, Wang J-L, Osmond CB (1987). Environmental effects on photosynthesis, nitrogen-use efficiency and metabolite pools in leaves of sun and shade plants. Plant Physiol 84:796-802.

Terashima I, Inoue Y (1985). Vertical gradient in photosynthetic properties of spinach chloroplasts dependent on intraleaf light environment. Plant Cell Physiol 26:781-785.

Terashima I, Saeiki T (1983). Light environment within a leaf. I. Optical properties of paradermal sections of *Camellia* leaves with special reference to difference in the optical properties of palisade and spongy tissue. Plant Cell Physiol 24:1493-1501.

Terashima I, Sakaguchi S, Hana N (1986). Intra-leaf and intracellular gradients in chloroplast ultrastructure of dorsiventral leaves illuminated from the adaxial or abaxial side during their development. Plant Cell Physiol 27:1023-1031.

Virgin I, Styring S, Anderson B (1988). Photosystem II disorganization and manganese release after photoinhibition of isolated spinach thylakoid membranes. FEBS Letts 233:408-412.

Weiss E, Berry JA (1987). Quantum efficiency of photosystem 2 in relation to "energy" dependent quenching of chlorophyll fluorescence. Biochim Biophys Acta 894: 198-208.

Yoda K (1974). Three-dimensional distribution of light intensity in a tropical rainforest of West Malaysia. Jpn J Ecol 24:247-254.

Photosynthesis, pages 19–44

PHOTOSYNTHESIS AND PLANT PRODUCTIVITY -- SCALING TO THE BIOSPHERE

H.A. Mooney and C.B. Field

Dept. of Biological Sciences, Stanford University, Stanford, CA 94305 (H.A.M.); Dept. of Plant Biology, Carnegie Institution of Washington, 290 Panama Street, Stanford, CA 94305 (C.B.F.)

We have a rather limited view of the productive capacity of the biosphere. The available estimates of annual carbon fixation, storage, and decomposition are imprecise and are based principally on geographical-ecological techniques. The large implications for climate regulation of the changing storage and flux rates of carbon make it important to obtain more precise estimates. In the following we review some of the emerging techniques for improving these estimates.

THE GLOBAL CARBON CYCLE

Unequivocal evidence that the atmosphere is increasing in carbon dioxide concentration at about one and a half parts per million per year (Trabalka, 1985) is generating enormous interest and activity in understanding and quantifying the carbon cycle of the Earth. Climate models project that a doubling of CO_2 concentration will lead to an increase in the mean annual global temperature of 1.5 to 5.5°C (Dickinson 1986). This is a temperature change comparable to that between the last glacial period and the present interglacial (Bolin, *et al.*, 1986). These concerns have resulted in a large effort to balance the total global carbon budget, including studies of all known pools and fluxes.

A recent summary of this knowledge is shown in Fig. 1. Here we concentrate on the part of the carbon cycle that is controlled by the terrestrial biota. The flux between the atmosphere and the terrestrial biota is large; about 15% of the atmospheric reservoir is exchanged per year. Carbon in the soils exceeds that in the atmosphere and in the living vegetation but is still a small fraction of that stored in the depths of the oceans. The flux of carbon from deforestation can be viewed as an acceleration of normal biological exchange of carbon between the vegetation and the atmosphere. Since this flux may be a significant

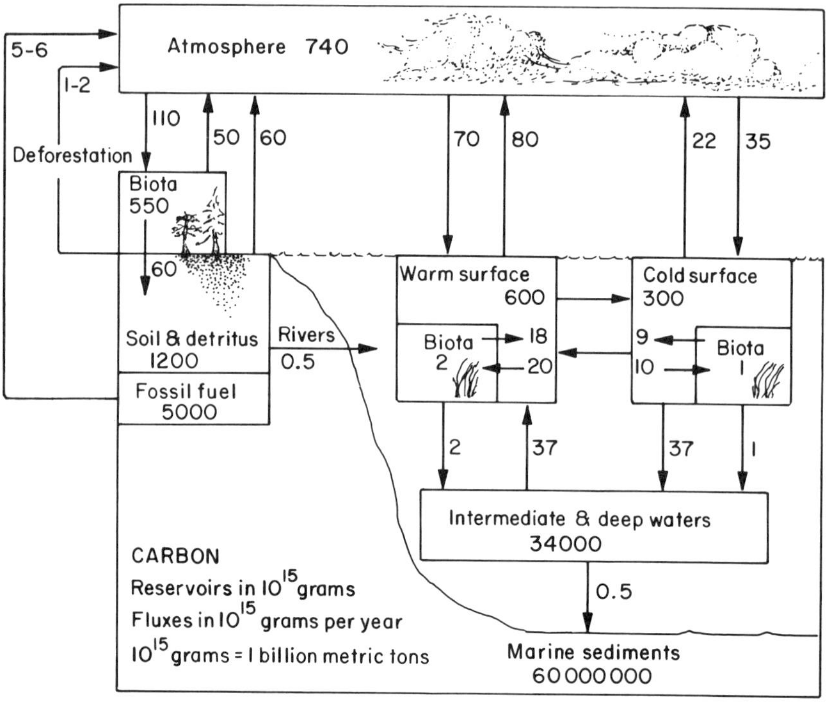

Figure 1. The global carbon cycle. Arrows denote fluxes in 10^{15} g year and boxes show carbon reservoirs in 10^{15} g (from Moore and Bolin, 1986/87).

contributor to the increasing atmospheric concentration, it has been examined in some detail (Richards, *et al.*, 1983). However, good estimates of the amount of CO_2 released through deforestation are, like all of the global pools and fluxes of carbon, difficult to obtain. New techniques increasingly enable us to employ more appropriate spatial and temporal sampling approaches than have been available for past studies. Further progress is needed, but new approaches available now allow us to monitor continuously, and certainly more accurately, our changing biosphere.

GLOBAL BIOMASS

Fundamental to any understanding of the global carbon budget is the inventory of carbon stocks, both live and dead. Estimates of global living biomass by Rodin and Bazilevich (1966) were approximately 1080 Pg C

(1 Pg = 10^{15} g = 1 billion metric tons). This estimate was evidently based on potential vegetation cover, without human influences (Olson, *et al.*, 1983). Whittaker and Likens (1975), in an independent assessment utilizing less extreme estimates of standing biomass for a given vegetation type give a value of 827 Pg of C (1837 Pg dry matter). Olson, *et al.* (1983) in a global survey utilizing modern data on biomass of various biomes place global carbon estimates at 561 Pg of C with boundaries on the high and low side of 665 and 461 Pg. The lowering of estimates through time is due to both incorporation of the effects of humans on reducing forest area and density, and the utilization of average rather than extreme community standing biomass data. All these values are derived from point samples of biomass multiplied by estimates of the global extent of a vegetation type, usually obtained from maps. In another approach, the distribution of vegetation types throughout the world has been categorized using a predictive climate-vegetation model developed by Holdridge (1947). One example of this approach shows how climate change resulting from increased atmospheric CO_2 would affect community distribution (Emanuel *et al.*, 1985).

Recently, Matthews (1983) digitized the world distribution of vegetation types, utilizing both maps and satellite imagery. This important effort will no doubt be refined as we get better techniques for assessing vegetation cover and biomass from satellite images.

GLOBAL PRODUCTIVITY

As Choudhury (1987) noted, there have been two basic ways of estimating the rate at which organic matter accumulates in the biosphere. One, exemplified by the work of Whittaker and Likens (1975), calculates the area covered by each biome type and multiplies that by an average productivity value for each, derived from point studies. As they note, "the means...are in some cases averages of published values; but in many cases they have been subjectively chosen as reasonable, intermediate values for a range indicated by a few field measurements". The summation of these values gives the global average. Estimates of the net primary productivity of the earth's terrestrial ecosystems vary from about 45 to 62 Pg yr^{-1} (Houghton *et al.*, 1985).

The other approach is to utilize climate-productivity algorithms such as those developed by Lieth (1975). Lieth constructed world maps of productivity based on precipitation, evapotranspiration, and length of growing season. This approach potentially allows delineating the spatial variability in productivity as finely as the available weather data.

REMOTE SENSING OF PRIMARY PRODUCTIVITY

A. Optical Vegetation Indices

It can be seen from the above that we do not have a very precise estimate of either the earth's stock of carbon or its changing dimensions. There are, however, new approaches for improving these estimates as indicated below.

Plants and soil differ dramatically in their spectral characteristics. Leaves have low reflectivity throughout the visible wave lengths, but high reflectivity in the near infrared. In contrast, the reflectivity of soil increases gradually from the blue through the near infrared. This difference makes it possible to use a ratio of infrared to visible reflectance or radiance to estimate several components of the vegetation status, including leaf area index (Jordan, 1969), forage biomass (Pearson and Miller, 1972), intensity of drought stress (Tucker *et al.*, 1980a), and predicted grain yield (Tucker *et al.*, 1980b). However, the ratio of infrared to visible reflectance is a non-linear index of parameters related to green-leaf biomass and becomes increasingly unsatisfactory for predicting plant biomass or leaf area index as leaf area index increases (Curran, 1981).

Monteith (1977) explored the hypothesis that the rate of crop growth is linearly related to radiation absorption in the visible waveband. Kumar and Monteith (1981) then developed theory predicting that a ratio of infrared to visible reflectances should vary linearly with visible radiation interception by a crop canopy. Since visible radiation interception is a non-linear function of leaf biomass or leaf area index but a potentially linear function of crop growth, spectral signals based on visible and infrared reflectances may be much better indices of canopy growth than of biomass or leaf area index.

Tucker and colleagues have been very successful in establishing the validity and utility of spectral indices combining near infrared and visible reflectances for predicting growth or primary productivity. Using a hand-held radiometer looking at wavebands of 0.65-0.70 and 0.775-0.825 μm, Tucker *et al.* (1981) found a strong correlation between end-of-season biomass of winter wheat and two spectral indices, the simple ratio ($SR=R_n/R_v$) and the normalized difference [$ND=(R_n-R_v)/(R_n+R_v)$], where R_n and R_v are radiance in the near infrared and visible, respectively. End-of-season biomass was most strongly correlated with spectral signals measured approximately one month before harvest or integrated over the entire growing season.

The success with a hand-held instrument stimulated further development of the technique for satellite radiometry. The principal data source to date has been the Advanced Very High Resolution Radiometer (AVHRR) aboard the NOAA 6 and 7 polar-orbiting satellites. AVHRR provides radiance data in the visible (0.55-0.68 μm) and near-infrared (0.73-1.1μm), updated several times during every 9-day orbital cycle, at a 1-km resolution for local areas and 4-km for the entire globe.

Utilizing the AVHHR data, Tucker *et al.* (1985) demonstrated a strong correlation between either maximum or time-integrated ND and growing-season biomass accumulation over a three year period characterized by nearly three-fold variation in precipitation and even greater variation in biomass production (Fig. 2). The strong correlations between ND (or NDVI, for normalized difference vegetation index) or SR and biomass accumulation in herbaceous systems suggest the hypothesis that these vegetation indices are generally valid predictors of primary productivity or of the time-integrated net photosynthesis of all the leaves in the canopy.

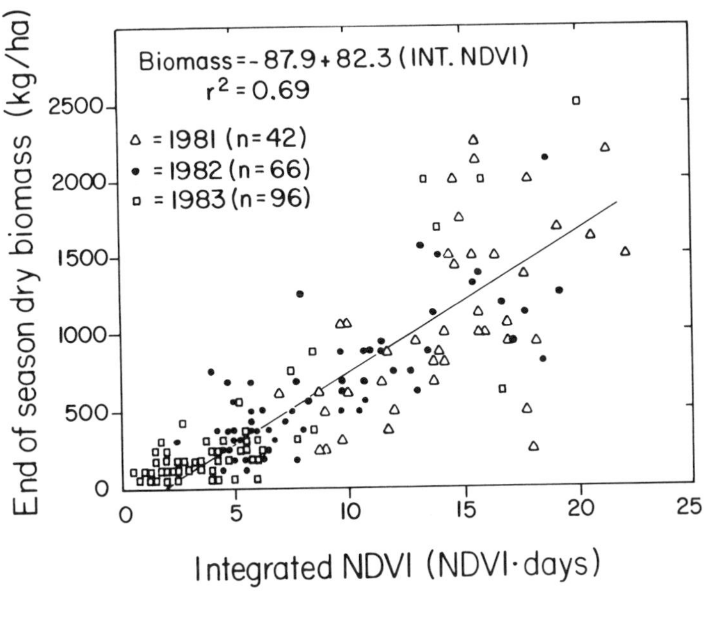

Figure 2. Growing season herbaceous biomass accumulation in the Senegalese Sahel during 1981-1983 as related to the integrated normalized difference vegetation index (NDVI) (from Tucker *et al.*, 1985).

Three lines of evidence support this hypothesis. These are (1) correlations between time-integrated NDVI or SR and end-of-season above-ground biomass in herbaceous systems, as discussed above; (2) correlations between NDVI and variation in atmospheric CO_2 on a global or broad latitudinal scale as discussed later (Tucker *et al.*, 1986; Fung *et al.*, 1987); and (3) theoretical studies combining models of light penetration, photosynthesis, and gas transport (Sellers, 1985, 1987; Choudhury, 1987). The accumulated evidence leaves little doubt that NDVI and SR are useful indicators of ecosystem photosynthesis in many systems and under a range of conditions. Sellers (1985, 1987) has also begun testing the hypothesis that NDVI provides a useful index of canopy conductance to water vapor.

Other evidence challenges the generality of the relationships between photosynthesis and the spectral indices. Singh (1988) used simulation models to show that much of the annual variation in NDVI at high latitudes may result from variation in the solar zenith angle, and found that NDVI is a decreasing function of zenith angle for angles greater than 30° (30° away from vertical). Choudhury (1987), applying sensitivity analysis to a simulation combining models of light availability, light penetration, and photosynthesis, found that the relationship between NDVI or SR and either absorbed visible radiation or photosynthesis is not always linear and not necessarily unique. The relationship between NDVI and photosynthesis may be quite sensitive to variation in soil reflectance and in photosynthetic pathway (C_3 vs. C_4).

Sellers' (1987) studies of the biophysical basis for the near linearity of the relationship between SR or NDVI and the fraction of photosynthetically active radiation absorbed (APAR) identify the dominant parameters and the prerequisites for linearity. Variation in the infrared reflectance dominates the relationship under normal field conditions because reflectance in the visible is typically very similar for soil and for an infinitely dense plant canopy (Sellers, 1987). The approximation to linearity is partly an intrinsic consequence of the large change in the scattering coefficient of leaves at the visible-infrared boundary and of the fact that APAR depends on a single pass through the vegetation canopy while infrared reflectance depends on a double pass (in and out). The extent of the linearity of the SR-APAR or NDVI-APAR relationship also depends on the choice of wavelengths in the visible and infrared. The wavelength choice for maximum linearity depends on the leaf angle distribution. AVHRR channels 1 and 2, the wavelengths used most frequently, are almost the best possible.

The mechanistic basis of the relationship between APAR and photosynthesis is incompletely known but clearly multifaceted, involving temporal, spatial, and physiological components. The temporal component to the relationship between APAR and photosynthesis arises because APAR is always measured, at least from satellite platforms, as an instantaneous value, while photosynthesis is a time integral. APAR is an index of the capacity for light absorption. Maximum photosynthesis under high light and optimal temperature and humidity may scale with the capacity for light absorption, but daily photosynthesis need not scale with photosynthetic capacity, especially in light-limited sites (Field, 1988). The tuning of photosynthetic capacity to the potential for daily carbon gain may be a result of evolution, molding photosynthetic characteristics to eliminate capacity rendered unusable by resource limitations (Mooney and Gulmon, 1979). However, little is known about the relationship between maximum photosynthetic capacity and daily carbon balance.

The spatial component of the APAR-photosynthesis relationship is relatively straightforward. Increasing or decreasing leaf area index tends to change APAR and photosynthesis in concert. Sellers (1985) concludes from simulations that both are similar, saturating functions of leaf area index over a broad range of leaf-angle distributions.

The physiological component of the APAR-photosynthesis relationship has direct and indirect aspects. Increasing the density of pigments directly increases the capacity for light absorption and the potential for photochemistry. However, photosynthetic capacity is not determined solely by the potential for photochemistry. Under normal conditions photosynthesis is partially limited by multiple processes, including carboxylation by rubisco, RuBP regeneration, and gas exchange through stomata (Woodrow *et al.*, 1987). The APAR-photosynthesis relationship depends on balanced responses of all components of photosynthesis to changes in the resource availability and stress. While the evidence for substantial balance is quite strong (Field and Mooney, 1986), the variance resulting from differences in photosynthetic pathway (C_3 vs. C_4) (Sage and Pearcy, 1987), habitat of origin (Evans, 1988), and distribution of photosynthetic potential through a canopy (Hirose and Werger, 1987; Gutschick and Wiegel, 1988) are not insignificant.

B. Microwave Vegetation Index

Choudhury *et al.* (1987), Choudhury and Tucker (1987), and Choudhury (1988) develop an index of vegetation quantity and function using data from the scanning multispectral microwave radiometer (SMMR) on the Nimbus-7 satellite. The quantity of interest is the intensity difference between horizontally and vertically polarized microwave radiation at 37 GHz (wavelength ~8mm). At the 50° angle of incidence of the SMMR, the horizontally polarized microwave signal from the soil is stronger than the vertical, typically by an amount sufficient to make the brightness temperature 25 to 30 K higher for the horizontally than for the vertically polarized radiation. Vegetation, in contrast, emits primarily unpolarized radiation and scatters and absorbs the polarized signal from the soil. Over heavily vegetated surfaces, the polarization difference is typically 5 to 10 K (Choudhury *et al.*, 1987). The polarization difference is most sensitive to vegetation effects when the cover is sparse, making the microwave vegetation index well suited to studies of arid and semi-arid regions and hence complementary to vegetation indices based on visible and near infrared signals, which tend to be most successful when the vegetation cover is high (Choudhury and Tucker, 1987).

The accuracy and range of applications for the microwave vegetation index are comparable to those of the normalized difference vegetation index from AVHRR data. The microwave polarization difference is curvilinearly related to the AVHRR NDVI and highly correlated with an inverse fit for data from several sites in Africa and the U.S. (Choudhury *et al.*, 1987). The microwave vegetation index is also highly correlated with the global average relative CO_2 concentration (the relative CO_2 concentration is the deviation from the detrended mean) (Choudhury and Tucker, 1987; Choudhury, 1988), as well as with the land-surface net primary productivity (Fig. 3) and the land-surface actual evaporation averaged over 5° latitude bands (Choudhury, 1988). The estimates of primary productivity lose sensitivity as primary productivity approaches values over 500 g m^{-2} yr^{-1}.

CO_2 EXCHANGE OF ENTIRE ECOSYSTEMS

A. Meteorological Measurements

As CO_2 is depleted around photosynthesizing tissue, it tends to be replenished by atmospheric processes that transport CO_2 from regions of higher to regions of lower concentration. Just as a product of a CO_2 differential and a forced flow gives the rate of photosynthesis for a leaf or plant in a chamber, a combination of a CO_2 gradient and a natural transport can yield photosynthesis for an unenclosed, outdoor community. The complexity of natural air movements insures that the combination is more detailed than a simple product of a concentration difference and a flow, but the basic idea is simple: the rate of CO_2 uptake by the vegetation must equal the rate of CO_2 supply from the ground below and the atmosphere above.

Three kinds of meteorological measurements have been widely used. All focus on turbulent flow as the dominant process transporting gases

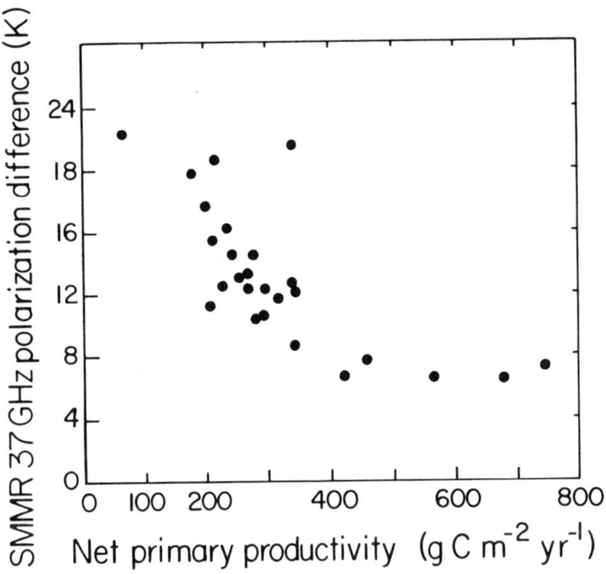

Figure 3. Correlation of primary productivity and polarization difference at 37 GHz averaged over 5° latitude bands, as measured by the scanning multichannel microwave radiometer on Nimbus-7 (from Choudhury, 1988).

into and out of the vegetation canopy, but they differ in the time scale over which the turbulence is considered and in the requirement for quantifying the atmosphere's tendency for vertical transport. Two of the methods, the mean-value or aerodynamic technique and the Bowen-ratio or energy-balance technique, are based on flux-gradient theory and the assumption that turbulent transfer is analogous to molecular diffusion. This assumption implies that flux is equal to the product of a mean vertical concentration gradient and an analog of a molecular diffusivity, termed a turbulent or eddy diffusivity (Thom, 1975; Kanemasu *et al.*, 1979). The mean-value method requires explicit specification of the turbulent diffusivities while the Bowen-ratio method requires only that the turbulent diffusivities for mass and energy are identical (Denmead and Bradley, 1985). The third major technique, the eddy correlation method (Denmead and Bradley, 1985; Baldocchi *et al.*, 1988), eliminates the need for time averaging the CO_2 gradient by accounting for the transport associated with each pulse of turbulent flow.

All of the meteorological methods for calculating CO_2 flux are subject to several limitations and sources of error. All work best for uniform vegetation covering a large expanse of horizontal surface. All require a substantial averaging period for accurate measurements [Baldocchi *et al.* (1988) estimate approximately one-half hour]. Even under good conditions, the accuracy of turbulent flux measurements is rarely better than ±10-20% (Baldocchi *et al.*, 1988). However, recent advances in sensor technology and in meteorological theory make it increasingly practical to obtain measurements that approach the maximum theoretical accuracy for vegetations with characteristics that are far from ideal.

The mean value method has the longest history of the three methods, but also has the most stringent requirements for high accuracy. The turbulent diffusivity for CO_2 must be estimated from that for another quantity, usually momentum, and the empirical formulae for interrelating turbulent diffusivities are imprecise. Further, both the mean-value and Bowen-ratio methods require very great sensor accuracy because the magnitude of the CO_2 gradient is often small. The Bowen-ratio method tends to be more reliable than the mean-value method, typically yielding errors of ±10-15% (Baldocchi *et al.*, 1988). Within canopies, where much of the flux occurs in sporadic pulses and where flux can be against the prevailing concentration gradient, both flux-gradient methods fail completely (Denmead and Bradley, 1985).

The eddy correlation method provides a direct measure of flux that can be used both above and within canopies. Rotating the analysis framework makes it possible to obtain accurate measurements on sloping sites. Eddy correlation measurements do require fast-response instruments and high sampling rates. Instrument responses should be faster than the period of the smallest eddies that are significant for transport, typically well below 1 s (Anderson *et al.*, 1986). However, digital low-pass recursive filtering can eliminate the need for storing large quantities of raw data, and increasing the height of the sensors relative to the height of the canopy can decrease the need for a high sampling rate (Baldocchi *et al.*, 1988).

B. Boundary Layer Measurements

Meteorological methods yield CO_2 exchange of vegetation units on a scale of several meters to a few kilometers, a scale large enough to encompass many sources of variation and to interface with satellite-based measures. However, even measurements at this scale need to be replicated extensively in order to provide an accurate picture of the photosynthesis of large landscape units. The aircraft-based eddy correlation approach of Desjardins *et al.* (1982, 1985) offers one option for extensive sampling. The boundary-layer method of Wofsy *et al.* (1988) offers another.

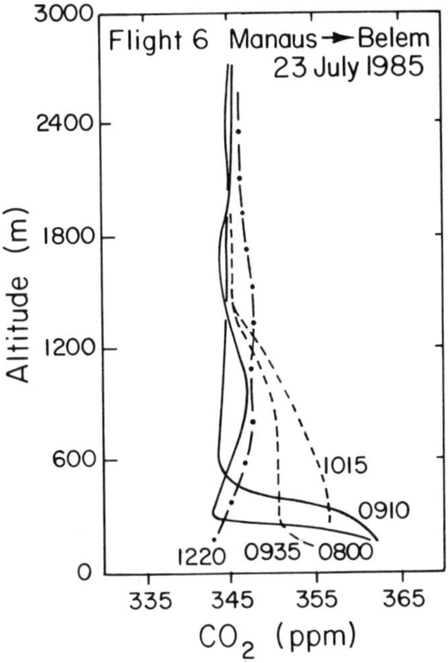

Figure 4. Morning CO_2 profiles over the Amazonian forest (from Wofsy *et al.*, 1988).

The basic idea of the boundary-layer method is that, under appropriate atmospheric conditions, regional CO_2 exchange is given by the rate of change of the CO_2 content of the air column extending from the ground to, or just above, the highest level ever reached by the planetary boundary layer. Conditions are appropriate when the atmosphere is stable or negatively buoyant, and when the height of the planetary boundary layer is stable or increasing. Under these conditions exchange between the free troposphere and the planetary boundary layer is limited largely to the gas incorporated into the boundary layer as its height increases (Matson and Harriss, 1988).

Flying an aircraft-mounted CO_2 analyzer over forests in the Amazon Basin, Wofsy *et al.* (1988) observed a discontinuity in the CO_2 content of the atmosphere at the boundary between the free troposphere and the planetary boundary layer. This was paralleled by similar discontinuities in the abundance of ozone and aerosols (Browell *et al.*, 1988) The planetary boundary layer was short at night and in the early morning and characterized by an excess of CO_2 relative to the free troposphere (Fig. 4). Through a typical morning, the CO_2 concentration decreased (as a

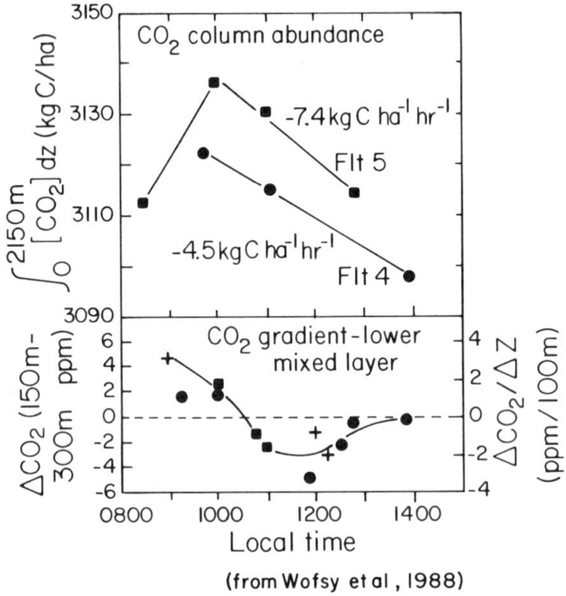

Figure 5. Changes in the total column abundance of carbon dioxide in the mixed layer above an Amazonian forest for two separate days (top). The slope gives the rate of net photosynthesis of the forest. The low value at 0830 shows the influence of the relic mixed layer. (Bottom panel) The difference in CO_2 concentration between 150 and 300 meters during the course of the two different days. The left hand axis is the total gradient from 150 to 300 m and the right hand axis ($\Delta CO_2/\Delta Z$) is the gradient per 100 m.

result of photosynthesis) and the height of the boundary layer increased (as a result of surface heating) (Fig. 4). The CO_2 content of the air column, integrated from the surface to 2150 m, also decreased through the morning (Fig. 5). The slope of the decrease is the whole-system CO_2 exchange.

Wofsy *et al.* (1988) found that photosynthesis estimates from the boundary-layer method are quite local, with CO_2 concentrations at 150 m responding to landscape features (e.g. forest, wetland, or river) over a few hundred meters. They measured midday CO_2 uptake of 9 kg C ha^{-1} hr^{-1} or 22 μmol CO_2 m^{-2} s^{-1} over forest and 1.6 kg C ha^{-1} hr^{-1} or 4 μmol CO_2 m^{-2} s^{-1} over wetlands.

C. Empirical Results

Table I shows midday rates of net photosynthesis of several vegetation types, measured with either remote sensing or a canopy chamber or

Table I. Maximum rates of whole-canopy photosynthesis for natural and agricultural systems, measured with several techniques. All of the data are from vegetations with complete cover and not subject to unusual water or temperature stress. The column "Incl. Soil Resp." indicates whether the reported rate is net flux downward from the atmosphere (which equals the sum of canopy CO_2 uptake and soil CO_2 release - indicated by Yes in the table) or only canopy CO_2 uptake (indicated by No in the table).

Vegetation type	Photosyn. Capacity (μmol m^{-2} s^{-1})	Method	Incl. Soil Resp.	Ref.
Temperate Forests				
Deciduous Forest (Tennessee)	23	Eddy correlation	No	Baldocchi *et al.*, 1987
Mixed Forest (Canada)	30	Eddy correlation	Yes	Desjardins *et al.*, 1985
Loblolly Pine	32	Flux gradient	Yes	Jarvis *et al.*, 1976
Loblolly Pine (South Carolina)	35	Flux gradient	Yes	Lorenz and Murphy 1985
White Pine (Canada)	23	Eddy correlation	Yes	Desjardins *et al.*, 1985
Scots Pine (Scotland)	27	Leaf chamber	No	Beadle *et al.*, 1985
Norway Spruce (Scotland)	23	Flux gradient	Yes	Jarvis *et al.*, 1976
Sitka Spruce (Scotland)	23	Flux gradient	Yes	Jarvis *et al.*, 1976
Sitka Spruce (Scotland)	21	Leaf chamber	No	Jarvis *et al.*, 1985
Tropical Forests				
Amazon Rainforest	22	Boundary layer	Yes	Wofsy *et al.*, 1988
Costa Rican Rainforest	21	Flux gradient	Yes	Allen and Lemon 1976
Malaysian Rainforest	28	Flux gradient	Yes	Aoki *et al.*, 1975
Rubber Plantation	41	Flux gradient	No	Monteney *et al.*, 1985
Grasslands				
Mixed Prairie (Canada)	13	Flux gradient	Yes	Ripley and Redmann 1976
Arctic Tundra (Alaska)	5	Flux gradient	Yes	Coyne and Kelly 1975
Perennial ryegrass (England)	32	Canopy chamber	No	Parsons *et al.*, 1983
Crops				
Alfalfa (France)	34	Flux gradient	Yes	Saugier 1970
Alfalfa (Nebraska)	31	Flux gradient	Yes	Baldocchi *et al.*, 1981a
Alfalfa (Canada)	27	Canopy chamber	No	Daley and McNeil 1987
Barley (England)	25	Flux gradient	No	Biscoe *et al.*, 1975
Corn (Japan)	63	Flux gradient	Yes	Uchijima 1976
Corn (Canada)	30	Eddy correlation	Yes	Desjardins *et al.*, 1982
Cotton (Mississippi)	27	Canopy chamber	Yes	Wells *et al.*, 1986
Rice (Japan)	38	Flux gradient	Yes	Uchijima 1976
Soybean (Nebraska)	35	Flux gradient	Yes	Baldocchi *et al.*, 1981b
Soybean (Florida)	38	Canopy chamber	Yes	Jones *et al.*, 1985
Soybean (Georgia)	22	Canopy chamber	Yes	Boerma and Ashley 1988
Sugar beet (England)	25	Flux gradient	Yes	Brown 1976
Sugar beet (Netherlands)	51	Leaf chamber	No	Spitters 1986
Sunflower (France)	37	Flux gradient	No	Saugier 1976
Wheat (Australia)	44	Flux gradient	No	Denmead 1976
Wheat (Colorado)	50	Canopy chamber	No	Morgan and Willis 1983

calculated from simulations based on leaf-level measurements. For a single vegetation type, variation among estimates based on different techniques is relatively small. In fact, variation among plant communities with full vegetation cover is relatively small. With the exception of the measurement for arctic tundra, a system characterized by low light flux and high organic carbon accumulation, the extremes differ by less than one order of magnitude, much less than the two order of magnitude variation among rates at the leaf level (Mooney and Gulmon 1979). Neither the photosynthetic capacity of individual leaves nor the photosynthetic pathway of the constituent species is a good predictor of the maximum photosynthesis of entire communities. For plant communities in which the photosynthetic capacity of individual leaves is high, whole-canopy maximum photosynthesis is only approximately the same as the capacity of a single leaf. In communities in which the photosynthetic capacities of individual leaves are low (e.g. conifers), the maximum photosynthesis of the canopy may be 5 times that of a single leaf (Lorenz and Murphy 1985).

One contributor to this relative uniformity of canopy photosynthetic capacity is that biophysical constraints on the efficiency of light utilization place an upper limit on canopy photosynthesis. Maximum inputs of photosynthetically active radiation are typically near 2000 μmol m^{-2} s^{-1} and the photon yield of CO_2 fixation in leaves in normal air and not subject to photoinhibition or other stresses is near 0.06 mol CO_2/mol photons (Pearcy and Ehleringer, 1984). Thus, canopy photosynthesis should never exceed 120 μmol m^{-2} s^{-1}. Respiration from the soil and from non-photosynthetic tissues makes real maxima lower. Another factor decreasing canopy photosynthesis is light saturation in individual leaves. The relative uniformity of canopy photosynthetic capacity argues that canopies comprised of leaves with low photosynthetic capacities effectively distribute the incident radiation over a large leaf area per unit of ground area. In canopies of high capacity leaves, leaf area indices may be lower, the distribution of radiation may be less efficient, photosynthetic capacities may be adjusted to the prevailing radiation environment (Field, 1983), or photosynthesis may be limited by diffusion resistance at the canopy level (Jarvis and McNaughton, 1986). The relative constancy of canopy photosynthetic capacity has important consequences for satellite remote sensing, implying that the important quantities to measure are more related to canopy cover and stress than to the photosynthetic capacity of individual leaves or the entire canopy.

GLOBAL METABOLISM

One of the most significant biospheric data sets of recent times is the time course of CO_2 concentration measured at the Mauna Loa Observatory in Hawaii. This record, extending back continuously to 1958, shows a gradual increase in mean concentration through time as well as a strong seasonal amplitude in concentration (Gammon *et al.*, 1985). The seasonal signal indicates the "metabolism" of the biosphere -- the changing balance between photosynthesis and decomposition. The strength of the signal is greatest at the higher northern latitudes, a consequence of both the large land mass and the strong seasonality in those regions.

Fung *et al.*, (1987) found, at mid-latitudes, a correspondence of the duration of high NDVI and the duration of the CO_2 drawdown (Fig. 6). The date of the maximum drawdown closely matches that of the peak NDVI. These workers further modeled seasonal ecosystem carbon flux. They estimated release of carbon utilizing published soil respiration value corrected with a temperature coefficient.

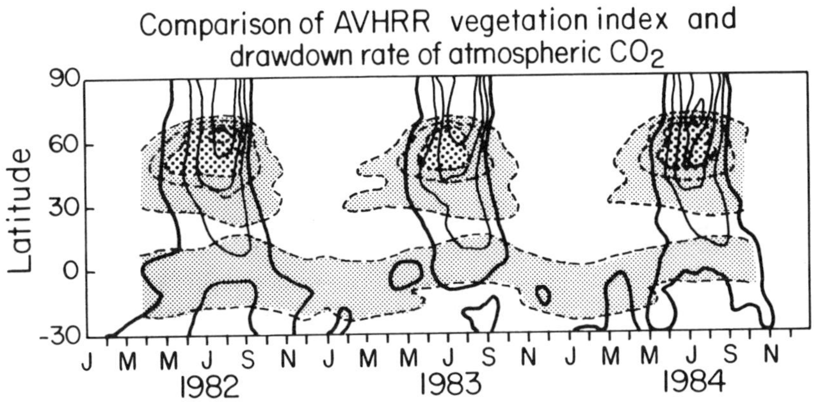

Figure 6. Comparison of seasonal variation in the NDVI and drawdown of CO_2. Solid lines indicate drawdown rates with the heavy solid line representing 0ppm/month. Contour intervals are -2ppm/month. Dashed lines are contours of averaged NDVI, spread at intervals of 0.05. Light shading is NDVI >0.05 and heavy shading NDVI >0.15 (from Fung *et al.*, 1987).

For uptake of carbon they employed published net primary productivity data for various vegetation types and scaled these values seasonally by the changing NDVI. The modelled values for four grid boxes of 4 degrees latitude and 5 degrees longitude are shown in Fig. 7. These simulations

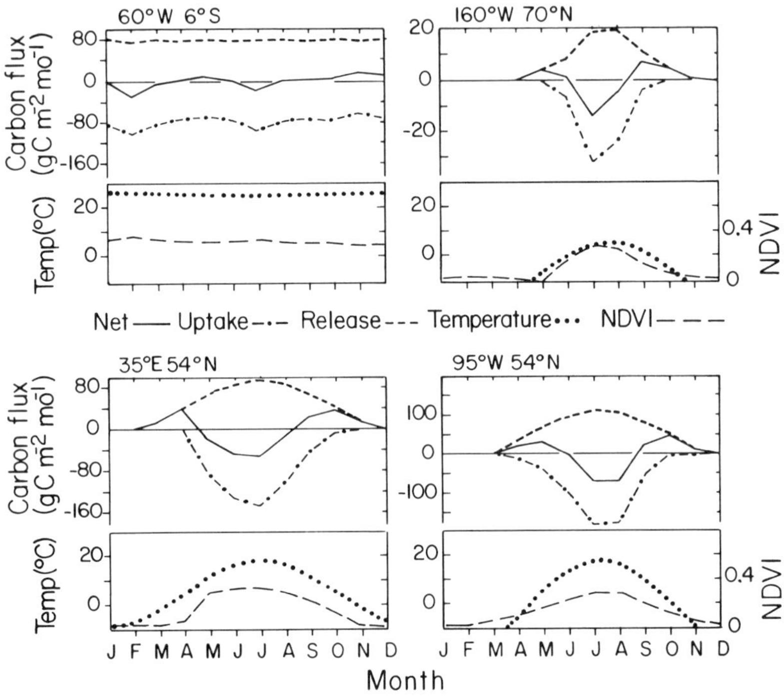

Figure 7. Simulations of seasonal CO_2 fluxes from tropical (top left), tundra (top right), temperate deciduous (bottom left) and evergreen (bottom right) forests (from Fung *et al.*, 1987).

show large fluctuations in the net CO_2 flux in the temperate latitudes and less in the tropics, fluctuations which correspond to the data on atmospheric drawdown.

Another indication of the role of the biota in modulating atmospheric CO_2 concentration is the difference in the growth rate of CO_2 in the atmosphere prior and post-El Niño. Gammon *et al.*, (1985) show that during the El Niño period in 1982, 5 X 10^{15} grams of C did not appear in the average yearly addition to the atmosphere (Fig. 8). Higher addition rates in the subsequent year indicated that the carbon had been temporarily stored. One possibility suggested from NDVI data is that there was greater than normal terrestrial productivity during 1982.

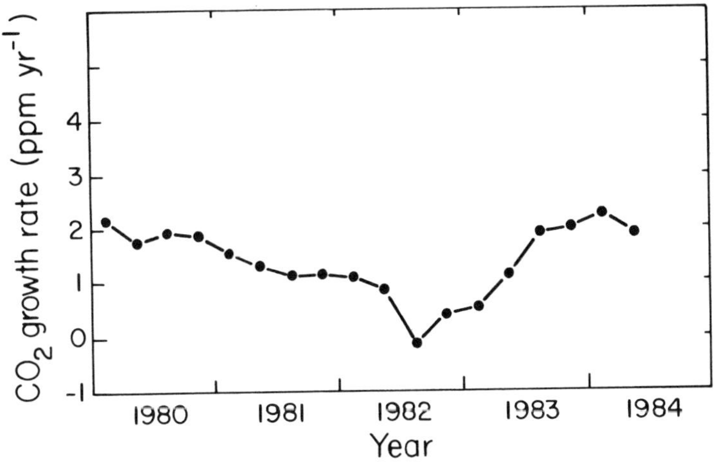

Figure 8. Yearly change in the CO_2 growth rate (from Gammon *et al.*, 1985).

INFLUENCES ON THE CARBON CYCLE

Humans are influencing the carbon cycle directly through the use of fossil fuels but also through deforestation. There has been considerable effort in recent years to quantify atmospheric inputs from these processes both currently and historically. As noted above, estimates of both living biomass and deforestation rates are quite variable. Estimates of forested areas in tropical regions, based on various sources, differ by a factor of two for certain regions (Molofsky *et al.*, 1986). Tropical deforestation estimates also differ substantially, varying from 3 to 30 X 10^6 ha yr^{-1} (Houghton *et al.*, 1985). The large variation in estimates is due to the variable data base as well as the interpretations of the status of the land. Satellite technology offers some promise of providing more accurate time-based estimates of forest cover and its conversion. The LANDSAT imagery, to a resolution of 80m for the multispectral scanner, is available going back to 1972, collecting data once every 16 days. Woodwell *et al.* (1983) indicate that it takes 1200 LANDSAT images to cover the land surface of the earth. In contrast the AVHRR images from the NOAA polar-orbiting satellites are available up to twice a day with a ground resolution of 1 km. Thus large scale patterns can be discerned with the use of AVHRR and the detailed patterns with LANDSAT.

Malingreau and Tucker (1988) utilized AVHRR data to assess disturbance and deforestation rates in Brazil. Their study shows both the problems and promise of these techniques. Over a five year period they were only able to retrieve, at most, 15 useful images in any given year because of cloudiness, smoke, and poor transmission of data. From these, however, they were able to put together a remarkable, quantitative record of vegetation change. They utilized the 3.5-3.9 μm temperature channel to discriminate between cleared and forested areas, using ground observations for verification. They showed that by 1985, 11% of the Brazilian state of Rondonia (with an area of 243,400 km^2) had been deforested (Fig. 9) and 36% was in some degree of disturbance (mix of marginal to radical conversions of the forest canopy). This all occurred in the span of 15 years. At the current rate of cutting, all the forest in Rondonia will be cleared by the turn of the century.

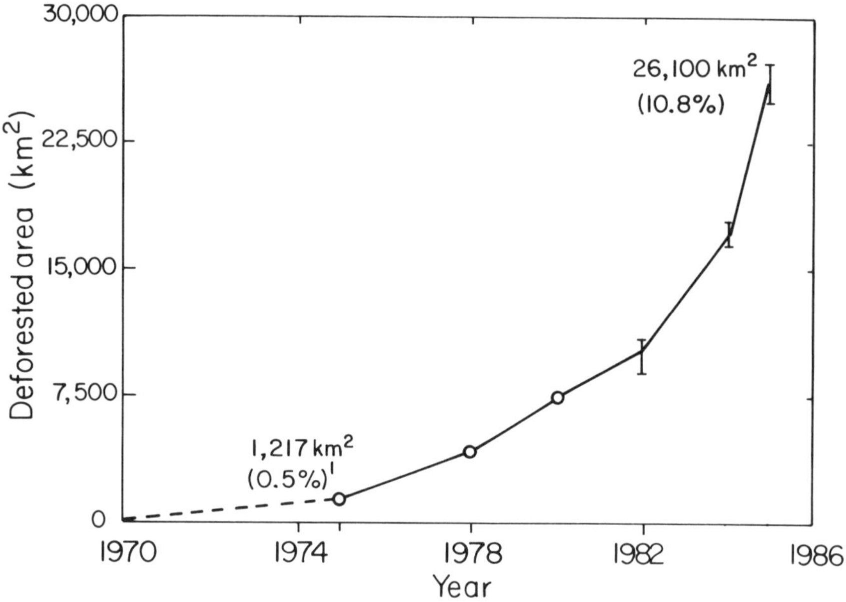

Figure 9. Time course of deforestation of the tropical forests of the state of Rondonia in Brazil. Data for 1975, 1978, and 1980 (open circles) are derived from LANDSAT images. Data for 1982, 1984, and 1985 (shown by brackets estimating the error determined by ground-based studies in 1985) are derived from the Advanced Very High Resolution Radiometer (AVHRR) on the NOAA satellites (from Malingreau and Tucker, 1988).

There is no doubt that we will be attaining more precise global assessments of land-use impacts on the biosphere in the future, utilizing satellite imagery. Although the estimates from records and land-based measurements have a large variance, they enable us to understand the general magnitude and history of various land use practices and how these impact the global carbon budget. Matthews (1983), utilizing the data base on global vegetation cover mentioned earlier, indicates that large areas of the globe that have been modified by human activities (Table II). Houghton *et al.* (1983) give a time dimension to these activities and relate them to changing values of carbon flux (Fig. 10). The shift in deforestation activity from North America to tropical regions, in the middle of the century, is striking. Houghton *et al.* (1985) estimate that the net release of carbon to the atmosphere through human activities in the past 100-220 years is 90-180 X 10^{15} g C. This is in the same range as estimates given by Siegenthaler and Oescher (1987) for total biospheric carbon release based on air bubbles in polar ice.

Table II. Estimates of past and present global vegetation cover (from Matthews, 1983).

	(Pre-agricultural) Vegetation data	Land cover data	(Present) Ecosystem reduction	
	Area (10^6 km^2)	Area (10^6 km^2)	Area (10^6 km^2)	Percent total ecosystem
Total forest	46.28	39.27	7.01	15.2
Tropical rainforest	12.77	12.29	0.48	3.8
Other forest	33.51	26.98	6.53	19.5
Woodland	15.23	13.10	2.13	13.8
Shrubland	12.99	12.12	0.87	6.7
Grassland	33.90	27.43	6.47	19.1
Tundra	7.34	7.34	0.0	0.0
Desert	15.82	15.57	0.25	1.6
Cultivation	0.93*	17.56	-16.63	--

*This figure is a residual land classification number and is not truly agriculture.

A BIOSPHERE SCRUBBER

It has been proposed that we make an effort to reverse the loss of carbon stores and carbon fixation capacity of the biosphere by an

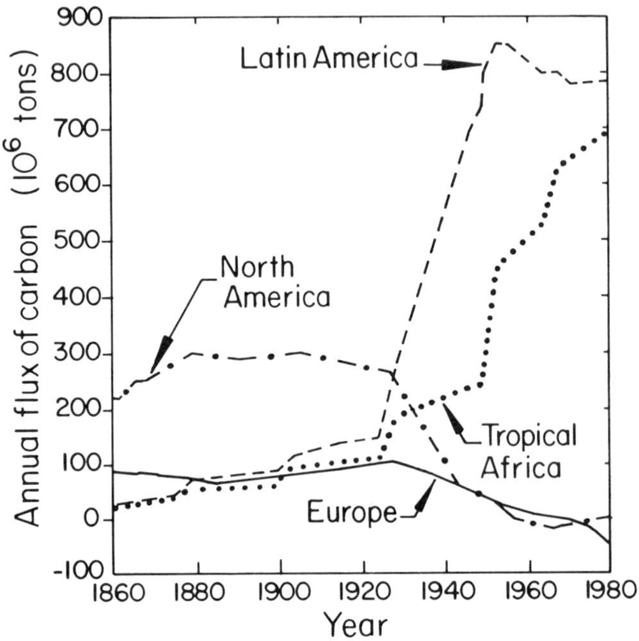

Figure 10. Carbon flux rates from several geographic areas through time (from Houghton *et al*., 1983).

intensive program of either enhancement of forest productivity or of reforestation. For reforestation to balance the 5 X 10^{15} g of carbon discharged into the atmosphere each year through fossil fuel combustion, we would have to plant fast-growing forests over an area equivalent to all that has been cleared to date or double the net annual yield of all existing closed forests (Marland, 1988). The costs of such an effort would be immense and not totally practical; however, Marland proposes that forestry can play a substantial role in alleviating the CO_2 problem (see Richards, *et al*., 1983).

SUMMARY

The fluxes and pools of biospheric carbon are changing as is evidenced by the increase in atmospheric concentration. There is, however, considerable uncertainty in the estimates of both pools and fluxes because the techniques available for their global measurement have been rather crude. Emerging techniques will give a better temporal and spatial resolution for these measurements and hence create the possibility of continuously measuring the earth's "metabolism".

ACKNOWLEDGMENTS

We are very grateful for the following persons for providing materials and help: K. Cromwell, D. Reichle, I. Fung, B. Choudhury, P. Vitousek, and E. Matthews. We express appreciation to the National Science Foundation (Harold Mooney, NSF BSR 86-14769; Chris Field, NSF BSR 87-17422, and to the A. W. Mellon Foundation for supporting studies related to this manuscript. This is CIW DPB publication number 1032.

REFERENCES

Allen LH, Lemon ER 1976. Carbon dioxide exchange and turbulence in a Costa Rican tropical rain forest. In Monteith JL (ed): "Vegetation and the Atmosphere,: London: Academic Press, pp 265-308.

Aoki M, Yabuki K, Koyama H 1975. Micrometeorology and assessment of primary production of a tropical rain forest in West Malaysia. J Agri Met (Japan) 31:115-124.

Anderson DE, Verma SB, Clement RJ, Baldocchi DD, Matt DR (1986). Turbulence spectra of CO_2, water vapor, temperature and velocity over a deciduous forest. Agric For Meteorol 38:81-99.

Baldocchi DD, Verma SB, Rosenberg NJ (1981a). Environmental effects of CO_2 flux and CO_2-water flux ratio of alfalfa. Agric Meteorol 24:175-184.

Baldocchi DD, Verma SB, Rosenberg NJ (1981b). Mass and energy exchanges of a soybean canopy under various environmental regimes. Agron J 73: 706-710.

Baldocchi DD, Hicks BB, Meyers TP 1988. Measuring biosphere-atmosphere exchanges of biologically-related gases with micrometeorological methods. Ecology 69:1331-1340.

Baldocchi DD, Verma SB, Anderson DE 1987. Canopy photosynthesis and water-use efficiency in a deciduous forest. J Appl Ecol 24:251-260.

Beadle CL, Talbot H, Neilson RE, Jarvis PG (1985). Stomatal conductance and photosynthesis in a mature Scots pine forest III. Variation in canopy conductance and canopy photosynthesis. J Appl Ecol 22: 587-595.

Biscoe PV, Scott RK, Monteith JL (1975). Barley and its environment. III Carbon budget of the stand. J Appl Ecol 12:269-291.

Boerma HR, Ashley DA (1988). Canopy photosynthesis and seed-fill duration in recently developed soybean cultivars and selected plant introductions. Crop Sci 28:137-140.

Bolin B, Doos BR, Jager J, Warrick RA (eds) 1986. "The Greenhouse Effect, Climatic Change, and Ecosystems." SCOPE 29: Chichester: John Wiley and Sons, 541 pp.

Browell EG, Gregory GL, Harriss RC, Kirchoof VWJH (1988). Tropospheric ozone and aerosol distributions across the Amazon basin. J Geophys Res 93:1431-1451.

Brown KW (1976). Sugar beet and potatoes. In Monteith JL (ed): "Vegetation and the atmosphere, Volume 2: Case studies," London: Academic Press, pp 65-86.

Choudhury BJ (1987). Relationships between vegetation indices, radiation absorption, and net photosynthesis evaluated by a sensitivity analysis. Remote Sensing Environ 22:209-233.

Choudhury BJ (1988). Relating Nimbus-7 37 GHz data to global land-surface evaporation, primary productivity and the atmospheric CO_2 concentration. Int J Remote Sensing 9:169-176.

Choudhury BJ, Tucker CJ (1987). Monitoring vegetation using Nimbus-7 37 GHz data: some empirical relations. Int J Remote Sensing 8:1085-1090.

Choudhury BJ, Tucker CJ, Golus RE, Newcomb WW (1987). Monitoring vegetation using Nimbus-7 scanning multichannel microwave radiometer's data. Int J Remote Sensing 8:533-538.

Coyne PI, Kelly JJ (1975). CO_2 exchange over the Alaskan arctic tundra: micrometeorological assessment by an aerodynamic method. J Appl Ecol 12:587-612.

Curran PJ (1981). Multispectral remote sensing for estimating biomass and productivity. In Smith H (ed): "Plants and the Daylight Spectrum," London: Academic Press, pp 65-99.

Daley PF, McNeil JN (1987). Canopy photosynthesis and dry matter partitioning of alfalfa infested by the alfalfa blotch leafminer [*Agromyza fontella* (Rondani)]. Can J Plant Sci 67:433-443.

Denmead OT (1976). Temperate Cereals. In Monteith JL (ed): "Vegetation and the Atmosphere, Volume 2: Case studies," London: Academic Press, pp 1-31.

Denmead OT, Bradley EF (1985.) Flux-gradient relationships in a forest canopy. In Hutchison BA, Hicks BB (eds): "The Forest-Atmosphere Interaction," Dordrecht: D Reidel Publishing Company, pp 421-442.

Desjardins RL, Brach EJ, Alvo P, Schuepp PH (1982). Aircraft monitoring of surface CO_2 exchange. Science 216:733-735.

Desjardins RL, MacPherson JL, Alvo P, Schuepp PH (1985). Measurements of turbulent heat and CO_2 exchange over forests from aircraft. In Hutchison BA, Hicks BB (eds): "The Forest-Atmosphere Interaction," Dordrecht: D Reidel Publishing Company, pp 645-658.

Dickinson RW (1986). How will climate change? In Bolin B, Doos BR, Jager J, Warrick RA (eds.): "The Greenhouse Effect, Climatic Change, and Ecosystems." SCOPE 29.: Chichester: John Wiley and Sons, pp 206-270.

Emanuel WR, Shugart HH, Stevenson MP (1985). Climate change and the broad scale distribution of terrestrial ecosystem complexes. Climate Change 7:29-43.

Evans JR (1988). Photosynthesis and nitrogen relationships in leaves of C3 plants. Oecologia (in press).

Field CB (1983). Allocating leaf nitrogen for the maximization of carbon gain: leaf age as a control on the allocation program. Oecologia 56:341-347.

Field C, Mooney HA (1986). The photosynthesis - nitrogen relationship in wild plants. In Givnish TJ (ed): "On the Economy of Plant Form and Function," Cambridge: Cambridge University Press, pp 25-55.

Field CB (1988). On the role of photosynthetic responses in constraining the habitat distribution of rainforest plants. Aust J Plant Physiol 15:343-358.

Fung IY, Tucker CJ, Prentice KC (1987). Application of advanced very high resolution radiometer vegetation index to study atmosphere-biosphere exchange of CO_2. J Geophys Res 92:2999-3015.

Gammon RH, Sundquist ET, Fraser PJ (1985). History of carbon dioxide in the atmosphere. In "Atmospheric Carbon Dioxide and the Global Carbon Cycle," US Department of Energy, DOE/ER-0239, Washington DC pp 25-62.

Gutschick VP, Wiegel FW (1988). Optimizing the canopy photosynthetic rate by patterns of investment in specific leaf mass. Am Nat 132:67-86.

Hirose T, Werger MA (1987). Maximizing daily canopy photosynthesis with respect to the leaf nitrogen allocation pattern in the canopy. Oecologia (Berlin) 72:520-526.

Holdridge LR (1947). Determination of world plant formations from simple climate data. Science 105:367-368.

Houghton RA, Hobbie JE, Melillo JM, Moore B, Peterson BJ, Shaver GR, Woodwell GM (1983). Changes in the carbon content of terrestrial biota and soils between 1860 and 1980: a net release of CO_2 to the atmosphere. Ecol Monog 53:235-262.

Houghton RA, Schlesinger WH, Brown S, Richards JF (1985). Carbon dioxide exchange between the atmosphere and terrestrial ecosystems. In Trabalka JR (ed): "Atmospheric Caron Dioxide and the Global Carbon Cycle," US Department of Energy, DOE/ER-0239, Washington DC, pp 11-140.

Jarvis PG, James GB, Landsberg JJ (1976). Coniferous forest. In Monteith JL (ed): "Vegetation and the Atmosphere, Volume 2: Case Studies," London: Academic Press, pp 171-240.

Jarvis PG, Miranda HS, Muetzelfeldt RI (1985). Modelling canopy exchanges of water vapor and carbon dioxide in coniferous forest plantations. n Hutchison BA, Hicks BB (eds): "The Forest-Atmosphere Interaction," Dordrecht: D Reidel Publishing Company, pp 521-542.

Jarvis PG, McNaughton KG (1986). Stomatal control of transpiration: scaling up from leaf to region. Adv Ecol Res 15:1-49.

Jones P, Allen LHJ, Jones JW (1985). Responses of soybean canopy photosynthesis and transpiration to whole-day temperature changes in different CO_2 environments. Agron J 77:242-249.

Jordan CF (1969). Derivation of leaf area index from light quality of the forest floor. Ecology 50:663-666.

Kanemasu ET, Wesley ML, Hicks BB, Heilman JL (1979). Techniques for calculating energy and mass fluxes. In Barfield BJ, Gerber JF (eds.): "Modification of the Aerial Environment of Plants," St Joseph, Michigan: Am Soc Agric Eng, pp 156-182.

Kumar M, Monteith JL (1981). Remote sensing of crop growth. In Smith H (ed): "Plants and the Daylight Spectrum," London: Academic Press, pp 133-144.

Lieth H (1975). Modeling the primary productivity of the world. In Lieth H, Whittaker RH (eds): "Primary Productivity of the Biosphere," New York: Springer-Verlag, pp 237-263.

Lorenz R, Murphy CE Jr (1985). Sulfur dioxide, carbon dioxide, and water vapor flux measurements utilizing a microprocessor-controlled data acquisition system in a pine plantation. In Hutchison BA, Hicks BB (eds): "The Forest-Atmosphere Interaction," Dordrecht: D Reidel Publishing, pp 133-147.

Malingreau J-P, Tucker CJ (1988). Large-scale deforestation in the southeastern Amazon Basin of Brazil. Ambio 17:49-55.

Marland G (1988). The prospect of solving the CO_2 problem through global reforestation. US Department of Energy, DOE/NBB-0082.

Matson PA, Harriss RC (1988). Prospects for aircraft-based gas exchange measurements in ecosystem studies. Ecology 69:1318-1325.

Matthews E (1983). Global vegetation and land use: new high-resolution data bases for climate studies. J Clim Appl Meteorol 22:474-487.

Molofsky J, Hall C, Myers N (1986). A Comparison of Tropical Forest Surveys. US Department of Energy, DOE/NBB-0078, 66 pp.

Monteith JL (1977). Climate and the efficiency of crop production in Britain. Phil Trans R Soc Lond B 281:277-294.

Monteney BA, Barbier JM, Bernos CM (1985). Determination of the energy exchanges of a forest type culture: *Hevea brasiliensis*. In Hutchison BA, Hicks BB (eds): "The Forest-Atmosphere Interaction," Dordrecht: D Reidel Publishing Company, pp 211-233.

Mooney HA, Gulmon SL (1979). Environmental and evolutionary constraints in the photosynthetic characteristics of higher plants. In Solbrig OT, Jain S, Johnson GB, Raven PH (eds): "Topics in Plant Population Biology," New York: Columbia University Press, pp 316-337.

Moore B, Bolin B (1986/87). The oceans, carbon dioxide, and global climate change. Oceanus 29:(4) 9-15.

Morgan JA, Willis WO (1983). Gas exchange and water relations of 'Olaf' spring wheat. Crop Sci 23:541-546.

Olson JS, Watts JA, Allison LJ (1983). Carbon in live vegetation of major world ecosystems. Oak Ridge National Laboratory, Environmental Sciences Division Publication No. 1997.

Parsons AJ, Leafe EL, Collett B, Stiles W (1983). The physiology of grass production under grazing. I. Characteristics of the leaf and canopy photosynthesis of continuously-grazed swards. J Appl Ecol 20:117-126.

Pearcy RW, Ehleringer J (1984). Comparative ecophysiology of C_3 and C_4 plants. Plant Cell Env 7:1-13.

Pearson RL, Miller LD (1972). Remote sensing of standing crop biomass for estimation of the productivity of the shortgrass prairie. In "Proceedings of the Eighth International Symposium on Remote Sensing of Environment," Ann Arbor: ERIM, pp 1357-1381.

Richards JF, Olson JS, Rotty RM (1983). Development of a Database for Carbon Dioxide Releases Resulting from Conversion of Land to Agricultural Uses. Oak Ridge Ass Univ, Oak Ridge, Tennessee.

Ripley EA, Redmann RE (1976). Grassland. In Monteith JL (ed): "Vegetation and the Atmosphere, Volume 2: Case Studies," London: Academic Press, pp 351-398.

Rodin LE, Bazilevich NI (1967). "Production and Mineral Cycling in Terrestrial Vegetation," Edinburgh, Scotland: Oliver and Boyd, 288 pp.

Sage RF, Pearcy RW (1987). The nitrogen use efficiency of C_3 and C_4 plants 1. Leaf nitrogen, growth, and biomass partitioning in *Chenopodium album* (L.) and *Amaranthus retroflexus* (L.) Plant Physiol 84:984-958.

Saugier B (1970). Transport turbulents de CO_2 et de vapeur d'eau au-dessers et a' l'ionterieur de la vegetacion. Methodes de mesure micrometeorologiques. Oecol Plant 5:179-223.

Saugier B (1976). Sunflower. In Monteith JL (ed): "Vegetation and the Atmosphere, Volume 2: Case Studies," London: Academic Press, pp 87-119.

Sellers PJ (1985). Canopy reflectance, photosynthesis, and transpiration. Int J Remote Sensing 6:1335-1372.

Sellers PJ (1987). Canopy reflectance, photosynthesis, and transpiration. II. The role of biophysics in the linearity of their interdependence. Remote Sens Environ 21:143-183.

Siegenthaler U, Oescher H (1987). Biospheric CO_2 emissions during the past 200 years reconstructed by deconvolution of ice core data. Tellus 39B:140-154.

Singh SM (1988). Simulation of solar zenith angle effect on global vegetation index (GVI) data. Int J Remote Sensing 9:237-248.

Spitters CJT (1986). Separating the diffuse and direct component of global radiation and its implications for modeling canopy photosynthesis. II. Calculation of canopy photosynthesis. AgRIC For Meteorol 38:231-242.

Thom AS (1976). Momentum, mass and heat exchange of plant communities. In Monteith JL (ed): "Vegetation and the Atmosphere, Volume 1: Principles," London: Academic Press, pp 57-109.

Trabalka JR (1985). Atmospheric Carbon Dioxide and the Global Carbon Cycle. US Department of Energy, DOE/ER-0239.

Tucker CJ, Elgin JH, McMurtrey JE (1980a). Relationship of crop radiance to alfalfa agronomic values. Int J Remote Sens 1:69-75.

Tucker CJ, Holben BN, Elgin JH, McMurtrey JE (1980b). Relationship of spectral data to grain yield variation. Photogram Eng Remote Sens 45:600-608.

Tucker CJ, Holben BN, Elgin JH, McMurtrey JE (1981). Remote sensing of total dry-matter accumulation in winter wheat. Remote Sensing Environ 11:171-189.

Tucker CJ, Vanpraet CL, Sharman MJ, van Ittersum G (1985). Satellite remote sensing of total dry matter production in the Sengalese Sahel. Remote Sensing Environ 13:17:233-249.

Tucker CJ, Fung IY, Keeling CD, Gammon RH (1986). Relationship between atmospheric CO_2 variations and a satellite-derived vegetation index. Nature 319:195-199.

Uchijima Z (1976). Maize and Rice. In Monteith JL (ed): "Vegetation and the Atmosphere, Volume 2: Case Studies," London: Academic Press, pp 33-64.

Wells R, Meredith WRJ, Williford JR (1986). Canopy photosynthesis and its relationship to plant productivity in near-isogenic cotton lines differing in leaf morphology. Plant Physiol 82:635-640.

Whittaker RH, Likens GE (1975). The biosphere and man. In Lieth H, Whittaker RH (eds): "Primary Productivity of the Biosphere. Ecological Studies 14," New York: Springer Verlag, pp 305-328.

Wofsy SC, Harriss RC, Kaplan WA (1988). Carbon dioxide in the atmosphere over the Amazon Basin. J Geophys Res 93 (D2):1377-1387.

Woodrow IE, Ball JT, Berry JA (1987). A general expression for the control of the rate of photosynthetic CO_2 fixation by stomata, the boundary layer and radiation exchange. In Biggins J (ed): "Progress in Photosynthesis Research," The Netherlands: Martinus Nijhoff, pp 225-228.

Woodwell GM, Hobbie JE, Houghton RA, Melillo JM, Peterson BJ, Shaver GR, Stone TA, Moore B, Park AB (1983). Deforestation Measured by LANDSAT: Steps Toward a Method. US Department of Energy, DOE/EV/10468.

Photosynthesis, pages 45–58

SOME VIEWPOINTS ON PHOTOSYNTHETIC RESPONSE AND ADAPTATION TO ENVIRONMENTAL STRESS

Olle Björkman

Carnegie Institution of Washington, Department of Plant Biology, 290 Panama Street, Stanford, CA 94305 USA

I suspect that most of us who are gathered at this Symposium in honor of Stacy French are studying photosynthesis simply because we would like to find out how this important process works. However, it would probably please us a great deal if we felt that our efforts might ultimately contribute to an improved plant productivity, either by providing a better basis for managing natural or cultivated plant communities, or for breeding more efficient crops. In any case, there is no doubt that this aspect is important to our sponsors.

In this context it may be appropriate to quote the following statements by Roger Gifford who, I am certain, has given a great deal of thought to the problem: "After several decades of effort there has, to my knowledge, been no cultivar released in which any photosynthetic attribute has been successfully incorporated to increase productivity.... Our assumption that by exerting selection pressure on photosynthetic attributes, or by applying mechanistic knowledge of photosynthesis in breeding will increase productivity has so far failed. Perhaps we should now assume that evolution has perfected photosynthesis for high productivity in resource-limited natural environments" (Gifford, 1987).

I agree with Roger Gifford's statement that attempts to increase crop productivity by applying mechanistic knowledge of photosynthesis has been unsuccessful so far. Perhaps it is also true that the photosynthetic characteristics of wild plants are already perfectly adapted to make maximum use of the resources available in the respective environments. If so, any attempts to improve the productivity of natural vegetation by modification of its photosynthetic characteristics seem doomed to fail. To a large degree the same argument may also apply to crop plants that are grown in climates for which the crop is well suited. However, the situation may be different when a crop is introduced into an environment to which it is ill-adapted; or when crops are subjected to unusually severe stress, as happened in much of the drought- and heat-stricken part of this country during the current growing season. Under these circumstances, modification of genetically determined photosynthetic characteristics may be one of the requirements to improve the fitness and productivity of the crop.

Recent progress in plant molecular genetics promises that we may be able to modify a crop's genetic constitution at will, without the constraints imposed by traditional plant breeding techniques. Genetic traits from quite unrelated species may be introduced into a crop and this will hopefully include desirable photosynthetic traits. However, an essential prerequisite for such endeavors to succeed obviously requires that we are able to identify the desirable photosynthetic traits that we wish to introduce. I believe that our present level of understanding of what constitutes a truly desirable photosynthetic trait in any particular field situation is quite rudimentary, and far from sufficient to provide the plant breeder with any clue as to what photosynthetic trait he should attempt to incorporate.

It seems to me that the failure in applying mechanistic knowledge of photosynthesis to crop improvement is in large part due to insufficient knowledge of the factors that underlie photosynthetic adaptation to an unfavorable environment. I will attempt to illustrate some of the problems encountered in trying to determine these factors.

SHADE TOLERANCE

I will first briefly address the factors that are associated with shade tolerance. The reason for doing so is that we probably know more about how plants respond and adjust to the unfavorable situation of having to maintain a net carbon balance even when the driving force itself is very weak, than we know about their adaptive response to any other environmental stress (for recent reviews, see Anderson and Osmond, 1987; Anderson *et al.*, 1988). Although plants differ greatly in the extent to which they can tolerate deep shade, I think it is safe to say that all plants have considerable capacity to respond in an adaptive manner when they are faced with a scarcity of light. Factors associated with shade tolerance are:

1) Leaves displayed so that maximum light interception per unit leaf area is obtained. This often includes active leaf movements.
2) The allocation of carbon, nitrogen, and other resources to leaf growth in relation to the growth of roots and other non-photosynthetic organs is increased.
3) The ratio of leaf area to leaf mass is increased.
4) The chlorophyll content per chloroplast volume is increased; this change is related to an increased grana stacking.

5) There is an increased ratio of chlorophyll to other chloroplast components such as Rubisco and other enzymes of carbon metabolism; to electron carriers such as cytochrome *f* and plastoquinone; and to the amount of the phosphorylation coupling factor CF_1.
6) The antenna size of PSII in increased.
7) Associated with these changes is a decreased rate of leaf respiration resulting in a low light compensation point.

All of these responses evidently result in increased light interception relative to the investment and maintenance costs. This increased light-harvesting efficiency occurs at the expense of enzymes and electron carriers, that is, at the expense of components that are required for effective use of *high* irradiances. This puts severe restriction on the ability of the plant to take advantage of periods of bright light. The high light-harvesting efficiency and early light saturation of photosynthesis could also result in a potentially harmful accumulation of excessive excitation energy at the reaction centers. I will return to this problem later in my talk.

It is evident that shade tolerance involves modifications at several levels of organization and it would be difficult to point out a specific single trait to select for, in the event that it would indeed be desirable to increase the shade tolerance of a species or cultivar. But at least we have some idea in what direction we should go. Perhaps we could start out by selecting for a high ratio of chlorophyll per chloroplast volume or extensive grana stacking.

As suggested by Gifford (1987) perhaps there has been a tendency to breed for unnecessarily high chlorophyll per chloroplast volume, not because of any wish to increase the shade-tolerance of the crop, but simply because dark-green leaves look healthy. This may be counterproductive to a plant growing in a high-irradiance environment since high chlorophyll content may provide little or no benefit. It would seem that the resources tied up in an over-investment in light-harvesting chlorophyll could be better used for chloroplast components that would increase the capacity for light-saturated photosynthesis. This would improve the return per investment in terms of daily carbon gain per biosynthetic cost, without increasing the total cost.

Photosynthetic capacity per unit leaf area can, of course, also be increased by increasing the total amount of chloroplast components per leaf area. However, it must be kept in mind that such an increase in photosynthetic rate per leaf area occurs at the expense of a smaller total leaf area per plant. The daily carbon gain by the plant is the product of photosynthetic rate per leaf area, *times* leaf area. Therefore, sooner or later a point will be reached at which the gain in photosynthesis from a further increase in photosynthetic capacity will be more than counterbalanced by the reduction in leaf area. This point will undoubtedly depend on the particular environment in which the plant is grown. It is therefore not surprising that selection for high photosynthetic rate per leaf area without taking the cost into account, has not been accompanied by an improved productivity. To my knowledge, the only photosynthetic trait than can increase the photosynthetic rate per leaf area in a high-irradiance environment without an increased investment cost or water use, is C-4 photosynthesis (e.g. Osmond *et al.*, 1980).

TOLERANCE TO DROUGHT AND EXCESSIVE IRRADIANCE

I will now turn to considering photosynthetic responses to drought, an environmental condition that on a world-wide scale is the most important limiting factor to productivity. Drought is commonly defined as the absence of rainfall for a period of time long enough to affect plant growth. Drought leads to water stress as the plants gradually deplete soil moisture. Many books have been published on the response and adaptation of plants to water stress and I will make no attempt to review this intensive field of research. It is of course impossible to consider photosynthetic responses out of context with the responses of other growth processes but I will try to highlight some of the problems relevant to photosynthetic function during drought.

Drought is usually accompanied by high irradiance, high temperature and low air humidity. By increasing the evaporative demand these factors exacerbate the stress imposed by the reduced availability of water. Conversely, drought can increase the excessive excitation energy at a given high irradiance by decreasing the rate with which excitation energy can be dissipated via photosynthesis. Drought tolerance may therefore be closely linked with the ability of the leaves to avoid or to tolerate excessive radiation. Many plant species are able to decrease the amount of excessive energy by decreasing light interception (Koller, 1986). In some species such as *Oxalis oregana* (Powles and Björkman, 1981), Siratro (Ludlow and Björkman, 1984), and probably many other species, light interception is reduced by active paraheliotropic leaf movements. In other species, leaves developed under excessive radiation (drought) are steeply inclined; and in some this steep leaf angle is combined with development of a highly reflective layer of leaf pubescence (Ehleringer *et al.*, 1976). These responses can cause strong attenuations of the total energy absorbed by the leaf; thus, not only is the photosynthetically active radiation decreased, but the buildup of excessively high leaf temperatures may also be avoided.

It is obvious that accumulation of excessive excitation energy would also be prevented if that portion of the energy which cannot be used in photosynthesis can be safely dissipated via another pathway in the light-harvesting antenna of chlorophyll molecules. I will discuss this question in some detail below in a more general context of how photosynthesis responds to excessive light for I believe that our new concepts can have far-reaching implications on our view of how plants cope with drought and other stress factors.

It has been known for many years that exposure to intense light of isolated chloroplasts, algal cells and single leaves, causes a reduction in the photon yield of photosynthesis measured after return to low light. Such high-light treatments can also cause a reduction in the rate of electron transport and a strong quenching of the maximum and variable fluorescence from photosystem II. These and other inhibitory effects of excessive light were generally attributed to damage to the PSII reaction centers, or to a site closely associated with these centers (for reviews, see Powles, 1984; Kyle, 1987). Thus the term photoinhibition, originally defined as a reduction in photosynthetic activity by excessive light, became more or less synonymous with photoinhibitory damage.

In this vein, Steve Powles and I subjected well-established field-grown plants of *Nerium oleander* to gradual water stress and followed the rate of electron transport and chlorophyll fluorescence characteristics (Björkman and Powles, 1984). Plants droughted in full sunlight exhibited a strong decline in electron transport activity with increased duration and intensity of the stress. Much smaller effects were observed when the plants were droughted in the shade. Moreover, as shown in Fig. 1, chlorophyll fluorescence, measured at dawn, declined as the water potential fell over a period of about 2 weeks. F_M is the fluorescence yield in the state when all the PSII reaction centers are closed; F_O is the yield when all centers are open. F_M fell steeply as the water stress increased, F_O was less affected. The curve labeled ϕ_{pc}, depicts the efficiency of PSII photochemistry. Clearly, this efficiency was strongly affected by the drought treatment. Although not shown in Fig. 1, no quenching of fluorescence could be detected when the plants were droughted in the shade. The responses of oleander leaves, droughted in the sun, were in all respects similar to those obtained when well-watered oleander plants, grown under a relatively low light regime, were exposed to bright light. So we concluded that, in both cases, we were observing an inhibition of photosynthetic activity by excessive excitation energy. I believe that this conclusion is still valid. However, we also concluded at the time that these responses were indicative of photoinhibitory damage. This interpretation was probably incorrect.

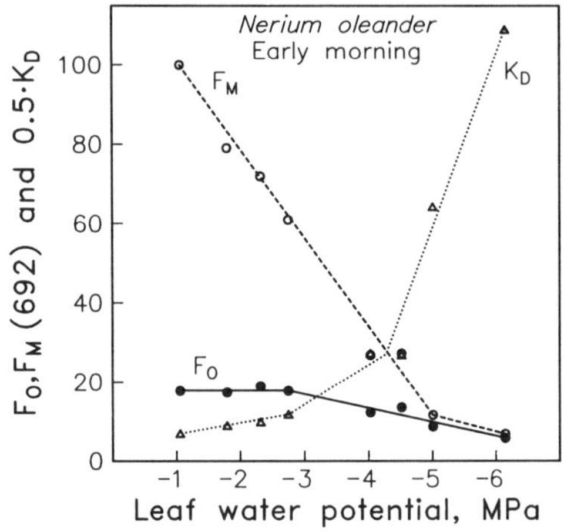

Figure 1. Effect of leaf water potential on 77K fluorescence in *Nerium oleander* leaves growing in direct sunlight. Measurements were made in the early morning before the leaves received direct radiation.

Subsequent studies provided strong evidence that the fluorescence quenching observed in leaves that had been exposed to excessive light was not necessarily caused by damage to the PSII reaction centers (Demmig and Björkman, 1987). We concluded that instead, this quenching reflects an increase in non-radiative dissipation of excitation energy which protects the PSII centers from damage by excessive energy.

Figure 2 serves to illustrate this point. The box on the top represents photosystem II in its ground state, that is, its non-excited state. When photons are absorbed, PSII becomes excited. This excited state is depicted in the lower box. PSII can then be de-excited and returned to the ground state via several competing processes. These are PSII photochemistry, energy transfer to photosystem 1, non-radiative dissipation, and chlorophyll fluorescence. The fraction dissipated via fluorescence is at most 2% of the total, so we can ignore its contribution, but fluorescence serves a convenient probe. As long as the rate of photosynthetic electron transport is high enough to keep up with the energy flow into PSII, almost all of the de-excitation occurs via electron transport. Little or no de-excitation occurs via non-radiative dissipation. But when the energy flow into PSII exceeds the electron transport rate, then the rate constant for non-radiative dissipation increases so that the excess energy is dissipated via this process. It does not matter whether the excess of excitation energy is brought about by increasing the light intensity or whether it is brought about by restricting the rate of electron transport. By diverting the excessive energy away from the PSII reaction centers, the increase in non-radiative dissipation protects these centers

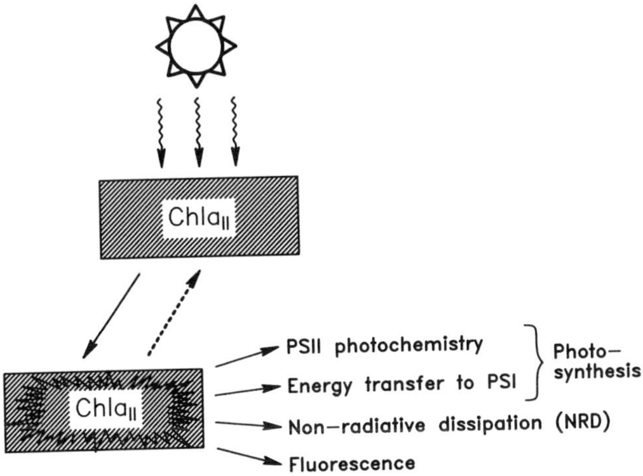

Figure 2. Processes of energy dissipation from photosystem II.

from harmful overexcitation. However, unless the rate constant for non-radiative dissipation rapidly returns to a low value when the light intensity is lowered to a rate-limiting level, we will find that the photon yield of photosynthesis will be reduced. In this case the photon yield is reduced not because of any damage but simply because some of the energy continues to be dissipated via the non-radiative process rather than being used to drive PSII photochemistry.

Let us return to Fig. 1 for a moment. The dotted line, labelled K_D, depicts the rise in the rate constant for non-radiative energy dissipation as the potential for accumulation of excess energy increased in these oleander leaves as the severity of water stress increased. In the present example, what at first sight appeared to reflect a stress-induced damage may instead turn out to reflect the operation of a regulatory and adaptive process. I think this is a good example of the problems we face in trying to interpret photosynthetic responses to stress.

Experimental data on the responses of plants to excessive irradiance in real field situations are sparse. In a recently published study on several species of mangroves growing in an open intertidal habitat on the coast of North Queensland, Australia it was found that leaves naturally exposed to excessive light had markedly depressed photon yields of O_2 evolution. Fluorescence yields were also severely quenched, showing that the efficiency of photosystem was reduced (Björkman *et al.*, 1988). Mangrove shade leaves had normal photon yields and fluorescence characteristics showing that energy conversion efficiency was unaffected by the high salinity or the associated low leaf water potentials ("physiological drought"). Shading of mangrove sun leaves resulted in a gradual increase in the photon yield and relaxation of fluorescence quenching but in most species more than 1 week was required to reach the values found in leaves that had developed in the shade. Such slow recovery could indicate that repair of photoinhibitory damage is involved. However, on the basis of the degree of quenching of F_O relative to the quenching of F_M in these sun leaves and during recovery we concluded that, like the situation in water-stressed oleander leaves, the reduced photon yields and the massive fluorescence quenching in mangrove sun leaves were more likely caused by a large, sustained increase in non-radiative energy dissipation. Adams (1988) arrived at similar conclusions regarding the responses to excessive light in the epiphytic fern *Pyrrosia confluence*, growing on the exposed edge of a subtropical rainforest as well as in another CAM species, the cactus *Opuntia stricta*, growing in a mixed *Eucalyptus* woodland. Both habitats are in New South Wales, Australia.

To illustrate how an important agricultural crop plant responds to a combination of high irradiance and drought, I will now discuss some of the recent results obtained in a cotton field in the San Joaquin Valley, California (Björkman *et al.*, 1989). Since this region is characterized by very high irradiance, high temperatures, and no significant rainfall during the cotton growing season, the crop is normally grown under irrigation. To study the response to drought we simply withheld irrigation. When this was done, midday leaf water potential gradually fell from about -1.0 to -3.0 MPa. Because of the diahelioptropic leaf movements exhibited by this species of cotton (*Gossypium hirsutum* L. cv. SJ2) the daily radiation receipt by exposed leaves approached 90 mol photon m^{-2}. This is approx. 40% higher than that received by horizontally-oriented leaves.

Fully sunlit leaves of irrigated plants had very high rates of net CO_2 uptake (ca. 44 μmol m^{-2} s^{-1}) throughout the growing season from mid-May to early October. The intrinsic photon yield of O_2 evolution (measured at limiting photon flux densities and 4.5 - 5% CO_2) of leaves developed in full sunlight also remained high and constant throughout the season with a mean value of 0.104 ± 0.006 O_2 per absorbed photon. Similarly, fluorescence yields measured in the early morning showed no signs of photoinhibitory effects. We therefore conclude that no photoinhibitory damage occurred even though these solar tracking leaves received extremely high irradiances.

Fig. 3 (left panel) shows the response of photosynthetic CO_2 fixation $P(CO_2)$ to increasing photon flux density in the field. Also shown are the light dependencies of the rate of photorespiration, calculated according to von Caemmerer and Farquhar (1981) and of P(ET) which is the sum of

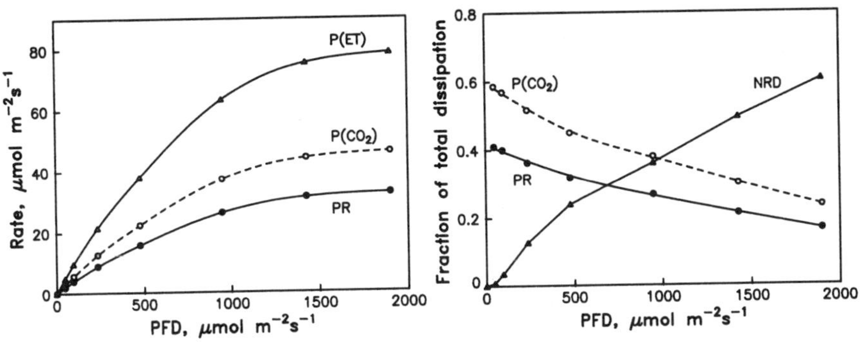

Figure 3. Response of photosynthesis, photorespiration and electron flow to absorbed photon flux density in cotton leaves in the field (left) and fraction of total excitation energy dissipated via photosynthesis, photorespiration, and non-radiative dissipation (right). After Björkman *et al*., 1989).

$P(CO_2)$ and PR. P(ET) is therefore a measure of the electron or energy flow via photosynthesis. Fig. 3 (right panel) shows the fractions of the total excitation energy that was dissipated via $P(CO_2)$ and PR. The curve labeled NRD depicts the fraction dissipated via non-radiative energy dissipation, determined by measuring the quenching of the variable fluorescence yield. That component of fluorescence quenching which saturates at low PFD levels and which probably reflects an increased energy transfer to photosystem 1 (see Fig. 2) was not included in these estimates of NRD. The results clearly show that the gradual decrease in the fraction of energy dissipated via $P(CO_2)$ and PR that takes place when PFD increases is almost quantitatively matched by a compensating increase in NRD. Our estimates show that in fully sunlit leaves of irrigated cotton, approx. 25% of the excitation energy was used for CO_2 fixation; an additional 19% was dissipated in photorespiration and the remainder, 56%, was dissipated via NRD. This dissipation of excessive energy prevented overreduction of the photosystem II centers. Estimates of the reduction state of these centers, using a fluorescence analysis approach, indicated that at most 27% of the centers were in the reduced state in fully sunlit cotton leaves.

So what happened when the cotton plants were subjected to drought? I will first briefly discuss the effect of mild water stress which was insufficient to cause any detectable decrease in midday leaf water potential. Two groups of plants were sown in well-preirrigated soil. The control plants received a second irrigation on Julian day 166, the stress-treatment plants did not. The above ground biomass gain of the unirrigated plants during the 25-day period from day 166-191 was only some 42% of the control plants. From these and other results we can calculate that the average relative growth rate during this period fell from approx. 10% per day in the control to 6.7% per day in the unirrigated plants. The larger portion of this reduction was clearly caused by a reduced rate of leaf expansion. The growth rate expressed on the basis of leaf mass (or area) was only about 10% lower in unirrigated as compared with irrigated plants. The rate of CO_2 fixation and leaf conductance to water vapor loss also were only about 10% lower. Evidently, the decreased leaf development together with the slightly lower leaf conductances slowed down whole plant transpiration rate so that leaf water potential remained the same as in the controls (-1.0 to 1.1 MPa). These results illustrate a response to drought that may be common knowledge among many crop scientists but seems to have received scant attention among plant physiologists (Bradford and Hsiao, 1982). It seems to me that a decreased rate of leaf development is a clearly adaptive response to a restriction of the water supply since it allows the existing leaves to photosynthesize at a nearly undiminished rate. The response would be especially favorable if, as often seems to be the case, the decreased allocation to leaf growth is associated with an increased allocation to root growth.

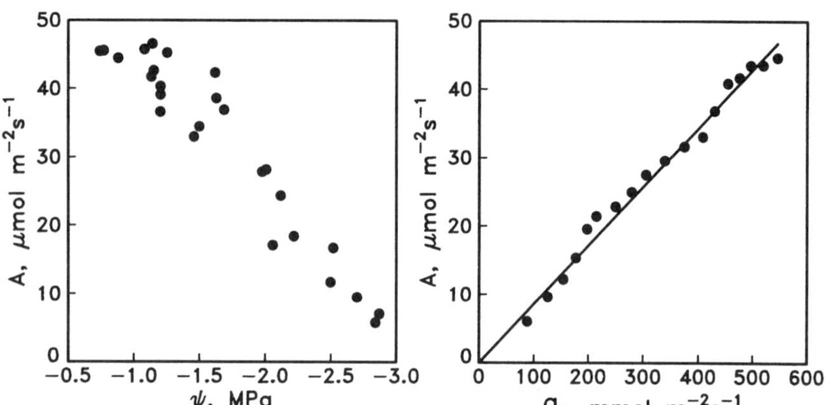

Figure 4. Effect of increasing water stress on midday rate of net CO_2 uptake (A) in sunlit cotton leaves in the field (left); and relationship between A and leaf conductance (g_L) in these leaves (right).

Fig. 4 (left) shows how the light-saturated rate of net CO_2 uptake (A) responded to continued drought over time. An essentially linear decrease in A from about 44 to about 7 μmol m^{-2} s^{-1} was obtained as the leaf potential (Ψ) decreased from approx. -1.25 MPa to -2.8 MPa. There was a parallel decrease in leaf conductance to water vapor loss so that a more or less constant slope is obtained when A is plotted *versus* g_L (Fig. 4, right). This means that the photosynthetic water use efficiency remained essentially constant as the intensity and duration of water stress gradually increased. Contrary to many statements in the water-stress literature a linear relationship between A and g_L does not indicate that non-stomatal components of photosynthesis were unaffected by drought or that photosynthesis was solely limited by stomatal conductance. If this were the case, then we would expect a convex curvilinear relationship in this plot (cf. Wong *et al.*, 1979), at least as long as stomatal apertures were reasonably uniform across the leaf surface. The relationship between A and g_L shown in Fig. 4 (right) would rather seem to indicate that stomatal aperture responded to a decreased capacity for CO_2 fixation and, as a result, the leaves maintained a more or less constant intercellular CO_2 concentration and photosynthetic water use efficiency. This may be another example of the operation of a regulatory mechanism that keeps the different component processes of photosynthesis in balance.

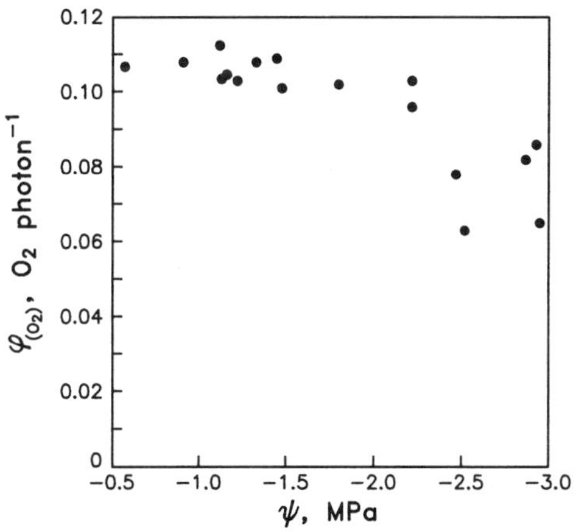

Figure 5. Effect of increasing water stress on the photon yield of light-limited, CO_2 saturated photosynthesis of cotton leaves exposed to high irradiance in the field.

Fig. 5 shows that despite the strong reductions in the rate of light-saturated photosynthesis the intrinsic photon yield (ϕO_2) remained essentially unaffected by water stress above leaf water potentials of -2.2 MPa. Significantly reduced photon yields were only found under quite severe water stress levels (< -2.4 MPa). At -2.8 MPa the mean photon yield was approx. 30% below that for leaves of irrigated plants. Fluorescence studies *in situ* also indicated that the efficiency of energy conversion by photosystem II was also significantly lower in leaves of severely stressed plants. While it is possible some photoinhibitory damage was present in these leaves it seems more likely that the reduction in this efficiency was caused by a sustained increase in the rate constant for non-radiative energy dissipation. This strongly resembles the situation in mangrove sun leaves and sun leaves of water-stressed oleander plants discussed earlier. In any event, the lowered efficiency of energy conversion found in these severely stressed cotton leaves is insufficient to exert any significant influence on the photosynthetic rate other than during the brief periods when the light intensity is sufficiently low to be rate-limiting.

Fluorescence analyses showed that increased water stress levels increased the fraction of energy dissipation via non-radiative dissipation. In plants growing under intermediate water stress levels (Ψ = -1.8 MPa) the net rate of CO_2 uptake in full sunlight was approx. 27 μmol m^{-2} s^{-1}. In this case, 18% of the excitation energy was used in CO_2 fixation and 16% was dissipated via photorespiration, raising the excess to 66% of the total. The cotton leaves compensated by increasing NRD to this percentage. The fraction of reduced PSII centers was higher than in leaves of irrigated plants but at most 38% were in the reduced state. In severely water stressed plants (Ψ = -2.8 MPa) only about 13% of the excitation energy was used for CO_2 fixation and we estimate that the same percentage was dissipated in photorespiration. Since the intrinsic photon yield and the efficiency of PSII photochemistry of these leaves were below normal, even after overnight recovery, it is difficult to obtain accurate values for NRD and the reduction state of the PSII centers. Rough estimates show that NRD increased still further perhaps to as much as 73% of the total and that less than one-half of the PSII centers were in the reduced state at peak solar radiation.

The molecular mechanism(s) of non-radiative energy dissipation still remains to be determined. This is now an object of active research in several laboratories including my own. Various mechanisms have been suggested, including the operation of a "futile" cycle of electrons around photosystem II (Schreiber and Neubauer, 1987). There is also increasing evidence in support of the involvement of specific carotenoid pigments of the thylakoid membranes ("xanthophyll cycle"). For an account of progress in these studies see article by Demmig-Adams (this Volume).

CONCLUDING REMARKS

In conclusion I don't think that we are yet ready to give much advice based on mechanistic knowledge of photosynthesis to a group of plant breeders or genetic engineers as to the photosynthetic modifications they should make to improve crop productivity in a stressful environment. Undoubtedly, knowledge of the mechanistic basis of plant responses to stress greatly improves our ability to determine whether a particular response is adaptive or whether it reflects a stress injury. I therefore believe that it is essential that experiments on the responses to stress in real field situations be closely linked with studies of the mechanistic basis of the response; conversely it is equally important that hypotheses regarding the adaptive significance of a particular molecular mechanism, derived from experiments in the laboratory, be adequately tested in the field. This will require a much closer interaction and collaboration among specialists in different areas of photosynthesis research and also a closer interaction between investigators concerned with basic and applied aspects of plant growth and productivity, than we have seen in the past.

REFERENCES

Adams III WW (1988). Photosynthetic acclimation and photoinhibition of terrestrial and epiphytic CAM tissues growing in full sunlight and deep shade. Aust J Plant Physiol 15:123-134.

Anderson JM, Osmond CB (1987). Sun-shade responses: compromises between acclimation and photoinhibition. In Kyle DJ, Osmond CB, Arntzen CJ (eds): "Photoinhibition," Elsevier, Amsterdam, New York, Oxford, pp 1-38.

Anderson JM, Chow WS, Goodchild DJ (1988). Thylakoid membrane organization in sun/shade acclimation. Aust J Plant Physiol 15:11-26.

Björkman O, Powles SB (1981). Leaf movement in the shade species *Oxalis oregana*. I. Responses to light level and quality. Carnegie Inst Washington Year Book 80:59-62.

Björkman O, Powles SB (1984). Inhibition of photosynthetic reactions under water stress: interaction with light level. Planta 161:490-504.

Björkman O, Demmig B, Andrews TJ (1988). Mangrove photosynthesis: Response to high-irradiance stress. Aust J Plant Physiol 15:43-61.

Björkman O, Schäfer C, Shih C, Thayer S (1989). Photosynthetic performance under high irradiance: Effect of drought stress on photosynthetic energy conversion in a cotton crop in the San Joaquin Valley, California (in preparation).

Bradford KJ, Hsiao TC (1982). Physiological responses to moderate water stress. In Lange OL, Nobel PS, Osmond CB, Ziegler H (eds): "Encyclopedia of Plant Physiology" NS Vol 12B, Springer, Berlin, Heidelberg, New York, pp. 263-324.

Demmig B, Björkman O (1987). Comparison of the effect of excessive light on chlorophyll fluorescence (77K) and photon yield of O_2 evolution in leaves of higher plants. Planta 171:171-184.

Ehleringer J, Björkman O, Mooney HA (1976). Leaf pubescence: Effects on absorptance and photosynthesis in a desert shrub. Science 192:376-377.

Gifford R (1987). Barriers to increasing crop productivity by genetic improvement in photosynthesis. In Biggins J (ed): "Progress in Photosynthesis Research," Vol 4, Martinus Nijhoff, Dordrecht, Boston, Lancaster, pp 377-384.

Koller D (1986). The control of leaf orientation by light. Photochem Photobiol 44:819-826.

Kyle DJ (1987). The biochemical basis for photoinhibition of photosystem II. In Kyle DJ, Osmond CB, Arntzen CJ (eds): "Photoinhibition," Elsevier, Amsterdam, New York, Oxford, pp 197-227.

Ludlow MM, Björkman O (1984). Paraheliotropic leaf movement in Siratro as a protective mechanism against drought-induced damage to primary photosynthetic reactions: damage by excessive light and heat. Planta 161:505-518.

Osmond CB, Björkman O, Anderson DJ (1980). "Physiological Processes in Plant Ecology." Ecological Studies no 36, Chapters 9 and 10, Springer, Berlin, Heidelberg, New York.

Powles SB (1984). Photoinhibition of photosynthesis induced by visible light. Annu Rev Plant Physiol 35:15-44.

Schreiber U, Neubauer C (1987). The polyphasic rise of chlorophyll fluorescence upon onset of strong illumination: II. Partial control by the photosystem II donor side and possible ways of interpretation. Z. Naturforschung 42C (in press).

von Caemmerer S, Farquhar G (1981). Some relationships between biochemistry of photosynthesis and the gas exchange of leaves. Planta 153:376-387.

Wong SC, Cowan IR, Farquhar GD (1979). Stomatal conductance correlates with photosynthetic capacity. Nature 282:424-426.

Photosynthesis, pages 59–70

PHOTOSYSTEM II AND OXYGEN EVOLUTION

Norio Murata and Mitsue Miyao

National Institute for Basic Biology, Myodaiji, Okazaki 444, Japan

INTRODUCTION

The photosystem II (PSII) complex is a complicated membrane protein and is composed of at least 20 kinds of protein components, chlorophyll (Chl), pheophytin (Pheo), carotenoids, lipids, quinones, and inorganic ions (see Fig. 1). It is located in the appressed region of the thylakoid membrane (Arntzen, 1978). Within the PSII complex, light energy is captured by the light-harvesting Chl, the captured light energy is used to

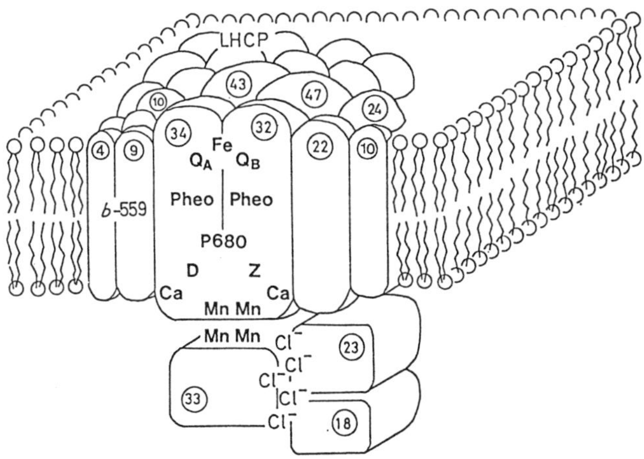

Figure 1. A model for the PSII complex of spinach. P680 and pheophytin (Pheo) are the electron donor and acceptor, respectively, of the photochemical reaction center. Q_A and Q_B are the primary and secondary quinone acceptors, respectively. Z and D are tyrosine residues of the 32-kDa and 34-kDa proteins, respectively, and Z acts as the electron donor to P680. LHCP is the light-harvesting Chl *a/b* protein. Numbers in circles refer to the estimated relative molecular masses of proteins. The proteins of 32 kDa and 34 kDa correspond to D1 and D2 proteins, respectively. Fe, Ca, Mn and Cl are inorganic components.

separate positive and negative charges at the photochemical reaction center (P680 and Pheo), and electrons are transported from water molecules to plastoquinone. As a result of the dehydrogenation of two water molecules, an oxygen molecule is evolved. The oxygen-evolving reaction is catalyzed by the Mn cluster composed of four Mn atoms, and requires calcium and chloride ions as essential inorganic factors (Miyao and Murata, 1987; Murata and Miyao, 1987). The characteristics and functions of some of the protein components in the PSII complex have been reviewed (Miyao and Murata, 1987; Murata and Miyao, 1987). In the present paper, we describe various types of PSII preparation, PSII extrinsic proteins, a new PSII gene, and lipids in the PSII complex.

PHOTOSYSTEM II COMPLEX

Recent biochemical techniques using detergents have made it possible to isolate various types of PSII complexes and sub-complexes from thylakoid membranes. The first success was the treatment of thylakoid membranes with Triton X-100 followed by differential centrifugation to isolate membranes which contain only PSII complexes but none or little of the other membrane-bound protein complexes, such as photosystem I, cytochrome b_6/f and CF_1/CF_0 complexes (Berthold *et al.*, 1981; Kuwabara and Murata, 1982b). This type of preparation is denoted as a PSII membrane. In addition to the protein components shown in Fig. 1, there are at least five protein components of molecular masses below 8 kDa (Ljungberg *et al.*, 1986; Murata *et al.*, 1988). This type of PSII preparation retains oxygen-evolution activity of 400-500 μmol oxygen/mg Chl/h (Berthold *et al.*, 1981; Kuwabara and Murata, 1982b).

A treatment of the PSII membranes with octylglucoside dissociates the PSII complex into sub-complexes and a following centrifugation isolates an oxygen-evolving PSII core complex which contains the 47-kDa, 43-kDa, 34-kDa and 32-kDa proteins, cytochrome *b*-559 (its subunits, 9 kDa and 4 kDa), and the 33-kDa extrinsic protein (Ikeuchi *et al.*, 1985). A similar type of complex can be prepared upon treatment of thylakoid membranes with digitonin followed by cation-exchange column chromatography (Tang and Satoh, 1985). The oxygen-evolving PSII core complex possesses oxygen-evolution activity as high as 1,200-1,500 μmol oxygen/mg Chl/h in the presence of sufficient concentrations of calcium and chloride ions and an artificial electron acceptor (Ikeuchi and Inoue, 1986). When the 33-kDa extrinsic protein is removed at high pH, the complex is no longer active in oxygen evolution, but contains the ability to undergo a photochemical charge separation (Tang and Satoh, 1985). This preparation is called a PSII core complex.

A treatment of the PSII membranes with Triton X-100 followed by cation-exchange column chromatography can isolate a smaller sub-complex of the photochemical reaction center, which is composed of 34-kDa and 32-kDa proteins and cytochrome *b*-559 (Nanba and Satoh, 1987). This preparation is active in transferring an electron from P680 to Pheo *a* at a quantum yield close to unity (Danielius *et al*., 1987), and is called a reaction center II complex.

In addition to the protein components there are low-molecular mass components such as pigments, lipids, quinones, and inorganic ions in the PSII preparations. Stoichiometry among these components is summarized in Table I. Striking differences among the PSII preparations appear in contents of lipid and Chl relative to P680, but not in those of plastoquinone and Pheo *a*.

Table I. Stoichiometry among low-molecular-mass components in spinach PSII preparations

Component	PSII[a,b] membranes	PSII[b-e] core	Reaction[f,g] center II
	Molar ratio		
P680	1	1	1
Pheophytin *a*	2	2	2
Cytochrome *b*-559	2	2	1
Plastoquinone	3	2	0
Mn	4	4	0
Ca	2	1	--
Chlorophyll	220	50	4-5
MGDG	90	3	1
DGDG	40	3	0
PG	15	3	0
SQDG	5	0	0

References: a, Murata *et al.* (1984); b, Shen *et al.* (1988); c, Ikeuchi *et al.* (1985); d, Omata *et al.* (1984); e, Yamada *et al.* (1985); f, Nanba and Satoh (1987); g, Satoh (1988). Data for lipid content, Murata N, Higashi S, Fujimura Y (unpublished).

EXTRINSIC PROTEINS

Among the three extrinsic proteins, the 33-kDa protein was first purified and characterized (Kuwabara and Murata, 1978; 1982a). Later, the 23-kDa and 18-kDa proteins were also purified and characterized (Jansson, 1984; Kuwabara and Murata, 1984). They are all hydrophilic proteins. One molecule each of the extrinsic proteins is bound to the PSII complex (Murata *et al.*, 1984).

The 23-kDa and 18-kDa proteins can be released from the PSII complex in the PSII membrane with 1.0 M NaCl (Miyao and Murata, 1983), and the 33-kDa protein with 1.0 M $CaCl_2$ (Ono and Inoue, 1983) or 2.6 M urea (Miyao and Murata, 1984). The PSII complex can be reconstituted by rebinding these extrinsic proteins to the intrinsic part of the PSII complex with saturation at the stoichiometric level (Miyao and Murata, 1987). The reconstitution with the extrinsic proteins is sequential; the rebinding occurs in the order of 33-kDa, then 23-kDa, and finally the 18-kDa protein (Miyao and Murata, 1987). By manipulating this reconstitution system, it is possible to study the functions of these proteins. The functions which have so far been elucidated are as follows: The 33-kDa protein (i) stabilizes the Mn cluster (Miyao and Murata, 1984; Kuwabara *et al.*, 1985), (ii) accelerates the S state transition from S_3 to S_0 (Miyao *et al.*, 1987), (iii) reduces the chloride requirement from 200 mM to 30 mM (Miyao and Murata, 1985), and (iv) provides a binding site for the 23-kDa protein (Murata *et al.*, 1983). The 23-kDa protein (i) stabilizes the calcium trap (Ghanotakis *et al.*, 1984; Miyao and Murata, 1986; 1987), (ii) reduces the chloride requirement from 30 mM to 10 mM (Miyao and Murata, 1985), and (iii) provides the binding site for the 18-kDa protein (Miyao and Murata, 1983). The 18-kDa protein sustains oxygen-evolution activity at chloride concentrations below 10 mM (Akabori *et al.*, 1984; Miyao and Murata, 1985).

Table II summarizes the dissociation constants and the number of binding sites for the extrinsic proteins. The binding of the 33-kDa protein to the intrinsic part of the PSII complex gives a dissociation constant of about 20 nM. The binding of the 23-kDa protein to the intrinsic part requires the presence of the 33-kDa protein for stoichiometric binding with a low dissociation constant of about 3 nM. When the 33-kDa protein is absent, the 23-kDa protein binds to the intrinsic part of the PSII complex with very low affinity and at a level much lower than the stoichiometry. The 18-kDa protein can stoichiometrically bind to the PSII complex supplemented with the 33-kDa protein but no 23-kDa protein with a very high dissociation constant, 160 nM. The dissociation constant is 50-fold reduced by the supplementation of the 23-kDa protein. This is consistent with a

Table II. Dissociation constant (K_d) between the extrinsic proteins and the PSII complex, and number of binding sites (n/RC II) of the extrinsic proteins on the PSII complex

Extrinsic protein	PSII complex supplemented with		K_d (nM)	n/RC II
33-kDa	--	--	16	0.8
23-kDa	--	--	100	0.3
23-kDa	--	33-kDa	2.8	0.9
18-kDa	--	33-kDa	160	0.8
18-kDa	23-kDa	33-kDa	3.5	1.0

Experiments were done with PSII membranes suspended in 10 mM NaCl.

previous result that the 23-kDa protein is necessary for the binding of the 18-kDa protein to the PSII complex (Miyao and Murata, 1983).

The three extrinsic proteins bind to the intrinsic part of the PSII complex within the lumen of the intact thylakoid membrane. The internal volume of the grana thylakoid can be calculated to be 0.001 μm^3, on the assumption that the thylakoid is a vesicle of disk shape with a diameter of 300 nm and an internal thickness of 10 nm (Kirk and Tilney-Bassett, 1978). The calculation indicates that when a concentration of the 33-kDa protein is equivalent to its dissociation constant, about 20 nM, only about 0.01 molecule of this protein is dissolved in the lumenal space. At concentrations of the 23-kDa and 18-kDa proteins equivalent to their dissociation constants, about 3 nM, only 0.002 molecule each of these proteins is dissolved in the lumenal space. This estimation suggests that the extrinsic proteins are bound to the PSII complex in the thylakoid membrane with very high affinity.

PHOTOSYSTEM II GENES IN THE CHLOROPLAST GENOME

Seven genes of PSII in the chloroplast genome have been well characterized; they are termed *psb*A, *psb*B, *psb*C, *psb*D, *psb*E, *psb*F, *psb*H (Sugiura, 1987). The *psb*A and *psb*D genes encode the reaction center 32-kDa (D1) and 34-kDa (D2) proteins, respectively. The *psb*B and *psb*C encode the 47-kDa and 43-kDa proteins of PSII core antenna. The *psb*E and *psb*F are the genes for the two subunits of 9 kDa and 4 kDa of the cytochrome *b*-559 apoprotein. The *psb*H encodes the 10-kDa phosphoprotein, the function of which is not known. Steinmetz *et al.*, (1986) reported another PSII gene (*psb*G according to them) in the maize chloroplast genome, but the presence of the protein synthesized by this gene in the PSII complex has not be demonstrated. Cushman *et al.*, (1988) reported a set of genes, termed *psb*I and *psb*J, in the *Euglena* chloroplast genome, which are located downstream from the *psb*E and *psb*F genes. Two genes corresponding to *psb*I and *psb*J of *Euglena* have also been found in the genome of a cyanobacterium (Cantrell and Bryant, 1987).

We have recently found a new PSII gene in the tobacco chloroplast genome (Murata *et al.*, 1988). When the low-molecular-mass polypeptides in spinach PSII membranes are analyzed by SDS-polyacrylamide gel electrophoresis, eight polypeptide bands emerge below the 9-kDa subunit of cytochrome *b*-559. Three of them correspond to already characterized proteins; the 10-kDa phosphoprotein (Farchaus and Dilley, 1986; Westhoff *et al.*, 1986), the 4-kDa subunit of cytochrome *b*-559 (Herrmann *et al.*, 1984; Widger *et al.*, 1985), and the hydrophilic 5-kDa protein (Ljungberg *et al.*, 1986; Murata *et al.*, 1987). When a partial amino acid sequence at the amino-terminus of one of these proteins having an apparent molecular mass of 2 kDa is compared with those deduced from the open-reading frames in the tobacco chloroplast genome (Shinozaki *et al.*, 1986), it is

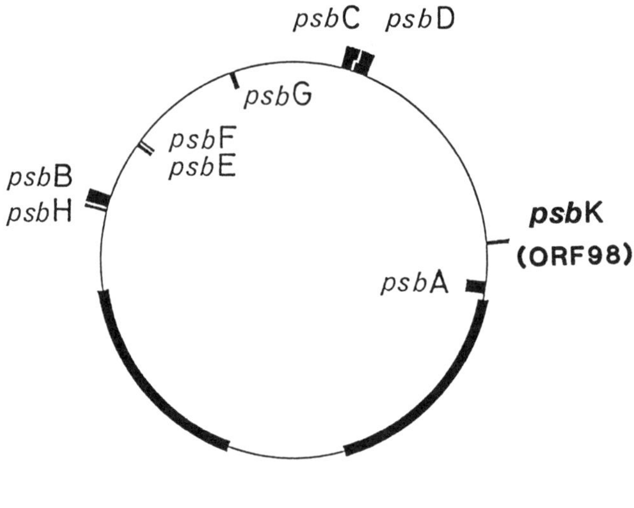

Figure 2. Location of PSII genes on the tobacco chloroplast genome. The inverted repeat is shown by broad lines. Genes indicated outside the circle are located on a strand transcribed counterclockwise and genes indicated inside are on the other strand transcribed clockwise.

identical with a part of ORF98. Figure 2 shows the location of the ORF98 in the tobacco chloroplast genome. It is located between the genes for tRNA (serine) and tRNA (glutamine) in the large single-copy region, and oriented in the direction opposite to the tRNA genes. The determined amino-acid sequence of the polypeptide matches the deduced amino-acid sequence from the 62nd to 74th residues of the ORF98. The putative polypeptide deduced from the ORF consists of 37 amino acid residues with a molecular mass of 4,285 Da. We designate this gene as *psb*K, and this polypeptide as the K polypeptide.

The *psb*K gene has two possible initiation codons, GTG from position 1 and ATG from position 112. If the first codon is the real initiation codon, the translation product consists of 98 amino-acid residues and a polypeptide of 61 amino-acid residues at the amino-terminus would be removed post-translationally. If the second codon is the real initiation codon, the precursor protein should consist of 61 amino-acid residues and 24 amino-acid residues are removed upon the processing of this precursor to the mature protein. The K polypeptide contains two basic amino-acid and two acidic amino-acid residues. Twenty-seven of 37 amino-acid residues are hydrophobic, which makes the K polypeptide a very hydrophobic protein. The hydropathy profile suggests that there is one membrane-spanning region in the sequence of the protein.

The role of the K polypeptide in PSII is not known. It is absent in the oxygen-evolving PSII core complex (Murata *et al.*, 1988), suggesting that it does not directly contribute to the photochemical reaction nor oxygen evolution. It may function as a structural and/or regulatory factor.

LIPIDS IN SPINACH PHOTOSYSTEM II

The major components of thylakoid membranes are proteins, glycerolipids, Chl and plastoquinone which amount to about 60%, 20%, 10%, and 3% in weight percent, respectively. The glycerolipids in the thylakoid membranes are monogalactosyldiacylglycerol (MGDG), digalactosyldiacylglycerol (DGDG), sulfoquinovosyldiacylglycerol (SQDG) and phosphatidylglycerol (PG). In order to examine the roles of these lipids in the photochemical and oxygen-evolving reactions, the lipid and fatty acid compositions in the different types of PSII preparation must be established. In the thylakoid membrane, in which the molar ratio of lipid to Chl is 2.1:1, the lipid composition is 60% MGDG, 30% DGDG, 4% SQDG and 6% PG. The lipid composition of PSII membranes is almost the same as that of thylakoid membranes, whereas the molar ratio of lipid to Chl is reduced to 0.7:1. Since the molar ratio of Chl to P680 is 220:1 (Murata *et al.*, 1984), the molar ratio of lipid to the reaction center is calculated to be 150:1. As shown in Table I, the PSII membranes contain 90 molecules of MGDG, 40 of DGDG, 5 of SQDG, and 15 of PG per one molecule of P680. In the oxygen-evolving PSII core complex, the lipid to Chl ratio is 0.19:1 and the lipid composition is 34% MGDG, 36% DGDG and 30% PG, but SQDG is lacking. Provided that the molar ratio of Chl to P680 in this type of preparation is 50:1 (Satoh, 1985), the molar ratio of MGDG, DGDG, PG, and P680 is calculated to be about 3:3:3:1. These findings suggest that SQDG is not essential to the photochemical reaction nor oxygen evolution. In the reaction center II complex, the ratio of lipid to [Chl + Pheo] was 0.16:1, and the lipid composition was 90% MGDG and 10% PG. Since the reaction center II complex contains 4 Chl and 2 Pheo molecules per P680 (Satoh, 1988), the molar ratio of MGDG to P680 is calculated to be 1:1. This suggests that the MGDG may participate in the organization of reaction center complex and/or the functioning of the photochemical reaction.

The fatty acid composition of total lipids in the various PSII preparations is presented in Table III. The PSII membranes are characterized by high contents of polyunsaturated fatty acids such as α-linolenic acid (18:3) and hexadecatrienoic acid (16:3) which amount to 70 and 10% of the total fatty acids, respectively. The oxygen-evolving PSII

Table III. Fatty acid composition of total lipids from spinach PSII preparations

<table>
<tr><th rowspan="2">Fatty acid</th><th colspan="3">Type of preparation</th></tr>
<tr><th>PSII membranes</th><th>PSII core</th><th>Reaction center II</th></tr>
<tr><td></td><td colspan="3">Mole percent</td></tr>
<tr><td>14:0</td><td>tr</td><td>3</td><td>7</td></tr>
<tr><td>16:0</td><td>9</td><td>15</td><td>34</td></tr>
<tr><td>16:1(9-cis)</td><td>1</td><td>3</td><td rowspan="2">9</td></tr>
<tr><td>16:1(3-trans)</td><td>4</td><td>12</td></tr>
<tr><td>16:3</td><td>9</td><td>7</td><td>2</td></tr>
<tr><td>18:0</td><td>1</td><td>3</td><td>12</td></tr>
<tr><td>18:1</td><td>1</td><td>4</td><td>10</td></tr>
<tr><td>18:2</td><td>6</td><td>6</td><td>3</td></tr>
<tr><td>18:3</td><td>70</td><td>45</td><td>22</td></tr>
</table>

core complex contains higher proportions of palmitic acid (16:0) and (3-trans)-hexadecenoic acid (16:1) but less of 18:3 if compared to PSII membranes. The fatty acid composition of the reaction center complex is characteristic of the saturation level. The saturated fatty acids such as myristic acid (14:0), 16:0 and stearic acid (18:0) amount to 50% of the total fatty acids, and the polyunsaturated acids 18:3 plus 16:3 amount to only 25%. Since the saturation of fatty acids in lipid rigidifies the lipid molecules, it can be concluded that the MGDG molecule in the reaction center complex is more rigid than the bulk lipid molecules in the PSII membranes. We speculate that such a rigid lipid molecule is essential for the assembly of the reaction center complex.

ACKNOWLEDGEMENTS

The authors are grateful to Dr. Julian J. Eaton-Rye, National Institute for Basic Biology, for helpful discussion and critical reading of the manuscript. This work is supported in part by a Grant-in-Aid for Scientific Research on Priority Areas of the Molecular Mechanism of Photoreception (63621003) from the Ministry of Education, Science and Culture, Japan, by a grant from Toray Science Foundation, Japan, and by the NIBB Program for Biomembrane Research.

REFERENCES

Akabori K, Imaoka A, Toyoshima Y (1984). The role of lipid and 17-kDa protein in enhancing the recovery of O_2 evolution in cholate-treated thylakoid membranes. FEBS Lett 173:36-40.

Arntzen CJ (1978). Dynamic structural features of chloroplast lamellae. Curr Top Bioenerg 8:111-160.

Berthold DA, Babcock GT, Yocum CF (1981). A highly resolved, oxygen-evolving photosystem II preparation from spinach thylakoid membranes. FEBS Lett 134:231-234.

Cantrell A, Bryant DA (1987). Molecular cloning and nucleotide sequences of the genes encoding cytochrome *b*-559 from the cyanelle genome of *Cyanophora paradoxa*. In Biggins J (ed): "Progress in Photosynthesis Research," Dordrecht: Martinus Nijhoff Publishers, Vol IV, pp 659-662.

Cushman JC, Christopher DA, Little MC, Hallick, RB, Price CA (1988). Organization of *psb*E, *psb*F, *orf*38 and *orf*42 gene loci on the *Euglena gracilis* chloroplast genome. Curr Genet 13:173-180.

Danielius RV, Satoh K, van Kan PJM, Plijter JJ, Nuijs AM, van Gorkom HJ (1987). The primary reaction of photosystem II in the D1-D2-cytochrome *b*-559 complex. FEBS Lett 213:241-244.

Farchaus J, Dilley RA (1986). Purification and partial sequence of the *M*r 10,000 phosphoprotein from spinach thylakoids. Arch Biochem Biophys 244:94-101.

Ghanotakis DF, Topper JN, Babcock GT, Yocum CF (1984). Water-soluble 17 and 23 kDa polypeptides restore oxygen evolution activity by creating a high-affinity binding site for Ca^{2+} on the oxidizing side of Photosystem II. FEBS Lett 170:169-173.

Herrmann RG, Alt J, Schiller B, Widger WR, Cramer WA (1984). Nucleotide sequence of the gene for apocytochrome *b*-559 on the spinach plastid chromosome: implications for the structure of the membrane protein. FEBS Lett 176:239-244.

Ikeuchi M, Yuasa M, Inoue Y (1985). Simple and discrete isolation of an O_2-evolving PS II reaction center complex retaining Mn and the extrinsic 33 kDa protein. FEBS Lett 185:316-322.

Ikeuchi M, Inoue Y (1986). Characterization of O_2 evolution by wheat photosystem II reaction center complex isolated by a simple method: disjunction of secondary acceptor quinone and enhanced Ca^{2+} demand. Arch Biochem Biophys 247:97-107.

Jansson C (1984). Proteins involved in photosynthetic oxygen evolution--isolation and characterization. In Sybesma C (ed): "Advances in Photosynthesis Research," The Hague: Martinus Nijhoff/Dr. W Junk Publishers, Vol I, pp 375-378.

Kirk JTO, Tilney-Bassett RAE (1978). "The Plastids. Their Chemistry, Structure, Growth and Inheritance." Amsterdam: Elsevier/North-Holland Biomedical Press, pp 155-189.

Kuwabara T, Murata N (1979). Purification and characterization of 33 kilodalton protein of spinach chloroplasts. Biochim Biophys Acta 581:228-236.

Kuwabara T, Murata N (1982a). An improved purification method and a further characterization of the 33 kilodalton protein of spinach chloroplasts. Biochim Biophys Acta 680:210-215.

Kuwabara T, Murata N (1982b). Inactivation of photosynthetic oxygen evolution and concomitant release of three polypeptides in the photosystem II particles of spinach chloroplasts. Plant Cell Physiol 23:533-539.

Kuwabara T, Murata N (1984). Chemical and physicochemical characterization of the proteins involved in the oxygen evolution system. In Sybesma C (ed): "Advances in Photosynthesis Research," The Hague: Martinus Nijhoff/Dr. W Junk Publishers, Vol I, pp 371-374.

Kuwabara T, Miyao M, Murata T, Murata N (1985). The function of 33-kDa protein in the photosynthetic oxygen-evolution system studied by reconstitution experiments. Biochim Biophys Acta 806:283-289.

Ljungberg U, Henrysson T, Rochester CP, Åkerlund H-E, Andersson B (1986). The presence of low-molecular-weight polypeptides in spinach Photosystem II core preparations. Isolation of a 5 kDa hydrophilic polypeptide. Biochim Biophys Acta 849:112-120.

Miyao M, Murata N (1983). Partial disintegration and reconstitution of the photosynthetic oxygen evolution system--Binding of 24 kilodalton and 18 kilodalton polypeptides. Biochim Biophys Acta 725:87-93.

Miyao M, Murata N (1984). Role of the 33-kDa polypeptide in preserving Mn in the photosynthetic oxygen-evolution system and its replacement by chloride ions. FEBS Lett 170:350-354.

Miyao M, Murata N (1985). The Cl^- effect on photosynthetic oxygen evolution: interaction of Cl^- with 18-kDa, 24-kDa and 33-kDa proteins. FEBS Lett 180:303-308.

Miyao M, Murata N (1986). Light-dependent inactivation of photosynthetic oxygen evolution during NaCl treatment of photosystem II particles: The role of the 24-kDa protein. Photosyn Res 10:489-496.

Miyao M, Murata N (1987). Photoinhibition of the oxygen-evolving complex of photosystem II. In Kyle DJ, Osmond CB, Arntzen CJ (eds): "Topics in Photosynthesis Vol 9, Photoinhibition," Amsterdam: Elsevier Science Publishers BV, pp 289-307.

Miyao M, Murata N, Lavorel J, Maison-Peteri B, Boussac A, Etienne A-L (1987). Effect of the 33-kDa protein on the S-state transitions in photosynthetic oxygen evolution. Biochim Biophys Acta 890:151-159.

Murata N, Miyao M, Kuwabara T (1983). Organization of the photosynthetic oxygen evolution system. In Inoue Y, Crofts AR, Govindjee, Murata N, Renger G, Satoh K (eds): "The oxygen evolving system of photosynthesis," Tokyo: Academic Press, pp 213-222.

Murata N, Miyao M, Omata T, Matsunami H, Kuwabara T (1984). Stoichiometry of components in the photosynthetic oxygen evolution system of photosystem II particles prepared with Triton X-100 from spinach chloroplasts. Biochim Biophys Acta 765:363-369.

Murata N, Miyao M (1987). Oxygen-evolving complex of photosystem II in higher plants. In Biggins J (ed): "Progress in Photosynthesis Research," Dordrecht: Martinus Nijhoff Publishers, Vol I, pp 453-462.

Murata N, Kajiura H, Fujimura Y, Miyao M, Murata T, Watanabe A, Shinozaki K (1987). Partial amino acid sequences of the proteins of pea and spinach photosystem II complex. In Biggins J (ed): "Progress in Photosynthesis Research," Dordrecht: Martinus Nijhoff Publishers, Vol I, pp 701-704.

Murata N, Miyao M, Hayashida N, Hidaka T, Sugiura M (1988). Identification of a new gene in the chloroplast genome encoding a low-molecular-mass polypeptide of photosystem II complex. FEBS Lett 235:283-288.

Nanba O, Satoh K (1987). Isolation of a photosystem II reaction center consisting of D-1 and D-2 polypeptides and cytochrome *b*-559. Proc Natl Acad Sci USA 84:109-112.

Omata T, Murata N, Satoh K (1984). Quinone and pheophytin in the photosynthetic reaction center II from spinach chloroplasts. Biochim Biophys Acta 765:403-405.

Ono T, Inoue Y (1983). Mn-preserving extraction of 33-, 24- and 16-kDa proteins from O_2-evolving PSII particles by divalent salt washing. FEBS Lett 164:255-260.

Satoh K (1985). Protein-pigments and photosystem II reaction center. Photochem Photobiol 42:845-853.

Satoh K (1988). Reality of P-680 chlorophyll protein -- Identification of the site of primary photochemistry in oxygenic photosynthesis. Physiol Plant 72:209-212.

Shen J-R, Satoh K, Katoh S (1988). Calcium content of oxygen-evolving Photosystem II preparations from higher plants. Effects of NaCl treatment. Biochim Biophys Acta 933:358-364.

Shinozaki K, Ohme M, Tanaka M, Wakasugi T, Hayashida N, Matsubayashi T, Zaita N, Chunwongse J, Obokata J, Yamaguchi-Shinozaki K, Ohto C, Torazawa K, Meng BY, Sugita M, Deno H, Kamogashira T, Yamada K, Kusuda J, Takaiwa F, Kato A, Tohdoh N, Shimada H, Sugiura M (1986). The complete nucleotide sequence of the tobacco chloroplast genome: its gene organization and expression. EMBO J 5:2043-2049.

Steinmetz AA, Castroviejo M, Sayre RT, Bogorad L (1986). Protein PSII-G, an additional component of photosystem II identified through its plastid gene in maize. J Biol Chem 261:2485-2488.

Sugiura M (1987). Structure and function of the tobacco chloroplast genome. Bot Mag Tokyo 100:407-436.

Tang X-S, Satoh K (1985). The oxygen-evolving photosystem II core complex. FEBS Lett 179:60-64.

Westhoff P, Farchaus JW, Herrmann RG (1986). The gene for the M_r 10,000 phosphoprotein associated with photosystem II is part of the *psb*B operon of the spinach plastid chromosome. Curr Genet 11:165-169.

Widger WR, Cramer WA, Hermodson M, Herrmann RG (1985). Evidence for a hetero-oligomeric structure of the chloroplast cytochrome *b*-559. FEBS Lett 191:186-190.

Yamada Y, Itoh N, Satoh K (1985). A versatile chromatographic procedure for purifying PS II reaction center complex from digitonin extracts of spinach thylakoids. Plant Cell Physiol 26:1263-1271.

Photosynthesis, pages 71–103

PHOTOSYSTEM II: FROM A FEMTOSECOND TO A MILLISECOND

Govindjee, and M.R. Wasielewski

Departments of Physiology and Biophysics and Plant Biology, University of Illinois at Urbana-Champaign, 289 Morrill Hall, 505 South Goodwin Avenue, Urbana, IL 61801 (G.), Chemistry Division, Argonne National Laboratory, Argonne, IL 60439 (M.R.W.).

ABSTRACT

We present here a minireview on the time sequence of events in water-plastoquinone oxido-reductase (*i.e.*, photosystem II, PSII) from a femtosecond to a millisecond. The minimal components of PSII include: three intrinsic polypeptides D-1 and D-2, and cytochrome (Cyt) *b*-559, and an extrinsic polypeptide of 33 kilodalton (kDa) mass (labeled as Y-33). The first act of photosynthesis, light absorption, occurs in chlorophyll (Chl) *a* molecules in about 1 femosecond (fs). Then, the following reactions take place. **(1)** The primary charge separation, *i.e.*, the formation of the oxidized reaction center Chl *a*, $P680^+$, and the reduced pheophytin, $Pheo^-$, has been shown to occur in reaction center preparation (Nanba and Satoh, 1987) with a lifetime of 3 ± 0.6 picoseconds (ps) (Wasielewski *et al.*, 1989), using 500 fs laser flashes. **(2)** Electron flow from $Pheo^-$ to the first plastoquinone electron acceptor, Q_A, has been estimated to occur in several hundreds of ps. **(3)** The re-reduction of $P680^+$ by the electron donor Z [suggested recently to be a tyrosine molecule (Debus *et al.*, 1988; Vermaas *et al.*, 1988, in press)] occurs in 25 to 300 nanoseconds (ns) (Brettel *et al.*, 1984). **(4)** The electron flow from the reduced Q_A, Q_A^-, to the second plastoquinone electron acceptor, Q_B, occurs in several hundreths of a microsecond (μs). Approximately, in the same time range, Z^+ is restored to Z by electron flow from the water oxidation complex (WOC) producing the **stable** charge-separated state $S_2Q_B^-$, where S_2 refers to the redox state of the WOC -- the dark-adapted state being S_1Q_B. Thus, after twelve orders of magnitude of time, a stable state is created. We shall not discuss here the steps of oxygen evolution, and the details of the production of plastoquinol. However, we shall review the possible role of bicarbonate (HCO_3^-) in the mechanism of plastoquinone reduction in plants and cyanobacteria (Blubaugh and Govindjee, 1988, in press). It is suggested that HCO_3^- functions to provide (a) a proper conformation to the reaction center by binding to the iron in the Q_A-Fe-Q_B complex, and (b) protons for plastoquinol formation by binding to an arginine/histidine pair on the D-1 protein of PSII.

INTRODUCTION

The function of Photosystem II (PSII) is to act as water-plastoquinone oxido-reductase (WPOR). Water is oxidized to O_2, and plastoquinone (PQ) is reduced to plastoquinol (PQH_2): $2H_2O + 2PQ + 4H^+$ (outside) + 4 $h\nu \rightarrow O_2 + 2\ PQH_2 + 4H^+$ (inside). This uphill transfer of electrons requires 4 light quanta ($h\nu$), which are used via the excitation of the reaction center (RC) chlorophyll (Chl) *a* of PSII, P680; the latter leads to the primary charge separation: P680 · Pheo + $h\nu \rightarrow$ P680* · Pheo $\rightarrow$ P680$^+$ · Pheo$^-$, where Pheo stands for pheophytin. This minireview is concerned solely with the light reaction and the electron transport in PSII. For earlier reviews, see Govindjee (1980, 1982, 1984), Velthuys (1980, 1987), van Gorkom (1985), Diner (1986), Mathis (1986), Ort (1986), and Andréasson and Vänngård (1988). The electron acceptor side of PSII has been discussed by Vermaas and Govindjee (1981), Crofts and Wraight (1983), and Govindjee and Eaton-Rye (1986), and the electron donor side by Ghanotakis and Yocum (1985), Govindjee *et al.* (1985), Renger and Govindjee (1985), Andersson and Åkerlund (1987), Babcock (1987), and Renger (1987).

Figure 1 shows a self-explanatory generalized model of the thylakoid membrane with the four major complexes: PSII, cytochrome b_6/f complex, PSI and the ATP synthase, whereas Fig. 2 shows a detailed current picture of the electron flow in PSII, along with the measured or estimated times of reactions. Electron carrier redox pairs are placed vertically according to their approximate known or estimated redox midpoint potential (Em,7). The main path of electron flow is as follows (alternate names of the intermediates are in the figure): H_2O/O_2 (Em,7, + 0.82V) $\rightarrow$ Mn^{3+}/Mn^{4+} $\rightarrow$ Z/Z^+ (+1.0V; tyrosine/tyrosine$^+$) $\rightarrow$ P680/P680$^+$ (+1.15V) $\xrightarrow{h\nu}$ P680/P680* (-0.7V) $\rightarrow$ Pheo/Pheo$^-$ (-0.6V) $\rightarrow$ Q_A/Q_A^- (0 to -30mV) $\rightarrow$ Q_B/Q_B^- $\rightarrow$ PQ/PQH_2 (+90mV). Oxygen and protons are released to the inside, and protons are picked up from the outside for the production of PQH_2. Here, the symbol "M" under Mn^{3+}/Mn^{4+} represents more than the Mn atoms: it is the charge-accumulating entity necessary for water oxidation and it may, therefore, include the necessary (Y-33, the extrinsic polypeptide with 33 kDa molecular mass) and stimulatory (regulatory) polypeptides (Y-24 and Y-18), chloride (Cl^-), and entities (proteins?) producing absorbance changes at 320nm (Y-320), etc. The redox state of "M" is referred to as the S-states, where S_0, S_1, S_2, S_3, and S_4 represent the five different states

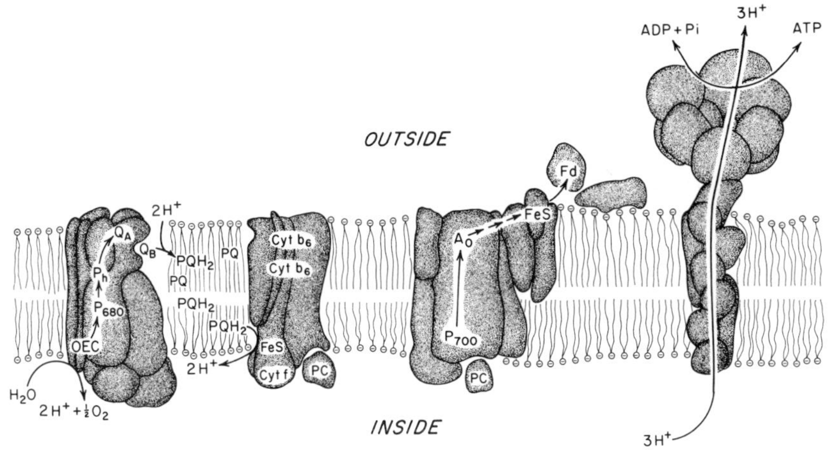

Figure 1. A schematic diagram of the four major protein complexes. From left to right: Photosystem II [OEC = oxygen evolving complex, P680 = reaction center chlorophyll *a*, Ph= functional pheophytin, Q_A = first plastoquinone electron acceptor (on D-2), Q_B = second plastoquinone electron acceptor (on D-1), PQ (PQH_2) = mobile plastoquinone (plastoquinol)]; Cytochrome b_6/f complex (cyt b_6 = cytochrome b_6, Cyt f = cytochrome f, FeS = Rieske Fe-S center, Pc = plastocyanin); Photosystem I (P700 = reaction center chlorophyll *a*, Ao = chlorophyll *a*, first electron acceptor, FeS = iron sulfur centers, Fd = ferredoxin); and the ATP synthase (the membrane portion is called CF_0; and the hydrophilic portion is called CF_1). Simplified by Donald R. Ort (1988, personal communication) from Govindjee and Eaton-Rye (1986). Courtesy of D. Ort.

with increasing positive equivalents on them. "Z" represents the entities that donate electrons directly to $P680^+$; the oxidized form of Z, Z^+, is described by its ESR signal labeled II very fast (IIvf); earlier, it was suggested to be an oxidized plastoquinol molecule. Now, new data suggest that it may be tyrosine-160 on the polypeptide D-1 of PSII (Debus *et al.*, 1988, Vermaas *et al.*, 1988). P680, which is most likely a dimer in its ground state, behaves as a monomer in its cationic and triplet form. In the current picture, it is attached to the α-helices of the D-1 and D-2 polypeptides of the reaction center, just as the reaction center bacteriorchlorophyll is attached to the "L" and "M" subunits of the bacterial reaction center (Michel and Deisenhofer, 1988). The photochemically active Pheo, one of the two pheophytin molecules in the reaction center complex, is a monomer and suggested to be located on the D-1 polypeptide (see Fig. 3). Q_A, the bound plastoquinone, may also be on D-2; by analogy, again, with photosynthetic bacteria, a non-heme iron (Fe^{2+}) sits in the middle of Q_A and the second bound plastoquinone Q_B, which is located on D-1. Q_B acts as a two-electron gate; it is very tightly

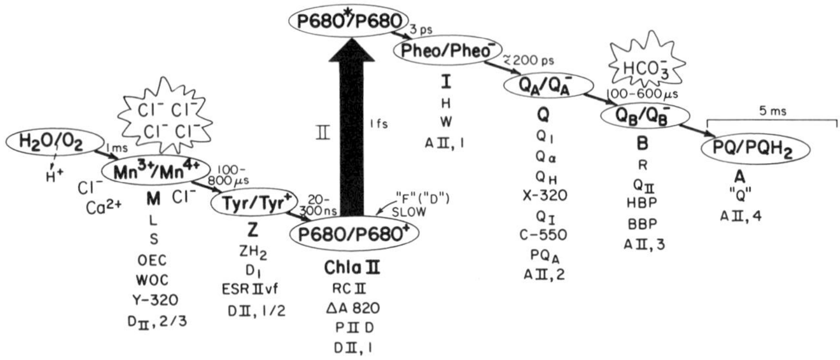

Figure 2. Electron transport scheme of Photosystem IIα. Ovals show major components; measured or estimated times involved in each step are also shown. Location of sites of Cl^- and HCO_3^- actions are highlighted. Alternate names of intermediates, used by various authors for the major components, are also shown (see text). Modified after Govindjee (1984) and Govindjee *et al.* (1985).

bound in its Q_B^- form; and it exchanges with a plastoquinone molecule only after it is doubly reduced (Crofts and Wraight 1983; Velthuys, 1987). Both the D-1 and D-2 are suggested to span the membrane 5 times (Trebst and Draber, 1986; Trebst, 1987). Herbicides (Trebst 1986) and bicarbonate (Blubaugh and Govindjee 1988) are suggested to bind to this protein on the side facing the stroma, whereas manganese and chloride, involved in O_2 evolution, are suggested to bind on the side facing the lumen (Coleman and Govindjee, 1987).

A working model for the organization of PSII (excluding the light harvesting Chl *a*/Chl *b* complex) is shown in Fig. 4. In most higher plants, active PSII is located in the "appressed" regions of the membrane; this is, perhaps, the PSIIα units (for a discussion of heterogeneity of PSII, see Vermaas and Govindjee, 1981; and Black *et al.*, 1986). We recognize a minimum of 7 polypeptides: (I) 47 kDa (CP-47), an internal Chl *a*-protein antenna complex; it fluoresces at 696nm at 77K; (II) 43 kDa (CP-43), a distal Chl *a*-protein antenna complex; it fluoresces at 685nm at 77K; (III) D-1: 32 kDa Q_B-binding lysine-rich protein to which herbicides can also bind (HBP) and that may also carry bicarbonate binding sites (BBP); (IV) D-2: 34 kDa Q_A-binding lysine-poor protein; (V) cytochrome *b*-559 having two polypeptides of 4 and 9 kDa molecular masses; (VI) Y-33: a 33 kDa extrinsic polypeptide, that can be released by high concentrations (1M) of $CaCl_2$, is involved in O_2 evolution; (VII) Y-24: a 24 kDa extrinsic polypeptide that can be released by low

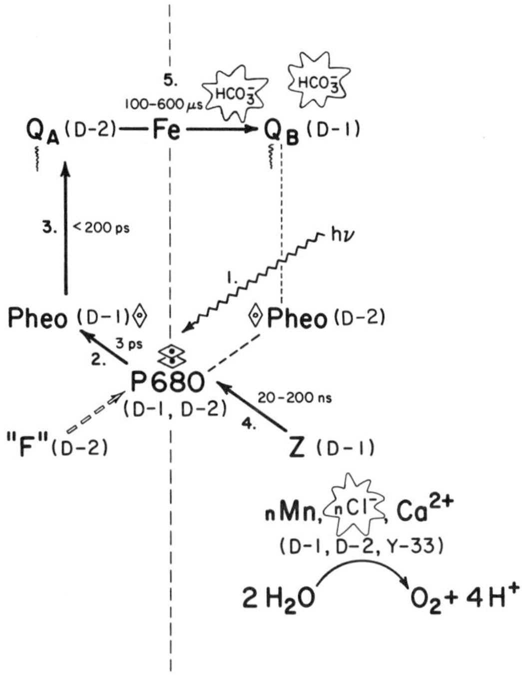

Figure 3. A simplified model of the arrangement of the various chromophores in the reaction center of photosystem IIα. P680 = reaction center chlorophyll *a*, Pheo = pheophytin, Q_A = first plastoquinone electron acceptor, Fe = iron, Q_B = second plastoquinone electron acceptor, Z = suggested to be a tyrosine; D-1 = the 32 kilodalton intrinsic protein, D-2 = the 34 kilodalton intrinsic protein, Y-33 = the 33 kilodalton extrinsic protein (that is not included in the Nanba-Satoh preparations). Cytochrome *b*-559, included in the Nanba-Satoh preparation, is not shown in the diagram. A slow electron donor "F" (or "D") on the protein D-2, identified as tyrosine, is shown. Measured or estimated approximate times for reactions are shown over the arrows that indicate the direction of electron flow.

concentrations of salts; it stimulates O_2 evolution and is required for Ca^{2+} and Cl^- binding to the membrane; and (VIII) Y-18: a 18 kDa extrinsic polypeptide that can also be released by low concentrations of salts; its function is less clear than that of Y-24. Cyanobacteria do not contain Y-18 and Y-24. The minimal system required, thus far, for O_2 evolution contains, at least, the first 6 polypeptides (Ghanotakis *et al.*, 1987). However, the reaction center (RC) preparation, capable of the primary, charge separation, is simpler and was shown to contain only three polypeptides: D-1, D-2, and cytochrome *b*-559; it contained 4 chlorophyll

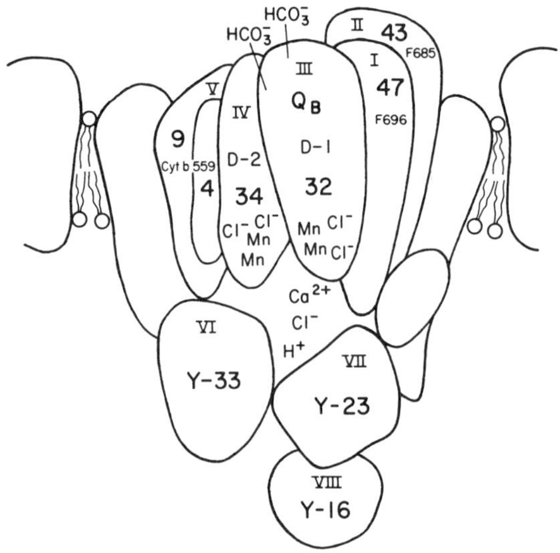

Figure 4. A possible architecture of the oxygen-evolving complex. Modified from Andersson and Åkerlund (1987). (See text for the symbols and the nomenclature.)

molecules (2 of which, presumably, form P680) and 2 pheophytin molecules (Nanba and Satoh, 1987). Unfortunately, the currently available RC preparations lack Q_A and Q_B. Although the role of cytochrome *b*-559 in the main electron flow has not been discovered, it has been shown to act as an electron donor at 77K or when normal-electron flow is blocked.

The major electrogenic event in PSII is suggested to be due to the electron flow from reduced pheophytin to Q_A. This is supported by electrically-induced reduction of Pheophytin by Q_A^- (Meiburg *et al.*, 1983); a direct kinetic measurement of membrane potential by a macroscopic electrode technique in PSII suggests that it is a fast reaction (<200 ps) (see Trissl and Kunze, 1985).

What follows is a discussion of the major components and their function in the electron transfer from the reaction center chlorophyll *a* P680 to pheophytin; pheophytin$^-$ to Q_A; and from Q_A^- to Q_B. We shall end with a discussion of the role of HCO_3^- in the reduction of plastoquinone.

PRIMARY REACTANTS AND REACTIONS: FEMTOSECONDS TO PICOSECONDS

Reactants.

P680. The existence of a reaction center Chl *a* in PSII, labeled as Chl *a* II, was first observed as an absorbance decrease at 690nm by Döring *et al.* (1967, 1969), and later by others (see reviews in Witt, 1975; Parson and Ke, 1982; Ke and Shuvalov, 1987). Floyd *et al.* (1971) measured these changes at 77K and labeled the responsible component as P680 by analogy with P700 of PSI. Since still earlier measurements in the 680-690nm region were shown to be artifacts of Chl *a* fluorescence (Rubinstein and Rabinowitch, 1963), it took many experiments and several years to accept this change to be due to an absorbance decrease (scc *e.g.*, Govindjee *ct al.*, 1970; Butler, 1972; Eckert *et al.*, 1984). Finally, measurements of absorbance increase in the 820nm region (van Gorkom *et al.*, 1975; Van Best and Mathis, 1978) established that the change was due to the oxidation of P680 to $P680^+$, not due to Chl *a* fluorescence increase. Chl *a* cations (Chla^+) show such a positive absorbance change (Fujita *et al.*, 1978). Although the 820nm change is not as specific to $P680^+$, it is easier to measure. It is now generally believed that: (a) P680 is a dimer in its ground state, but the positive charge in the cation ($P680^+$) and the triplet (^{3}P680) states are localized on a monomer (see *e.g.*, Diner, 1986; Den Blanken *et al.*, 1983); and (b) the redox potential (Em,7) of P680/$P680^+$ is 1.1V (Jursinic and Govindjee, 1977; Klimov *et al.*, 1980).

Recovery of P680 from $P680^+$. The kinetics of the decay of $P680^+$ to P680 are rather complex. Several components in the range of 200μs, 35μs, 5μs, 250 ns and 20-30 ns, have been observed. These may represent electron donation by electron donors and/or by reduced electron acceptors available for back reactions. The fastest (20-250 ns components representing 50-60% of the total signal) must be due to the main routes of electron flow from Z to $P680^+$. These reactions have been measured by:

(1) Absorbance changes at 820nm (see *e.g.* van Best and Mathis, 1978, who measured a 30 ns component in dark-adapted thylakoids; Brettel *et al.*, 1984; also see Schlödder *et al.*, 1985) show that the kinetics of the reduction of $P680^+$ depend upon the redox state of WOC; it is fastest when the WOC is in S_0 and S_1 states (20 ns), and slowest (50 and 260 ns; two phases of equal amplitude) in S_2 or S_3 state. This suggests that the coulombic force of the positive charge on the WOC slows the electron donation from Z to $P680^+$. Furthermore, existence of two donors Z_1 and Z_2 was suggested. (For an earlier discussion of Z_1 and Z_2, see Govindjee and Jursinic, 1979.)

(2) Chlorophyll *a* fluorescence increase. Mauzerall (1972) observed a Chl *a* fluorescence rise time of 20-30 ns in dark-adapted algae; Butler (1972) explained these data to be due to the disappearance of $P680^+$ which acts as a chlorophyll *a* quencher. Sonneveld *et al.* (1979), in an elegant analysis of Chl *a* fluorescence rise during a flash excitation at different light intensities, established that in dark-adapted thylakoids, this rise is best explained by a 20-40 ns reduction time of $P680^+$ to P680. However, in preilluminated samples, this rise was 400 ns. Whether these differences are due simply to different charged states of WOC, or, also due to different electron donors, (*e.g.* Z_1 and Z_2, see *e.g.* Bouges-Bocquet, 1980) is not yet known. Just as in kinetic data on absorbance changes, fluorescence changes also show microsecond components (see *e.g.* Jursinic and Govindjee, 1977).

Pheophytin. Pheophytin is a chlorophyll molecule in which magnesium in the center has been replaced by two hydrogens. Klimov *et al.* (1977) showed that $Pheo^-$ can be accumulated in PSII if $P680^+$ is kept reduced by an external donor and Q_A is chemically reduced prior to illumination (see review by Klimov and Krasnovsky, 1981). Pheophytin *a* is different from chlorophyll *a* in several respects: (a) it absorbs more weakly in the red and more strongly at 505 and 535nm than Chl *a* (Fujita *et al.*, 1980); (b) *in vitro*, it can be reduced more easily compared to Chl *a* (Fajer *et al.*, 1980); (c) its midpoint redox potential is -0.6V, less negative than that of Chl *a* (Klimov *et al.*, 1979; Rutherford *et al.*, 1981); pheophytin exists in a monomer form (Klimov *et al.*, 1980) in PSII, but there are 2 pheophytin molecules per reaction center (Murata *et al.*, 1986; Omato *et al.*, 1984). As suggested in Fig. 2, Pheo must sit quite close to P680 as no membrane potential seems to be observed.

Reaction Center Preparation

In order to measure the primary photochemical reaction, *i.e.*, the charge separation, one needs to have a purified, stable reaction center preparation that is free of antenna Chl *a* molecules. Nanba and Satoh (1987) isolated a complex that contains only three proteins: D_1, D_2 and cytochrome *b*-559 (also see Satoh *et al.*, 1987; Barber *et al.*, 1987; Marder *et al.*, 1988; Seibert *et al.*, 1988). This preparation contains only 4 or 5 Chl *a*, 2 Pheo, and a single β-carotene. Since Pheo *a* could be reduced by illumination, it was suggested that this complex is analogous to the reaction center from photosynthetic bacteria [see *e.g.* that the latter contains 4 Bchl and 2 B Pheo molecules (Deisenhofer *et al.*, 1984)]. Unfortunately, the Nanba-Satoh preparation lacks the bound quinones Q_A and Q_B, and has not been crystallized for structural and functional studies. However, herbicide binding sites are shown to exist (Giardi *et al.*, 1988). Akabori *et al.* (1988) have, however, prepared fractions that contain the bound quinones but their method is much more involved. Marder *et al.* (1987) have demonstrated that D-1 and D-2, in their preparation, are indeed the products of appropriate genes; and Chapman *et al.* (1988) have measured electron flow properties of these preparations. Although electron flow from $Pheo^-$ to Q_A cannot be measured, these preparations are highly suitable for measurements on primary charge separation. We shall review below our results, obtained in collaboration with Doug Johnson (of Argonne) and Mike Seibert (of SERI) (Wasielewski *et al.*, 1989). A major reason for our success was the availability of a preparation that is stable and remains optically clear for an entire day at 4°C under intense illumination (10^5 laser pulses). The stability and clarity were obtained (H. McTavish, R. Picorel and M. Seibert, 1988) through the following: (1) reaction center was concentrated with polyethylene glycol (a stabilizer); (2) addition of low concentration (final 0.04%) of Triton X-100 keeps the material from aggregating; (3) the material was kept under anaerobic conditions by the addition of glucose, catalase, and glucose oxidase.

Photochemistry

The primary reactions can be written as:

$$^1\text{Chl}a + h\nu \xrightarrow{1} {}^1\text{Chl}a^* \text{ (light absorption; creation of excited state)}$$

$$^1\text{Chl}a^* + \text{P680} \xrightarrow{2} {}^1\text{Chl}a + {}^1\text{P680}^* \text{ (creation of singlet excited state of P680)}$$

$$^1\text{P680}^* + \text{Pheo} \xrightarrow{3} {}^1\text{P680}^+ + \text{Pheo}^- \text{ (primary charge separation)}$$

$$^1\text{P680}^+ + \text{Pheo}^- \xrightarrow{4} \text{P680} + \text{Pheo} \text{ (charge recombination, or loss of charges by other mechanisms)}$$

$$^1\text{P680}^+ + \text{Pheo}^- \xrightarrow{5} {}^3\text{P680} + \text{Pheo} \text{ (triplet state formation)}$$

The act of light absorption is the most rapid reaction and is estimated from the interaction of light with Chl *a* molecule. For example, Chl *a* goes into first singlet excited state by absorbing red light (680nm). The 680nm light oscillates with a frequency of 4.41×10^{14} cycles s^{-1}. Thus, one transition (or cycle) occurs in 2.5 x 10^{-15}s or 2.5 femtoseconds. This, then, is the approximate time of excitation (reaction 1). The excitation energy transfer in a reaction center must also be very rapid, considering the rise time for $^1P680^*$ production to be within the 500 fs instrument function (reaction 2). Wasielewski *et al.* (1989) have measured the lifetime of reaction 3 to be 3.0 ± 0.6 picoseconds in stable reaction center II preparations (to be discussed in details). The lifetime of $P680^+$ - $Pheo^-$ (reaction 4) had been measured in Nanba and Satoh preparations to be 32 nanoseconds (ns) by Takahashi *et al.* (1987) and Danielius *et al.* (1987). Hansson *et al.* (1988) have reported that the lifetime of $P680^+$ - $Pheo^-$ depends upon the number of antenna Chl *a* remaining attached to the reaction center. Finally Okamura *et al.* (1986) have shown the formation of the triplet state of P680, and have established the radical pair nature of its precursor through its spin polarization characteristics. Nuijs *et al.* (1986a,b) have studied primary reactions, using a 35ps laser, in photosystem II particles enriched in reaction centers (P680/80 Chl). This study, however, does not provide direct answers to the steps involved in reaction 3 above. We discuss below the newest finding, that is, the first direct measurement of the kinetics of charge separation in a stable reaction center of plant photosynthesis, using 500 fs laser flashes.

Charge Separation in Photosystem II Reaction Center

Wasielewski *et al.* (1989) have provided the first direct measurement of charge separation in stabilized PSII reaction centers with a 500 fs resolution. Using 610nm exciting flashes, they observed the following: (1) an instrument, limited (0.5 ± 0.4 ps) decrease in absorbance at 670nm and 485nm indicating the formation of the singlet excited state of P680 ($^1P680^*$); (2) an increase in absorbance at 820nm (due to the formation of $P680^+$) with a time constant (1/e) of 3.0 ± 0.6 ps (Table I); (3) a decrease in absorbance at 670nm and an increase in absorbance at 485nm or at 538nm reflecting the formation of $P680^+$ $Pheophytin^-$ with a time constant also of approximately 3 to 4 ps; and (4) a decrease in absorbance at 650nm (at the isosbestic point for the $P680^+$ $Pheo^-$ - P680 Pheo changes), reflecting the decay of $^1P680^*$ with a time constant of 2.6 ± 0.6 ps that matches, within experimental error, the formation of $P680^+$ (Fig. 5). Furthermore, it was shown that when Pheo was kept reduced by the addition of sodium dithionite, methylviologen and light, only the absorbance changes due to the formation of $^1P680^*$ were observed; the absorbance increase at 820nm was eliminated confirming that the observed changes are indeed due to the charge separation.

The above conclusion was further supported by quantitative comparisons of the absorbance changes measured at various wavelengths (Table II). For example, the expected ratio of absorbance at 800nm (due

Table I. Summary of Rise and Decay Components (From data of Wasielewski *et al.*, 1989)

λ,nm	absorbance	time	components
820	increase	3.0 ± 0.6ps	$P680^+$ formation
670	decrease	0.5 ± 0.4ps	Instrument-limited, $P680^*$ formation
		3.3 ± 0.4ps	$P680^+$ and $Pheo^-$ formation
650	decrease	2.6 ± 0.6ps	Decay of $P680^*$ (isosbestic for both $P680^+$ and $Pheo^-$)
538	increase	2.8 ± 0.6ps	$P680^+$ and $Pheo^-$ formation
485	decrease	0.5 ± 0.4ps	Instrument-limited, $P680^*$ formation
	increase	3.6 ± 0.8ps	$P680^+$ and $Pheo^-$ formation

to $P680^+Pheo^-$) and that at 538nm trough (due to $Pheo^-$) of 5 (2 x 10^4 M^{-1} cm^{-1}/4 x 10^3 $M^{-1}cm^{-1}$) was matched by the same observed ratio (ΔOD at 800nm = 0.035/ΔOD at 538nm trough = 0.007). The same conclusion was obtained from the ratio of absorbance changes at 674nm to that at 800nm, where the extinction coefficient ratio was 4.0 (8 x 10^4 $M^{-1}cm^{-1}$ at 674nm/ 2 x 10^4 M^{-1} cm^{-1} at 800nm) and the measured absorbance ratio was 4.3 (OD 674nm = 0.15/OD at 800nm = 0.035). Wasielewski *et al.* (1989) further concluded that: (1) no other intermediate exists between P680 and pheophytin at times > 500 fs, and (2) the quantum yield of charge separation is about 1.0.

ELECTRON FLOW FROM PHEOPHYTIN TO Q_A: PICOSECONDS TO A NANOSECOND

Since the back reaction between $Pheo^-$ and $P680^+$ has been measured to be 30ns (Takahashi *et al.*, 1987; Danielius *et al.*, 1987) in Satoh and Nanba's D-1, D-2, cyt *b*-559 preparations, and, is expected to be 2-4ns in intact PSII, the forward electron flow from $Pheo^-$ to Q_A is expected to be many orders of magnitude faster. Based on the measurements of the lifetime of chlorophyll *a* fluorescence of 180 to 300 ps from PSII in algae

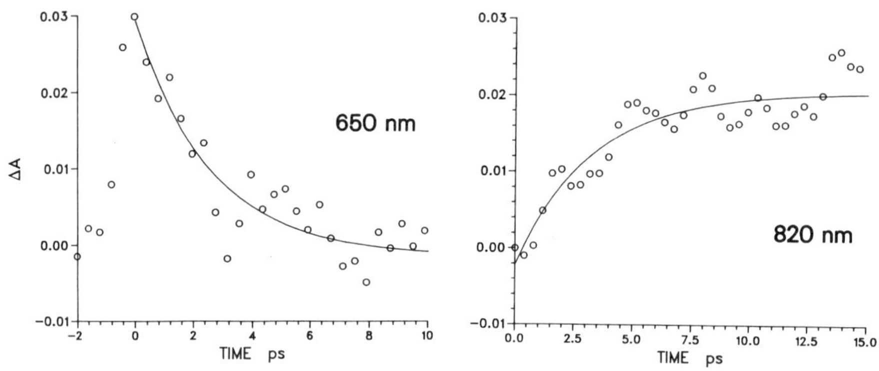

Figure 5. Transient absorption changes at 820nm (due to formation of $P680^+$) and at 650nm (due to decay of $^1Chla^*$) for photosystem II reaction centers following a 100 μJ, 500 fs laser flash at 610nm. Left: 650nm change; right: 820nm change. (After Wasielewski *et al*., 1989.)

when the reaction centers are fully open, the reduction time of Q_A is suggested to be <300 ps (see Moya *et al*., 1986; Holzwarth, 1987, for reviews on lifetime of fluorescence measurements). Our own (J. Fenton, N. S. Rao, E. Gratton, 1982, presented at the Midwest Photosynthesis Congress, Argonne, IL, USA) unpublished measurements of 200 ps in spinach thylakoids is in agreement with this conclusion. Since $P680^+$ to P680 reaction occurs in 20-30ns ($P680^+$ is a quencher of Chl *a* fluorescence), the reduction time of Q_A to Q_A^- cannot be measured by fluorescence. However, the electron flow from $Pheo^-$ to Q_A can be measured by absorbance changes. Nuijs *et al.* (1986a), from measurements with 35ps resolution, concluded that the electron transfer from $Pheo^-$ to Q_A occurs with a time constant of 270ps in spinach PSII preparations enriched in P680 (^{1}P680/80 Chl *a*). On the other hand, Schatz *et al.* (1987), using *Synechococcus* PSII preparations enriched in P680 (also ^{1}P680/80 Chl *a*), estimated that the electron transfer from $Pheo^-$ to a subsequent quinone electron acceptor takes place in ~ 500ps. Furthermore, the quantum yield of photoreduction of Q_A has been estimated to be 0.9 (see Thielen and van Gorkom, 1981).

Since electron flow from $Pheo^-$ to Q_A is assumed to traverse a large portion of the thylakoid membrane, a membrane potential ($\Delta\Psi$) is expected to be produced. Measurements of Trissl and Kunze (1985) by macroelectrodes suggest that $\Delta\Psi$ in PSII occurs in the hundreds of picosecond range, and the flash-induced reduction of pheophytin is not

Table II. Magnitudes of Absorbance Changes (From Data of Wasielewski *et al.*, 1989)

λ,nm	Component	Extinction coefficient M^{-1} cm^{-1}	ΔOD, observed
800	$P680^+$	10^4	
	$Pheo^-$	10^4	
	$P680^+$ + $Pheo^-$	2 x 10^4	0.035
538 trough	$Pheo^-$	4 x 10^3	0.007
Ratio 800/538	--	5	5
674	$P680^+$ + $Pheo^-$	8 x 10^4	0.15
Ratio 674/800	--	4	4.3

electrogenic. This confirms the picture that P680 and pheophytin are on the same side of the membrane and Q_A is on the other side.

ELECTRON FLOW OUT OF Q_A-: THE Q_A TO Q_B REACTION: NANOSECONDS TO A MILLISECOND

Reactants.

Q_A. The major electron acceptor Q_A is a plastoquinone permanently bound on the D_2 protein. There is a tremendous heterogeneity among the quinone acceptors in PSII (see *e.g.* Vermaas and Govindjee, 1981; Black *et al.*, 1985; Diner, 1986; Rich and Moss, 1987). Since it is the second acceptor of PSII, the first being pheophytin, it is also referred to as AII,2. Originally, it was called "Q" for quencher of fluorescence (Duysens and Sweers, 1963; see a still earlier suggestion by Kautsky *et al.*, 1960), and is equivalent to Q_1 of P. Joliot, and, most likely, has a Em,7 of O mV (see *e.g.* Horton and Croze, 1979; Malkin, 1982). Its operating potential could be as low as -130mV. It is assigned to be equivalent to Q_H (a Q high potential). (For an earlier discussion of Q_1 and Q_2, see R. Govindjee *et al.*, 1970.)

When Q_A is reduced to the semiquinone Q_A^-, a characteristic absorbance change at 320nm is observed, and is labeled as X-320 (discovered by Stiehl and Witt (1968). (See Schatz and van Gorkom, 1985, for a recent UV spectra.) Although the protein (or a group) near Q_A may become protonated (Rutherford and Zimmerman, 1984), no evidence of protonation of Q_A^- is observed even at pH4.0 (van Gorkom, 1974).

Q_A^- interacts with pheophytin$^-$ as well as a non-heme iron (see *e.g.* Klimov *et al.*, 1980). Although there is no apparent reason for believing that the antenna complexes CP-43 and CP-47 are involved in any chemistry, deVitry *et al.* (1984) have observed that both these complexes were necessary for Q_A reduction.

All the permanently bound acceptors in the so-called PSIIα are also referred to as Qα and those in PSIIβ as Qβ. There is the possibility that the low potential Q, Q_L, may be equivalent to Q_2 (Joliot and Joliot, 1981), or Xa (Eckert and Renger, 1980). Neither Q_2 nor Xa is a quinone. The so-called Q_{400} that has been shown to be Fe (see Pterouleas and Diner, 1986) of the Q_A-Fe-Q_B complex cannot be equivalent to Q_L. No membrane potential ($\Delta\Psi$; C-550) is associated with Q_2^- or X_a^-, but Q_A^- production shows $\Delta\Psi$, and, an associated C-550 change that involves pheophytin and carotene.

In view of the above, it is not clear whether there exists only one or two tightly bound iron-quinones per reaction center in PSII. Brettel *et al.* (1985) find evidence for only one, but Evans and Ford (1986) claim that there are two tightly-bound iron quinones. Relationship to the PSIIα and PSIIβ centers must be examined before we can reach any conclusions.

Q_B. There has not been much discussion of heterogeneity of Q_B except that there are two types of centers: "B type" (*i.e.*, those that go through a two electron gate, see explanation below) and those that are "non-B type" (*i.e.* those that do not go through a two electron gate) (see Lavergne, 1982). PSIIβ centers also do not go through the two electron gate (Thielen and van Gorkom, 1981). It is assumed that some non-B type centers exist in PSIIα centers also. Q_B is bound to the 32 kDa lysine-free protein; it is also a herbicide binding protein, as mentioned in the Introduction. We have suggested that it also binds two HCO_3^- ions per polypeptide (see Blubaugh and Govindjee, 1988). Binding of the herbicide atrazine has been shown by photoaffinity labeling of ^{14}C azido-atrazine (Gardner, 1981; Satoh *et al.*, 1983). Herbicides (inhibitors) compete with Q_B binding by binding on the apoprotein; they are suggested to displace Q_B (Velthuys, 1981; Wraight, 1981). Herbicide resistance has been traced to alterations in single amino acids in the Q_B apoprotein (see *e.g.* Erickson *et al.*, 1985; Johanningmeier *et al.*, 1987).

Oettmeier and Soll (1983) and Vermaas *et al.* (1984) have shown apparent competition of short-chain plastoquinone analogs with ^{14}C-DCMU and ^{14}C-atrazine, respectively. Lavergne (1982) has clearly shown that binding of DCMU to centers with Q_B^- is much lower than with Q_B; furthermore, Lavergne (1982) has also shown competition of Q_B with benzoquinone. Recently, Wydrzynski and Inoue (1987) showed that artificial quinones can easily replace Q_B and modify the properties of the PSII acceptor site.

It is of interest to us to mention that effects on the D-1 protein, usually on the acceptor side by mutation or by inhibitors, can show effects on the donor side of PSII (*e.g.* Golbeck and Warden, 1985 for salicylaldoxime; Metz *et al.*, 1986 for herbicides; and Warden and Csatorday, 1987 for linolenic acid).

Electron Flow from Q_A^- to Q_B: Electrons are rapidly transferred out of Q_A^- with a halftime of 200 to 600 μs (Forbush and Kok, 1968; Mauzerall, 1972; Zankel, 1973). Bowes and Crofts (1980) measured this time as a function of flash number and have obtained evidence that $Q_A^-Q_B \rightarrow Q_AQ_B^-$ occurs in 200 to 400 μs and $Q_A^- Q_B^- \rightarrow Q_AQ_B^{2-}$ in 600-800 μs. The slowness of the second reaction may be due to electrostatic repulsion of the negative charge. In the presence of herbicides, that displace Q_B, electron flow out of Q_A^- is blocked and takes seconds.

As hinted above, Q_B is a two electron acceptor and acts as a two electron gate. Bouges-Bocquet (1973) and Velthuys and Amesz (1974), independently, reported evidence for the existence of what they have called "B" or "R", which they suggested accepts electrons from $Q^-(Q_A^-)$. Bouges-Bocquet (1973) measured the reduction of methylviologen (an electron acceptor for photosystem I only) in PSI light, and found that the number of electrons reaching methylviologen was higher after the 2 and 4 than after 1 and 3 preilluminating flashes in PSII; thus, it was suggested that a component B is reduced by the first PSII flash to B^- and then to B^{2-} by a second flash and only when B^{2-} was formed were electrons transferred to the PQ pool, and then to PSI. Velthuys and Amesz (1974) first blocked the electron donor side of PSII, added artificial electron donors to donate electrons to Z^+ or $P680^+$, exposed the samples to a series of light flashes, and then measured DCMU-induced Chl *a* fluorescence increase (ΔF) by a weak measuring beam. They observed a binary oscillation in F with peaks after odd number of flashes. This was explained as follows: after flash 1, Q^-R is converted to QR^-; DCMU blocks further electron flow, equilibrium is shifted to Q^-R and fluorescence is high as Q^- is a condition for high fluorescence. After flash 2, Q^-R^- is converted to QR^{2-}, electrons move out to the PQ pool, QR is produced; DCMU addition gives low fluorescence since the system is in low fluorescent state as Q is a condition for low fluorescence. Thus, the concept of the "two electron gate" was born. "B" or "R" has now been renamed Q_B. The important concept as to how electrons are transferred from Q_B to the PQ pool was suggested by Velthuys (1981): In this concept, after Q_B^{2-} is formed, it exchanges with a PQ molecule on the Q_B protein. The reactions can now be written as (see *e.g.* Stein *et al.*, 1984):

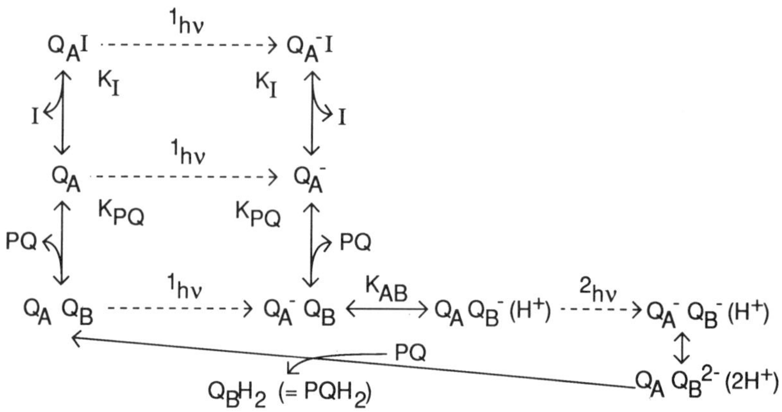

Here, I = inhibitor, Ks stands for equilibrium constants and PQ for plastoquinone.

The ratio of Q_B:Q_B^- in dark-adapted isolated thylakoids is approximately 7 : 3 (see *e.g.* Wollman, 1978). However, in intact chloroplasts (Govindjee, T. Ono, and Y. Inoue, unpublished, 1983), leaves (Rutherford *et al.*, 1984), and in higher plant cells grown in tissue culture (Xu *et al.*, 1988) it is 1:1. Figure 6 shows that the decay of Chl *a* fluorescence, reflecting Q_A^- to Q_A decay, is identical after both flash 1 and 2 (Xu *et al.*, 1988) suggesting 1:1 ratio for Q_B:Q_B^- in dark-adapted soybean cells.

ROLE OF BICARBONATE IN PLASTOQUINONE REDUCTION

Warburg and Krippahl (1958, 1960) (also see Franck, 1945) discovered that CO_2 (or HCO_3^-) stimulates electron flow in the Hill reaction. Warburg (see *e.g.*, his paper in 1964) considered this phenomenon as evidence for CO_2 being the source of O_2 in photosynthesis. Although Metzner (1975) and Stemler (1982) have favored an important role of CO_2 (HCO_3^-) on the electron donor, *i.e.* the O_2 evolving side of PSII, we have

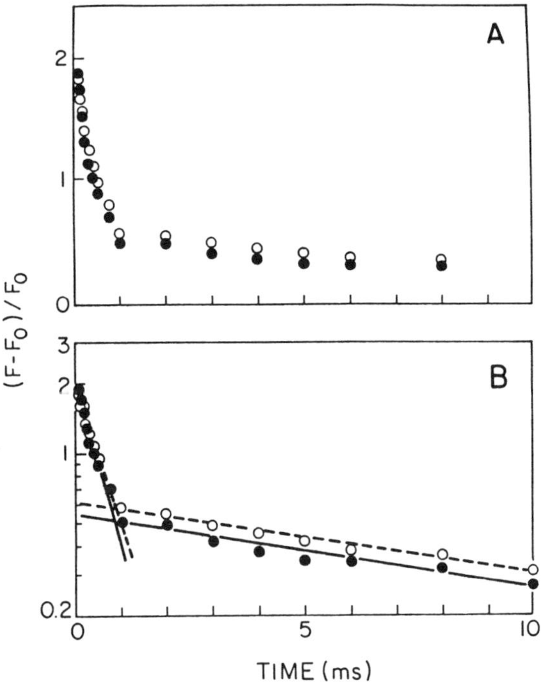

Figure 6. (A) Decay of variable chlorophyll *a* fluorescence ($(F-F_0)/F_0$) after the first (•) and second (∘) actinic flash, spaced 1 second apart, in soybean cells. (B) The same data on a semilogrithmic scale. F_0 is the fluorescence yield taken with the measuring light when all Q_A are oxidized and F is the fluorescence yield after the actinic flash. Absence of difference in data between flash 1 and 2 indicates that the ratio of Q_B:Q_B^- is 1 in dark-adapted intact higher plant cells. (After Xu *et al.*, 1988.)

consistently obtained evidence for a major effect of CO_2 (HCO_3^-) on the electron acceptor, *i.e.*, the plastoquinone reduction side of PSII (see reviews by Govindjee and van Rensen, 1978; Vermaas and Govindjee, 1981, 1982; van Rensen and Snel, 1985; Govindjee and Eaton-Rye, 1986; Blubaugh and Govindjee, 1988). We believe that HCO_3^- *is* required for the electron flow and the protons needed for plastoquinone reduction in PSII.

Our current picture is: CO_2 is the diffusing and HCO_3^- is the active species (see *e.g.*, Sarojini and Govindjee[a], 1981; and Blubaugh and Govindjee, 1986). At the pH of the stroma, we expect enough (220μM) HCO_3^-, much higher than the estimated 35-60μM dissociation constant, to be present at ambient air levels. We consider that two tight HCO_3^- binding sites exist on the Q_B-apoprotein, discussed above: (1) on the Fe of the Q_A-Fe-Q_B complex, as suggested by Michel and Deisenhofer

[a]Initially, the faster kinetics with CO_2 was misinterpreted for it being the active species, rather than the diffusing species.

(1988); and (2) on one of the arginines, near the Q_B binding, as suggested by Shipman (1981). Perhaps, the binding to the Fe in the Q_B protein places the PSII reaction in its proper conformation (allosteric effector) so that Q_B can efficiently accept electrons from Q_A^-, and Q_B^{2-} can efficiently exchange with the PQ pool. On the other hand, the function of the second HCO_3^- may be to bring protons needed for the reduction of the plastoquinone (for details, see Blubaugh and Govindjee, 1988).

A binding to Fe is supported by the experiments of Vermaas and Rutherford (1984), who observed a large formate-induced (*i.e.* HCO_3^- removal) increase in the $Q_A^-Fe^{2+}$ complex EPR signal at g = 1.82. A role of HCO_3^- in protonation, that may not involve the Fe site, was suggested first by experiments of Khanna *et al.* (1980) and later by Govindjee, J.J. van Rensen, J.F.H. Snel and W. Tonk (unpublished, 1983; see Govindjee and Eaton-Rye, 1986). The latter authors showed that a binary oscillation, superimposed on a period of 4 oscillation, of H^+ release to the lumen was abolished by HCO_3^- depletion (confirmed, reversibility of the effect shown, and extended by van Rensen *et al.*, 1988). These experiments may not provide any clue as to any direct effect of HCO_3^- on the protonation reactions *per se*, since the dips at uneven flashes may not be due to protonation of Q_B^{2-} but due to increased H^+ release during PQH_2 oxidation at even flashes. On the other hand, pH dependence of electron flow from Q_A^- to Q_B, as measured by Chl *a* fluorescence decay, shows remarkable differences with and without HCO_3^- present. In HCO_3^--depleted samples, Eaton-Rye and Govindjee (1988) observed that at pH 7.5, two turnovers of the Q_B "two-electron gate" were necessary before the maximum slowing down of Q_A^- to Q_B electron flow could be seen on subsequent flashes. By contrast, at pH 6.5, the slowed time was maximum only after one turnover of Q_B (*i.e.*, after two flashes). However, the maximum slowed time was much greater at pH 7.5 than at pH 6.5. After 1 or 2 flashes, however, the situation was reversed, *i.e.*, the decay was greater at pH 6.5 than at pH 7.5 (Fig. 7). Although complicated, the phenomenon can be understood by considering that there are two effects of HCO_3^- depletion: one, a conformational change and the other a protonation event. Whether the second is a result of the first or whether they are two separate effects remains to be determined. As a result of conformational change, binding of plastoquinone that forms Q_B is suggested to be decreased leading to changes in rate of $Q_A^-Q_B \rightarrow Q_AQ_B^-$ reaction. The effect is pH dependent, being faster at alkaline pH. It is likely that this effect involves changes at the Fe site. However, after higher flash numbers, either the conformational effect exposes another protonation effect, or a new separate protonation effect becomes observable. At alkaline pH: absence of HCO_3^- leads to a larger lack of H^+s for the protonation of plastoquinone sites, thus, slowing down the Q_A^- to $Q_B^{(-)}$ reactions. At

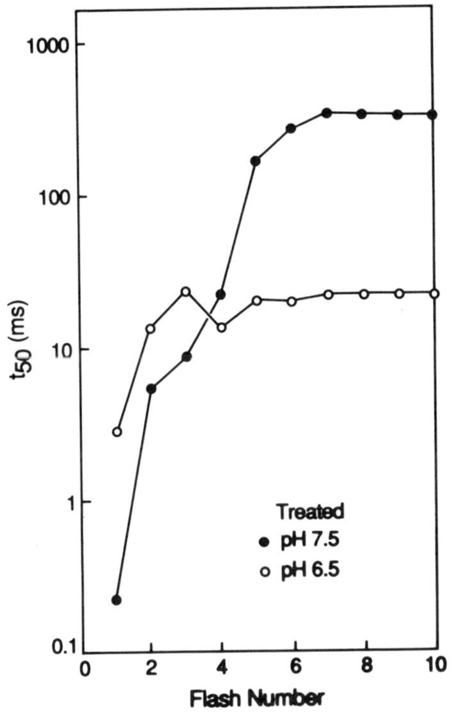

Figure 7. Plot of the times (t_{50}) at which (Q_A^-) is 50% of maximum [Q_A^-] in HCO_3^--depleted (treated) membranes with measurements at pH 6.5 and pH 7.5. At pH 7.5, (Q_A^-) decay is much slower than at pH 6.5 after four flashes suggesting the role of HCO_3^- in protonation reactions. The flash frequency was 1 Hz. (After Eaton-Rye and Govindjee, 1988.)

acidic pH, however, HCO_3^- has a lesser role to play due to the greater availability of protons. This explains the larger slowing down at pH 7.5 than at pH 6.5.

The concept of two binding sites was further supported by the data of Blubaugh and Govindjee (1988b) who showed by kinetic analysis that the residual activity in HCO_3^--depleted thylakoids was due to endogenously bound HCO_3^-: the amount of Hill activity restored to HCO_3^--depleted thylakoids by a half-saturating [HCO_3^-] was non-linear with respect to [Chl]. Thus, the essentiality of HCO_3^- was suggested. After corrections were made for endogenous [HCO_3^-], the activity versus [HCO_3^-] no longer followed Michelis-Menten kinetics, and suggested the existence of two cooperative binding sites ($n > 1.4$).

A detailed model of how HCO_3^- may function as a H^+ donor was discussed by Blubaugh and Govindjee (1988) (see Fig. 8). It is proposed that when Q_B (PQ) is reduced to Q_B^-, the pKa of a nearby histidine shifts

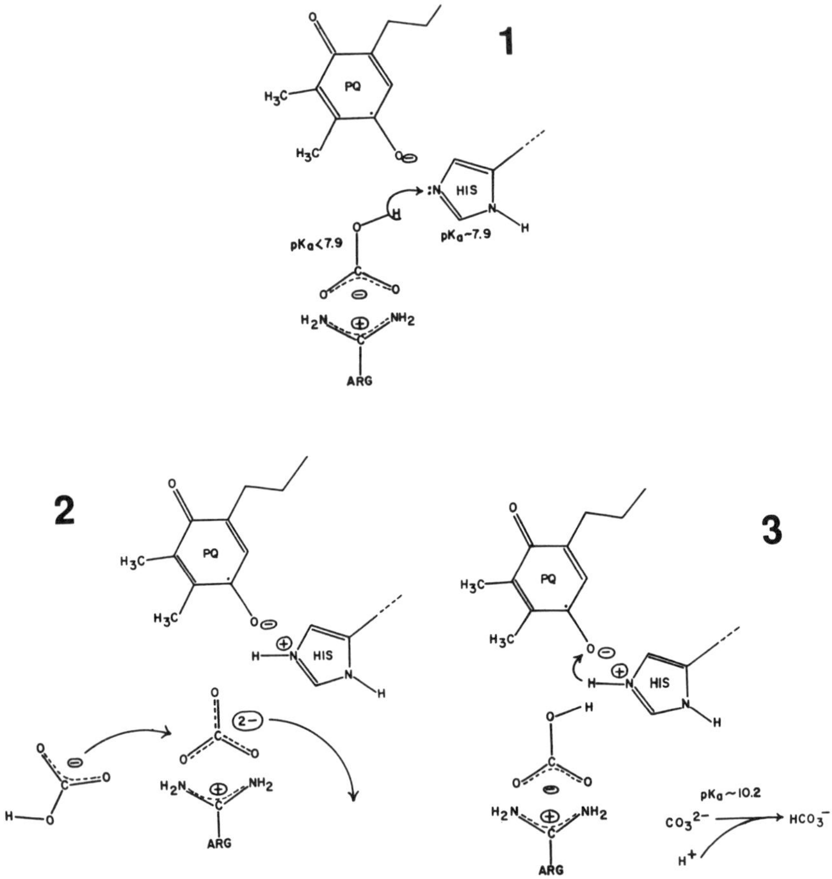

Figure 8. A working model for the involvement of HCO_3^- in the protonation of Q_B^-. 1: Bicarbonate sits on an arginine (ARG), and shares its proton with a nearby histidine (HIS); PQ stands for Q_B. 2: HCO_3^- replaces CO_3^{2-}, a H^+ has already been delivered to HIS. 3: Proton on HIS provides the positive charge needed to stabilize Q_B^- formed after a light reaction. The CO_3^{2-} is converted to HCO_3^- by the uptake of a H^+ from the surrounding (see text for details). (After Blubaugh and Govindjee, 1988.)

from 6.4 to 7.9 (see Crofts *et al.*, 1984) due to repulsive interaction between the negative charge on Q_B^- and the lone pair of electrons on histidine. The pKa of the HCO_3^- bound to a neighboring arginine is presumed to change from 10.2 to 7.9 due to electron withdrawal by arginine. Thus, a H^+ is shared between HCO_3^- and histidine. This H^+ ends up on histidine as another HCO_3^- replaces the CO_3^{2-} group holding the H^+ close to the histidine. This ensures irreversibility of H^+ transfer from HCO_3^- to histidine. Since the pKa of free CO_3^{2-} is 10.2, reprotonation is rapid. Finally, steric effects between the OH of HCO_3^- and H on the histidine favor the transfer of H^+ from histidine to near Q_B^- or Q_B^{2-}. We favor the possibility that arginine 257 and histidine 252 on D-1 protein may be the most likely pairs involved in protonation of Q_B^- or Q_B^{2-}. Thus, they are our targets of site-directed mutagenesis in order to test the above hypothesis.

At present, this test can be done on a transformable cyanobacterium *Synechocystis* PCC 6803. It, thus, became necessary to check for the existence of bicarbonate effect in cyanobacteria. The only published work on this effect on cyanobacteria showed a negative result (van Rensen and Vermaas, 1981). However, results of J. Cao and Govindjee

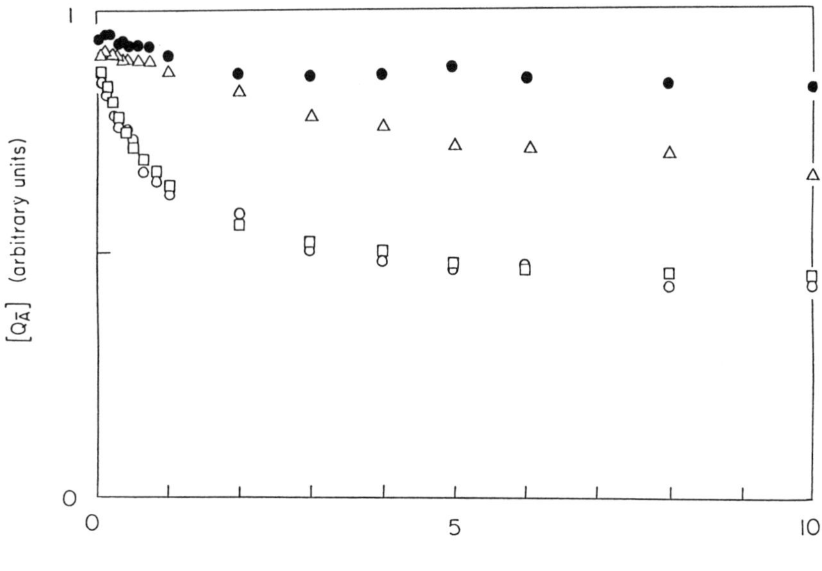

Figure 9. Decay of (Q_A^-), calculated from chlorophyll *a* fluorescence decay, in thylakoids from *Synechocystis* PCC 6803. Open circles: Control; triangles: depleted of HCO_3^- (25mM sodium formate); squares: restored with 5mM HCO_3^- addition; solid circles: control with 5μM diuron. Note that HCO_3^- depletion reversibly blocks the decay of Q_A^- just as diuron does. (Data of J. Cao and Govindjee, 1988, unpublished.)

(unpublished, 1988) show the existence of HCO_3^- effect in both thylakoids and intact cells of *Synechocystis* PCC 6803. Figure 9 shows that the decay of Q_A^- after the third flash is dramatically slowed down upon bicarbonate depletion approaching the rate of decay in herbicide treated samples.

Photosynthetic bacteria, that have quite a different amino acid sequence on their electron acceptor side, do not show any significant bicarbonate effect (R. J. Shopes, D. Blubaugh, Govindjee and C. Wraight, unpublished, 1987). As noted earlier, Michel and Deisenhofer (1988) have suggested that in PSII, HCO_3^- may replace the M-252 glutamate that provides the fifth ligand to the Fe. The possible relationship between the bacterial and plant systems as regards the second HCO_3^- binding site (an arginine) has not yet been explored.

A major question about the HCO_3^- effect is its role *in vivo*. A hint of this CO_2 effect was provided for wheat leaves by Gerbaud and André (1980). Govindjee, J. F. H. Snel and J. J. S. van Rensen (1983, unpublished; cited by Govindjee and Eaton-Rye, 1986) showed that bicarbonate depletion reversibly decreases the parabenzoquinone Hill reaction by several fold and produces a Chl *a* fluorescence transient that mimics DCMU-treatment in *Scenedesmus* cells. Garab *et al.* (1983) showed similar changes in Chl *a* fluorescence transient in leaves when air or CO_2 was replaced with N_2. Mende and Wiessner (1985) showed that in *Chlamydobotrys stellata*, but not in *Chlorella fusca*, replacement of air with CO_2-free-air first led to a decreased fluorescence suggestive of a blockage on the electron donor or water side, followed by an increase in fluorescence suggestive of a block on the electron acceptor side. Unfortunately, the parallel measurement on O_2 evolution was not done in identical conditions and perceived correlation between the two measurements was not established. Ireland *et al.* (1987), using wheat and maize leaves, showed that decreasing [CO_2] from already a low level (when CO_2 fixation was negligible) to a very low level caused changes that reflected a block between Q_A^- and Q_B. Garab *et al.* (1988) showed that removal of CO_2 from the leaf atmosphere produced thermoluminescence bands indicating a preponderance of Q_A^- over Q_B^- clearly showing that the absence of CO_2 blocked electron flow from Q_A^- to Q_B. Ireland *et al.* (1987) working on wheat and maize, and Fatma El-Shintinawy and Govindjee (paper presented at the annual meeting of the American Society of Plant Physiology, 1988, Reno) working on spinach leaves showed that, when methylviologen is used as an electron acceptor (anaerobic conditions), absence of HCO_3^- caused a faster fluorescence rise indicating a block in electron flow from Q_A^- to Q_B. Addition of HCO_3^- reversed the effect. Interestingly, El-Shintinawy and Govindjee observed that if the spinach leaves were infiltrated for a very short time (*e.g.*, 10s) with 100 mM sodium formate, both O_2 evolution and Chl *a* fluorescence levels decreased, apparently, suggesting a block on the electron donor side of PSII (Fig. 10). However, a longer infiltration (*e.g.*, minutes) showed a dramatic decrease in O_2 evolution and a dramatic increase in the rate of Chl *a* fluorescence rise indicating a block on the electron acceptor side. It appears that the HCO_3^- pool associated with the first effect must be more loosely bound than that associated with the second effect. Our data on leaves, although reminiscent of Mende and Wiessner's data on *Chlamydobotrys*, cannot be interpreted, however, as showing a block on the electron donor side because a block between pheophytin and Q_A (a site close to the Fe site) would also give the same result since reduced pheophytin is a quencher of Chl *a* fluorescence.

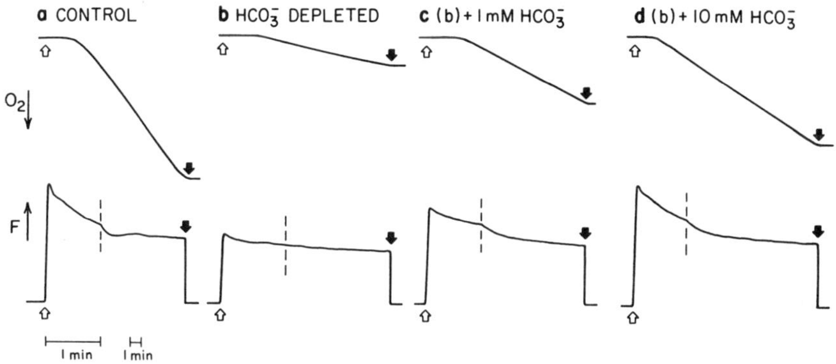

Figure 10. Parallel measurements on the rate of oxygen evolution (top curves) and chlorophyll a fluorescence changes in spinach leaf discs. (a): control; (b) infiltrated with 100mM sodium formate for 10s; (c) : (b) restored with 1mM HCO_3^-; (d) : (b) restored with 10mM HCO_3^-. Infiltration of leaf discs with 100mM sodium formate for longer time (> 1 min) shows elimination of O_2 evolution and a faster fluorescence rise to maximum fluorescence (data not shown). (After F. El-Shintinawy and Govindjee, 1988, presented at the 1988 Americal Society of Plant Physiology Meeting, Reno, Nevada.)

It appears to us that when HCO_3^- is completely removed, electron flow from Q_A^- to the plastoquinone pool is almost completely blocked after 1 or 2 turnovers of the two electron gate, (see Vermaas and Govindjee, 1982; Govindjee, 1984; Govindjee *et al.*, 1984; J. Cao and Govindjee, unpublished, 1988). Experiments are in progress to further understand the mechanism of bicarbonate effect *in vivo*.

ACKNOWLEDGEMENT

Govindjee is thankful to Professor Donald Ort for his continued support and to an Interdisciplinary McKnight Grant to the University of Illinois. Work at ANL is supported by the Division of Chemical Sciences, Office of Basic Energy Sciences of the Department of Energy under contract W-31-109-ENG-38.

REFERENCES

Akabori K, Tsukamoto H, Tsukihara J, Nagatsuka T, Motokawa O, Toyoshima Y (1988). Disintegration and reconstitution of photosystem II reaction center core complex I. Preparation and characterization of three different types of sub-complex. Biochim Biophys Acta 932:345-357.

Andersson B, Åkerlund H-E (1987). Proteins of the oxygen-evolving complex. Topics in Photosynthesis 8:379-420.

Andréasson L-E, Vänngård T (1988). Electron transport in photosystem I and II. Ann Rev Plant Physiol 39:379-411.

Babcock GT (1987). The photosynthetic oxygen-evolving process. In Amesz J.(ed): "Photosynthesis," Amsterdam: Elsevier Science Publishers BV Biomed Div, pp 125-158.

Barber J, Chapman DJ, Telfer A (1987). Characterisation of a photosystem II reaction centre isolated from the chloroplasts of *Pisum sativum*. FEBS Lett 220:67-73.

Black MT, Brearley TH, Horton P (1986). Heterogeneity in chloroplast photosystem II. Photosyn Res 8:193-207.

Blubaugh D, Govindjee (1986). Bicarbonate, not CO_2, is the species required for the stimulation of photosystem II electron transport. Biochim Biophys Acta 848:147-152.

Blubaugh DJ, Govindjee (1988a). The molecular mechanism of the bicarbonate effect at the plastoquinone reductase site of photosynthesis. Photosyn Res, 19:85-128.

Blubaugh DJ, Govindjee (1988b). Kinetics of the bicarbonate effect and the number of HCO_3^- binding sites in thylakoid membranes. Biochim Biophys Acta, 936:208-214.

Bouges-Bocquet B (1980). Kinetic models for the electron donors of photosystem II of photosynthesis. Biochim Biophys Acta 594:85-103.

Brettel K, Schlödder E, Witt HT (1984). Nanosecond reduction kinetics of photooxidized chlorophyll a_{II} (P680) in single flashes as a probe for the electron pathway, H^+ release and charge accumulation in the O_2-evolving complex. Biochim Biophys Acta 766:403-428.

Brettel K, Schlödder E, Witt HT (1985). Evidence for only one stable electron acceptor in the reaction center of photosystem II in spinach chloroplasts. Photobiochem Photobiophys 9:205-213.

Butler WL (1972a). The relationship between P-680 and C-550. Biophys J 12:851-857.

Butler WL (1972b). On the primary nature of fluorescence yield changes associated with photosynthesis. Proc Natl Acad Sci USA 69:3420-3422.

Chapman DJ, Gounaris K, Barber J(1988). Electron transport properties of the isolated D-1-D-2-cytochrome b559 photosystem two reaction centre, submitted.

Coleman WJ, Govindjee (1987). A model for the mechanism of chloride activation of oxygen evolution in photosystem II. Photosyn Res 13:199-223.

Crofts AR, Wraight C (1983). The electrochemical domain of photosynthesis. Biochim Biophys Acta 726:149-185.

Crofts AR, Robinson HH, Snozzi M (1984). Reactions of quinones at catalytic sites; a diffusional role in H-transfer. In Sybesma C (ed): "Advances in Photosynthesis Research," The Hague: Martinus Nijhoff, pp 461-468.

Danielius RV, Satoh K, vanKan PJM, Plitzer JJ, Nuijs AM, van Gorkom HJ (1987). The primary reaction of photosystem II in the D1-D2-cytochrome b-559 complex. FEBS Lett 213:241-244.

Debus RJ, Barry BA, Babcock GT, McIntosh L (1988). Site-directed mutagenesis identifies a tyrosine radical in the photosynthetic oxygen-evolving system. Proc Natl Acad Sci USA 85:427-430.

Deisenhofer J, Epp O, Miki K, Huber R, Michel H (1984). X-ray structure analysis of a membrane protein complex: electron density map at 3Å resolution and a model of the chromophores of the photosynthetic reaction center from *Rhodopseudomonas viridis*. J Mol Biol 180:385-398.

Den Blanken HJ, Hoff AJ, Jongenelis APJM, Diner BA (1983). High resolution triplet-minus-singlet absorbance difference spectrum of photosystem II particles. FEBS Lett 157:21-27.

de Vitry C, Wollman FA, Delepelaire P (1984). Function of the polypeptides of the photosystem II reaction center in *Chlamydomonas reinhardtii*: localization of the primary events. Biochim Biophys Acta 767:415-422.

Diner B (1986). The reaction center of photosystem II. Encyclopedia of Plant Physiology New Series 19:422-436.

Döring G, Stiehl HH, Witt HT (1967). A second chlorophyll reaction in the electron transport chain of photosynthesis: registration by the repetitive excitation technique. Z Naturforsch B22:639-644.

Döring G, Renger G, Vater J, Witt HT (1969). Properties of the photoactive chlorophyll a_{II} in photosynthesis. Z Naturforsch B24:1139-1143.

Duysens LNM, Sweers HE (1963). Mechanism of two photochemical reactions in algae as studied by means of fluorescence. In Jpn Soc Plant Physiol (eds): "Microalogae and Photosynthetic Bacteria," Tokyo: University of Tokyo Press, pp 353-372.

Eaton-Rye JJ, Govindjee (1988). Electron transfer through the quinone acceptor complex of photosystem II in bicarbonate-depleted spinach thylakoid membranes as a function of actinic flash number and frequency. Biochim Biophys Acta, 935:237-257.

Eckert HJ, Renger G (1980). Photochemistry of the reaction centers of system II under repetitive flash group excitation in isolated chloroplasts. Photochem Photobiol 31:501-511.

Eckert HJ, Renger G, Witt HT (1984). Reduction kinetics of the photo-oxidized chlorophyll a_{II} in the nanosecond range. Measurements of the absorption changes at 688nm under repetitive flash excitation. FEBS Lett 167:316-320.

Erickson JM, Rahire M, Rochaix J-D, Mets L(1985). Herbicide resistance and cross resistance: changes at three distinct sites in the herbicide resistant protein. Science 228:204-207.

Evans MCW, Ford RC (1986). Evidence for two tightly bound iron quinones in the electron acceptor complex of photosystem II. FEBS Lett 195:290-294.

Fajer J, Davies MS, Forman A, Klimov VV, Dolan E, Ke B (1980). Primary electron acceptors in plant photosynthesis. J Amer Chem Soc 102:7143-7145.

Floyd RA, Chance B, DeVault D (1971). Low temperature photo-induced reactions in green leaves and chloroplasts. Biochim Biophys Acta 226:103-112.

Franck J (1945). Photosynthetic activity of isolated chloroplasts. Rev Mod Phys 17:112-119.

Fujita I, Davis MS, Fajer J (1978). Anion radicals of pheophytin a and chlorophyll *a*. Their role in the primary charge separation of plant photosynthesis. J Amer Chem Soc 100:6280-6282.

Garab G, Sanchez Burgos AA, Zimanyi L, Faludi-Daniel A (1983). Effect of CO_2 on the energization of thylakoids in leaves of higher plants. FEBS Lett 154:323-327.

Garab Gy, Rozsa Zs, Govindjee (1988). Carbon-dioxide affects charge accumulation in leaves: measurements by thermoluminiscence. Naturwiss, 75:517-519.

Gardner G (1981). Azidoatrazine: photoaffinity label for the site of triazine herbicide action in chloroplasts. Science 211:937-940.

Gerbaud A and André M (1980). Effect of CO_2, O_2 and light on photosynthesis and photorespiration. Plant Physiol 66: 1032-1036.

Ghanotakis DF, Yocum CF (1985). Polypeptides of photosystem II and their role in oxygen evolution. Photosyn Res 7:97-114.

Ghanotakis DF, Waggoner CM, Bowlby NR, Demetriou DM, Babcock GT, Yocum CF (1987). Comparative structural and catalytic properties of oxygen-evolving photosystem II preparations. Photosyn Res 14:211-227.

Giardi MT, Marder JB, Barber J (1988). Herbicide binding to the isolated photosystem two reaction centre. Biochim Biophys Acta, submitted.

Golbeck JH, Warden JT (1985). Site of salicyladoxime interaction with photosystem II. Photosyn Res 6:371-380.

Govindjee (1980). The oxygen evolving system of photosynthesis. Plant Biochem J (India) Sircar Memorial Volume: 7-30.

Govindjee (ed.) (1982). Photosynthesis, Volume I. Energy conversion by Plants and Bacteria. New York: Academic Press, pp 195-272; 331-506.

Govindjee (1984). Photosystem II: the oxygen evolving system of photosynthesis. In Sybesma C (ed): "Advances in Photosynthesis Research, Vol 1," The Hague: Martinus Nijhoff/Dr W Junk Publishers, pp 227-238.

Govindjee, Eaton-Rye JJS (1986). Electron transfer through photosystem II acceptors: interaction with anions. Photosyn Res 10:365-379.

Govindjee, van Rensen JJS (1978). Bicarbonate effect on the electron flow in isolated broken chloroplasts. Biochim Biophys Acta 505:183-213.

Govindjee, Döring G, Govindjee R (1976). The active chlorophyll a_{II} in suspensions of lyophilized and Tris-washed chloroplasts. Biochim Biophys Acta 205:303-306.

Govindjee, Nakatani HY, Rutherford AW, Inoue Y (1984). Evidence from thermoluminescence for bicarbonate action on the recombination reactions involving the secondary electron acceptor of photosystem II. Biochim Biophys Acta 766:416-423.

Govindjee, Kambara T, Coleman W (1985). The electron donor side of photosystem II: the oxygen evolving complex. Photochem Photobiol 42:1187-210.

Hansson Ö, Duranton J, Mathis P (1988). Yield and lifetime of the primary radical pair in preparations of photosystem II with different antenna size. Biochim Biophys Acta 932:91-96.

Holzwarth AR (1987). Picosecond fluorescence spectroscopy and energy transfer in photosynthetic antenna pigments. Topics in Photosynthesis 8:95-157.

Horton P, Croze E (1979). Characterization of two quenchers of chlorophyll fluorescence with different midpoint oxidation-reduction potentials in chloroplasts. Biochim Biophys Acta 545:188-201.

Ireland CR, Baker NR, Long SP (1987). Evidence for a physiological role of CO_2 in the regulation of photosynthetic electron transport in intact leaves. Biochim Biophys Acta 893:434-443.

Johanningmeier U, Bodner U, Wildner GF (1987). A new mutation in the gene coding for the herbicide protein in *Chlamydomonas*. FEBS Lett 211:221-224.

Joliot P, Joliot A (1981). A photosystem II electron acceptor which is not a plastoquinone. FEBS Lett 134:155-158.

Jursinic P, Govindjee (1977a). Temperature dependence of delayed light emission in the 6 to 340 μs range after a single flash in chloroplasts. Photochem Photobiol 26:617-628.

Jursinic P, Govindjee (1977b). The rise in chlorophyll *a* fluorescence yield and decay in delayed light emission in Tris-washed chloroplasts in the 6-100 μs time range after an excitation flash. Biochim Biophys Acta 461:253-267.

Kautsky H, Apel W, Amman H (1960). Chlorophyll Fluorescenz und Kohlensauerassimilation XII. Mitteilung die Fluorescenzkurve und die Photochemie der Pflanzen. Biochem Z 332:277-292.

Ke B, Shuvalov VA (1987). Picosecond absorption spectroscopy in photosynthesis and primary electron transfer processes. Topics in Photosynthesis 8:31-93.

Khanna R, Wagner R, Junge W, Govindjee (1980). Effects of CO_2-depletion on proton uptake and release in thylakoid membranes. FEBS Lett 121:222-224.

Klimov VV, Krasnovsky AA (1981). Pheophytin as the primary electron acceptor in photosystem 2 reaction centres. Photosynthetica 15:592-609.

Klimov VV, Klevanik AV, Shuvalov VA, Krasnovsky AA (1977). Reduction of pheophytin in the primary light reaction of photosystem II. FEBS Lett 82:183-186.

Klimov VV, Allakhverdiev SI, Demeter S, Krasnovsky AA (1979). Photoreduction of pheophytin in photosystem 2 of chloroplasts with respect to the redox potential of the medium. (in Russian) Doklady Akad Nauk SSSR 249:227-230.

Klimov VV, Dolan E, Shaw ER, Ke B (1980). Interaction between the intermediary electron acceptor (pheophytin) and a possible plastoquinone-iron complex in photosystem II reaction centers. Proc Natl Acad Sci USA 77:7227-7231.

Lavergne J (1982a). Mode of action of 3-(3,4 dichlorophenyl)-1,1-dimethylurea. Evidence that the inhibitor competes with plastoquinone for binding to a common site on the acceptor side of photosystem II. Biochim Biophys Acta 682:345-353.

Lavergne J (1982b). Two types of primary acceptors in chloroplasts. Photosystem II. I. Different recombination properties II. Reduction in two successive photoacts. Photobiochem Photobiophys 3:257-285.

Malkin R (1982). Redox properties and functional aspects of electron carriers in chloroplast photosynthesis. Topics in Photosynthesis 4:1-47.

Marder JB, Chapman DJ, Telfer A, Nixon PJ, Barber J (1987). Identification of psbA and psbD gene products, D_1, and D_2, as reaction center proteins of photosystem 2. Plant Mol Biol 9:325-333.

Marder JB, Telfer A, Barber J (1988). The D-1 polypeptide subunit of the photosystem II reaction center has a phosphorylation site at its amino terminus. Biochim Biophys Acta 932:362-365.

Mathis P (1986). Structural aspects of vectorial electron transfer in photosynthetic reaction centers. Photosyn Res 8:97-111.

Mauzerall D (1972). Light-induced fluorescence changes in *Chlorella*, and the primary photoreactions for the production of oxygen. Proc Natl Acad Sci USA 69:1358-1362.

Meiburg RF, van Gorkom HJ, van Dorssen RJ (1983). Excitation trapping and charge separation in photosystem II in the presence of an electric field. Biochim Biophys Acta 724:352-358.

Mende D, Wiessner W (1985). Bicarbonate *in vivo* requirement of photosystem II in the green alga *Chlamydobotrys stellata.* J Plant Physiol 118:259-266.

Metz JG, Pakrasi H, Seibert M, Arntzen CJ (1986). Evidence for a dual function of the herbicide binding D-1 protein in Photosystem II. FEBS Lett 205:269-274.

Metzner H (1975). Role of CO_2 in water decomposition in photosynthesis? A critical consideration. J Theoret Biol 51:201-231.

Michel H, Deisenhofer J (1988). Relevance of the photosynthetic reaction center from purple bacteria to the structure of photosystem II. Biochemistry 27:1-7.

Moya I, Sebban P, Haehnel W (1986). Lifetime of excited states and quantum yield of chlorophyll *a* fluorescence *in vivo.* In Govindjee, Amesz J, Fork DC, (eds): "Light Emission by Plants and Bacteria," Orlando: Academic Press, pp 161-190.

Murata N, Arakai S, Fujita Y, Suzuki K, Kuwabara T, Mathis P (1986). Stoichio- metric determination of pheophytin in photosystem II of oxygenic photosynthesis. Photosyn Res 9:63-70.

Nanba O, Satoh K (1987). Isolation of a photosystem II reaction center consisting of D-1 and D-2 polypeptides and cytochrome *b*-559. Proc Natl Acad Sci USA 84:109-112.

Nuijs AM, van Gorkom HJ, Plijter JJ, Duysens LNM (1986a). Primary charge separation and excitation of chlorophyll *a* in photosystem II particles from spinach as studied by picosecond absorption difference spectroscopy. Biochim Biophys Acta 848:167-175.

Nuijs AM, Shuvalov VA, van Gorkom HJ, Plijter JJ, Duysens LNM (1986b). Picosecond absorbance difference spectroscopy on the primary reactions and the antenna excited states in photosystem I particles. Biochim Biophys Acta 850:310-318.

Oettmeier W, Soll HJ (1983). Competition between plastoquinone and 3-(3,4-dichlorophenyl),-1,1-dimethylurea at the acceptor side of photosystem II. Biochim Biophys Acta 724:287-290.

Okamura MY, Satoh K, Isaacson RA, Feher G (1987). Evidence of the primary charge separation in D_1 D_2 complex of photosystem II from spinach: EPR of the triplet state. In Biggins J (ed): "Progress in Photosynthesis Research, Proceedings of the 7th International Conference on Photosynthesis, August, 1986, Vol I," The Hague: Martinus Nijhoff, pp 379-391.

Omato T, Murata N, Satoh K (1984). Quinone and pheophytin in the photosynthetic reaction center II from spinach chloroplasts. Biochim Biophys Acta 765:403-405.

Ort D (1986). Energy transduction in oxygenic photosynthesis: an overview of structure and mechanism. Encyclopedia of Plant Physiology New Series 19:143-196.

Parson WW, Ke B (1982). Primary photochemical reactions. In Govindjee (ed): "Photosynthesis I. Energy conversion by Plants and Bacteria," New York: Academic Press, pp 331-385.

Petrouleas V, Diner BA (1986). Identification of Q_{400}, a high potential electron acceptor of photosystem II, with the iron of the quinone-iron acceptor complex. Biochim Biophys Acta 893:126-137.

Renger G (1987). Mechanistic aspects of photosynthetic water cleavage. Photosynthetica 21:203-224.

Renger G, Govindjee (1985). The mechanism of photosynthetic water oxidation. Photosyn Res 6:33-55.

Rich PR, Moss DA (1987). The reactions of quinones in higher plant photosynthesis. Topics in Photosynthesis 8:421-445.

Rubinstein D, Rabinowitch E (1963). Fluorescence and absorption changes in *Chlorella* exposed to strong light: the red band. Science 142:681-682.

Rutherford AW, Zimmerman JL (1984). A new EPR signal attributed to the primary plastosemiquinone acceptor in photosystem II. Biochim Biophys Acta 767:168-175.

Rutherford AW, Mullet JE, Crofts AR (1981). Measurement of the midpoint potential of the pheophytin acceptor of photosystem II. FEBS Lett 123:235-237.

Rutherford AW, Govindjee, Inoue Y (1984). Charge accumulation and photochemistry in leaves studied by thermoluminescence and delayed light emission. Proc Natl Acad Sci USA 81:1107-1111.

Sarojini G, Govindjee (1981) Is CO_2 an active species in stimulating the Hill reaction in thylakoid membranes? In Akoyunoglou G (ed): "Photosynthesis II. Electron transport and photophosphorylation," Philadelphia: Balaban International Science Service, pp 143-149.

Sarojini G, Daniell H, Vermaas WFJ (1981). Site of electron acceptance by 3,6 -dichloro-2,5-dimethoxy p-benzoquinone and its relation to the bicarbonate effect on photosynthetic electron transport. Biochem Biophys Res Commun 102:944-951.

Satoh K, Nakatani HY, Steinback KE, Watson J, Arntzen CJ (1983). Polypeptide composition of a photosystem II core complex: presence of a herbicide-binding protein. Biochim Biophys Acta 724:142-150.

Satoh K, Fujii Y, Aoshima T, Tado T (1987). Immunological identification of the polypeptide bands in the SDS polyacrylamide gel electrophoresis of photosystem II preparations. FEBS Lett 216:7-10.

Schatz GH, van Gorkom HJ (1985). Absorbance difference spectra upon charge transfer to secondary donors and acceptors in photosystem II. Biochim Biophys Acta 810:283-294.

Schatz GH, Brock H, Holzwarth AR (1987). Picosecond kinetics of fluorescence and absorbance changes in photosystem II particles excited at low photon density. Proc. Natl Acad Sci USA 84: 8414-8418.

Schlödder E, Brettel K, Witt HT (1985). Relation between microsecond reduction kinetics of photooxidized chlorophyll *a*II (P-680) in photosystem II particles from Synechococcus sp. Biochim Biophys Acta 808: 123-131.

Seibert M, Picorel R, Rubin AB, Connolly JS (1988). Spectral, photophysical and stability properties of isolated photosystem II reaction center. Plant Physiol, 87:303-306.

Shipman LL (1981). Theoretical study of the binding site and mode of action for photosystem II herbicides. J Theoret Biol 90:123-148.

Sonneveld A, Rademaker H, Duysens LNM (1979). Chlorophyll *a* fluorescence as a monitor of nanosecond reduction of the photooxidized primary donor P-680$^+$ of photosystem II. Biochim Biophys Acta 548:536-551.

Stemler A (1982). The functional role of bicarbonate in photosynthetic light reaction II. In Govindjee (ed): "Photosynthesis. Vol II. Development, Carbon Metabolism, and Plant Productivity," New York: Academic Press, pp 513-558.

Stein RR, Castellvi AL, Bogacz JP, Wraight CR (1984). Herbicide-quinone competition in the acceptor complex of photosynthetic reaction centers from *Rhodopseudomonas sphaeroides*: a bacterial model for PSII herbicide activity in plants. J Cell Biochem 24:243-259.

Stiehl HH, Witt H (1968). Die kurzeitigen ultravioletten Differenzspektren bei der Photosynthese. Z. Naturforschg 23b:220-224.

Takahashi Y, Hansson Ö, Mathis P, Satoh K (1987). Primary radical pair in the photosystem II reaction center. Biochim Biophys Acta 893:49-59.

Thielen APGM, van Gorkom HJ (1981a). Energy transfer and quantum yield in photosystem II. Biochim Biophys Acta 637:439-446.

Thielen APGM, van Gorkom HJ (1981b). Redox potentials of electron acceptors in Photosystem IIα and IIβ. FEBS Lett 129:205-209.

Trebst A (1986). The topology of the plastoquinone binding peptides of photosystem II in the thylakoid membrane. Z Naturforschg 41C:240-245.

Trebst A (1987). The three-dimensional structure of the herbicide binding niche on the reaction center polypeptides of photosystem II. Z Naturforschg 24C:742-750.

Trebst A, Draber W (1986). Inhibitors of photosystem II and the topology of the herbicide and Q_B binding polypeptide in the thylakoid membrane. Photosyn Res 10:381-392.

Trissl H.-W, Kunze, V (1985). Primary electrogenic reactions in the chloroplasts probed by picosecond flash-induced dielectric polarization. Biochim Biophys Acta 806:136-144.

Trissl HW, Kunze U, Junge W (1982). Extremely fast photoelectric signals from suspensions of broken chloroplasts and of isolated chromatophores. Biochim Biophys Acta 682: 364-377.

van Best JA, Mathis P (1978). Kinetics of the oxidized primary electron donor of photosystem II in spinach chloroplasts and in *Chlorella* cells in the microsecond and nanosecond time ranges following flash excitation. Biochim Biophys Acta 503:178-188.

van Gorkom HJ (1974). Identification of the reduced primary electron acceptor of photosystem II as a bound semiquinone anion. Biochim Biophys Acta 347:439-442.

van Gorkom HJ (1985). Electron transfer in Photosystem II. Photosyn Res 6:97-112.

van Gorkom HJ, Pulles MPJ, Wessels JSC (1975). Light-induced changes of absorbance and electron spin resonance in small photosystem II particles. Biochim Biophys Acta 408:331-339.

van Rensen JJS, Snel JFH (1985). Regulation of photosynthetic electron transport by bicarbonate, formate and herbicides in isolated broken and intact chloroplasts. Photosyn Res 6:231-246.

van Rensen JJS, Tonk WJ, De Bruijn SM (1988). Involvement of bicarbonate in the protonation of the secondary quinone electron acceptor of photosystem II via the non-haem iron of the quinone-iron acceptor complex. FEBS Lett 226:347-351.

Velthuys B (1980). Mechanisms of electron flow in photosystem II and towards photosystem I. Ann Rev Plant Physiol 31:545-567.

Velthuys BR (1981). Electron dependent competition between plastoquinone and inhibitors for binding to photosystem II. FEBS Lett 126:277-281.

Velthuys B (1987). Photosystem II reaction center. Topics in Photosynthesis 8:341-378.

Vermaas WFJ, Govindjee (1981a). Unique role(s) of carbon dioxide and bicarbonate in the photosynthetic electron transport system. Proc Indian Natl Sci Acad B47:581-605.

Vermaas WFJ, Govindjee (1981b). The acceptor side of photosystem II in photosynthesis. Photochem Photobiol 34:775-793.

Vermaas WFJ, Govindjee (1982a). Bicarbonate or CO_2 as a requirement for efficient electron transport on the acceptor side of photosystem II. In Govindjee (ed): "Photosynthesis. Vol II. Development, Carbon Metabolism and Plant Productivity," New York: Academic Press, pp 541-558.

Vermaas WFJ, Govindjee (1982b). Bicarbonate effects on chlorophyll *a* fluorescence transients in the presence and absence of diuron. Biochim Biophys Acta 680:202-209.

Vermaas WFJ, Rutherford AW (1984). EPR measurements on the effect of HCO_3^- and triazine resistance on the acceptor side of photosystem II. FEBS Lett. 175:243-248.

Vermaas WFJ, Renger G, Arntzen CJ (1984). Herbicide/quinone binding interactions in photosystem II. Z. Naturforschg 39c:368-373.

Vermaas WFJ, Rutherford AW, Hansson Ö (1988). Site-directed mutagenesis in photosystem II of the cyanobacterium *Synechocystis* sp. PCC 6803: the donor D is a tyrosine residue in the D_2 Protein. Proc Natl Acad Sci USA, in press.

Warburg O (1964). Prefatory chapter. Annu Rev Biochem 33:1-18.

Warburg O, Krippahl G (1958). Hill-Reaktionen. Z Naturforsch 13b:509-514.

Warburg O, Krippahl G (1960). Notwindigkeit der Kohlensaure für die Chinon und Ferricyanid-Reactionen in grünen Grana. Z Naturforsch 14b:712-724.

Warden JT, Csatorday K (1987). On the mechanism of linolenic acid inhibition in photosystem II. Biochim Biophys Acta 890:215-223.

Wasielewski MR, Johnson DG, Seibert M, Govindjee (1989). Determination of the primary charge separation rate in isolated photosystem II reaction centers with 500 femtosecond time resolution. Proc Natl Acad Sci USA, 86:524-528.

Witt HT (1975). Primary acts of energy conservation in the functional membrane of photosynthesis. In Govindjee (ed): "Bioenergetics of Photosynthesis," New York: Academic Press, pp 493-554.

Wollman F-A (1978). Determination and modification of the redox state of the secondary acceptor of photosystem II in the dark. Biochim Biophys Acta 503:263-273.

Wraight CA (1981). Oxidation-reduction physical chemistry of the acceptor quinone complex in bacterial photosynthetic reaction centers: evidence for a new model of herbicide activity. Israel J. Chem. 21:348-351.

Wydrzynski T, Inoue Y (1987). Modified photosystem II acceptor side properties upon replacement of the quinone at the Q_B site with 2,5-dimethyl-p-benzoquinone and phenyl-p-benzoquinone. Biochim Biophys Acta 893:33-42.

Xu C, Rogers SMD, Goldstein C, Widholm JM, Govindjee (1988). Fluorescence characteristics of photoautotrophic soybean cells. Photosyn Res, in press.

Photosynthesis, pages 105–122

TEMPERATURE-DEPENDENT STEADY-STATE AND PICOSECOND KINETIC FLUORESCENCE MEASUREMENTS OF A PHOTOSYSTEM I PREPARATION FROM SPINACH

Ishita Mukerji and Kenneth Sauer

Department of Chemistry and Chemical Biodynamics Division, Lawrence Berkeley Laboratory, University of California, Berkeley, CA 94720 USA

ABSTRACT

The fluorescence properties of a photosystem I (PSI) preparation from spinach containing approximately 200 chlorophyll (Chl) per reaction center were investigated. The preparation, characterized both spectroscopically and biochemically, contained the peripheral light harvesting antenna associated with PSI. In this study steady-state fluorescence measurements were performed as a function of temperature. An emission maximum at 690nm and a long wavelength shoulder from 710 to 740nm were observed. The fluorescence yield at 690nm is temperature independent, while the yield of the long wavelength shoulder increases dramatically with decreasing temperature. Additionally, kinetic measurements using the technique of single photon counting were done at room temperature and 77K. At 295K a four-component fit was needed to describe the fluorescence decay; whereas at 77K, an additional 40-50 ps rise component indicative of fluorescence induction was necessary. The kinetic measurements demonstrate that the increase in yield as a function of decreasing temperature arises primarily from an increase in fluorescence decay component amplitude rather than an increase in lifetime. The dependence of the yield of the long wavelength emission on excitation wavelength indicates that an integral relationship exists between the chlorophyll *b* (Chl *b*) contained in the peripheral light harvesting antenna and the long wavelength emission.

INTRODUCTION

Intact PSI complexes have been isolated with different amounts of core and peripheral antenna pigments associated with the reaction center (Mullet *et al.*, 1980; Lam *et al.*, 1984). A peripheral light harvesting complex (LHCPI) containing Chl *b* is associated with PSI. This light harvesting complex contributes significantly to the long wavelength emission characteristic of PSI at low temperature (Haworth *et al.*, 1983). The core antenna and reaction center are associated with two high molecular mass peptides, approximately 62 and 58 kDa, which bind the primary electron donors and acceptors and 100 Chl *a* molecules (Fish and Bogorad, 1986; Golbeck *et al.*, 1987; Malkin, 1986). The LHCPI complex contains polypeptides in the 20-24 kDa range. This complex consists of 100 Chl *a* and Chl *b* molecules in a ratio of approximately 4:1. All of the Chl *b* in PSI is found in the LHCPI complex (Mullet *et al.*, 1980; Lam *et al.*, 1984; Malkin, 1986; Ortiz *et al.*, 1984). In this study the role of Chl *b* in stimulating long wavelength emission is specifically addressed.

The dynamics of energy transfer between antenna pigments and reaction centers of photosynthetic systems has motivated many spectroscopic studies in recent years (Bose, 1982; Holzwarth, 1987b). The occurrence of uphill energy transfer to the reaction center from a long wavelength-absorbing pigment (Butler *et al.*, 1979) causes PSI to be a particularly interesting system for study. Time-resolved emission spectroscopy using the method of single-photon counting is a useful tool for unravelling the mechanisms of energy transfer within photosynthetic systems. Analysis of fluorescence decay kinetics can provide information relating to structure, dynamics of energy transfer and photochemical trapping within a photosynthetic unit. Recently, single-photon counting measurements have led Owens *et al.* (1987b) to conclude that a linear relation exists between trapping times and antenna size, indicating that in PSI core preparations excitation migration is trap limited. Although many time-resolved fluorescence studies have been done on chloroplasts at low temperature (Butler *et al.*, 1979; Reisberg *et al.*, 1982; Wittmershaus *et al.*, 1985; Mimuro *et al.*, 1987; Mimuro, 1988; Beddard *et al.*, 1979; Gulotty *et al.*, 1985; Holzwarth *et al.*, 1984) and on PSI reaction-center-enriched preparations (Mimuro *et al.*, 1987; Mimuro, 1988; Searle *et al.*, 1988; Owens *et al.*, 1987a; Owens *et al.*, 1988; Wittmershaus, 1987), comparatively few studies have been done on PSI preparations containing more than 100 Chl/reaction center at low temperature (Mimuro *et al.*, 1987; Mimuro, 1988; Wittmershaus, 1987). In this study all measurements were performed on a PSI preparation from spinach containing 200 Chl/P700. Steady state fluorescence and time-resolved emission spectroscopy were done at room temperature and at 77K. Fluorescence decays at 295K were adequately described by 4 components: <30 ps, 80-100 ps, 270-330 ps, and 1.3-1.8 ns. An emission maximum was observed at 690nm at room temperature with a strong shoulder from 710 to 740nm. At 77K the emission maximum is red-shifted to 740nm, and the fluorescence decay kinetics was deconvolved to 5 components: 40-50 ps rise, and decays of 250-280 ps , 1.0-1.1 ns , 2.4-2.5 ns and 4.8-5.3 ns. The fluorescence yield at 740nm increases about 20 fold with decreasing temperature, arising primarily from an increase in the time-resolved component amplitudes rather than lifetime. Additionally, the ratio of fluorescence yield at 735nm (F735) to the yield at 690nm (F690) increases when the excitation is at 650nm relative to excitation at 670nm. This effect is observed at both 295K and at 77K. This enhanced yield at longer wavelengths reflects an increase in the amplitude of the 90 ps component at room temperature and the longer lived components at 77K. Steady state measurements confirm the results obtained in the kinetic studies. Because 650nm is the absorption maximum for Chl *b* (Clayton, 1986), we conclude that Chl *b* plays an integral role in stimulating long wavelength emission in PSI.

MATERIALS AND METHODS

The PSI complex (PSI-200) containing the reaction center and a full complement of the antenna was isolated from spinach according to the procedure described by Mullet *et al.* (Mullet *et al.*, 1980) with some modifications. All procedures were carried out at 4°C with limited light exposure. Chloroplasts were prepared by grinding market spinach leaves in buffer [0.3M sucrose, 0.01M NaCl, 5.0mM $MgCl_2$ and 50.0mM N-tris(hydroxymethyl)methyl glycine (Tricine), pH=7.8], filtering through 4 layers of cheesecloth, followed by centrifuge spin (1000xg; 5 min). Chloroplasts were resuspended (50.0 mM sorbitol, 0.75mM EDTA, pH=7.8) and then concentrated by centrifugation (11,000xg; 5 min). This step was repeated. Resulting pellets were resuspended (50.0 mM sorbitol, 5.0 mM Tricine, pH=7.8) and then centrifuged (19,000xg; 15 min) to remove soluble components. Pellets were then diluted with ice cold H_2O to a Chl concentration of 0.75-0.8 mg/ml with a Triton-X-100 concentration of 1.0-1.2% (w/v). These conditions differ depending on the nature of the spinach and the season. This detergent-chloroplast mixture was shaken for 1 h at room temperature in a water bath protected from light and then centrifuged (44,000xg; 30 min). The resulting supernatant was immediately loaded onto a 0.1-1.0M linear sucrose density gradient containing 0.02% Triton-X-100 and centrifuged (100,000xg; 16 h). The band of interest occurred at the interface between the 2.0M sucrose cushion and the linear gradient.

Chl concentration and Chl *a/b* ratios were determined as described by Arnon (1949). Polypeptide composition was determined by SDS-PAGE; separating gels were 12.5% acrylamide and stacking gels were 6.0% acrylamide. Absorption spectra were obtained using a Varian 2300 spectrophotometer in the UV-Vis mode. The amount of Chl per P700 was determined by light-induced absorption changes using an Aminco DW-2 spectrophotometer with side illumination from a tungsten lamp filtered by a Corning 4-96 color filter. Absorption changes were monitored at 699nm with an EMI GENCOM 9684 (S1 type) photomultiplier tube through a 700nm narrow bandpass filter using $\Delta\varepsilon=64mM^{-1}cm^{-1}$ (Hiyama and Ke, 1972). Samples were suspended at 10μg Chl/ml in 50mM Tris, pH=7.8 and 5mM ascorbate. Steady state fluorescence measurements were performed using a SPEX Fluorolog 2 model 212 spectrofluorimeter with an RCA: C31034A photomultiplier tube. All emission scans were corrected for emission monochromator and photomultiplier tube wavelength dependence. Excitation scans were corrected for lamp and monochromator wavelength dependence. Low temperature fluorescence studies were done using a liquid nitrogen-filled cylindrical glass dewar. Samples were diluted to <0.1 absorbance at 675nm and were frozen to a cracked glass in 60% glycerol, 40% 50.0mM Tricine, pH=7.8 in a glass NMR tube. Variable temperature measurements were done using an Oxford cryostat.

Fluorescence decays were measured by time-correlated single photon counting. Picosecond pulses were provided by a tunable cavity-dumped dye laser (Spectra Physics 375, 341) synchronously pumped by an Ar ion laser (SP 171). DCM was used as the lasing dye. Pulses were issued at a repetition rate of 823 kHz. Emission wavelength was selected by a double monochromator (Instruments, SA) with a 2nm bandwidth and detected by an ITT microchannel plate detector (model F4129). The instrument response function, measured by scattering the excitation beam with a solution of non-dairy creamer in H_2O, had a FWHM of 140-150 ps. Deconvolution of the decays with the instrument response function gives a time resolution of 20 ps. Fluorescence decays, following a short flash of exciting light were described in terms of a sum of simultaneous exponential decays

$$F(t)=\Sigma\alpha_i e^{-t/\tau_i}$$

where F is the fluorescence intensity, a_i the amplitude and τ_i the lifetime of the individual components. Quality of the fit was judged by the value of the reduced χ^2 (1.0-1.15) and from a plot of the weighted residuals.

Both single exponential analysis and global analysis (Holzwarth, 1987a) were applied to the decays. When global analysis was used, a set of decays was measured at 5nm intervals under the same conditions, and the lifetimes were constrained to be constant for the corresponding decay components at all wavelengths. This method of analysis is useful for indicating inadequacies in the model function, as well as for reducing the total number of parameters. It therefore improves the precision and stability of each parameter. The results from the global analysis were consistent with those obtained by analyzing the decays individually with unconstrained lifetimes. Consistency with respect to lifetime and reduced χ^2 values was monitored specifically. The decay component amplitudes at each wavelength obtained from the global analysis were corrected for data collection time, excitation intensity, and wavelength dependence of the detector. Decay-associated spectra refer to corrected amplitudes plotted as a function of emission wavelength.

RESULTS

The PSI preparation used for these spectroscopic studies is stable and photochemically active. It contains approximately 200 Chl/ P700 and has a Chl *a/b* ratio of 8:1. The presence of Chl *b* is confirmed by prominent shoulders at 472nm and 650nm (Clayton, 1980) in the absorption spectrum (data not shown). SDS-PAGE was used to characterize the polypeptide composition. Reaction center and peripheral antenna polypeptides were observed, as well as low molecular weight polypeptides which possibly bind Fe-S centers (Høj and Moller, 1986). Due to the large antenna complement and the relatively low amounts of detergents used for isolation, PSI-200 represents a preparation which resembles PSI *in vivo* (Mullet *et al.*, 1980, Lam *et al.*,1984).

Fluorescence Temperature Dependence

PSI-200 at room temperature exhibits an emission maximum at 690nm with a shoulder from 710 to 740nm; at 77K the emission maximum is red-shifted to 738nm with a shoulder at 690nm (Mullet *et al.*, 1980; Haworth *et al.*, 1983). In Fig. 1A it can be seen that the intensity of F735 increases approximately 20 fold as a function of decreasing temperature; whereas F690 remains relatively small. Changes in the apparent intensity of F690 result in part from instabilities in the glass at temperatures between 77K and 250K. The maximum of the long-wavelength emission becomes progressively red-shifted as the temperature is decreased (Fig. 1B), which is suggestive that more than one component is contributing to this fluorescence. This temperature-dependent fluorescence behavior is not a property of Chl *a in vitro* (Goedheer, 1964). We conclude that for the PSI complex the fluorescence behavior is directly related to changes in excitation quenching and energy transfer dynamics.

Additionally, the relative fluorescence intensity of the long wavelength shoulder is dependent on excitation wavelength. At 295K excitation was done at 435, 472, 650 and 670nm. The highest fluorescence yield was observed with excitation at 435nm. Excitation spectra confirm that Chl *b* particularly sensitizes long wavelength emission at both 295K and 77K (Fig. 2A). Excitation at the Chl *b* absorption maxima enhances the relative fluorescence yield of the long wavelength

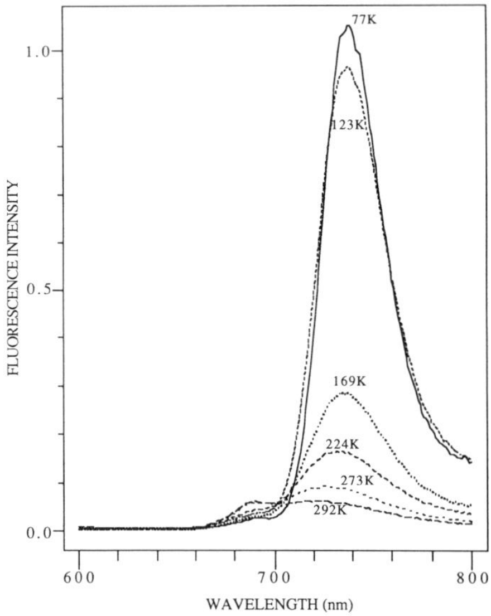

Figure 1A. Temperature dependence of PSI-200 fluorescence. Steady-state fluorescence emission spectra of PSI-200 at 77K, 123K, 169K, 224K, 273K, 292K. Samples originally in 1 M sucrose were diluted 5 fold in 40% 50.0mM Tricine, pH = 7.8, 60% glycerol. Excitation was at 435nm. All spectra were taken on the same sample.

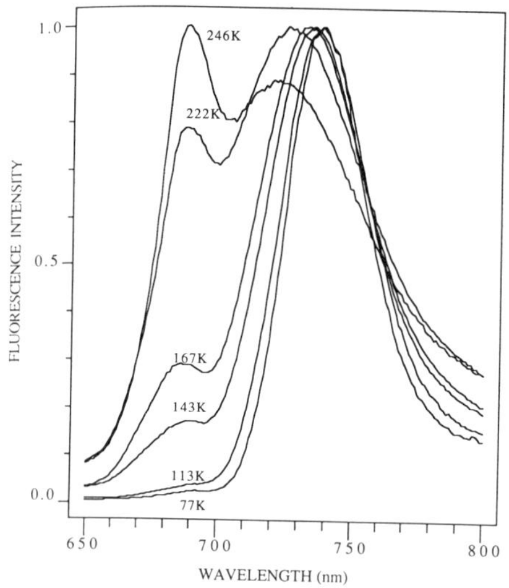

Figure 1B. Temperature dependence of PSI-200 fluorescence. Steady-state fluorescence measurements of PSI-200 at 77K, 113K, 143K, 167K, 222K, 246K. Sample preparation is the same as in Fig. 1A. Excitation done at 435nm on different samples. Spectra are normalized at their emission maxima.

shoulder with respect to emission at 690nm. When excitation is at 472nm, there is an unknown amount of contribution from carotenoid absorption; however, carotenoids do not contribute to the absorption at 650nm (Fig. 2B). This enhanced yield was observed also at 77K (data not shown).

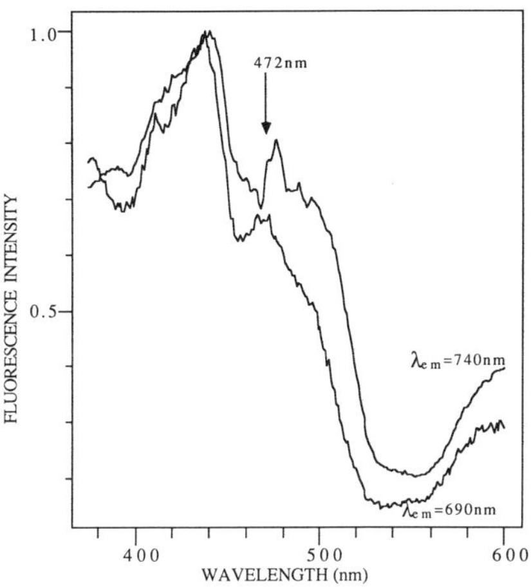

Figure 2A. Steady-state excitation spectra at 77K with emission at 690nm or 740nm. Samples frozen to a cracked glass in 60% glycerol/40% 50.0mM Tricine, pH = 7.8.

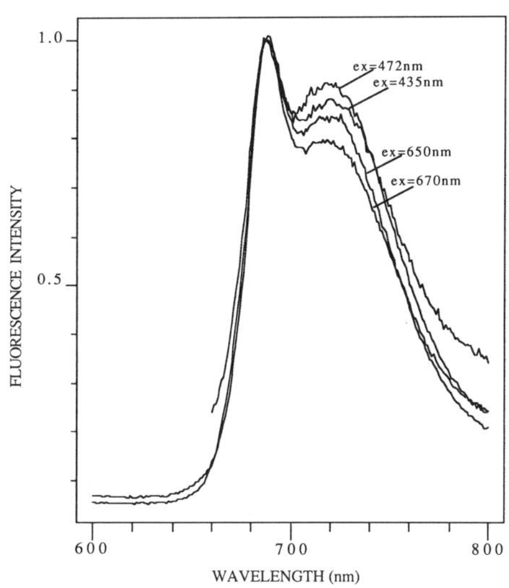

Figure 2B) PSI-200 Fluorescence dependence on excitation wavelength. Steady-state emission spectra at 295K, normalized to emission maxima. Excitation at 435nm, 472nm, 650nm or 670nm. Samples diluted 5-10 fold in 50.0 mM Tricine, pH = 7.8 from 1M sucrose.

Picosecond Fluorescence Kinetics

The time-resolved fluorescence decay measurements on PSI-200 excited at 650nm or 670nm at 295K were fit to a sum of four exponentials (Fig. 3). Although Owens *et al.* (1988) report decays which fit to 3 components for smaller PSI complexes, we find that the PSI-200 complex with an extended antenna requires an additional component. Our data were taken over the emission wavelength range 685-740nm. The data obtained were analyzed globally using both a 3-exponential and a 4-exponential model. The weighted residuals of the fits are shown in Fig. 4A and 4B. It is apparent from the plot of the weighted residuals and a comparison of the χ^2 values that the 4-exponential model best satisfies the data at room temperature.

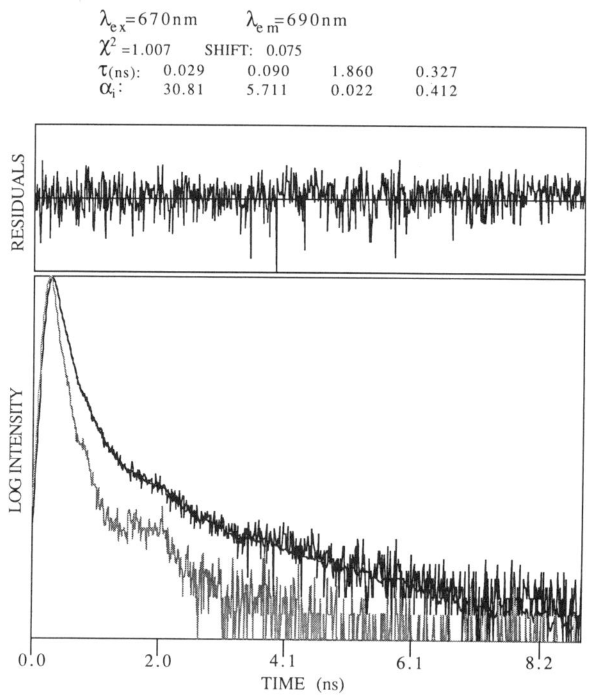

Figure 3. Fluorescence decay of PSI-200 at room temperature. Below: Measured fluorescence decay with calculated fit and instrument response function. 10,000 counts were collected in peak channel. Above: Weighted residuals. Decay deconvolved to four components. Sample diluted in 50.0mM Tricine, pH = 7.8, to 10μg Chl/ml. λ_{ex} = 670nm. λ_{em} = 690nm.

In contrast to Owens *et al.* (1988) we find that the lifetime of the shortest decay component is relatively independent of emission wavelength, regardless of excitation at 650nm or at 675nm. The single

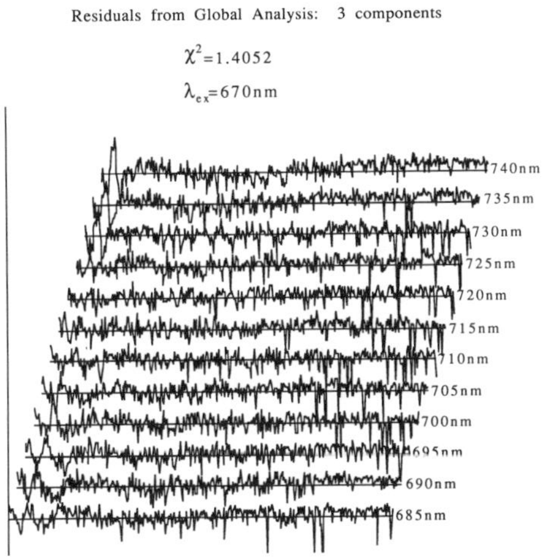

Figure 4A. Weighted residuals from global analysis of a set of decays measured on PSI-200. λ_{ex}=670nm. Emission range from 685nm to 735nm. Sample preparation as in Fig. 3. Residuals from 3 exponential model. χ^2=1.40.

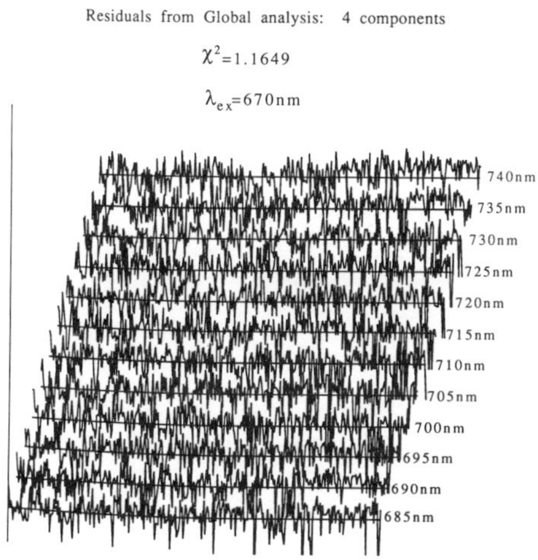

Figure 4B. Weighted residuals from global analysis of a set of decays measured on PSI-200. λ_{ex}=670nm. Emission range from 685nm to 735nm. Sample preparation as in Fig. 3. Residuals from 4 exponential model. χ^2=1.16. Scale expanded by a factor of 2 relative to Fig. 4A.

wavelength analysis lifetimes are independent of emission wavelength within the experimental uncertainty (± 10%) and are in good agreement with the lifetimes obtained from global analysis (see Table I).

Table 1. Decay-associated spectra obtained from global analysis of fluorescence kinetics from PSI-200.

Kinetic components	1	2	3	4	5
Temperature = 295K					
Global analysis lifetimes (ns)	0.018[a] ±0.005	0.085[b] ±0.010	0.30 ±0.04	1.53 ±0.19	
Emission maximum (nm)	695	730	--	<685	
Temperature = 77K					
Global analysis lifetimes (ns)	0.045[c] ±0.005	0.27 ±0.02	1.05 ±0.07	2.48 ±0.06	4.99 ±0.21
Emission maximum (nm)	695 (pos. amp.) 735 (neg. amp.)	720	725-730 (broad)	750	<680

Average lifetimes obtained from single wavelength analyses:
[a] 0.024 ns
[b] 0.102 ns
[c] 0.049 ns

Fluorescence yields for λ_{ex}=675nm, generated from global analysis, are plotted vs. wavelength in Fig. 5 (top). The fluorescence yield of each component is defined as:

$$\phi_i = \alpha_i \tau_i$$

The <30 ps component constitutes approximately 90% of the amplitude, corresponding to 70% of the fluorescence yield at its emission maximum at 695nm. This component, although at the limit of our time resolution, dominates the decay profile owing to its large amplitude. The 80-100 ps component contributes approximately 60% of the overall yield at its emission maximum at 730nm. The corrected yield arising from the middle component, 270-330 ps, is relatively wavelength independent and contributes 10% to the yield. The slowest component, 1.3-1.8 ns appears to have an emission maximum wavelength shorter than 685nm and adds at most 4% to the total fluorescence yield.

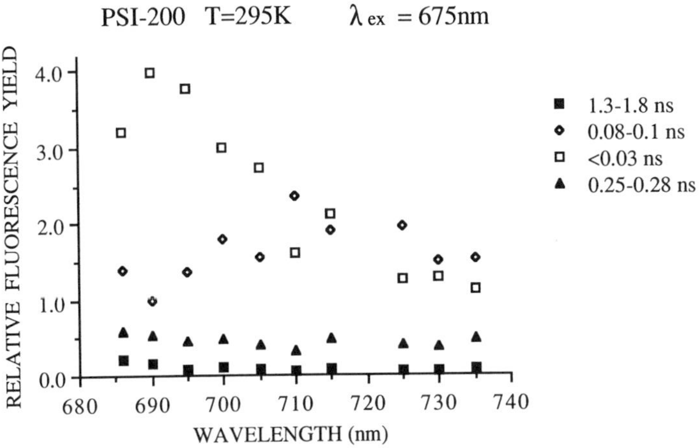

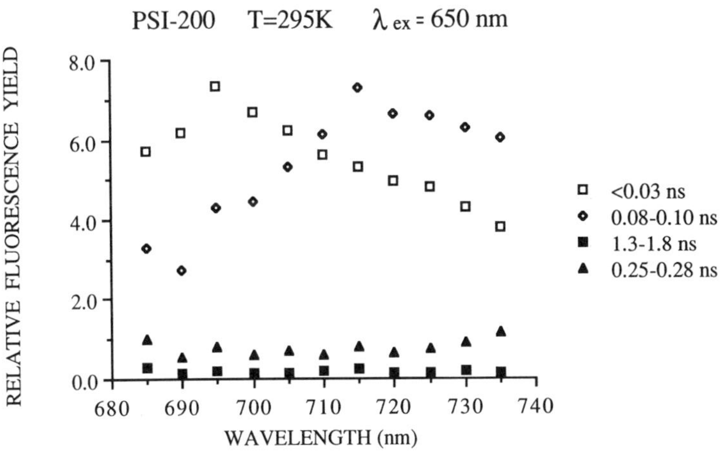

Figure 5. Decay associated spectra of the components determined by a 4-exponential global analysis. Data were taken at 295K. Fluorescence yield of each component is plotted as a function of emission wavelength. Top) λ_{ex}=675nm. Bottom) λ_{ex}=650nm.

Fig. 5 (bottom) depicts a decay-associated spectrum from global analysis, where λ_{ex}=650nm. The ratio of the fluorescence yields $\phi 730/\phi 690$

for the 80-100 ps component increases when excitation is at 650nm relative to that for excitation at 670nm. This enhancement of the long wavelength fluorescence through Chl *b* excitation is seen in both the kinetic measurements and the steady state measurements. Our data are suggestive that Chl *b* particularly sensitizes long wavelength emission, reflected by the increase in relative yield of the 80-100 ps decay component.

Fig. 6 (top) shows a decay associated spectrum of PSI-200 at 77K with λ_{ex}=670nm. The decays were analyzed by global analysis using a 5 component model. From global analysis it was determined that a 4 exponential model was not sufficient to describe the data. The lifetime of each of the components is 2-3 times longer than those observed at room temperature. In the slower components the dominant effect of lowering the temperature appears to be an increase in amplitude, whereas in the faster components the amplitudes are relatively unaffected.

The fastest component, 50 ps, comprises about 95% of the amplitude at 695nm. This component also exhibits a negative amplitude, indicative of fluorescence induction, at wavelengths >720nm. The maximum of the negative amplitude occurs at 735nm. Additionally, a fast decay of 250-280 ps is observed that has an emission maximum at 720nm and contributes 55% of the amplitude at that wavelength. The slowest decay component, 4.8-5.3 ns, has an emission maximum at wavelengths <680nm and contributes at most 3% to the overall amplitude. The 1.0-1.1 ns component has a broad maximum at 725-730nm and contributes about 40% to the total amplitude. The 2.4-2.5 ns decay peaks at 750nm, comprising approximately 70% of the total amplitude. It is apparent from the decay-associated spectra (Fig. 6, top and bottom) that excitation at 650nm increases the amplitudes of the slower decay components (0.25 ns, 1.1 ns, 2.4 ns) at longer wavelengths relative to the amplitude of the fast decay at 695nm.

DISCUSSION

Mimuro *et al.* have made time-resolved spectroscopic measurements on PSI complexes similar in antenna size and isolation procedure to the PSI-200 preparation (Mimuro *et al.*, 1987; Mimuro, 1988). The lifetimes obtained in the present work agree with the 4 components they used to deconvolute room temperature data (Mimuro *et al.*, 1987). Measurements at 77K (Mimuro, 1988) were deconvoluted to 7 components, including three designated F720, F735 and F747. The emission peaks of the slower decay components obtained in our study correlate with the 3 longest wavelength components identified by Mimuro.

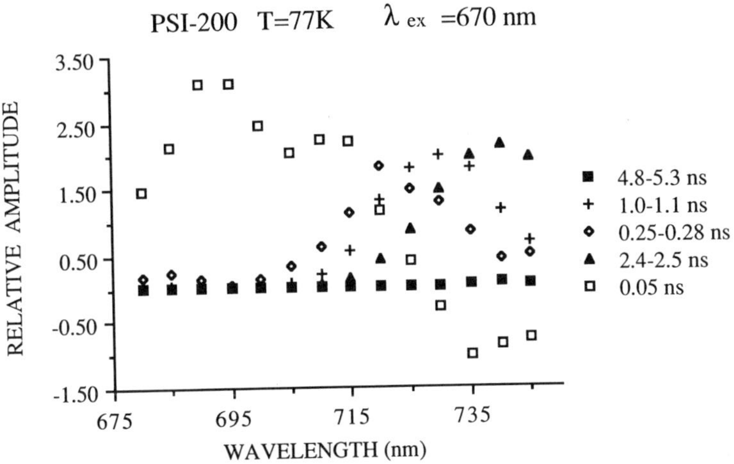

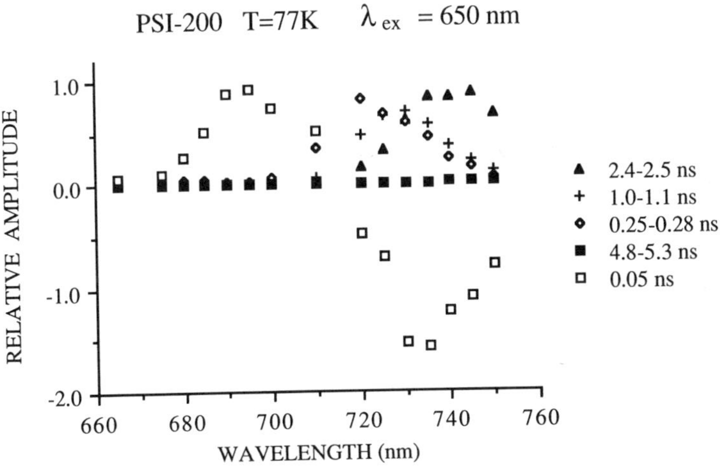

Figure 6. Decay-associated spectra determined from a 5 component global analysis. Data were taken at 77K. The amplitude of each component is plotted as a function of emission wavelength. Top) λ_{ex}=670nm. Bottom) λ_{ex}=650nm.

Wittmershaus (1987) reports lifetimes of 11 ps and 89 ps for PSI particles using a streak camera to observe room temperature emission at 685nm; at 77K lifetimes of 12 ps and 305 ps are reported (Wittmershaus *et al.*, 1985). The latter data were fit to only 2 exponential decay components. The effect of temperature on the amplitudes of these components was not addressed.

Our results confirm and extend those that have been previously reported. The lifetimes presented in this study correlate well with those of Mimuro and Wittmershaus. Additionally, our results support the biochemical data (Mullet *et al.*, 1980; Haworth *et al.*, 1983) which are suggestive that the longest wavelength emission in PSI arises primarily from the peripheral light harvesting antenna. We have observed both at 295K and at 77K that excitation at the absorption maxima of Chl *b* enhances fluorescence at 735nm. As Chl *b* is found only in LHCPI in PSI, this complex appears to be directly connected to the longest wavelength emission observed at both RT and 77K. This previously unreported result has been confirmed by both steady state and time-resolved emission studies.

Additionally, we observe from kinetic measurements at 77K that the exponential decay component amplitudes are the dominant factor for the dramatic increase in yield. We believe that the increase in amplitude reflects an increase in photon capture cross-section, caused by a decrease in the rate of photochemical trapping at lower temperatures. The fluorescence decay lifetimes increase by a factor of 3; however, the yield increases by approximately 20 fold between 290 and 80K. The majority of the fluorescence yield increase appears to arise from the increased amplitudes of the two longer lived decay components which peak at 730nm and 740nm. (Fig. 7) Studies of the temperature dependence of a PSI complex (CPI) with a smaller antenna indicated that, in that case also, the increased yield arises primarily from an increase in amplitude of the individual decay components (Tabbutt, 1987).

We propose that the kinetic component 1, (F690, see Table I) arises from the core antenna, while component 2 (F720) stems from an internal antenna. At 295K Chl *b* excitation increases the yield of this component, implying that the peripheral light harvesting antenna transfers energy to this antenna complex. Excitation transfer from F735 to F720 is not resolved in our measurements, as we observe no rise time. We assign the slowest decay component to dissociated Chl or to antennae which lack a functional trap. Such slowly decaying fluorescence amounts to no more than 3% of the total amplitude. We attribute kinetic components 3 (F735) and 4 at 77K to the peripheral light harvesting antenna.

We suggest a model (Fig. 8) for energy transfer within PSI which attempts to explain both the increase in lifetime and amplitude at lower temperatures. Searle *et al.* (1988) have proposed a model for a smaller antenna system; we have modified and extended this model to encompass the system used in this study. Our model consists of 3 distinct antenna

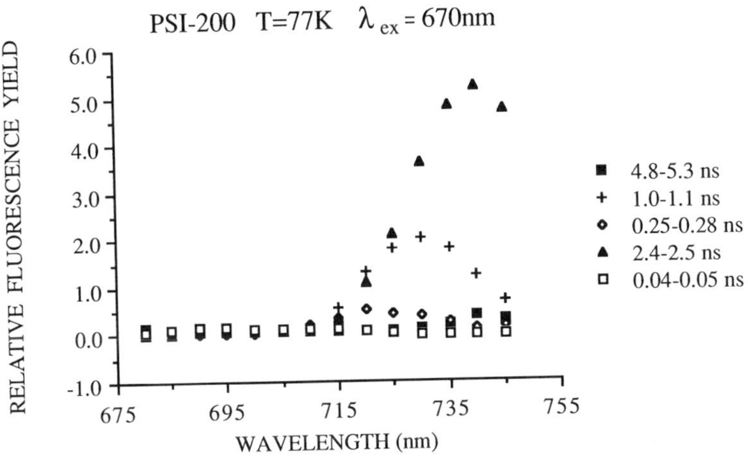

Figure 7. Decay-associated spectra of the components determined by a 5-exponential global analysis at 77K. The fluorescence yield of each component is plotted as a function of emission wavelength. λ_{ex}=670nm.

forms: F690, F720 and F735 and a photochemical trap, P700. We propose that the efficiency of the trapping reaction P700→P700$^+$ decreases with decreasing temperature. Additionally, excitation transfer from F735 to F720 and from F720 to P700 are both endothermic and temperature dependent. Fluorescence is a decay path for excited F720 and F735. We assume that most of the excitation arises from absorption by the main antenna Chl molecules and visits the reaction center before reaching F720 or F735. Excited F690 decays by temperature-independent transfer to P700 or by fluorescence. The increase in lifetime at lower temperatures reflects the decrease in rate constant for the trapping reaction. Consequently, excitation visiting P700, which would normally be trapped, is now diverted to F720 and F735, increasing the effective photon-capture cross-section of these two components. The increased cross-section results in the increase in amplitude of the corresponding components observed in the time-resolved fluorescence relaxation. Other de-excitation paths such as triplet formation and internal conversion to the ground state exist for P700*; however, the effect of temperature on the yield of these reactions with respect to the fluorescence yield has not been reported. Further experiments need to be done to verify that the trapping reaction is indeed temperature dependent.

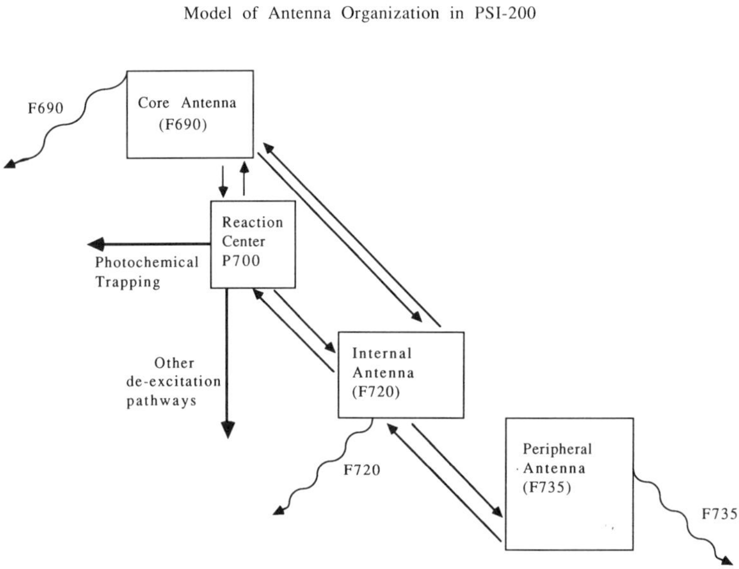

Figure 8. Proposed model of antenna organization and excitation transfer dynamics in PSI-200.

ACKNOWLEDGEMENTS

We would like to thank Dr. Alfred Holzwarth of the Max Planck Institute of Radiation Chemistry, Mülheim/Ruhr for generously supplying the data analysis programs. This research was supported by the Director, Office of Basic Energy Services, Division of Biological Energy Research of the Department of Energy under contract DE-AC03-76SF00098.

REFERENCES

Arnon DI (1949). Copper enzymes in isolated chloroplasts. Polyphenoloxidase in *Beta vulgaris*. Plant Physiol 24:1-15.

Beddard GS, Fleming GR, Porter G, Searle GFW, Synowiec JA (1979). The fluorescence decay kinetics of *in vivo* chlorophyll measured using low intensity excitation. Biochim Biophys Acta 545:165-174.

Bose S (1982). Chlorophyll fluorescence in green plants and energy transfer pathways in photosynthesis. Photochem Photobiol 36:725-731.

Butler WL, Tredwell CJ, Malkin R, Barber J (1979). The relationship between the lifetime and yield of the 735nm fluorescence of chloroplasts at low temperature. Biochim Biophys Acta 545:309-315.

Clayton RK (1980). "Photosynthesis: Physical mechanisms and Chemical Patterns." Cambridge: Cambridge University Press, p.38.

Fish LE, Bogorad L (1986). Identification and analysis of the maize P700 chlorophyll *a* apoproteins PSI-A1 and PSI-A2 by high pressure liquid chromatography analysis and partial sequence determination. J Biol Chem 261:8134-8139.

Goedheer JC (1964). Fluorescence bands and chlorophyll *a* forms. Biochim Biophys Acta 88:304-317.

Golbeck JH, Parrett KG, McDermott AE (1987). Photosystem I charge separation in the absence of centers A and B. III. Biochemical characterization of a reaction center particle containing P700 and F_X. Biochim Biophys Acta 893:149-160.

Gulotty RJ, Mets L, Alberte RS, Fleming GR (1985). Picosecond fluorescence study of photosynthetic mutants of *Chlamydomonas reinhardii*: Origin of the fluorescence decay kinetics of chloroplasts. Photochem Photobiol 41:487-496.

Haworth P, Watson JL, Arntzen CJ (1983). The detection, isolation and characterization of a light-harvesting complex which is specifically associated with photosystem I. Biochim Biophys Acta 724:151-158.

Hiyama T, Ke B (1972). Difference spectra and extinction coefficients of P700. Biochim Biophys Acta 267:160-171.

Holzwarth AR (1987a). A model for the functional antenna organization and energy distribution in the photosynthetic apparatus of higher plants and green algae. In Biggins J (ed): "Progress in Photosynthesis Research," Vol. I, The Hague: Martinus Nijhoff, pp 53-60.

Holzwarth AR (1987b). Picosecond fluorescence spectroscopy and energy transfer in photosynthetic antenna pigments. In Barber J (ed): "The Light Reactions," Amsterdam: Elsevier, pp 95-157.

Holzwarth AR, Haehnel W, Wendler J, Suter G, Ratajczak R (1984). Picosecond fluorescence kinetics and energy transfer in antennae chlorophylls of green algae and membrane fractions of thylakoids. In Sybesma C (ed): "Advances in Photosynthesis Research," Vol I, The Hague: Martinus Nijhoff, pp 73-76.

Høj PB, Moller BL (1986). The 110 kDa reaction center protein of photosystem I, P700-chlorophyll *a*-protein 1, is an iron-sulfur protein. J Biol Chem 261:14292-14300.

Lam E, Ortiz W, Malkin R (1984). Chlorophyll *a/b* proteins of photosystem I. FEBS Lett 168:10-14.

Malkin R (1986). A consideration of the organization of chloroplast photosystem I. Photosynthesis Res 10:197-201.

Mimuro M (1988). Analysis of excitation energy transfer in thylakoid membranes by the time-resolved fluorescence spectra. In Scheer H, Schneider S (eds): "Photosynthetic Light Harvesting Systems," Berlin: W. de Gruyter, pp 589-600.

Mimuro M, Yamazaki I, Tamai N, Yamazaki T, Fujita Y (1987). Analysis of the excitation energy transfer in spinach chloroplasts at room temperature: Identification of the component bands by the time-resolved fluorescence spectrum and by convolution of the decay kinetics. In Kobayashi T (ed): "Primary Processes in Photobiology," Berlin: Springer, pp 23-32.

Mullet JE, Burke JJ, Arntzen CJ (1980). Chlorophyll proteins of photosystem I. Plant Physiol 65:814-822.

Ortiz W, Lam E, Ghirardi M, Malkin R (1984). Antenna function of a chlorophyll *a/b* protein complex of photosystem I. Biochim Biophys Acta 766:505-509.

Owens TG, Webb SP, Alberte RS, Mets L, Fleming GR (1988). Antenna structure and excitation dynamics in photosystem I. I. Studies of detergent-isolated photosystem I preparations using time-resolved fluorescence analysis. Biophys J 53:733-745.

Owens TG, Webb SP, Eads DD, Alberte RS, Mets L, Fleming GR (1987a). Time-resolved fluorescence decay kinetics in photosystem I. Experimental estimates of charge separation and energy transfer rates. In Biggins J (ed): "Progress in Photosynthesis Research," Vol I, The Hague: Martinus Nijhoff, pp 83-86.

Owens TG, Webb SP, Mets L, Alberte RS, Fleming GR (1987b). Antenna size dependence of fluorescence decay in the core antenna of photosystem I: Estimates of charge separation and energy transfer rates. Proc Natl Acad Sci USA 84:1532-1536.

Reisberg P, Nairn JA, Sauer K (1982). Picosecond fluorescence kinetics in spinach chloroplasts at low temperature. Photochem Photobiol 36:657-661.

Searle GFW, Tamkivi R, van Hoek A, Schaafsma TJ (1988). Temperature dependence of antennae chlorophyll fluorescence kinetics in photosystem I reaction centre protein. J Chem Soc, Faraday Trans 2 84:315-327.

Tabbutt S (1987). Temperature dependence of CPI fluorescence. In "Spectroscopic Studies of Energy Transfer in Photosynthetic Reaction Centers of Higher Plants," PhD. Thesis Berkeley: Lawrence Berkeley Laboratory Report (LBL-24017) pp 105-165.

Wittmershaus B (1987). Measurements and kinetic modeling of picosecond time-resolved fluorescence from photosystem I and chloroplasts. In Biggins J (ed): "Progress in Photosynthesis Research," Vol. I, The Hague: Martinus Nijhoff, pp. 75-82.

Wittmershaus B, Nordlund TM, Knox WH, Knox RS, Geacintov NE, Breton J (1985). Picosecond studies at 77K of energy transfer in chloroplasts at low and high excitation intensities. Biochim Biophys Acta 806:93-106.

Photosynthesis, pages 123–133

BIOPHYSICAL STUDIES OF PHOTOSYNTHESIS: CAPTURE AND STORAGE OF QUANTUM ENERGY

J. Amesz

Department of Biophysics, Huygens Laboratory, University of Leiden, 2300 RA Leiden, The Netherlands

INTRODUCTION

One of the pioneering efforts of Stacy French in photosynthesis research was directed at the function of the so-called antenna pigments, which are involved in the capture of light energy and the transfer of energy to the reaction center. Various ingenious apparatuses were built at his direction to study the optical properties of photosynthetic material and he was the first to recognize that the absorption spectra of algae and higher plant chloroplasts are much more complicated than those of the constituent chlorophylls *in vitro*. One of the methods which French introduced in the study of photosynthetic pigments is that of differential spectroscopy (French, 1958). This method is still widely used to analyze absorption spectra. However, today one does not need a specially constructed apparatus anymore, but merely an appropriate computer program to register not only first, but also second and even fourth derivatives of absorption spectra.

Due to the combined efforts of biophysicists and biochemists we now know that the optical properties of the antenna pigments reflect a highly ordered, but in many cases quite complicated structure, built up of pigment-protein complexes. Normally, these complexes are contained in the photosynthetic membrane, together with the reaction center, but in some organisms, like green bacteria, cyanobacteria and red algae most of the antenna is situated outside the membrane in chlorosomes or phycobilisomes. Excitation energy absorbed by the antenna pigments is rapidly transferred to the reaction center where the primary charge separation takes place: the transfer of an electron from an excited chlorophyll or bacteriochlorophyll, the so-called primary electron donor P, to an acceptor molecule.

Much of our knowledge about these processes and the structures in which they occur is based on studies with photosynthetic bacteria. In this contribution I shall discuss some of the results of these studies, with special emphasis on the green photosynthetic bacteria and the recently discovered heliobacteria. These organisms have been less extensively studied than the well-known purple bacteria, but from the scientific point of view they are not less interesting.

ANTENNA SYSTEMS

The various groups of photosynthetic organisms display a large variety in structure and pigment composition of antenna systems. In some cases, like green plants, red algae, cyanobacteria and purple bacteria, these have been fairly thoroughly characterized, but the antenna systems of the green bacteria and the recently discovered heliobacteria have only been partially explored.

The main antenna of green bacteria is contained in the chlorosomes, oblong bodies, approximately 100 nm long and several tens of nm wide, which contain high concentrations of bacteriochlorophyll *c* or the related pigments bacteriochlorophyll d or e, and are attached to the cytoplasmic membrane. They are responsible for the main absorbance bands near 430 and 700 - 750 nm of the cells and give the bacteria their green (or brownish) color. Optical studies suggest that the structures of chlorosomes of green sulfur and non-sulfur bacteria are similar; for some species the main proteins, which may bind the bacteriochlorophyll *c*, have been sequenced and various degrees of homologies were found (Zuber, 1987,1988; Gerola *et al*., 1988). In addition to bacteriochlorophyll *c*, the chlorosomes contain small amounts (1 - 4 %) of bacteriochlorophyll *a*, absorbing near 800 nm (Fig. 1), which presumably acts as an intermediate in the energy transfer from bacteriochlorophyll *c* to the membrane (van Dorssen and Amesz, 1988; van Dorssen, 1988).

With respect to the structure of the cytoplasmic membrane a distinction must be made between the green sulfur bacteria and the gliding green (or green non-sulfur) bacteria. The only species of the second group which has been studied extensively is *Chloroflexus aurantiacus*. The membrane contains a pigment-protein complex, B808-866, which contains bacteriochlorophyll *a* and which has a similar structure as the antenna complexes of purple bacteria (Zuber, 1987). In addition, the membrane contains a reaction center complex which can be isolated as a separate entity (Pierson and Thornber, 1983) as with purple bacteria.

The structure of the membrane of green sulfur bacteria appears to be quite different. Again the main pigment is bacteriochlorophyll *a*, but in the green sulfur bacteria most of this pigment is contained in the so-called soluble bacteriochlorophyll *a* protein, which is attached to rather

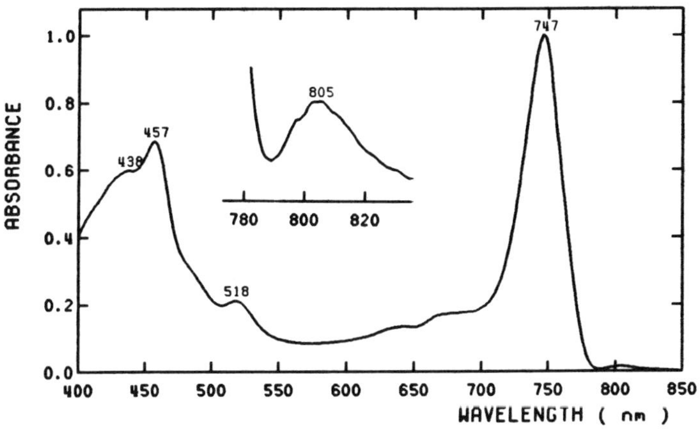

Figure 1. Absorption spectrum, measured at 4 K, of isolated chlorosomes from the green sulfur bacterium *Prosthecochloris aestuarii* (van Dorssen, 1988). The band at 747 nm is due to bacteriochlorophyll *c*; the band at 805 nm, shown on a 20-fold expanded scale in the insert, is attributed to bacteriochlorophyll *a*.

than embedded in the membrane and which may serve as a "base-plate" to connect the chlorosome to the membrane. Its three-dimensional structure has been determined by X-ray analysis of the crystalline protein (Tronrud *et al.*, 1986). The complex was found to consist of three subunits, each of which contains seven bacteriochlorophyll *a* molecules encaged in a protein which has mainly a β-sheet structure.

Upon removal of the soluble protein a so-called core complex is obtained (Vasmel *et al.*, 1983; Hurt and Hauska, 1984) which contains about 20 bacteriochlorophyll *a* molecules per reaction center, together with approximately 15 bacteriochlorophylls *c*. This bacteriochlorophyll *c* is more lipophilic than the chlorosomal bacteriochlorophyll *c* and is esterified with phytol, rather than farnesol (Braumann *et al.*, 1986). A separate reaction center has not been isolated so far from green sulfur bacteria.

Little is known about the structure of the antenna of heliobacteria. The main bacteriochlorophyll present in these bacteria is the newly discovered bacteriochlorophyll *g* (Gest and Favinger, 1983; Brockmann and Lipinski, 1983). Low temperature spectra of cytoplasmic membranes of *Heliobacterium chlorum* (Fig. 2) show the presence of three main spectral types of bacteriochlorophyll *g*, BChl *g* 778, BChl *g* 793 and BChl *g* 808, which can be resolved in the second derivative spectrum (van Dorssen *et al.*, 1985). Upon excitation with blue or yellow light, the

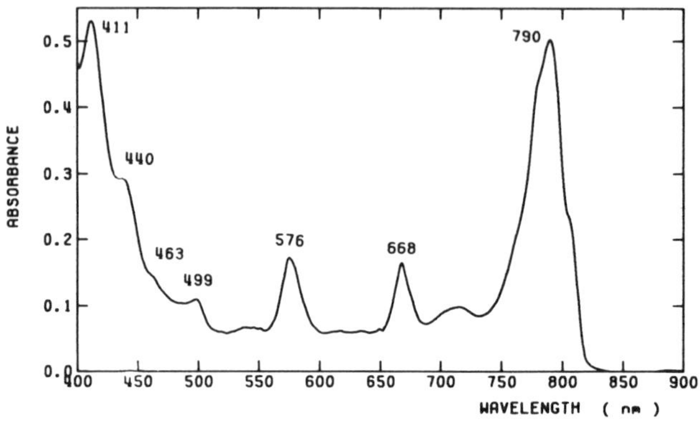

Figure 2. Absorption spectrum at 4 K of membranes of *Heliobacterium chlorum* (van Dorssen *et al.*, 1985).

fluorescence emission spectrum shows a maximum at 813 nm at room temperature and at 819 nm at 4 K. Apparently most or all of the emission comes from the long-wave pigment BChl *g* 808. In agreement with this, it was found that at 15 K short laser flashes induced a bleaching of the long-wave antenna component only, which indicates that excitations on the other bacteriochlorophyll species are rapidly, i.e. within a few tens of picoseconds, transferred to BChl *g* 808 (van Kan *et al.*, 1988). The fluorescence excitation spectrum shows that the transfer efficiency is close to 100% (van Dorssen *et al.*, 1985).

THE REACTION CENTER

The most thoroughly investigated reaction center is that of purple non-sulfur bacteria of group α (according to the recent classification scheme). Methods have been devised to prepare purified preparations of reaction center complexes from *Rhodobacter sphaeroides* and *Rhodopseudomonas viridis* and other species of this group after detergent solubilization of the intracytoplasmic membranes (Amesz and Knaff, 1988). The amino acid sequences of the constituent polypepides have been determined and the photochemical properties of the isolated reaction centers have been extensively studied by flash spectroscopy and various other means. Detailed pictures of the three-dimensional structure of the reaction center complexes of *Rps. viridis* and *Rb. sphaeroides* have been obtained by X-ray analysis of the crystalline complexes (Deisenhofer *et al.*, 1984; Allen *et al.*, 1987).

The so-called L and M peptides of the reaction center of these and other species show a fairly high degree of homology (Parson, 1987). Both peptides have five membrane spanning α-helices, whereas the third one, the H subunit, has a more polar character and traverses the membrane only once. The reaction center contains 4 bacteriochlorophylls *a* (in *Rb. sphaeroides* and most other purple bacteria) or bacteriochlorophylls *b* (in *Rps. viridis*), 2 bacteriopheophytins, and 2 quinones (ubiquinone or menaquinone) arranged in two chains in an approximately symmetrical fashion (see Fig. 3). Electron transport occurs from the primary electron donor P870 or P960 via the bacteriopheophytin to the quinone on the L chain (Q_A) and hence to the other quinone (Parson, 1987).

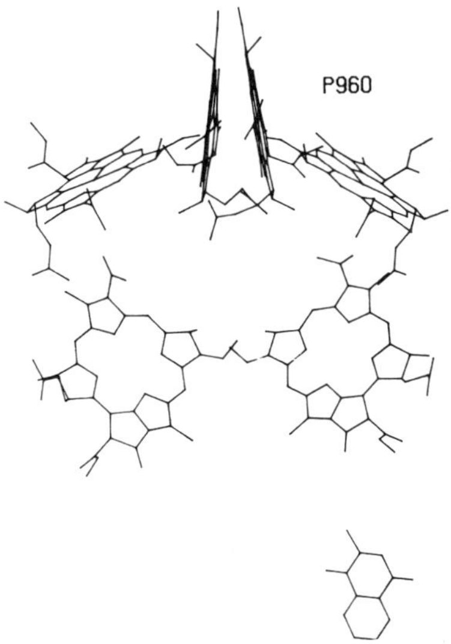

Figure 3. Arrangement of the chromophores and of menaquinone (Q_A) in the crystalline reaction center of *Rhodopseudomonas viridis*. Reproduced from Wasielewski and Tiede (1986).

The reaction center of the green non-sulfur bacterium *Cfl. aurantiacus* has a structure which is very similar to that of purple bacteria as indicated by an analysis of its optical properties (Vasmel *et al.*, 1986). This similarity also extends to the two constituent peptides, the primary structures of which show homology to the L and M subunits of purple bacteria (Ovchinnikov *et al.*, 1988a,b). There is various evidence now that the reaction center of photosystem II of plants also resembles that of purple bacteria (Nanba and Satoh, 1987).

Much less is known about the reaction centers of green sulfur bacteria and heliobacteria, but the available evidence indicates that they are structurally and functionally different from those of purple bacteria and suggests that they may form a second group, together with the reaction center of photosystem I. The primary electron donor, P840 in green sulfur bacteria, P798 in heliobacteria, is presumably a dimer of bacteriochlorophyll *a* and *g*, respectively (Amesz and Knaff, 1988; Prince *et al.*, 1985). Flash spectroscopic evidence indicates that in both groups the primary electron transfer reaction consists of the transfer of an electron to a pigment absorbing at 670 nm (Nuijs *et al.*, 1985a, b). For green sulfur bacteria this pigment has been identified as bacteriochlorophyll *c* or a closely related pigment (Braumann *et al.*, 1986); the electron acceptor in heliobacteria has not yet been identified, but may well be quite similar.

Fig. 4 shows the difference spectrum of *Heliobacterium chlorum* induced by 35 ps laser flashes. The spectrum shows a number of bleachings in the red and near-infrared region. A bleaching around 800 nm which decayed in roughly 200 ps was also observed when the reaction centers had been closed before the flash, and may be ascribed to excited antenna bacteriochlorophyll *g*. However, with open reaction centers additional bleachings at 670 and 800 nm were observed. That near 800 nm has a lifetime of several ms (Smit *et al.*, 1987) and is due to photooxidation of the primary electron donor P798 whereas the bleaching at 670 nm, attributed to the primary electron acceptor, reversed in approximately 500 ps, presumably by transfer of the electron to a second-

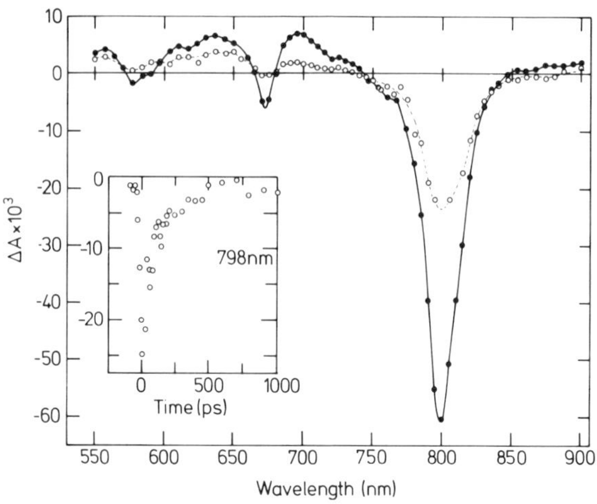

Figure 4. Absorbance difference spectrum of membranes of *Heliobacterium chlorum* measured during a 35 ps laser flash (Nuijs *et al.*, 1985b). Solid circles: with open reaction centers; open circles: with closed reaction centers. Insert: kinetics at 798 nm with closed reaction centers.

ary acceptor (Nuijs *et al.*, 1985b). The identity of this secondary electron acceptor is not known; although there is evidence that the reaction center of *H. chlorum*, like that of green sulfur bacteria, contains one or more iron-sulfur centers, the role of these in electron transport is not clear yet (Smit *et al.*, 1987).

Efficient photooxidation of P798 was also observed at cryogenic temperatures. Fig. 5 shows the difference spectrum, measured at 5 K. The main bleaching in the near-infrared was now located at 794 nm, and the bleaching reversed with a time constant of 2.3 ms, presumably by a back reaction with a secondary electron acceptor.

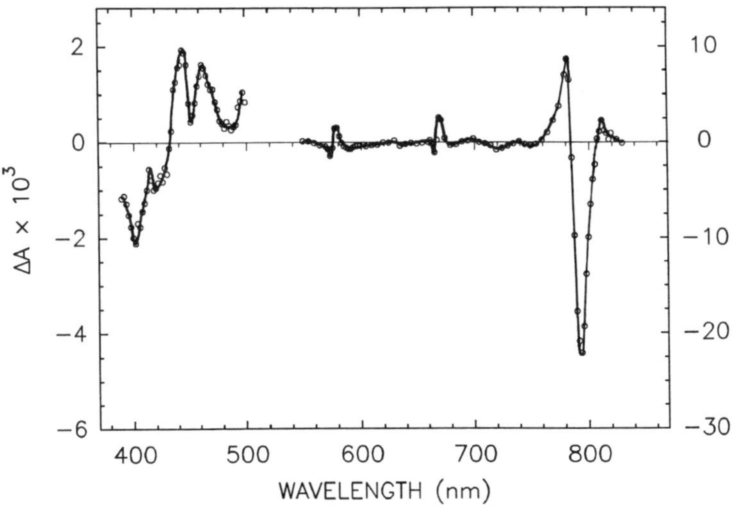

Figure 5. Absorbance difference spectrum due to photooxidation of P798 at 5 K. The lefthand scale applies to the region below 500 nm (H.W.J. Smit, unpublished).

ENERGY TRANSFER IN *Heliobacterium chlorum*

The presence of the long-wavelength component BChl *g* 808 in the antenna of *H. chlorum* raises questions about the pathway of energy transfer from the antenna to the reaction center. Energy absorbed by the other bacteriochlorophylls *g* is rapidly transferred to BChl *g* 808. Thus, one may envisage two pathways of energy transfer to the reaction center: (i) Energy transfer occurs from BChl *g* 793 directly to the reaction center at a high enough rate to compete effectively with transfer to BChl *g* 808 at low temperature, or (ii) this energy transfer occurs via BChl *g* 808, in which case its efficiency might be expected to be strongly reduced at low temperature, due to lack of thermal energy to overcome the energy difference between excited BChl *g* 808 and P798. We therefore studied the relative efficiencies for light absorbed by the various spectral forms of bacteriochlorophyll *g* to bring about P798 photooxidation.

The results of such an experiment are shown in Fig. 6, where the extent of P798 oxidation by a flash is plotted as a function of the absorbed intensity for three different filter combinations. The straight lines were obtained by means of a linear regression technique for the

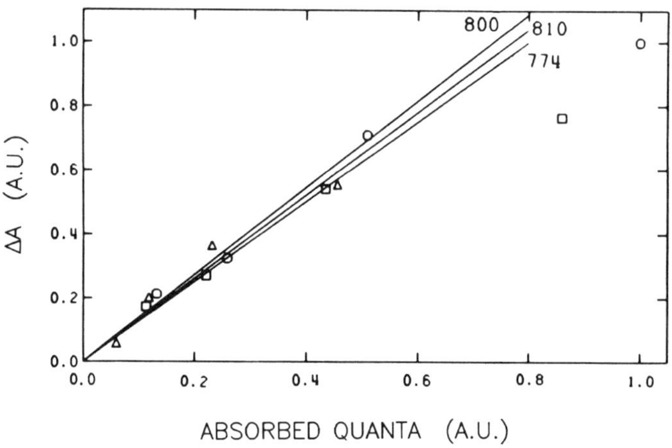

Figure 6. Relative amount of P798 photooxidized at 5 K as a function of absorbed flash energy (van Dorssen, 1988). Squares: 774 nm; circles: 800 nm; triangles: 810 nm (see text).

regions where the response appeared to be linear with intensity. The three lines appear to coincide within the error of measurement. Light transmitted by the 810 nm filter was about equally absorbed by BChl *g* 793 and BChl *g* 808, whereas for the 774 and 800 nm filters approximately 70 % was absorbed by BChl *g* 778 and BChl *g* 793, respectively. This indicates that no significant decrease of the photochemical activity occurs upon excitation of BChl *g* 808. Fig. 7 shows that the efficiency of photochemistry was roughly the same at 5 K and at room temperature. Thus we conclude that energy transfer to the reaction center occurs via BChl *g* 808, and that the efficiency of this transfer is quite high, even at 5 K.

The mechanism of this energy transfer is not clear at present. "Uphill" transfer from BChl *g* 808 to P798 would be expected to occur with very low efficiency, because of the extremely small overlap of the emission band of the former with the absorption band of the latter pigment.

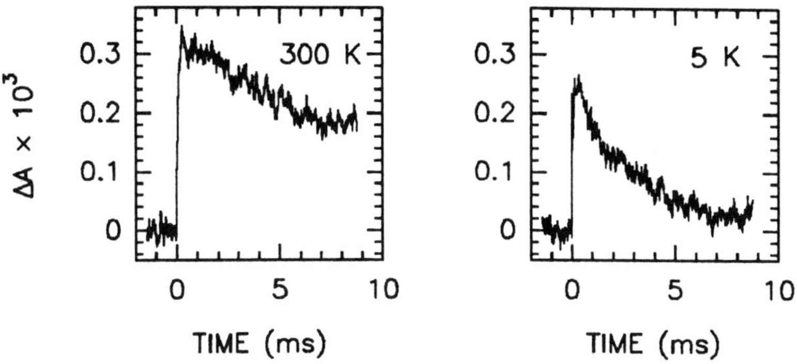

Figure 7. Kinetics of absorbance changes at 460 nm at 300 K and at 5 K induced by a non-saturating flash at 810 nm.

To avoid this difficulty, one could postulate the existence of a (weak) low energy band of the reaction center, which is not observed in the absorbance difference spectrum of Fig. 5. In this connection the recent model of Fischer and Scherer (1987) for charge separation in purple bacteria may be of interest. This model involves a charge transfer state of the "accessory" bacteriochlorophyll and bacteriopheophytin as the first intermediate in the primary reaction. Provided such a transition has sufficient oscillator strength, it might be excited directly by the antenna system, so that no excitation of P798 would be needed.

REFERENCES

Allen JP, Feher G, Yeates TG, Komiya H, Rees DC (1987). Structure of the reaction center from *Rhodobacter sphaeroides* R-26: The cofactors. Proc Natl Acad Sci USA 84: 5730-5734.

Amesz J, Knaff DB (1988). Molecular mechanism of bacterial photosynthesis. In Zehnder AJB (ed): "Biology of Anaerobes," New York: John Wiley, pp 113-178.

Braumann T, Vasmel H, Grimme LH, Amesz J (1986). Pigment composition of the photosynthetic membrane and reaction center of the green bacterium *Prosthecochloris aestuarii.* Biochim Biophys Acta 848:83-91.

Brockmann H, Lipinski, A (1983). Bacteriochlorophyll *g*. A new bacteriochlorophyll from *Heliobacterium chlorum.* Arch Microbiol 136:17-19.

Deisenhofer J, Epp O, Miki K, Huber R, Michel H (1984). Structure analysis of a membrane protein complex. Electron density map at 3 Å resolution and a model of the chromophores of the photosynthetic reaction center from *Rhodopseudomonas viridis*. J Mol Biol 180:385-398.

Fischer SF, Scherer POJ (1987). On the early charge separation and recombination processes in bacterial reaction centers. Chem Phys 115:151-158.

French CS (1958). Various forms of chlorophyll *a* in plants. Brookhaven Symp Biol 11:65-73.

Gerola PD, Høyrup P, Knudsen J, Roepstorff P, Olson JM (1988). The bacteriochlorophyll *c*-binding protein from chlorosomes of *Chlorobium limicola* f. *thiosulfatophilum*. In Olson JM, Ormerod JG, Amesz J, Stackebrandt D, Trüper HG (eds): "Green Photosynthetic Bacteria," New York: Plenum Press, pp 43-52.

Gest H and Favinger JL (1983). *Heliobacterium chlorum*, an anoxygenic brownish-green photosynthetic bacterium containing a "new" form of bacteriochlorophyll. Arch Microbiol 136:11-16.

Hurt EC, Hauska G (1984). Purification of membrane-bound cytochromes and a photoactive P840 protein complex of the green sulfur bacterium *Chlorobium limicola* f. *thiosulfatophilum*. FEBS Lett 168:149-154.

Nanba O and Satoh K (1987). Isolation of a photosystem II reaction center consisting of D-1 and D-2 polypeptides and cytochrome b-559. Proc Natl Acad Sci USA 84:109-112.

Nuijs AM, Vasmel H, Joppe HLP, Duysens LNM, Amesz J (1985a). Excited states and primary charge separation in the pigment system of the green photosynthetic bacterium *Prosthecochloris aestuarii* as studied by picosecond absorbance difference spectroscopy. Biochim Biophys Acta 807:24-34.

Nuijs AM, van Dorssen RJ, Duysens LNM, Amesz J (1985b). Excited states and primary photochemical reactions in the photosynthetic bacterium *Heliobacterium chlorum*. Proc Natl Acad Sci USA 82:6865-6868.

Ovchinnikov YuA, Abdulaev NG, Zolotarev AS, Schmukler BE, Zargarov AA, Kutuzov MA, Telezhinskaya IN, Levina NB (1988a). Photosynthetic reaction centre of *Chloroflexus aurantiacus*. I. Primary structure of L-subunit. FEBS Lett 231:237-242.

Ovchinnikov YuA, Abdulaev NG, Schmuckler BE, Zargarov AA, Kutuzov MA, Telezhinskaya IN, Levina NB, Zolotarev AS (1988b). Photosynthetic reaction centre of *Chloroflexus aurantiacus*. II. Primary structure of M-subunit. FEBS Lett 232:364-368.

Parson WW (1987). The bacterial reaction center. In Amesz J (ed): "Photosynthesis," Amsterdam: Elsevier, pp. 43-61.

Pierson BK, Thornber JP (1983). Isolation and spectral characterization of photochemical reaction centers from the thermophilic green bacterium *Chloroflexus aurantiacus* strain J-10-fl. Proc Natl Acad Sci USA 80:80-84.

Prince RC, Gest H, Blankenship RE (1985). Thermodynamic properties of the photochemical reaction center of *Heliobacterium chlorum*. Biochim Biophys Acta 810:377-384.

Smit HWJ, Amesz J and van der Hoeven MFR (1987). Electron transport and triplet formation in membranes of the photosynthetic bacterium *Heliobacterium chlorum*. Biochim Biophys Acta 893:232-240.

Tronrud DE, Schmid, MF, Matthews BW (1986). X-ray structure and amino acid sequence of a bacteriochlorophyll *a* protein from *Prosthecochloris aestuarii* refined at 1.9 Å resolution. J Mol Biol 188:443-454.

Van Dorssen RJ, Vasmel H, Amesz J (1985). Antenna organization and energy transfer in membranes of *Heliobacterium chlorum*. Biochim Biophys Acta 809:199-203.

Van Dorssen RJ, Amesz J (1988). Pigment organization and energy transfer in the green photosynthetic bacterium *Chloroflexus aurantiacus*. III. Energy transfer in whole cells. Photosynth Res 15:177-189.

Van Dorssen RJ (1988). Structure and function of light-harvesting complexes in photosynthesis. Doctoral thesis, University of Leiden.

Van Kan PJM, van Dorssen RJ, Aartsma TJ, Amesz J (1988). Primary photochemistry of *Heliobacterium chlorum*. In "Proc Vth Intern Symp on Ultrafast Phenomena," Vilnius 1987, Singapore: World Scientific Publ. Co., in the press.

Vasmel H, Swarthoff T, Kramer HJM, Amesz J (1983). Isolation and properties of a pigment-protein complex associated with the reaction center of the green photosynthetic sulfur bacterium *Prosthecochloris aestuarii*. Biochim Biophys Acta 725:361-367.

Vasmel H, Amesz J, Hoff AJ (1986). Analysis by exciton theory of the optical properties of the reaction center of *Chloroflexus aurantiacus*. Biochim Biophys Acta 852:159-168.

Wasielewski MR, Tiede DM (1986). Sub-picosecond measurements of primary electron transfer in *Rhodopseudomonas viridis* reaction centers using near-infrared excitation. FEBS Lett 204:368-372.

Zuber H (1987). Structure and function of light-harvesting pigment-protein complexes. In Amesz J (ed): "Photosynthesis," Amsterdam: Elsevier, pp 233-271.

Zuber H (1988). Structural studies on the antenna complexes and polypeptides of *Chloroflexus aurantiacus* and other green photosynthetic bacteria. In Olson JM, Ormerod JG, Amesz J, Stackebrandt D, Trüper HG (eds): "Green Photosynthetic Bacteria," New York: Plenum Press, pp. 53-55.

Photosynthesis, pages 135–148

BACTERIAL REACTION CENTERS

William W. Parson

Department of Biochemistry, University of Washington, Seattle, WA 98195

The bacterial reaction center has been at the focus of some remarkable developments in molecular biophysics. Because it was the first integral membrane protein whose structure was solved to high resolution by X-ray crystallography, its cluster of transmembrane α-helices has become a structural paradigm for a large and important group of proteins. The solution of the crystal structure has revealed how the bacteriochlorophylls, bacteriopheophytins, and quinones are arranged and how they interact with the amino acid side chains of the protein. As this wealth of detailed information on molecular structure has been emerging, new spectrophotometric techniques also have made it possible to explore molecular dynamics on time scales of fractions of picoseconds. By exciting preparations of purified reaction centers with short flashes of light, one can examine the kinetics of electron transfer reactions with high temporal resolution over a wide range of conditions, such as at very low temperatures or in the presence of external electrical or magnetic fields. The reaction center thus has become a rich source of information on how the rates of electron transfer reactions depend on the positions, orientations and nuclear motions of the reactants, and on the energies of the initial and final states. In addition, new computational techniques have opened these same questions to theoretical approaches based on explicit microscopic models of the protein and the electron carriers. It also has become possible to approach a quantum mechanical understanding of how the reaction center's complex absorption spectrum emerges from interactions among the pigments.

In this talk I shall summarize the main structural features of reaction centers of the purple photosynthetic bacteria, and describe some of the recent experimental and theoretical work on the mechanism of the initial electron transfer reactions.

Figure 1 is a drawing of the reaction center of *Rhodopseudomonas viridis*, based on the crystallographic studies of Deisenhofer *et al.* (1985) and Michel *et al.* (1986). Chang *et al.* (1986) and Allen *et al.* (1987a, b)

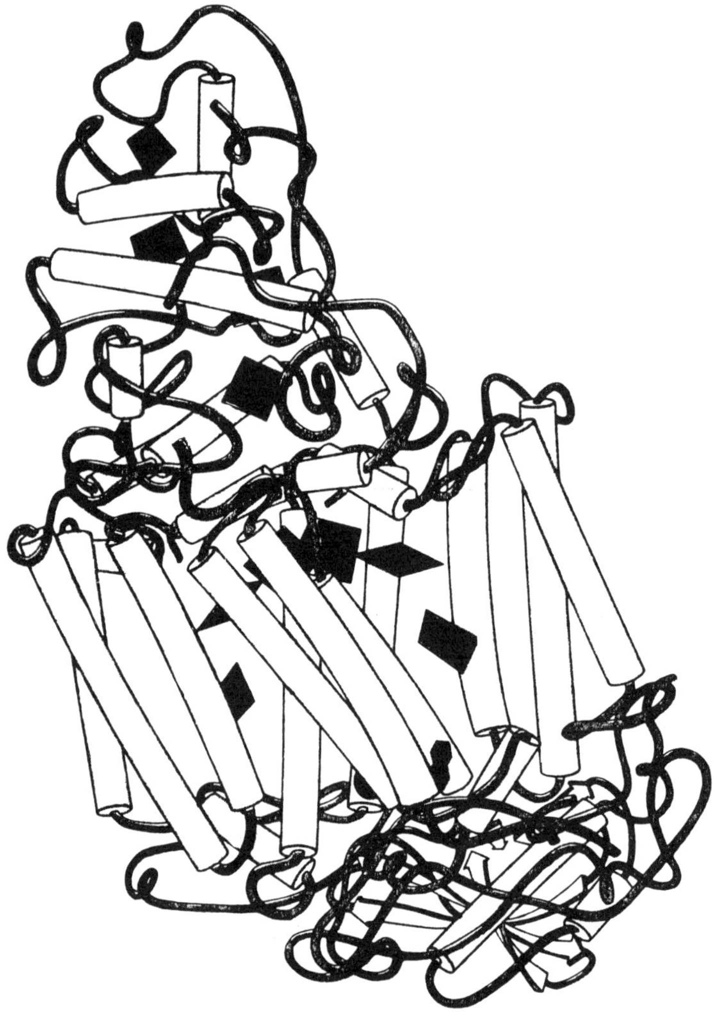

Figure 1. The reaction center of *Rps. viridis.* α-Helices are represented as cylinders, β-structure as flat arrows, and random coils as thin tubes. The chromophores are indicated with dark shading. The cytochrome subunit is at the top. After a drawing by J. Richardson (from Parson W (1988) "Photosynthesis and Other Reactions Involving Light," In Zubay G (ed): "Biochemistry," New York: Macmillan).

have found that the structure is very similar in *Rhodobacter sphaeroides*, except that the reaction centers of this species lack the cytochrome subunit that binds to the periplasmic side of the complex in *Rps. viridis*. The core of the structure is a set of eleven α-helices that pass back and forth across the bacterial membrane. Five of these α-helices come from one polypeptide subunit (L) and five from another (M). A third subunit (H) contributes the eleventh transmembrane helix, along with most of the globular structure on the cytoplasmic side of the membrane. Bound to the LM complex are four molecules of bacteriochlorophyll, two molecules of bacteriopheophytin, two quinones and one atom of non-heme iron (Fig. 2). Two of the bacteriochlorophylls are located very close together, and it is this special pair (P) that releases an electron when the reaction center is excited with light. The other bacteriochlorophylls (B_L and B_M) and the bacteriopheophytins (H_L and H_M) form two arms that stretch out in parallel from P toward the quinones. B_L and H_L are bound primarily to subunit L; B_M and H_M, to M. An axis of rotational pseudosymmetry extends from P to the iron atom, which is located approximately midway between the two quinones. A 180° rotation around this axis interchanges the positions of the pigments on the L and M sides, including the two bacteriochlorophylls of P, and also interchanges many corresponding elements of the protein. Although the pigment composition was known before the crystal structure was solved, and spectroscopic studies had suggested the general features of the special dimer, the highly symmetrical arrangement of the pigments was a surprise.

The Mg atoms of the bacteriochlorophylls are attached to histidyl residues of the protein. The other amino acids surrounding the pigments are generally nonpolar, and include a large number of residues with aromatic side chains. A hydroxylic amino acid generally is in position to form a hydrogen bond to the acetyl group of ring I. The ligands of the iron are four histidyl nitrogens and a glutamyl carboxyl group.

When the reaction center is excited with light, P transfers an electron to the bacteriopheophytin on the L side (H_L), which then reduces the quinone on the same side, Q_A (Fig. 2). Q_A in turn passes an electron to the other quinone, Q_B. The electron given up by P is replaced by one from a *c*-type cytochrome. In *Rps. viridis*, this is the bound cytochrome with four heme groups that is shown in Fig. 1; in *Rb. sphaeroides*, it is the soluble cytochrome-c_2. Excitation with a second photon sends another electron along the same path, reducing Q_B to the level of the quinol, Q_BH_2. Q_BH_2 then dissociates from the reaction center and diffuses to the cytochrome-bc_1 complex, where its reoxidation is coupled to the proton translocation that ultimately drives the formation of ATP.

The kinetics of these electron transfer steps can be measured by following changes in the reaction center's optical absorption spectrum.

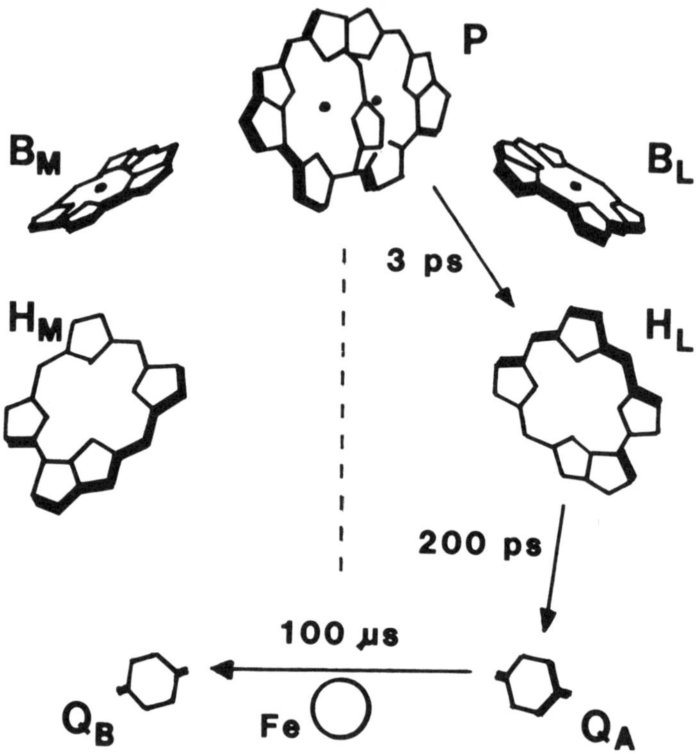

Figure 2. Arrangement of the bacteriochlorophylls (P, B_L and B_M), bacteriopheophytins (H_L and H_M), and quinones (Q_A and Q_B) in the reaction center of *Rb. sphaeroides.* Approximate time constants are indicated for the initial electron transfer steps at room temperature.

Figure 3 diagrams the type of apparatus typically used for such measurements. Picosecond or femtosecond flashes of 600-nm light first are generated by a mode-locked dye laser system. The picosecond pulse train is emitted continuously at a high repetition rate (82 MHz in our apparatus), but the pulses are very weak and generally need to be amplified by a factor of about a million before they can be used. The pulses are amplified by sending them through a series of dye cells that are pumped by a frequency-doubled Nd-YAG laser operating at a much lower repetition rate (10 Hz). After amplification, part of each pulse is used to excite the sample. The remainder of the pulse is focused into a liquid such as water to generate light of other wavelengths. The picosecond continuum then can be used to probe the absorption spectrum of the sample at a very precise time after the excitation. To do this, one

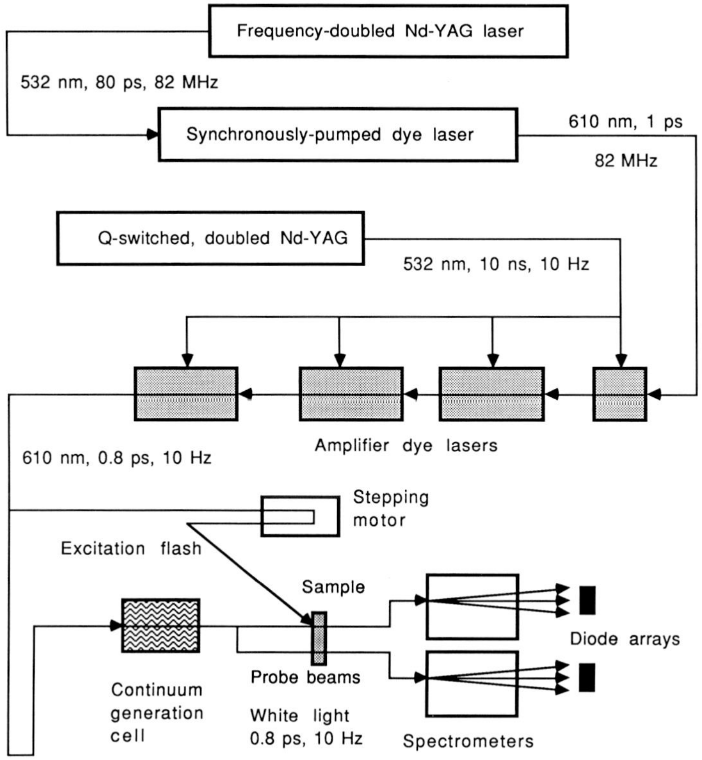

Figure 3. Apparatus for measuring light-induced absorbance changes.

delays the probe pulse relative to the excitation by adjusting the relative lengths of the optical paths that the pulses have to traverse before they pass through the sample. The delay path includes a prism that can be moved by a stepping motor controlled by a computer. The probe pulses are split into two beams, only one of which passes through the region of the sample that is excited; the other beam passes through an unexcited region of the sample and serves as a reference. After traversing the sample, the probe beams proceed to spectrometers, where they are dispersed and registered by two separate arrays of photodiodes. This allows one to measure the absorbance changes over a spectrum covering about 200 nm on each flash. The signal-to-noise ratio is improved by averaging the spectrum over hundreds or thousands of flashes.

Excitation of P causes a bleaching of the reaction center's strong absorption band at long wavelengths (865 nm in *Rb. sphaeroides* or 960 nm in *Rps. viridis*). Simultaneously, one begins to see emission of light in the 920-nm or 1-μ region, where the excited singlet state (P*) fluoresces. By measuring the decay kinetics of the emission, and the rise of absorbance changes associated with the formation of the oxidized dimer (P^+) or the reduced bacteriopheophytin (H_L^-), one can follow the initial electron transfer step in which P* releases an electron (Woodbury *et al.*, 1985; Martin *et al.*, 1986; Becker *et al.*, 1986). Figure 4 shows a measurement made with *Rb. sphaeroides* reaction centers at room temperature. The time constant of the electron transfer is about 3 ps. Essentially the same kinetics have been measured in *Rps. viridis* (Breton *et al.*, 1986, 1988) and *Rb. capsulatus* (Kirmaier C, and Holten D, personal communication). The reaction is slightly slower in reaction centers of *Chloroflexus aurantiacus*, in which B_M, the bacteriochlorophyll molecule neighboring P on the M side, is replaced by bacteriopheophytin (Becker *et al.*, 1986). Remarkably, the reaction increases in speed if one lowers the temperature (Woodbury *et al.*, 1985; Fleming *et al.*, 1988). With *Rps. viridis* at 4.2 K, the time constant is about 0.7 ps (Fleming *et al.*, 1988), which probably makes this the fastest chemical reaction ever measured.

Absorbance changes reflecting electron transfer from H_L^- to Q_A occur on the time scale of about 200 ps at room temperature (Rockley *et al.*, 1975; Kaufmann *et al.*, 1975; Kirmaier *et al.*, 1985a, b; Shuvalov *et al.*, 1978). Like the initial electron transfer step, this reaction increases in speed with decreasing temperature. At temperatures below 100 K, the time constant is about 100 ps (Kirmaier *et al.*, 1985b). The back-reaction between Q_A^- and P^+ also speeds up at low temperatures (Parson, 1966). Electron transfer from Q_A^- to Q_B behaves more conventionally, slowing down with decreasing temperature (Parson, 1969), but it also can occur at cryogenic temperatures under certain conditions (Kleinfeld *et al.*, 1984).

Measurements of the absorbance changes in the 540-nm region show that only the bacteriopheophytin on the L side of the reaction center (H_L) undergoes reduction to any significant extent in these initial steps (Rockley *et al.*, 1975; Kaufmann *et al.*, 1975; Kirmaier *et al.*, 1985a, b). The two bacteriopheophytins have distinct absorption bands in this region, and only one of the bands bleaches when the reaction center is excited with a single flash. Although H_M also can accept an electron if the reaction center is illuminated continuously under strongly reducing conditions (Thornber *et al.*, 1980; Robert *et al.*, 1985), this process has a much lower quantum yield. Considering the structural symmetry of the reaction center, the strong preference for electron transfer to H_L is puzzling. The most obvious structural difference between the two sides is that the ketone group on ring V of H_L appears to form a hydrogen bond to a glutamic acid residue. This could stabilize the $P^+H_L^-$ radical-pair slightly with respect to $P^+H_M^-$. However, there also are numerous other minor structural differences that could be important (Yeates *et al.*, 1987;

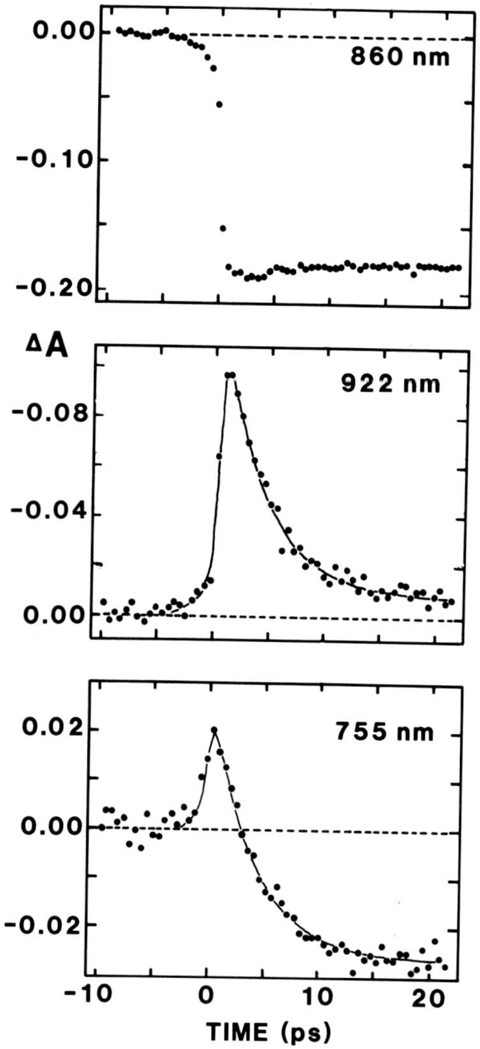

Figure 4. Kinetics of the initial electron transfer step in *Rb. sphaeroides* reaction centers. *Top:* Bleaching of the 860-nm absorption band of P. *Middle:* Transient appearance and decay of stimulated emission of light (fluorescence) from the excited singlet state P*. *Bottom:* Absorbance changes at 755 nm. The initial absorbance increase here probably reflects the formation of P*; the slower absorbance decrease reflects the reduction of H_L. From Woodbury *et al.* (1985).

Fischer and Scherer, 1987; Michel-Beyerle *et al.*, 1988; Warshel *et al.*, 1988). These include small differences in the distances between the electron carriers on the two sides, and differences in electrostatic interactions of the radicals with the surrounding protein.

The crystal structure shows that H_L is about 13 Å from P, edge-to-edge, and that one of the other bacteriochlorophylls, B_L, sits in between the two (Fig. 2). It is most unlikely that an electron could jump directly from P to H_L in 3 ps without any involvement of the intermediate bacteriochlorophyll: theoretical arguments indicate that a direct reaction over this distance would be slower by many orders of magnitude (Fischer and Scherer, 1987; Marcus, 1987; Michel-Beyerle *et al.*, 1988; Warshel *et al.*, 1988). A more plausible mechanism would be for an electron first to hop from P* to B_L, creating a $P^+B_L^-$ radical-pair, and then move from B_L^- to H_L in a second step:

$$P^* \, B_L \, H_L \rightarrow P^+ \, B_L^- \, H_L \rightarrow P^+ \, B_L \, H_L^-$$

Another possibility would be for an electron to jump from B_L to H_L, forming $B_L^+H_L^-$ in the first step, and for B_L^+ to oxidize P in the second step:

$$P^* \, B_L \, H_L \rightarrow P \, B_L^+ \, H_L^- \rightarrow P^+ \, B_L \, H_L^-$$

The individual steps in either of these mechanisms occur between molecules that are essentially in contact, and they could be very fast. Both mechanisms assume that the intermediate radical-pair, $P^+B_L^-$ or $B_L^+H_L^-$, lies below P* in energy, but above the final state, $P^+H_L^-$.

That B_L interacts in some manner with both P and H_L is clear from the observation that its absorption spectrum is perturbed by either the oxidation of P or the reduction of H_L (Shuvalov and Asadov, 1979; Shuvalov and Parson, 1981). However, recent attempts to detect the formation of either $P^+B_L^-$ or $B_L^+H_L^-$ as an intermediate in the electron transfer reaction have given negative results (Woodbury *et al.*, 1985; Martin *et al.*, 1986; Breton *et al.*, 1986, 1988; Paschenko *et al.*, 1985; Becker *et al.*, 1986; Kirmaier C, and Holten D, personal communication). If such an intermediate were present in a significant concentration, one might expect to see a transient bleaching of the absorption band of B_L near 800 nm in *Rb. sphaeroides*, or near 830 nm in *Rps. viridis*. This bleaching is not seen. When the excitation wavelength is chosen to excite P directly, there actually is a small absorbance increase, which probably reflects the formation of the excited state P*. (A short-lived bleaching does occur in these regions if the flash is absorbed by B_L and B_M instead of by P, but this appears to be an artifact due to excitation with more than one photon; its kinetics suggest that it is unrelated to the normal reaction mechanism.)

There are several possible explanations of these observations. In a two-step mechanism, the intermediate state might be undetectable if the rate constant for the second step is much greater than that for the first. Considering the signal-to-noise ratio in the experimental data, $P^+B_L^-$ or $B_L^+H_L^-$ probably could have been resolved if it decayed any less than 10 times faster than it formed, but it could not have been detected if the ratio of the rate constants were greater than 50 (Fleming *et al.*, 1988; Breton *et al.*, 1988). A ratio of 50 would imply that the time constant for the faster step is 1/50th of the time constant that is measured experimentally, or about 15 fs for *Rps. viridis* at 4 K. Whether the reaction could be *this* fast is not clear.

A second possible view is that B_L facilitates the reaction by a mechanism called "superexchange". Superexchange involves a quantum mechanical mixing of $P^+B_L^-$ or $B_L^+H_L^-$ with P*. Such a mixing can occur even if the radical-pair states lie above P* in energy, but it decreases in strength as the two states move apart. The rate of the reaction is proportional to the contribution that the radical-pair makes to the initial state. If the energies of P* and the radical-pair are equal, so that the two states come into resonance, this contribution could approach 50%. However, it seems unlikely that the contribution is higher than 10%, because then the formation of B_L^- or B_L^+ probably would be evident from the initial absorbance changes.

The energy gaps that separate the different states of the system are critical in the two-step mechanisms as well as in the superexchange mechanism. At the actual instant of the electron transfer, the energies of the initial and final states must be the same. This is because nuclear motions are too slow to remove any energy from the system during the electron's transit. If the overall reaction is exothermic, the transfer of energy to the surroundings must occur through nuclear relaxations that follow the electron transfer. One can diagram this by drawing the potential energy surfaces for the initial and final states as shown in Fig. 5. The abscissa in this diagram represents the nuclear coordinates of the system. Electron transfer occurs at the intersection of the two surfaces, where the energy difference between the two states is zero and where the nuclear coordinates for the two states are the same. The finding that the electron transfer reactions increase in speed with decreasing temperature suggests that the two potential energy surfaces cross near the minimum on the reactant surface, so that no activation energy is needed to reach the crossing point. However, Fig. 5 represents a static picture of the system. The actual energy gap between the initial and final states fluctuates with

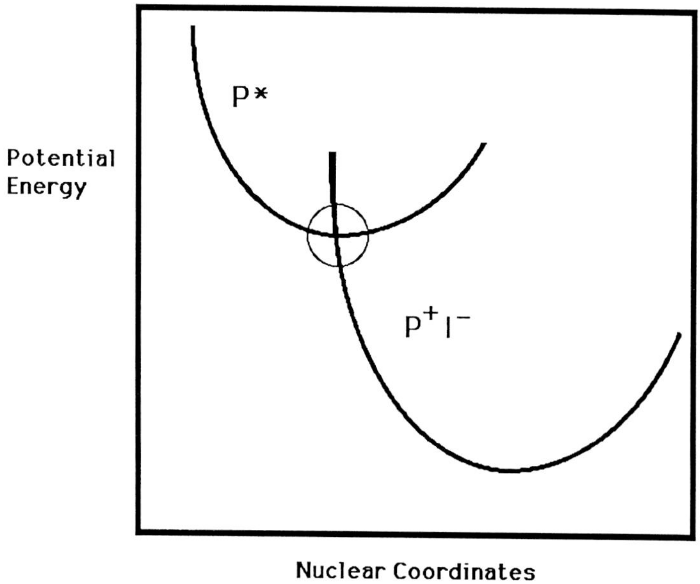

Figure 5. Schematic potential energy surfaces for the initial and final states of an "activationless" electron transfer reaction. The parabolas represent multidimensional energy surfaces that depend on the coordinates of many atoms of the protein, electron carriers and solvent. Electron transfer can occur where the two surfaces intersect.

time as a result of nuclear motions of the protein, and electron transfer events occur only when the gap passes through zero.

Major problems arise in trying to calculate molecular orbital energies for molecules as large as bacteriochlorophyll. Rather than trying to do this directly by a quantum mechanical treatment, we have adopted an alternative approach that makes use of the measured midpoint redox potentials for oxidation and reduction of the electron carriers in solution (Creighton *et al.*, 1988). The difference between the redox potentials of bacteriochlorophyll and bacteriopheophytin, for example, gives the standard free energy change for transferring an electron from bacteriochlorophyll to bacteriopheophytin, when the two molecules are at infinite distance from each other and are surrounded by solvent. The free energy change can be equated to the molecular orbital energy difference, plus the difference between the solvation free energies of the radicals and the neutral molecules:

$$\Delta G_{solution} = -(E_{m,BPh} - E_{m,BChl})F \qquad (1)$$

$$= \Delta E_{orbital} + \Delta G_{solv(sol'n)} \qquad (2)$$

(F in eq. 1 is the Faraday constant.) The free energy change for the same electron transfer reaction in the reaction center would involve essentially the same orbital energy difference, but different electrostatic interactions of the radicals with each other and with the surrounding protein:

$$\Delta G_{RC} = \Delta E_{orbital} + \Delta G_{elec(RC)} \qquad (3)$$

Techniques have been developed recently for calculating electrostatic energies in solution and proteins (Warshel and Russell, 1984; Warshel *et al.*, 1986). By combining these calculated energies with the measured redox potentials, one can obtain the free energy change for the electron transfer reaction in the reaction center, without having to calculate the orbital energy difference:

$$\Delta G_{RC} = \Delta G_{solution} + \Delta G_{elec(RC)} - \Delta G_{solv(sol'n)} \qquad (4)$$

The redox potentials and $\Delta G_{solv(sol'n)}$ represent averages over a long period of time. $\Delta G_{elec(RC)}$ also could represent such an average. On a short time scale, however, the electrostatic energy fluctuates as movements of the protein change the position and orientations of the electron carriers with respect to each other or with respect to charged or polarizable atoms nearby. By calculating the electrostatic energies as part of a molecular dynamics simulation, one can explore the time dependence of the free energy gaps that control the rate of electron transfer (Creighton *et al.*, 1988). The results of such a calculation (Fig. 6, *top*) suggest that the energy gap between P* and $P^+B_L^-$ goes through zero on the order of once or twice per picosecond.

Each time the energy gap for an electron transfer reaction passes through zero, there is a possibility for an electron to move from the donor to the acceptor. The quantum mechanical interaction energy that determines the probability of the transition depends on the molecular orbital overlap of the two molecules (Parson *et al.*, 1987; Fischer and Scherer, 1987; Michel-Beyerle *et al.*, 1988; Warshel *et al.*, 1988; Creighton *et al.*, 1988). Because the electron carriers in the reaction center are such large, complex molecules, it is not yet possible to calculate the interaction energies with much confidence. However, with estimates that appear to be reasonable, the calculated rate of electron transfer for the

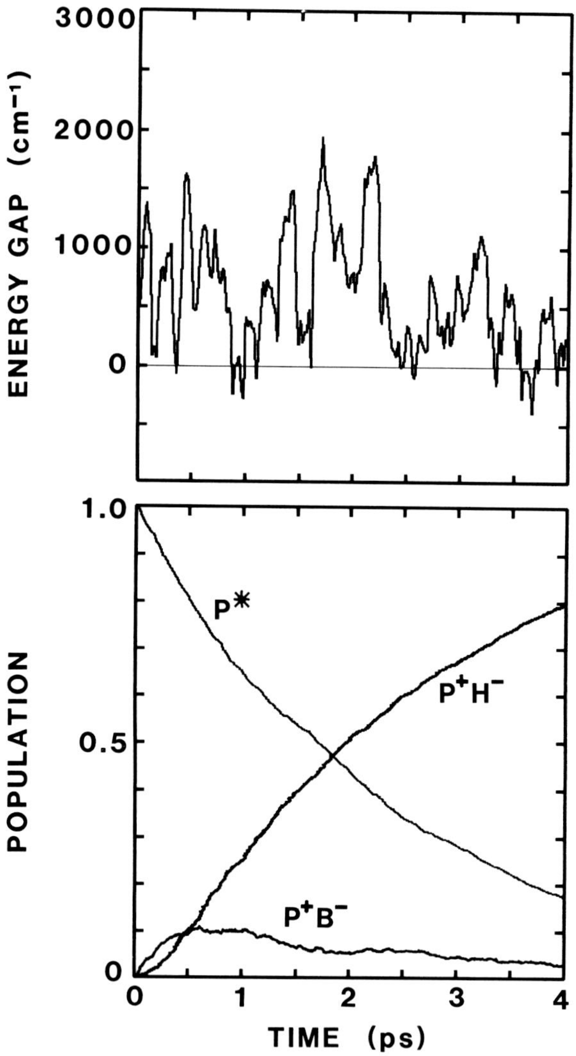

Figure 6. *Top:* Calculated energy difference between P* and $P^+B_L^-$ in *Rb. sphaeroides* reaction centers at 100 K. *Bottom:* Calculated time-dependence of P*, $P^+B_L^-$ and $P^+H_L^-$. From Parson *et al.* (1987) and Creighton *et al.* (1988).

two-step mechanism via $P^+B_L^-$ agrees reasonably well with the rate measured experimentally (Creighton *et al.*, 1988). The results of such a simulation are shown in the lower panel of Fig. 6.

We thus appear to be approaching a detailed molecular understanding of how the bacterial reaction center captures the energy of light. One of the rewards of work on the reaction centers of purple bacteria has been the increasing realization that many of the structural and functional features that are found here are remarkably similar to those of the reaction centers of cyanobacteria, algae and higher plants (Trebst, 1986; Michel and Deisenhofer, 1988; Nanba and Satoh, 1987). These similarities are most striking on the reducing side of Photosystem II, where the initial electron transfer steps from P680 to a pheophytin and from the pheophytin to a quinone occur with almost exactly the same kinetics as the corresponding reactions in the purple bacteria. The similarities do not extend to the oxygen-evolving complex of Photosystem II, or to the iron-sulfur centers that act as early electron acceptors in Photosystem I. Elucidation of these parts of the photosynthetic apparatus remains as a challenge for the future.

Acknowledgements: It is a pleasure to acknowledge the numerous contributions of M. Becker, S. Creighton, D. Middendorf, J. Norris, A. Warshel and N. Woodbury, and support from the National Science Foundation.

REFERENCES

Allen JP, Feher G, Yeates TO, Komiya H, Rees D (1987a). Proc. Natl. Acad. Sci. U. S. A. 84: 5730-5734.

Allen JP, Feher G, Yeates TO, Komiya H, Rees D (1987b). Proc. Natl. Acad. Sci. U. S. A. 84: 6162-6166.

Becker M, Middendorf D, Woodbury NW, Parson WW, Blankenship RE (1986). In Fleming GR, Siegman A (eds): "Ultrafast Phenomena V," Berlin: Springer-Verlag, pp 374-378.

Breton J, Martin JL, Migus A, Antonetti A, Orsag A (1986). Proc. Natl. Acad. Sci. U.S.A. 83: 5121-5125.

Breton J, Martin J-L, Fleming GR, Lanbry J-C (1988). Biochem. in press.

Chang C-H, Tiede D, Tang J, Smith U, Norris J, Schiffer M (1986). FEBS Lett. 186: 201-203.

Creighton S, Hwang J-K, Warshel A, Parson WW, Norris J (1988). Biochem. 27: 774-781.

Deisenhofer J, Epp O, Miki K, Huber R, Michel H (1985). Nature 318: 618-624.

Fischer SF, Scherer POJ (1987). Chem. Phys. 115: 151-158.

Fleming GR, Martin JL, Breton, J (1988). Nature, in press.

Kaufmann KJ, Dutton PL, Netzel TL, Leigh JS, Rentzepis PM (1975). Science 188: 1301-1304.

Kirmaier C, Holten D, Parson WW (1985a). Biochim. Biophys. Acta 810: 33-48.

Kirmaier C, Holten D, Parson WW (1985b). Biochim. Biophys. Acta 810: 49-61.

Kleinfeld D, Okamura MY, Feher G (1984). Biochem. 23: 5780-5786.

Marcus R (1987). Chem. Phys. Lett. 133: 471-477.

Martin JL, Breton J, Hoff AJ, Migus A, and Antonetti A (1986). Proc. Natl. Acad. Sci. U.S.A. 83: 957-961.

Michel H, Deisenhofer J (1988). Biochem. 27: 1-7.

Michel H, Epp O, Deisenhofer J (1986). EMBO J. 5: 2445-2451.

Michel-Beyerle ME., Plato M, Deisenhofer J, Michel H, Bixon M, Jortner J (1988). Biochim. Biophys. Acta 932: 52-70.

Nanba O, Satoh K (1987). Proc. Natl. Acad. Sci. U. S. A. 84: 109-112.

Woodbury NW, Becker M, Middendorf D, Parson WW (1985). Biochem. 24: 7516-7521.

Parson WW (1966). Biochim. Biophys. Acta 131: 154-172.

Parson WW (1969). Biochim. Biophys. Acta 189: 384-396.

Parson WW, Creighton S, Warshel, A (1987). In Kobayashi T (ed): "Primary Reactions in Photobiology," Berlin: Springer-Verlag, pp 43-51.

Paschenko VZ, Korvatovskii BN, Kononenko AA, Chamorovsky SK, Rubin AB (1985) Febs. Lett. 191: 245-248.

Robert B, Lutz M, Tiede DM (1985). FEBS Lett. 83: 326-330.

Rockley M, Windsor MW, Cogdell RJ, Parson WW (1975). Proc. Natl. Acad. Sci. U.S.A. 72: 2251-2255.

Shuvalov VA, Klevanik AV, Sharkov AV, Matveetz YA, Krukov PG (1978). FEBS Lett. 91: 135-139.

Shuvalov VA., Asadov AA (1979). Biochim. Biophys. Acta 545: 296-308.

Shuvalov VA, Parson WW (1981). Proc. Natl. Acad. Sci. U.S.A. 78: 957-961.

Thornber JP, Cogdel, RJ, Seftor REB., Webster GD (1980). Biochim. Biophys. Acta 593: 60-75.

Trebst AZ (1986). Naturforsch. 40c: 237-241.

Warshel A, Russell S (1984). Q. Rev. Biophys. 17: 283-422.

Warshel A, Sussman F, King G (1986). Biochem. 25: 8368-8372.

Warshel A, Creighton S, Parson WW (1988). J. Phys. Chem. 92: 2696-2701.

Yeates TO, Komiya H, Rees DC, Allen JP, Feher G (1987). Proc. Natl. Acad. Sci. U.S.A. 84: 6438-6442.

Photosynthesis, pages 149–159

RUBISCO: MECHANISMS AND THEIR POSSIBLE CONSTRAINTS ON SUBSTRATE SPECIFICITY

John Pierce

Agricultural Products Department, Experimental Station, Building 402, Room 2253, E.I. Du Pont de Nemours & Company, Inc., Wilmington, Delaware 19880-0402, (302) 695-1685

The bifunctional catalysis of both the carboxylation and oxygenation of RuBP[1] by Rubisco constitutes the metabolic interface between the major processes of photosynthesis and photorespiration. This single enzyme modulates the production of the starting metabolites for both photosynthesis (phosphoglycerate) and photorespiration (phosphoglycolate) as a result of the competition between CO_2 and O_2 for the enzyme's active site. The results of this competition determine, in large part, the metabolic profile of autotrophic organisms. As a result, the relative substrate specificity of Rubisco is a key factor in autotrophic metabolism. In this article, I will review recent advances in our understanding of the mechanistic determinants of Rubisco substrate specificity, and forward speculations concerning possible mechanistic constraints on its variation. More extensive, related information on Rubisco and its role in metabolism is available in recently published reviews (Andrews and Lorimer, 1987; Ellis and Gray, 1986; Gutteridge and Gatenby, 1987; Pierce, 1986; Pierce, 1988).

NATURAL VARIATION IN RUBISCO SUBSTRATE SPECIFICITY

In a broad survey of Rubisco from different sources, Jordan & Ogren (Jordan and Ogren, 1981; Jordan and Ogren, 1983) showed that enzymes from different natural sources have intrinsically different relative specificities for CO_2 and O_2, and that the relative substrate specificity is simply related to the kinetic constants of Rubisco.

---

[1] Abbreviations: Rubisco, ribulose 1,5-bisphosphate carboxylase/oxygenase (E.C. 4.1.1.39); RuBP, ribulose 1,5-bisphosphate; *keto*-CABP, 2'-carboxy-3-*keto*-arabinitol 1,5-bisphosphate.

$$\frac{v_c}{v_o} = \frac{V_c/K_c}{V_o/K_o} \cdot \frac{[CO_2]}{[O_2]} = s_{rel} \cdot \frac{[CO_2]}{[O_2]}.$$

That is, the relative rates of carboxylation and oxygenation (v_c/v_o) are proportional to the ratio of concentrations of the gaseous substrates, and the proportionality constant, s_{rel}, is simply the ratio of V_{max}/K_m values for the two reactions. When these substrates are at equal concentrations, the ratio of carboxylation to oxygenation is equal to s_{rel}. s_{rel} varied from ~10–80 over the extent of the survey. These numbers correspond to enzymes that spend 70 and 20% of their time, respectively, functioning as oxygenases at air levels of CO_2 and O_2. The value of s_{rel} appears to be under environmental selection. For example, Rubisco from photosynthetic anaerobes have the lowest values (s_{rel} ~10–15), a finding consistent with their anaerobic lifestyle. Rather higher values are found when one proceeds to the cyanobacteria (s_{rel} ~50). This higher number, while consistent with an aerobic lifestyle, would be too low to allow efficient carbon fixation were it not for the adaptive ability of the cyanobacteria to concentrate CO_2 internally (for a review, see Lucas and Berry, 1985). Finally, the highest values (s_{rel} ~80) are found with higher plant enzymes, which would appear to be under the most stringent selection.

A ROLE FOR SMALL SUBUNITS?

How might we account for these variations in s_{rel}? In comparing the large differences between the enzymes from photosynthetic anaerobes and cyanobacteria or higher plants, it is also important to note that these enzymes have different structures. The anaerobic organisms (e.g., *Rhodospirillum rubrum*) have a dimeric Rubisco (L_2) comprising two identical, active subunits. The cyanobacteria and higher plant enzymes have a more complicated L_8S_8 structure, and it is a common speculation that the small subunit might have a role in increasing s_{rel}. However, the function of the small subunit in enzymatic catalysis remains obscure even after a number of careful and innovative studies by Andrews and co-workers. For instance, the small subunits can be substantially (>97%) removed from cyanobacterial Rubisco (Andrews and Abel, 1981; Andrews and Ballment, 1983). The remaining L_8 core appeared to be inactive, but could be rendered once again competent for catalysis by addition of isolated small subunits. In an interesting study, Andrews & Lorimer re-constituted the cyanobacterial L_8 core with either its own small subunits, or heterologous small subunits from spinach Rubisco (Andrews and Lorimer, 1985). In both cases the reconstituted enzyme had the same s_{rel}, equivalent to the normal cyanobacterial enzyme. That is, substrate specificity appeared to be determined by the large subunit. This has been dramatically re-emphasized by recent observations. (S. Gutteridge, personal communication). By expressing the gene for a cyanobacterial large subunit (but not the small subunit gene) in *E. coli* , Gutteridge obtained a cyanobacterial L_8 fraction manifestly devoid of small subunits. Careful analysis showed that this L_8 enzyme still retained some carboxylase activity (~0.5% of the L_8S_8 activity). Further measurement indicated that s_{rel} for this L_8 core was identical to that of the WT enzyme. It is possible, of course, that as large subunit structures evolved to produce enzymes with higher s_{rel}, they may have come to require an accommodating interaction with small subunits. However, it seems clear that the small subunit has no essential catalytic involvement.

CATALYSIS AND ACTIVATION

The variation in s_{rel} observed in nature is ultimately a consequence of kinetic differences between the various Rubisco's. A great deal of kinetic and mechanistic information on Rubisco has been accumulated, and it is instructive to review this information in a way that might uncover possible mechanistic insights into the determinants of substrate specificity.

Analysis of the partial reactions of Rubisco has led to a fairly detailed understanding of the enzyme's reaction mechanism (Fig. 1). The catalytic intermediates shown in this figure have all been demonstrated to exist by a variety of experimental means, and their kinetic fates have been determined. In brief summary, the reactions of Rubisco are ordered. Surprisingly, neither of the gaseous substrates bind to the enzyme at concentrations orders of magnitude above their K_m values, regardless of whether RuBP or an RuBP substrate analogue is present (Pierce *et al.*, 1986b). Rather, after binding RuBP the enzyme commences the reaction by catalyzing the enolization of RuBP (Gutteridge *et al.*, 1984; Saver and Knowles, 1982). The partial reaction of enolization occurs quite readily in the virtual absence of CO_2 and O_2 (Gutteridge *et al.*, 1984; Pierce *et al.*, 1986b), and for the experimentally amenable *Rhodospirillum rubrum* enzyme it could be shown that enolization in the near absence of CO_2 proceeds at a rate in excess of the maximum rate of overall catalysis (Pierce *et al.*, 1986b). Subsequently, either CO_2 or O_2 condenses with *enol*-RuBP to form, respectively, a six-carbon carboxylated intermediate or a five-carbon peroxy intermediate. (Although the intermediate of the oxygenase reaction has not been isolated, there is suggestive spectrometric evidence for its existence. (Branden *et al.*, 1984b)) By rapidly denaturing Rubisco during catalysis, the six-carbon intermediate, *keto*-CABP, can be isolated (Jaworowski and Rose, 1985). When this intermediate is added back to fresh enzyme, the enzyme catalyzes both its hydration (Lorimer *et*

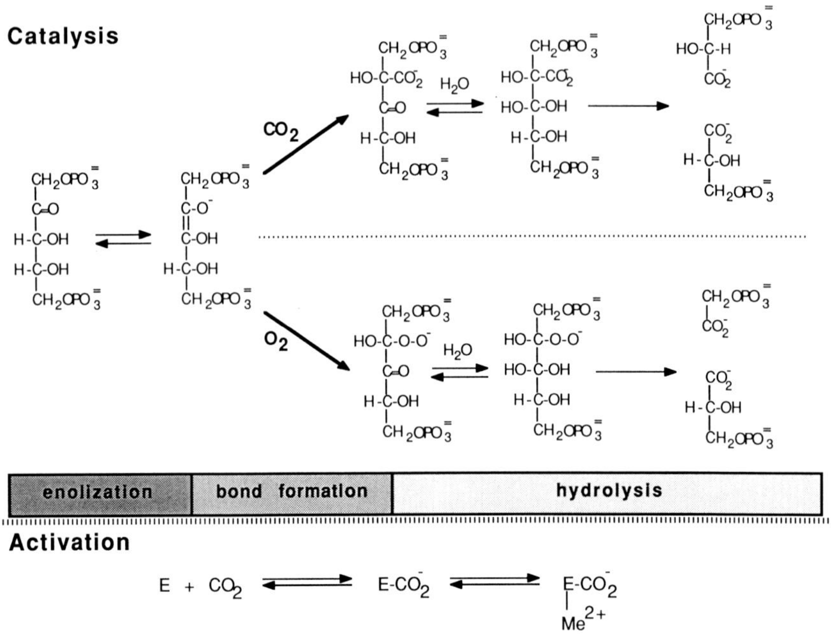

Figure 1. The catalytic and activation mechanisms of Rubisco.

al., 1986; Pierce *et al.*, 1986a) and subsequent cleavage to phosphoglycerate (Jaworowski and Rose, 1985; Pierce *et al.*, 1986a).

An additional complication associated with the enzymology of Rubisco can be introduced by noting that the enzyme can catalyze none of the above reactions without first being activated for catalysis. This activation (Fig. 1) renders the enzyme catalytically competent by reaction with CO_2 to form a carbamate residue on an evolutionarily conserved lysine side chain (Lorimer, 1981; Lorimer *et al.*, 1976). Subsequent condensation with a divalent metal (usually Mg^{2+}) provides the active ternary complex (Lorimer *et al.*, 1976). This complex is quite stable during catalysis, and the activating CO_2 molecule remains distinct from the substrate CO_2 molecule that is ultimately condensed with RuBP (Miziorko, 1979). The activating metal ion is intimately associated with catalysis. EPR (Branden *et al.*, 1984a; Branden *et al.*, 1984b; Miziorko and Sealy, 1984; Styring and Branden, 1985) and NMR (Pierce and Reddy, 1986) studies have revealed that the metal ion binds very closely to the activating carbamate moeity and that it stabilizes the reaction intermediates by direct complexation (Fig. 2).

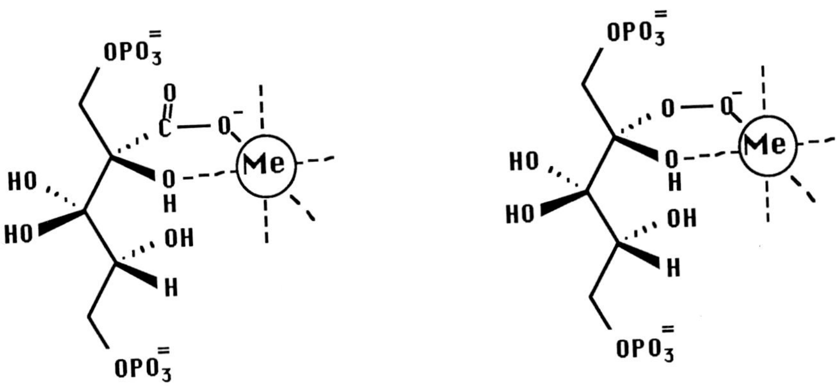

Figure 2. Stabilization of reaction intermediates by complexation with the catalytically essential metal atom.

MECHANISM AND SPECIFICITY

Additional experiments have related the above mechanistic results to the question of enzyme specificity. When enzymes from different natural sources and with widely differing relative specificities for CO_2 and O_2 were presented with *keto*-CABP, it was found that, regardless of the gaseous substrate specificity, essentially all of the intermediate was converted to phosphoglycerate (Pierce *et al.*, 1986a). That is, the intermediate did not appreciably revert to *enol*-RuBP and CO_2. Analogous arguments based on chemical precedence suggest that the same is true for the oxygenase intermediate (Pierce *et al.*, 1986a). Since these intermediates are committed to product formation, alterations in substrate specificity must occur prior to their formation. Similarly, since the enolization of RuBP is common to both reactions, specificity must be exerted after this initial step. Taken together, these results indicate that different enzymes exert their influence on relative substrate specificity at the step of gaseous substrate reaction with *enol*-RuBP.

So it is, that the formation of the stabilized reaction intermediates is a focal point in reaction specificity. The close structural similarities between these intermediates (Fig. 2) suggests that the different relative specificities of different enzymes may be determined by very subtle factors. The manifestation of substrate specificity during the formation of the stabilized reaction intermediates poses a dilemma, since enzymes with different relative substrate specificities obviously differentiate between CO_2 and O_2. A possible thermodynamic explanation can be proposed. While CO_2 and O_2 do not bind to enzyme in the free or RuBP bound form (Pierce *et al.*, 1986b), it may still be that the *enol*-RuBP-enzyme complex is somehow capable of differentially binding these substrates. While there is no experimental evidence to support this contention (and indeed a variety of circumstantial evidence argues against it (Christeller *et al.*, 1976; Pierce *et al.*, 1986b; Van Dyk and Schloss, 1986)), the possibility remains that differential binding could play a role. Perhaps a more likely explanation can be found in kinetic arguments.

It would appear that any preferential stabilization of the transition state leading to the carboxylated intermediate would allow for an increased specificity for CO_2. Although the structural differences between the carboxylated and oxygenated intermediates are quite small (Fig. 2), the extra carboxyl oxygen atom on *keto*-CABP could conceivably provide enough of a binding energy through interaction with an amino acid side chain. A successful crystallographic analysis of spinach Rubisco containing an enzyme-bound reaction intermediate analogue would be particularly illuminating in this regard. As an alternate kinetic speculation, it might also be that different enzymes differentially constrain *enol*-RuBP so that it is more or less reactive as a nucleophile. Under these circumstances, specificity could be exerted. In this view, an enzyme of low relative specificity would generate a reactive enol that is less discriminating towards CO_2 and O_2. A highly specific enzyme would generate a less reactive enol. How an enzyme might generate these differentially reactive enols is, however, completely obscure.

AN INEVITABLE CONSTRAINT?

Although we understand quite a bit about the mechanisms involved in Rubisco specificity, this understanding does not explain why higher plant enzymes do not have higher specificities than they do. Even quite small increases in the relative substrate specificity of a higher plant enzyme would be expected to provide phenotypic advantage, and the available evidence would appear to indicate that the enzyme's specificity is indeed under environmental selection. Nevertheless, high oxygenase activities persist in the higher plant enzymes and, we are left to conclude that, for some reason, further increases in the enzyme's relative specificity for CO_2 are unlikely. As pointed out by Somerville, *et al.* (1984), this apparent inability to accommodate further change may be an example of what Gould and Lewontin (1979), term an "inherent architectural constraint in the structure of life." While this viewpoint is agreeable in a logically formal sense, it is essentially a euphemistic restatement of our inability to understand the specificities observed in higher plants. We are still left wanting a detailed reason for why higher plant enzymes have not adapted further. The enzymatic mechanisms outlined above certainly provide no clear barrier to further enhancements in specificity. However, recent evidence from a number of laboratories has provided a possible basis for speculations concerning an alternate limitation to enzyme specificity, and it is to this evidence that we now turn.

TIME-DEPENDENT CATALYSIS BY HIGHER PLANT RUBISCO

A curious characteristic of higher plant enzymes is shown in Fig. 3. When purified, activated enzyme is exposed to RuBP and CO_2, its catalytic capacity decreases with time in a first order fashion to a final steady rate. These time-dependent kinetics, which are much less pronounced and difficult to observe in cyanobacterial and bacterial

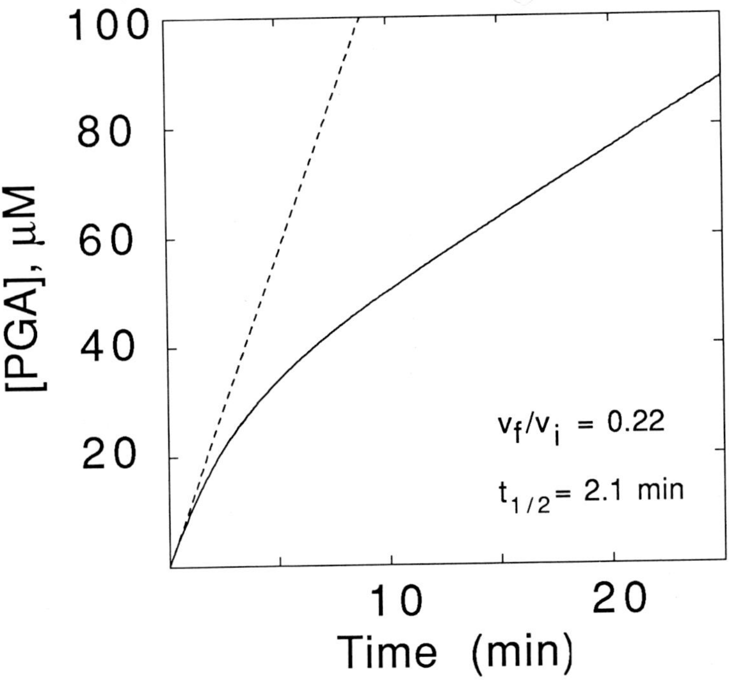

Figure 3. Time-dependent catalysis by spinach Rubisco. Fully carbamylated, active enzyme was added to a reaction mixture containing CO_2, RuBP, and Mg^{2+}. The dashed line is tangent to the observed initial rate.

enzymes, have been reported by a number of workers (Laing and Christeller, 1976; McCurry *et al.*, 1981; Mott and Berry, 1986; Sicher *et al.*, 1981). A variety of explanations for this phenomenon have been made. The presence of slow, tight binding inhibitors in RuBP preparations could certainly account for such behavior (Paech *et al.*, 1978), but more recent experiments using either carefully prepared substrate devoid of these inhibitors (T.J. Andrews, personal communication) or RuBP generated, *in situ*, from ribose 5-P by the action of ribose phosphate isomerase and phosphoribulokinase (J. Pierce, unpublished observations; Mott and Berry, 1986) still revealed the time-dependent loss in activity. An alternative explanation (McCurry *et al.*, 1981; Mott and Berry, 1986) invokes the well known, slow, tight binding of RuBP to decarbamylated enzyme (Jordan and Chollet, 1985; Jordan and Ogren, 1981). RuBP binds so tenaciously (K_i ~ 10 nM) to this enzyme form that it would appear to be able to perturb the activational equilibria and in so doing, sequester an initially carbamylated and active enzyme as an inactive complex (Fig. 4). In support of this, a careful study (Mott and Berry, 1986) showed that the tight binding of RuBP to the decarbamylated enzyme was strongly pH dependent, and that as the pH was raised above 8.6, both the tight binding of RuBP and the non-linear kinetics were eliminated. While this correlation was impressive, direct measurements have shown that the amount of carbamylated enzyme remains unchanged even after the reaction has proceeded to its final steady rate (J. Pierce & G.H. Lorimer, unpublished observations; T.J. Andrews & D. Edmonson, personal communication).

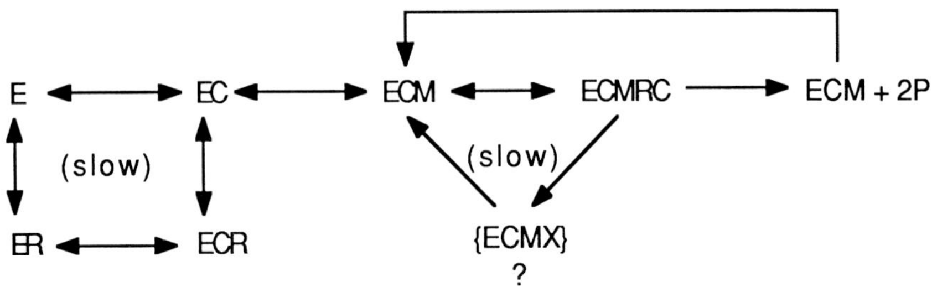

Figure 4. Equilibrium manifolds for activation and catalysis. E, enzyme; EC, carbamylated enzyme; ECM, ternary complex of carbamylated enzyme with Mg^{2+}; R, RuBP; C, CO_2; P, phosphoglycerate. The slow interconversion of decarbamylated and carbamylated enzyme forms in the presence or absence of RuBP is well established. The further possibility of improper substrate binding or inhibitor generation during catalytic turnover (suggested here by the ECMX form) is speculative.

RUBISCO ACTIVASE, AN ACCESSORY PROTEIN INVOLVED IN RUBISCO ACTIVATION.

Yet more complicating experiments have been reported based on studies of a mutant of *Arabidopsis thaliana* (Somerville *et al.*, 1982). In this mutant (termed *rca*), the carbamylation level of freshly isolated Rubisco was seriously deficient, even though the enzyme could be readily activated by subsequent incubation with CO_2 and Mg^{2+}. It was then shown that two stromal polypeptides were absent in the *rca* mutant (Salvucci *et al.*, 1985). Subsequent progress has been rapid. A protein, termed Rubisco activase, has been purified from spinach chloroplasts (Salvucci *et al.*, 1987) and shown to accelerate the rate and extent of activation of initially inactive Rubisco-RuBP complexes. The gene for this protein has been cloned, sequenced, and expressed in *E. coli* (Werneke *et al.*, 1988a; Werneke *et al.*, 1988b). Both the natural and *E. coli* derived enzymes have been shown to accelerate the activational equilibria of Rubisco in the presence of ATP. The detailed reasons for the presence of this protein in higher plants are still only poorly understood, but recent experiments are suggestive.

When Rubisco activase and ATP were included in a reaction with fully carbamylated Rubisco, the time-dependent loss of activity that is observed with Rubisco alone was dramatically reduced (Robinson and Portis, 1989). Furthermore, if the Rubisco reaction was allowed to proceed in the absence of Rubisco activase until the final, reduced steady state rate was achieved, subsequent addition of Rubisco activase and ATP resulted in the rapid, and nearly complete recovery of enzyme activity. Finally, when the Rubisco reaction was allowed to proceed to its final, steady state rate in the absence of Rubisco activase, and the Rubisco recovered from assay constituents by rapid gel filtration, its enzymatic activity remained low, but could be subsequently recovered by treatment with alkaline phosphatase (Robinson and Portis, 1989). Although these authors carefully allow that their results could be explained by a tight binding inhibitor that contaminated their RuBP preparation, it might also be due to unproductive (improper) binding of RuBP to a carbamylated enzyme form. A more intriguing explanation is that Rubisco may infrequently catalyze an alternate reaction which produces a slowly dissociating enzyme inhibitor (Fig. 4). Improper, catalytic misfiring by enzymes is a well-documented phenomenon, and the relative rates of catalysis and inactivation of Rubisco ($k_{cat} \sim 3\ s^{-1}$; $k_{inact} = 5.5 \times 10^{-3}\ s^{-1}$ (c.f., Fig. 3)) suggest that the alternate catalysis would occur once for every 500-1000 catalytic turnovers. Regardless of the specific details (improper substrate binding, catalytic misfiring, etc.) of the phenomenon, if time-dependent inactivation proves to be an intrinsic property of higher plant Rubisco, we would have a situation wherein the enzyme takes part in its own demise, and an accessory protein (Rubisco activase) would be required to alleviate this obstruction. The enhancement of the exchange rates of various ligands bound to Rubisco appears to be the hallmark of Rubisco activase. Its ability to accelerate and enhance the activation of non-carbamylated Rubisco as well as its ability to prevent inactivation of carbamylated Rubisco may be seen as manifestations of this basic property.

A MECHANISTIC CONSTRAINT ON FURTHER INCREASES IN S_{REL}?

Although higher plant Rubisco shares a large number of common features with other Rubisco's, it also has distinctive features. The higher plant enzyme has the highest relative substrate specificity; it catalyzes a time-dependent reaction; it associates with Rubisco activase, a protein which may have important regulatory effects on both chloroplast physiology and Rubisco catalysis. It remains to be seen, of course, whether these distinctive features are related in any causal manner. However, if we speculate that Rubisco acquired time-dependent kinetics as a result of mutations that allowed for higher specificity, then we have a clear view of a possible mechanistic constraint to further increases in specificity. There would also be a clear requirement for an accessory agent to ameliorate the effects of this time-dependence. Whether these speculations have any basis in fact will have to await the judgment of further studies on the specific details of the time-dependent kinetics and the mode of action of Rubisco activase.

Acknowledgments: I would like to thank my colleagues for sharing with me their unpublished observations and for stimulating discussions, especially T.J. Andrews, J.A. Berry, S. Gutteridge, G.H. Lorimer, and A.R. Portis.

REFERENCES

Andrews TJ, Abel KM (1981). J Biol Chem 256:8445-8451.
Andrews TJ, Ballment B (1983). J Biol Chem 258:7514-7518.
Andrews TJ, Lorimer GH (1985). J Biol Chem 260:4632-4636.
Andrews TJ, Lorimer GH (1987). In "The Biochemistry of Plants, Vol 10.", Academic Press. pp 131-218.
Branden R, Nilsson T, Styring S (1984a). Biochemistry 23:4373-4378.
Branden R, Nilsson T, Styring S (1984b). Biochemistry 23:4378-4382.
Christeller JT, Laing WA, Troughton JH (1976). Plant Physiol 57:580-582.
Ellis RJ, Gray JC, eds (1986). "Ribulose Bisphosphate Carboxylase-Oxygenase," The Royal Society, London, pp. 1-165.
Gould SJ, Lewontin RC (1979). Proc R Soc London, Ser. B 205:581-598.

Gutteridge S, Gatenby AA (1987). Oxford Surveys of Plant Molecular & Cell Biology 4:95-135.
Gutteridge S, Parry MAJ, Schmidt CNG, Feeney J (1984). FEBS Lett 170:355-359.
Jaworowski A, Rose IA (1985) J Biol Chem 260:944-948.
Jordan DB, Chollet R (1983). J Biol Chem 258:13752-13758.
Jordan DB, Chollet R (1985). Arch Biochem Biophys 236:487-496.
Jordan DB, Ogren WL (1981). Nature 291:513-515.
Jordan DB, Ogren WL (1983). Arch Biochem Biophys 227:425-433.
Laing WA, Christeller JT (1976). Biochem J 159:563-570.
Lorimer GH (1981). Biochemistry 20:1236-1240.
Lorimer GH, Andrews TJ, Pierce J, Schloss JV (1986). Phil Trans R Soc London Ser. B 313:397-407.
Lorimer GH, Badger MR, Andrews TJ (1976). Biochemistry 15:529-536.
Lucas WJ, Berry JA, eds (1985). "Inorganic carbon uptake by aquatic photosynthetic organisms," Waverly Press, Baltimore, pp. 1-480.
McCurry SD, Pierce J, Tolbert NE, Orme-Johnson WH (1981). J Biol Chem 256:6623-6628.
Miziorko HM (1979). J Biol Chem 254:270-272.
Miziorko HM, Sealy RC (1984). Biochemistry 23:479-485.
Mott KA, Berry JA (1986). Plant Physiol 82:77-82.
Paech C, Pierce J, McCurry SD, Tolbert NE (1978). Biochem Biophys Res Commun 83:1084-1092.
Pierce J (1986). Plant Physiol 81:943-945.
Pierce J (1988). Physiol Plantarum 72:690-698.
Pierce J, Andrews TJ, Lorimer GH (1986a). J Biol Chem 261:10248-10256.
Pierce J, Lorimer GH, Reddy GS (1986b). Biochemistry 25:1636-1644.
Pierce J, Reddy GS (1986). Arch Biochem Biophys 245:483-493.
Robinson SP, Portis AR (1989). Plant Physiol, in press.
Salvucci ME, Portis AR, Ogren WL (1985). Photosynth Res 7:191-203.
Salvucci ME, Werneke JM, Ogren WL, Portis AR (1987). Plant Physiol 84:930-936.
Saver BG, Knowles JR (1982). Biochemistry 21:5398-5403.
Sicher RC, Hatch AL, Stumpf DK, Jensen RG (1981). Plant Physiol 68:252-255.
Somerville CR, Fitchen J, Somerville S, McIntosh L, Nargang F (1984). in "Advances in Gene Technology: Molecular Genetics of Plants and Animals" (K. Downey, R.W. Voellmy, J. Schultz, & F. Ahmad, eds.), Academic Press, New York, pp. 295-309.
Somerville CR, Portis AR, Ogren WL (1982). Plant Physiol 70:381-387.
Styring S, Branden R (1985). Biochemistry 24:6011-6019.
Werneke JM, Chatfield JM, Ogren WL (1988a). Plant Physiol, in press.
Werneke JM, Zielinski RE, Ogren WL (1988b). Proc Natl Acad Sci, USA 85:787-791.
Van Dyk DE, Schloss JE (1986). Biochemistry 25:5145-5156.

Photosynthesis, pages 161–182

DYNAMIC MOLECULAR ORGANIZATION OF THE PLANT THYLAKOID MEMBRANE

Jan M. Anderson* and William W. Thomson+

*CSIRO, Division of Plant Industry, GPO Box 1600, Canberra ACT 2601, Australia, and +Department of Botany and Plant Sciences, University of California, Riverside, CA 92521, USA

INTRODUCTION

An ultimate understanding of the molecular mechanisms of photosynthetic energy transduction requires the elucidation of the molecular organization of the pigments, proteins, and lipids of the chloroplast thylakoid membranes. The photosynthetic apparatus is composed of several discrete multiprotein complexes located within the fluid bilayer of the thylakoid membranes (Fig. 1). The many chlorophyll and carotenoid molecules that are engaged in light-harvesting are associated with specific proteins. These include the chlorophyll α-proteins that, together with the reaction centres of each photosystem (PS), form the core PSII and PSI complexes, and the heterogeneous light-harvesting chlorophyll *a/b*-proteins of PSII (LHCII) and PSI (LHCI). The photosystem II complex, the cytochrome b/f complex, and the photosystem I complex are linked to each other through the mobile electron carriers, the plastoquinone pool, plastocyanin, and ferredoxin to transport electrons from H_20 to $NADP^+$. The protons deposited in the lumen, from water splitting and the PQ pool via the cytochrome b/f complex, generate electrochemical energy for ATP synthesis by ATP synthase.

This highly specific, asymmetric arrangement of protein complexes consisting of hydrophobic, membrane-spanning α-helices linked by the more hydrophilic amino acid sequences at the membrane surfaces imparts critical order to the arrangement of the pigment molecules and redox components. Further, the vast array of fixed surface charges in the protein and lipid domains at the outer and inner membrane surfaces not only gives rise to an overall asymmetry of electrostatic charges, but also means that small changes in different domains at the membrane surfaces will have a profound effect on the conformation of both the proteins and lipids. This important aspect of the transverse arrangement of the lipids and proteins of thylakoid membranes is reviewed elsewhere (Murphy, 1986; Anderson, 1987).

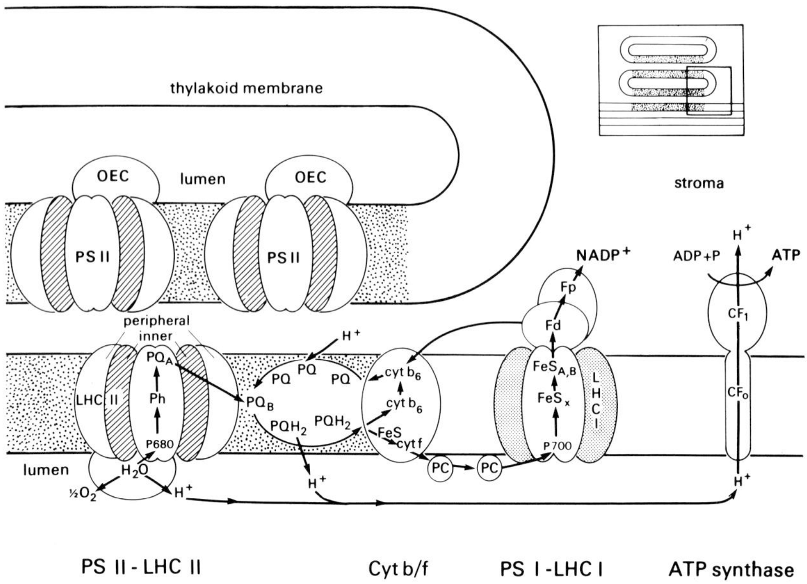

Figure 1. Molecular organization of the components of plant thylakoid membranes participating in light-harvesting, linear electron transport and proton translocation. The continuous thylakoid membrane network is structurally differentiated into appressed and nonappressed regions.

LATERAL ORGANIZATION OF THYLAKOID COMPLEXES.

The single thylakoid membrane continuum of mature higher plant chloroplasts shows a striking structural differentiation into grana stacks (whose flattened discs or compartments are closely appressed together in the partition region) and the stroma-exposed thylakoids that interconnect the grana (Weier 1961; Weier and Thomson, 1962; Paolillo, 1970). Thus, the membrane continuum forms an elaborate network in which stroma thylakoids interconnect several grana thylakoids in different grana stacks so that the inner lumenal space is continuous. This structural differentiation is directly related to function.

Early membrane fractionation studies with plant thylakoids suggested a nonuniform distribution of function between the appressed and non-appressed regions. It was suggested by Sane *et al.* (1970), and widely assumed during the 1970's, that PSII and PSI were located in close association in the appressed grana thylakoids to participate in linear electron transport, while some PSI was also present in the stroma thylakoids for cyclic electron transport. It was thought that equal amounts of PSII and PSI reaction centres shared excitation energy from a common pool of LHCII which was in contact with both photosystems. However, this picture slowly changed. Åkerlund *et al.* (1976) showed that inside-out vesicles derived from grana partitions were highly enriched in PSII, and depleted in PSI. Stroma thylakoids also had some PSII (Armond and Arntzen, 1977). Following analyses of the chlorophyll-proteins which showed that most PSI complexes were present in stroma thylakoid vesicles with very little in appressed, inside-out vesicles, Andersson and Anderson (1980) proposed that PSI was located only in the nonappressed membranes, and hence was laterally segregated from most PSII and LHCII. An heterogeneity was also found in PSII with interconnected PSIIα units having much larger light-harvesting antennae than PSIIβ units (Thielen and van Gorkhom, 1981). Further, PSIIα was exclusively located in the appressed membranes and PSIIβ in the nonappressed membranes (Anderson and Melis, 1983). Many structural studies (Staehelin and Arntzen, 1983; Haehnel, 1984) also support this concept of lateral heterogeneity (Fig. 1). Most of the core PSII complexes together with LHCII (PSIIα units) are located in the appressed membranes where they are laterally segregated from the PSI complexes, ATP synthase, and fewer PSIIβ units which are exclusively located in nonappressed domains (Fig. 1).

This lateral heterogeneity has important consequences. If indeed PSI were only in the nonappressed membranes there would be limited sharing of excitation energy between PSII and PSI; hence spillover would be curtailed (Anderson, 1981). Moreover, laterally segregated photosystems require mobile electron carriers to link them. To establish which mobile electron carriers were needed for long-range diffusion between the appressed and nonappressed domains, it became important to establish if the cytochrome b/f complex was present in both appressed and nonappressed domains (Anderson, 1982) or located in the fret region connecting the appressed and nonappressed membranes (Barber, 1983; Ghirardi and Melis, 1983).

While the concept of lateral heterogeneity in the distribution of thylakoid complexes illustrated in Fig. 1 had been generally accepted, the large body of evidence derived from biochemical and structural studies was indirect. Fortunately, immuno-electron microscope studies enable individual thylakoid components or complexes to be visualised directly. In 1976, Miller and Staehelin had already shown conclusively that ATP synthase was confined to nonappressed membranes. More recently, PSII complex (Vallon *et al.*, 1985; 1987), the oxygen evolving complex (Goodchild *et al.*, 1985b; Vallon *et al.*, 1986), and LHCII (Anderson and

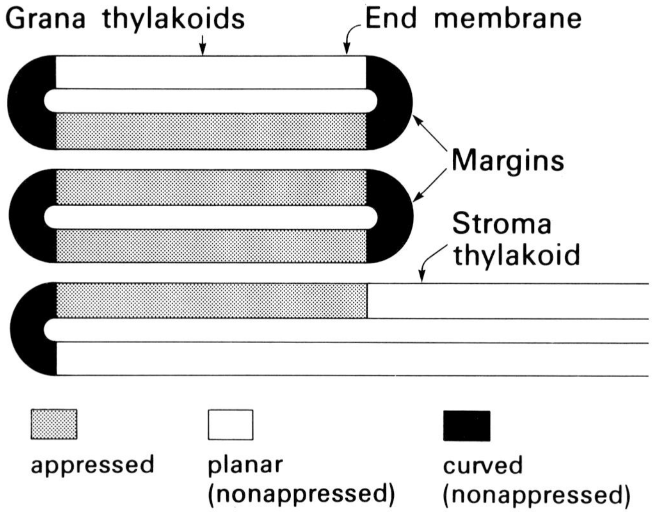

Figure 2. Diagramatic representation of appressed grana partitions and the nonappressed membranes which form two structural domains: the curved grana margins and the planar grana end membranes and stroma thylakoids.

Goodchild, 1987) have all been shown to be mainly present in appressed membranes. Importantly, PSI complex is exclusively located in non-appressed domains (Vallon *et al.*, 1986; Anderson and Goodchild, 1987). Indeed, the cytochrome b/f complex is the only one that is evenly distributed between the two membrane regions (Allred and Staehelin, 1985; Goodchild *et al.*, 1985a; Shaw and Henwood, 1985).

THE THYLAKOID MARGINS

Until recently most interest in the structural differentiation of the thylakoid membrane network has focussed on the distinction between the appressed regions where the membranes are in closer proximity than 4 nm, and the nonappressed regions whose outer surfaces are in direct contact with the stroma. However, the nonappressed membranes have two structurally distinct domains (Fig. 2). The end grana membranes and the stroma thylakoids are flat and planar, while the margins around the circumference of the granal compartments are highly curved. At various points around the circumference, the curved marginal domain is interrupted by the fret regions which connect the appressed grana membranes and the stroma thylakoids.

The marginal domains have been rather neglected. Murphy (1982) proposed that the margins would be enriched in neutral galactolipids. He suggested that the cone-shaped monogalactosyldiglycerides would be mainly located in the inner bilayer half, and the wedge-shaped digalactosyldiglycerides in the outer bilayer half, and proteins would be excluded from this curved region. Of more importance here is that Murphy (1982) has estimated that marginal membranes could account for approximately 11% of the total membrane surface area of grana. In reaching this figure, he assumed that the grana were 700-1400 nM in width and the thylakoids were approximately 25 nM in thickness. However, as he notes, if the thylakoids are smaller in width, e.g. 400 nM, and thicker, e.g. 35 nM, as they often are, then as much as 32% of the total surface area of granal membranes would be in the margins. Even if this figure is reduced somewhat by subtracting the area occupied by frets, the highly curved margins represent a significantly large, specialized domain of the total granal-fretwork system.

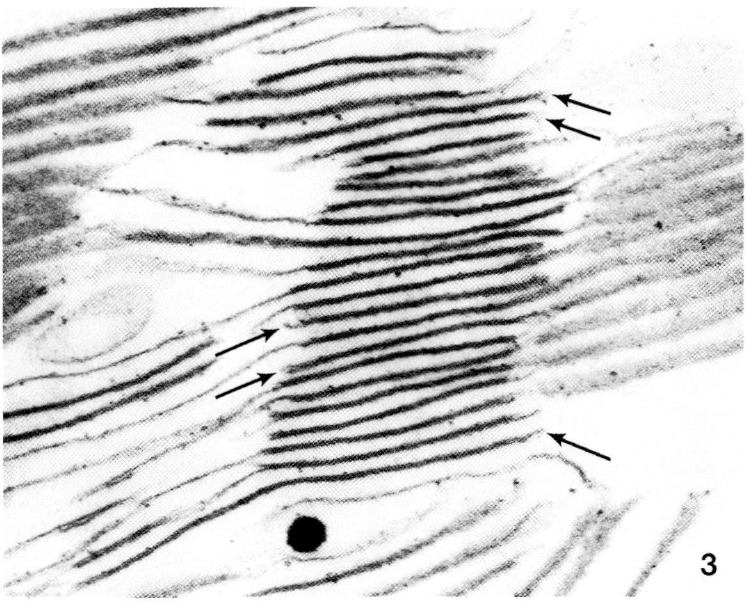

3

Figure 3. The granal fretwork system of an isolated spinach chloroplast after treatment with Tween 20. Note the absence of marginal regions, arrows. X = 93,000.

Thomson and Moeller (1982) showed that low concentrations of the detergent, Tween 20, preferentially solubilised the marginal membranes leaving intact grana partitions and stroma regions (Fig. 3). Later analyses of the Tween-solubilised supernatant showed the presence of proteins of the cytochrome b/f complex, PSI complex, and ATP synthase, but no LHCII (Webber *et al.*, 1988). Following this Tween 20 solubilisation, differential centrifugation enabled the appressed granal and nonappressed stroma membranes fractions to be isolated. Although only a small amount of total chlorophyll (5%) was isolated in the initial Tween-solubilised fraction, representing the marginal membranes, this distinct membrane region contains protein complexes, but no LHCII.

The results (Webber *et al.*, 1987) showing that the grana margins have some protein complexes (and not lacking proteins as Murphy (1982; 1986) suggested) are confirmed by immunocytochemical studies. Both ATP synthase (Miller and Staehelin, 1986) and the cytochrome b/f complex (Allred and Staehelin, 1985; Goodchild and Anderson, 1985a) were seen at the junctions between appressed and nonappressed regions as well as along the stroma-exposed membranes. In transverse sections, the gold particles attached to antibodies raised against the 68 kDa Chl *a*-apoprotein of PSI, are seen more frequently at the fret regions connecting appressed to nonappressed membranes, than along the stroma-exposed membranes (Anderson and Goodchild, 1987; Vallon *et al.*, 1986, 1987). Indeed using "on grid" labelling of grana discs, Vallon *et al.* (1986) showed that whereas the cytochrome b/f complex was uniformly distributed over the entire grana disc, the PSI complexes were found mainly around the periphery. We conclude that there is an heterogeneity in the location of protein complexes within the nonappressed membrane regions, with some cytochrome b/f complex, PSI complex, and ATP synthase being also located in grana margins.

The proposed localization of some cytochrome b/f complexes, PSI complexes and ATP synthase in the grana margins, has profound consequences for our current understanding of photosynthesis.

LINEAR ELECTRON TRANSPORT AT LIGHT SATURATION

To date, the model proposed with PSI located only in the planar, nonappressed membrane domains posed problems for linear electron transport at light saturation. Whitmarsh (1987) estimated that the minimum distance that plastocyanin would need to diffuse from the cytochrome b/f complexes of the appressed membranes to PSI out in the stroma-exposed membranes would be 100 nm, giving it a minimum lateral diffusion coefficient of 1.2 x 10^{-8} $cm^{-2}s^{-1}$. This might be adequate, but diffusion would be limited by both encounters with the many protein complexes of the appressed membranes and the narrow passage through the interconnecting fret regions. Haehnel (1984) argued that such long-range diffusion would be very difficult for the large plastocyanin molecules located in the very narrow lumen. Despite these difficulties, the lateral mobility of plastocyanin and plastoquinone *in vivo* is not a problem. We propose that the localisation of a substantial amount of the total PSI complexes in the grana margins means that the grana stacks themselves will be the main site for linear electron transport. Now plastoquinone molecules will be the main electron carriers from PSII to the cytochrome b/f complexes in the margins; plastocyanin will be very close to the marginal PSI. It is agreed that the dominant rate-limiting step of linear electron transport at light saturation is the oxidation of the plastoquinol pool with a half-time of 15 ms (Witt, 1971). The imbalance of PSII reaction centres to those of PSI reaction centres of the grana margins is not a problem when the light is not limiting. As pointed out (Atta-Asafo-Adjei and Dilley, 1985), regardless of the location of PSI relative to PSII, the turnover capacity of PSI is very high. A turnover number of 2.88 ms^{-1} for P700-PC-cyt f is about 100-fold greater than that of the rate-limiting step of linear electron transport. Hence, providing the light is not limiting, fewer PSI complexes will be able to receive electrons from many more PSII units. Further, protons will be able to be rapidly translocated via the marginal ATP synthase.

ELECTRON TRANSPORT AT LIGHT-LIMITING CONDITIONS

Recent measurements of quantum yield (Björkman and Demmig, 1987; Evans, 1987a) confirm earlier results and show a remarkable constancy of quantum yields of 0.11 mol O_2 per mol of photons for many C_3 sun and shade species, irrespective of growth irradiance.

These constant and very high values for quantum yield have important consequences under light-limiting conditions:

(1) The effective absorptive cross-section of the two photosystems of all plants must be equal.

(2) The number of pigments that each photosystem contains must be comparable (Evans, 1987a).

(3) The amount of PSI participating exclusively in cyclic electron transport must be drastically curtailed.

(4) These high quantum yields do not mean that the reaction centres must be present in a 1:1 stoichiometry. It is consistent, however, with the effective utilization of 8-9 quanta for oxygen evolution, involving the shuttling of mobile plastoquinonol from PSII complexes to cytochrome b/f complexes, and then plastocyanin to nearby PSI in the grana margins, and further-away PSI in the nonappressed membranes, since now there would be time for plastocyanin to diffuse over greater distances than at light

saturation. Hence, under light-limiting conditions, both the margin- and stroma-exposed-located PSI complexes will participate in linear electron transport.

(5) Since the marginal PSI complexes will be in proximity to some PSIIα units, the LHCII of the latter will be able to transfer light excitation to PSI, thereby redressing the imbalance of pigment content in PSII versus PSI. The dilemma posed by having more light-harvesting pigments in PSII in shade and low-irradiance plants than in sun or high-irradiance plants has hitherto made it difficult to comprehend constant high quantum yields. This is no longer a problem with some PSI located closer to PSII than had been appreciated.

We conclude that the grana stacks will be the site of non-cyclic electron transport with plastocyanin transferring electrons to PSI in margins, in preference to PSI in the stroma thylakoids (Anderson, 1982). The stroma thylakoids will be the site of cyclic electron transport at high irradiance, and the domain for the insertion of newly-synthesised membrane components and the assembly of multisubunit complexes (see below).

This picture of the molecular organization of thylakoid membranes (Fig. 1), being an instant snapshot in time, does not portray their highly dynamic nature. Clearly, there is a continuum of organization and function from appressed to nonappressed thylakoids within the chloroplast, to different chloroplasts both across and along the leaf, to leaves within and between species (Anderson and Osmond, 1987). This highly plastic organization of chloroplasts is a necessity for most plants which cannot move to escape from their surroundings, although some species have limited movements of leaves and chloroplasts to ameliorate their light environment. Most plants face considerable momentary, daily, and seasonal fluctuations in light quality and quantity, as well as changes in temperature, water, nutrition availability and other environmental factors. Hence, adaptation at the molecular level is vitally important for plants. Two broad adaptations occur both of which contribute to the dynamic nature of thylakoid membrane organization. These are the rapid, short-term regulatory changes (seconds to minutes) that involve a reorganization of existing membrane components. On the other hand, there are the slower changes (hours to days) brought about by changes in the rates of synthesis or degradation of thylakoid components, which in turn influence membrane organization, structure and function.

SHORT-TERM REGULATIONS

Plants and leaves growing in different light environments regulate their relative amounts of light-harvesting pigment-protein complexes, electron transport complexes, and ATP synthase, as well as the levels of the mobile electron transport carriers. These changes in composition lead to long-term morphological and functional changes in thylakoids (discussed later) which are ineffective in coping with the sudden fluctuations (ms to min) in light intensity and quality that may be due to changes in leaf movements, sun flecks, cloud cover or sun angle. The dynamic, rapid responses elicited by such environmental fluctuations alter the existing molecular organization of some thylakoid components, thereby influencing their light-harvesting and electron transfer characteristics, and indeed photosynthetic capacities. These responses permit the basic thylakoid functions (in relation to the amounts of $NADP^+$ and ATP, and the stromal environment) to be tuned to metabolic demands of the cell, light intensity, light quality, and temperature, as well as the availability of H_20, CO_2 and nutrients (Horton, 1985, 1987).

The rapid regulatory mechanism of State I/State II transitions ensures the even distribution of excitation energy between PSII and PSI, thereby maintaining optimum quantum efficiency of non-cyclic electron transport in leaves in response to variations in incident light (Chow *et al.*, 1981; Canaani *et al.*, 1983). It is believed that these State I/State II transitions are related to the reversible phosphorylation of peripheral LHCII and its subsequent lateral migration between appressed and nonappressed regions (Bennett, 1983; Barber, 1985). Thylakoids contain a kinase which phosphorylates threonine residues near the amino-terminal region of the apoproteins of peripheral LHCII. Too much light into PSII rapidly over-reduces the plastoquinone pool, which activates the kinase to phosphorylate preferentially the 25 kDa apoprotein of LHCII (Larsson and Andersson, 1985). The introduction of the negatively-charged phosphate groups into LHCII (which must form a major part of the protein mass of PSII units exposed at the stroma-facing surface) forces the phosphorylated LHCII to dissociate from the core PSII units, as well as forcing the edges of the grana stacks to become less firmly appressed. The phosphorylated peripheral LHCII complexes then migrate laterally in the membrane plane from the PSII-rich appressed regions to the PSI-rich nonappressed regions, leaving behind the core PSII units (Fig. 4). This redeployment strategy decreases the absorptive cross-section of PSII and increases that of PSI. Conversely, when the plastoquinone pool is over-oxidised by PSI light, the phosphorylated peripheral LHCII units are dephosphorylated and they laterally migrate back to appressed membrane regions where they re-associate with core PSII complexes to give PSIIα units. Such movements are very rapid (s to min). Under conditions of strong phosphorylation *in vivo*, both the phosphorylated LHCII and phosphorylated core PSII migrate out from the nonappressed membranes. Although evidence is not available for this reversible phosphorylation of peripheral LHCII *in vivo* the hypothesis is attractive: direct structural evidence should be sought.

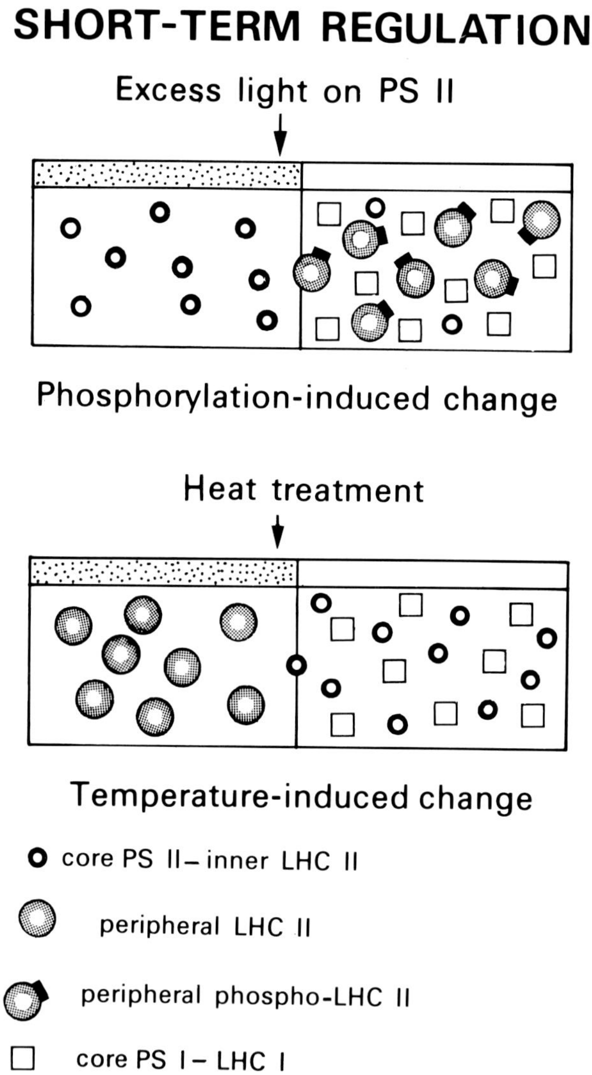

Figure 4. A scheme showing the rearrangements of membrane components between the appressed (left) and nonappressed (right) regions elicited by phosphorylation and brief heat treatment.

Interestingly, dynamic reversible changes in membrane organization have also been observed when isolated thylakoids are briefly heated. Exposure for 5 min to increasing temperatures in the range of 35-45°C, led to a progressive destacking of the thylakoid membrane network, accompanied by a different pattern of freeze-fracture particle distribution (Gounaris *et al.*, 1984). Indeed, brief heat treatments cause a dissociation of peripheral LHCII from core PSII units, but the ensuing lateral migration is quite different from that occurring with phosphorylation (Sundby and Andersson,1985; Larsson *et al.*, 1987). The core PSII units migrate out from the appressed membrane regions, while the LHCII units remain behind (Fig. 4). This reversible reorganization caused by brief heat treatments also decreases the absorptive cross-section of PSII, but "spillover" in the nonappressed regions is not possible, since excitation energy is not transferred between core PSII and core PSI units. This heat-induced response may be another strategy to minimise over-excitation and consequent photodamage to PSII, as high light is often accompanied by excess heat (Sundby and Andersson, 1985).

It is likely that these short-term regulatory responses have physiological significance and form part of a network of rapid reversible changes in the localisation of membrane components between the appressed and nonappressed domains. Accompanying these lateral migrations, will be the array of energy dissipative processes where changes in (Δ)pH, (Δ)g, (Δ)E and Mg++. cause more subtle conformational changes of the pigment beds of both photosystems, particularly that of PSII.

TRAFFICKING OF PROTEINS BETWEEN APPRESSED AND NONAPPRESSED REGIONS

The thylakoid complexes comprise both nuclear and chloroplast-encoded subunits. As the 70S thylakoid-bound ribosomes are too bulky to be accommodated within the grana partitions, it is likely that the chloroplast-encoded polypeptides are inserted only in nonappressed membranes. This may also be the case for nuclear-encoded polypeptides whose movements from the inner chloroplast envelope to the thylakoids are not characterised. Either these precursor polypeptides undergo transit through the stroma as soluble precursors, or they are transported in vesicles (derived from the inner envelope membrane) which function as carriers. Schmidt and Mishkind (1986) suggest that the vesicles thought to be formed for translocation of the lipid products of the biosynthetic pathways of the inner envelope membrane (Douce *et al.*, 1984) could also carry nuclear-encoded polypeptides, for example, the apoproteins of LHCII, to the thylakoids. If this latter process takes place these vesicles would also need to be incorporated into nonappressed regions.

In any event, most of the PSII units and some of the cytochrome b/f complexes are destined to reside in appressed regions. Since the core proteins of PSII, the D1 protein, the D2 protein, and cytochrome b-559, as well as the two Chl *a*-proteins (47 and 43 kDa aproproteins) are encoded by the chloroplast genome, there will need to be much lateral shuttling of newly-sythesized polypeptides from the nonappressed regions, through the narrow frets, to the appressed grana membranes. Although this dynamic lateral shuttling is not fully characterized, Mattoo and Edelman (1987) have traced events associated with the trafficking of the D1 protein. The D1 precursor protein (33.5 kDa), which is synthesised by the psbA gene, is the major light-driven translation product of the chloroplast genome (Ellis, 1981). Indeed the D1 protein is 40-80 times more rapidly turned-over than other thylakoid proteins. The D1 precursor protein is inserted exclusively in the nonappressed stroma thylakoids (Mattoo and Edelman, 1987) (Fig. 5). Then the carboxyl-terminal processing to the mature D1 protein (32 kDa) occurs also in nonappressed membranes with a half-time of 3-6 min. After processing, the D1 protein is rapidly translocated along the membrane to the

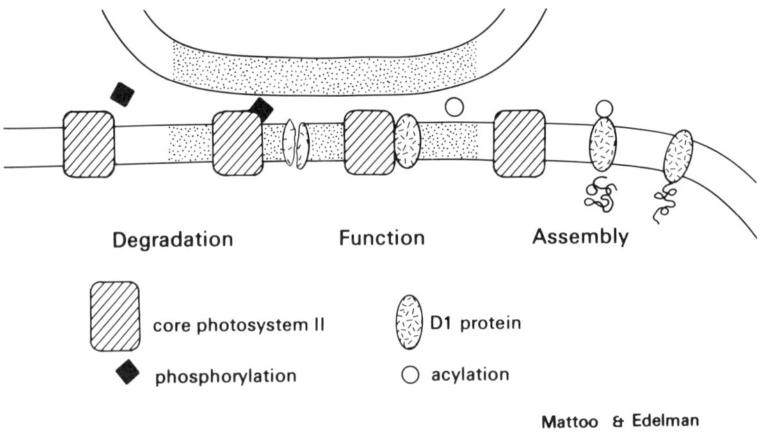

Figure 5. A scheme showing the lateral migration of the D1 protein between appressed and nonappressed membrane regions during turnover of PSII (Mattoo and Edelman, 1987; drawn by B. Andersson).

appressed grana membranes with a half-time of 6-12 min; thereafter it remains in the appressed membranes as an active PSII unit for 6-12 hours. Superimposed on this translocation is another event which involves a post-translational modification of the processed D1 protein: the D1 protein is acylated with palmitic acid (half-time of 3-6 min). Then deacylation is rapidly demonstrated in the appressed grana thylakoids.

Mattoo and Edelman (1987) suggest this protein modification might occur in the nonappressed region and serve as a tag for its rapid translocation to appressed domains. Callahan *et al.* (1987) suggest that the acylated protein is translocated to the appressed regions prior to being incorporated into the 32 kDa protein-depleted PSII units. Alternatively, the D1 protein-depleted units might themselves be translocated to the nonappressed regions, the acylated D1 protein become incorporated, and then the repaired PSII units be transported back to an appressed region.

The light intensity-dependent degradation of the D1 protein in *Spirodela oligorrhiza* occurs within the grana thylakoids to form a 23.5 kDa membrane-bound catabolic component (Greenberg *et al.*, 1987). Significantly, the loop-out region between the α-helices IV and V of the D1 protein contains the cleavage site. A 14 amino acid sequence (Arg 225 to Arg 238) contains the so-called PEST region, a region identified as the cleavage site in other eukaryotic proteins which are primed for rapid degradation (Rogers *et al.*, 1986). These elegant studies of Edelman and colleagues demonstrate that rapid dynamic changes in protein location are associated with synthesis of the D1 protein and perhaps other proteins. Callahan *et al.* (1987) have shown that the LHCII apoproteins, as well as other thylakoid proteins, are also post-translationally acylated with palmitic acid in the light. Since the post-translational acylation of viral and mammalian proteins have been implicated in processes such as protein processing and transport, assembly of multi-protein complexes or anchoring proteins into the membranes, it is likely that the light-stimulated post-translational acylation of thylakoid proteins will be significant in the dynamic shuttling of proteins and protein complexes between appressed and nonappressed regions. The maintenance and modulation of mature chloroplast thylakoids, necessitated by stress damage and long-term acclimation, clearly requires much dynamic shuttling.

ACCLIMATION OF PLANTS TO IRRADIANCE

The chloroplasts of plants adapted to low irradiance or shade have many more thylakoid membranes with huge granal stacks compared to sun or high irradiance-acclimated plants (Goodchild *et al.*, 1972). Given the marked lateral heterogeneity in the distribution of the multiprotein thylakoid complexes between appressed and nonappressed domains, these striking structural changes must mean changes in the relative amounts of thylakoid complexes and mobile electron transport carriers, and consequently in their function. Shade and low-irradiance-acclimated plants have lower Chl *a*/Chl *b* ratios; less of their total chlorophyll is associated with core PSII and core PSI units, and more with LHCII. Since LHCII mediates membrane stacking (Staehelin and Arntzen, 1983), the greater content of LHCII correlates with more appressed to nonappressed membranes in shade compared to sun plants.

Both the optimisation and balancing of light-harvesting between the photosystems are achieved by variations in the stoichiometries of P680 and P700, as well as in the light-harvesting antenna size of PSII (Anderson *et al.*, 1988). It is well-established that the Chl/P700 ratios of most plants are unaffected by the light environment in which the plants were grown (Björkman, 1981; Anderson and Osmond, 1987). Both shade and low irradiance-acclimated plants, however, have higher Chl/P680 ratios compared to sun plants (Wild *et al.*, 1986; Chow and Hope, 1987; Evans, 1987b; Anderson *et al.*, 1988): consequently, their PSII/PSI reaction centre ratios are also lower. Significantly, shade plants have fewer PSII units with larger light-harvesting antennae relative to PSI to allow for maximal light-harvesting under low irradiance. Since high rates of photosynthesis are not required, however, their lower rates of photosynthetic capacity are sustained by lower amounts of the cytochrome b/f complex, mobile electron carriers, ATP synthase and Rubisco on a chlorophyll basis.

On the other hand, under high irradiance where light is not limiting, the amounts of cytochrome b/f complex, mobile electron transport carriers, and ATP synthase per unit chlorophyll are increased so that the overall rates of electron transport and photophosphorylation are much greater. Both sun and high-irradiance-acclimated plants, have more PSII units with smaller light-harvesting antennae relative to PSI (Anderson *et al*,. 1988). The smaller number of pigment molecules associated with each P680 offsets the increased number of PSII centres, so that the total light-harvesting capacity of PSII balances that of PSI despite their PSII/PSI ratio being greater than 1 (Anderson *et al.*, 1988).

The coordinated and extremely well-regulated allocation of resources, especially proteins, to the processes of light-harvesting, electron transport, proton translocation, and carbon assimilation that are involved in acclimation are illustrated by dynamic transfer experiments (Davis *et al.*, 1986; Chow *et al.*, 1987a, b). When peas (previously acclimated to low irradiance) were transferred to a higher irradiance, the doubling in photosynthesis at CO_2 and light saturation is indeed closely associated with changes in most of the components of the photosynthetic apparatus. These changes include Chl *a*/Chl *b* ratios, and hence amount of chlorophyll-proteins, P680, cytochrome f, ATP synthase, and Rubisco, while the total Chl and Chl/P700 ratios are unchanged (Fig. 6). Although one week is required for the maximal changes to be induced, substantial increases take place in the 10 h light period following the initial transfer. While the acclimation of many sun plants to growth over a wide range of controlled irradiances has been studied (Björkman, 1981; Anderson and

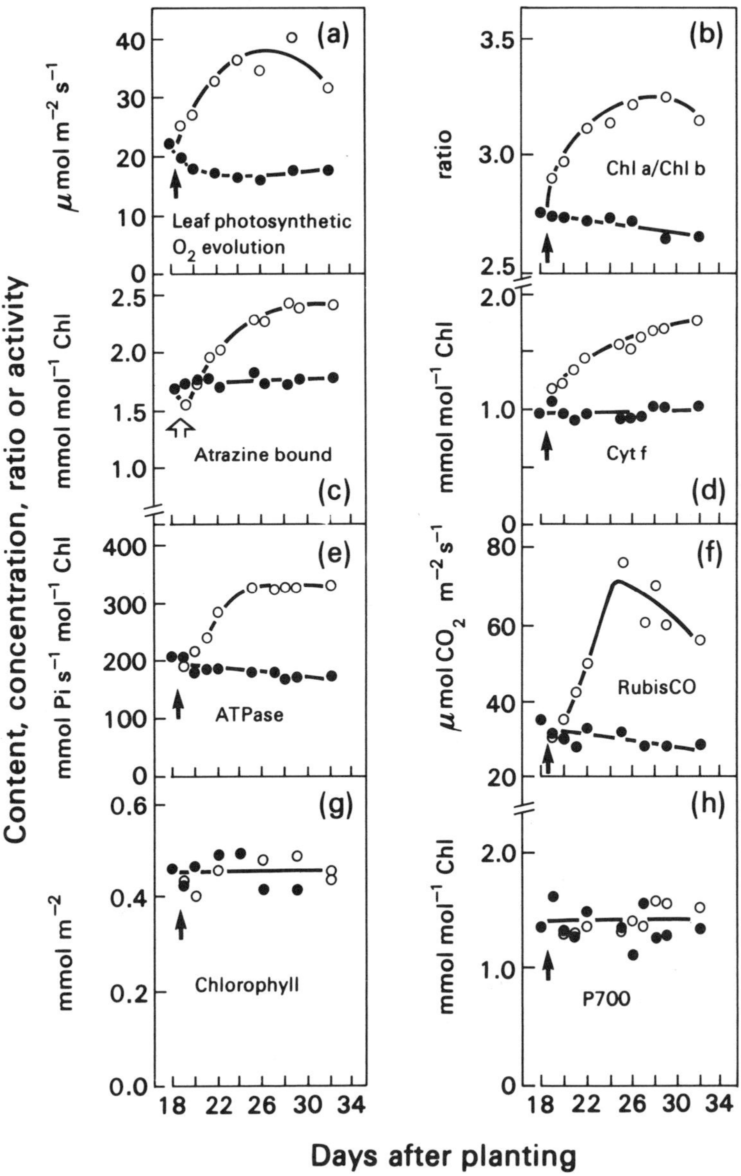

Figure 6. Acclimation of the photosynthetic apparatus of peas grown at low irradiance (60 μmol photons $m^{-2}s^{-1}$) for 19 days, and after transfer to high irradiance (390 μmol photons $m^{-2}s^{-1}$). •, low irradiance peas; O, acclimation to high irradiance (Chow and Anderson, 1987a,b).

Osmond, 1987), that of shade plants has been rather neglected. *Alocasia macrorrhiza* has sometimes been assumed to be an archetypal obligate shade species, yet it can grow in open sunny areas with apparently reduced quantum efficiency (Pearcy, 1987). Surprisingly, Chow *et al.* (1988) showed that *Alocasia* is capable of massive growth in high light environments. At the maximum growth irradiance studied, the photosynthetic capacity of *Alocasia* leaves (measured as the CO_2- and light-saturated rate of 0_2 evolution per unit leaf area) was very high (Fig. 7), and 4-fold greater than that sustained at the lowest growth irradiance studied. Concomitantly, with the increased photosynthetic capacities obtained with increasing irradiance during growth, the Rubisco activity per unit leaf area also increased markedly, while latent ATPase activity, cytochrome f and PSII centres (measured as atrazine binding) on a chlorophyll basis, increased to a lesser extent. While the Chl *a*/Chl *b* ratio increased substantially with increasing growth irradiance, the chlorophyll content per unit leaf area declined slightly; the Chl/P700 ratio was independent of growth irradiance (Fig. 3). These results show that given favorable nutrients and water supply, *Alocasia* is capable of rapid growth in high irradiance; the photosynthetic capacity attained was comparable with that of spinach.

Accompanying these coordinated changes in the composition and hence function of *Alocasia* chloroplasts, there were also marked changes in leaf anatomy and chloroplast ultrastructure (Chow *et al.*, 1988). At the low growth irradiance, the palisade cell chloroplasts which were preferentially located adjacent to the distal perclinal cell walls had very large granal stacks. At high irradiance, however, the palisade chloroplasts were mainly adjacent to the anticlinal cell walls; they had both smaller granal stacks and larger stromal space.

CONCLUDING REMARKS

Undoubtedly the thylakoid membranes of plant chloroplasts form a highly dynamic membrane network. We suggest that some cytochrome b/f complexes, PSI complexes, and ATP synthase are located in the curved grana margins, closer to PSIIα units than those located in the planar nonappressed membranes. Lateral heterogeneity in the distribution of the thylakoid supramolecular protein complexes between appressed and nonappressed domains is an operational definition. The many regulatory events occur within each chloroplast with varying time-frames. These events range from subtle conformational changes of the complexes which are associated with various energy dissipative processes to lateral migration of some components between appressed and nonappressed regions, that are involved in the regulation of excitation energy between the photosystems, and in repair mechanisms to stress-damaged proteins or

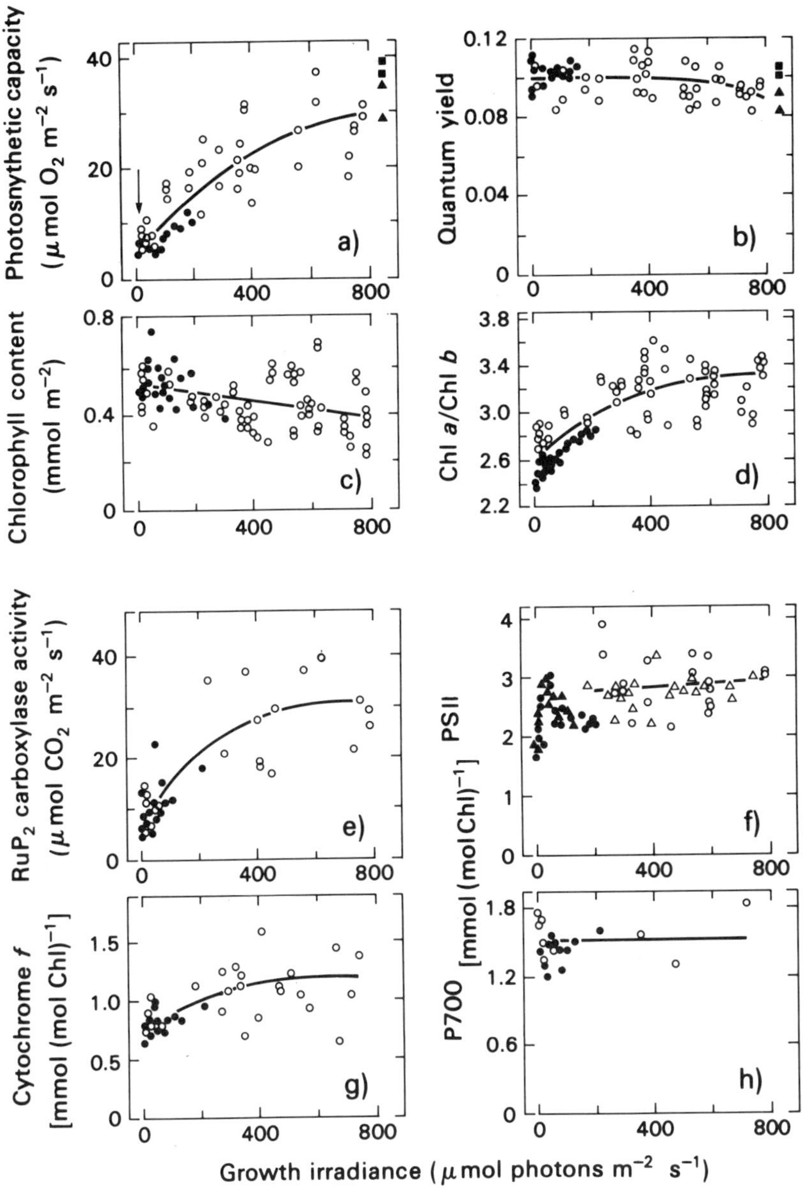

Figure 7. Composition and function of *Alocasia* leaves grown under varying irradiances of incandescent (•) or fluorescent with incandescent light (O). Quantum yields (b) are measured as mol O_2 (mol photons absorbed)$^{-1}$ (Chow *et al*., 1988).

pigments. Finally, on a longer-time scale, protein and lipid synthesis and degradation are involved in acclimation. This multiplicity of events requires a gigantic regulatory network involving light regulators, protein kinases, acylases, and so on which coordinate many scenarios with multiple players.

ACKNOWLEDGEMENTS

We are most grateful to our collaborators, B. Andersson, W.S. Chow, D.J. Goodchild, A. Melis, K.A. Platt-Aloia, and A.W. Webber who have contributed to our research and discussions. Grant support to WWT from NSF (DCB 8607765) is also acknowledged.

REFERENCES

Åkerlund H-E, Andersson B, Albertsson P-A (1976). Isolation of Photosystem II enriched membrane vesicles from spinach chloroplasts by phase partition. Biochim Biophys Acta 449:525-535.

Allred DR, Staehelin LA (1985). Lateral distribution of the cytochrome b6/f and coupling factor ATP synthetase complexes of chloroplast thylakoid membranes. Plant Physiol 78:199-202.

Anderson JM (1981). Consequences of spatial separation of Photosystem 1 and 2 in thylakoid membranes of higher plant chloroplasts. FEBS Lett 124:1-10.

Anderson JM (1982). Distribution of the cytochromes of spinach chloroplasts between the appressed membranes of grana stacks and stroma-exposed thylakoid regions. FEBS Lett 138:62-66.

Anderson JM (1987). Molecular organization of thylakoid membranes. In Amesz J (ed) "New Comprehensive Biochemistry Vol 15: Photosynthesis," Amsterdam: Elsevier, pp 273-297.

Anderson JM, Chow WS, Goodchild DJ (1988). Thylakoid membrane organization in sun/shade acclimation. Aust J Plant Physiol 15:11-21

Anderson JM, Goodchild DJ (1987). Lateral heterogeneity of thylakoid complexes and the transverse arrangement of the light-harvesting chlorophyll a/b-proteins of photosystem II of thylakoid membranes. Chem Scrip 27B:181-188.

Anderson JM, Melis A (1983). Localization of different photosystems in separate regions of chloroplast membranes. Proc Natl Acad Sci USA 80:745-749.

Anderson JM, Osmond CB (1987). Shade-sun responses: compromises between acclimation and photoinhibition. In Kyle DJ, Osmond CB, Arntzen CJ (eds) "Topics in Photosynthesis Vol 9: Photoinhibition," Amsterdam: Elsevier, pp 1-38.

Andersson B, Anderson JM (1980). Lateral heterogeneity in the distribution of chlorophyll-protein complexes of the thylakoid membranes of spinach chloroplasts. Biochim Biophys Acta 593:427-440.

Armond PA, Arntzen CJ (1977). Localization and characterization of photosystem II in grana and stroma lamellae. Plant Physiol 59:398-404.

Atta-Asafo-Adjei E, Dilley RA (1985). Plastocyanin stimulation of whole chain and photosystem I electron transport in inside-out thylakoid vesicles. Arch Biochem Biophys 243:660-667.

Barber J (1983). Photosynthetic electron transport in relation to thylakoid membrane composition and function. Plant Cell Envir 6:311-322.

Barber J (1985). Thylakoid membrane structure and organization of electron transport components. In Barber J, Baker NR (eds) "Topics in Photosynthesis: Photosynthetic Mechanisms and the Environment," Amsterdam: Elsevier, pp 91-134.

Bennett J (1983). Regulation of photosynthesis by reversible phosphorylation of the light-harvesting chlorophyll a/b-protein. Biochem J 212:1-13.

Björkman O (1981). Responses to different quantum flux densities. In Lange OL, Nobel PS, Osmond CB, Ziegler H (eds) "Encylopedia of Plant Physiology: Physiological Plant Ecology 1, New Series Vol 12A," Berlin: Springer-Verlag, pp 57-107.

Björkman O, Demmig B (1987). Photon yield of 0_2 evolution and chlorophyll fluorescence characteristics at 77 K among vascular plants of diverse origins. Planta 170:489-504.

Callahan FE, Edelman M, Mattoo AK (1987). Post-translational acylation and intra-thylakoid translocation of specific chloroplast proteins. In Biggin J (ed) "Progress in Photosynthesis Research Vol 3," Dordrecht: Martinus Nijhoff, pp 799-802.

Canaani O, Barber J, Malkin S (1984). Evidence that phosphorylation and dephosphorylation regulate the distribution of excitation energy between the two photosystems of photosynthesis *in vivo*: Photoacoustic and fluorimetric study of an intact leaf. Proc Natl Acad Sci USA 81:1614-1618.

Chow WS, Anderson JM (1987a). Photosynthetic responses of *Pisum sativum* to an increase in irradiance during growth. I. Photosynthetic activities. Aust J Plant Physiol 14:1-8.

Chow WS, Anderson JM (1987b). Photosynthetic responses of *Pisum sativum* to an increase in irradiance during growth. II. Thylakoid membrane components. Aust J Plant Physiol 14:9-19.

Chow WS, Hope AB (1987). The stoichiometries of supramolecular complexes in thylakoid membranes from spinach chloroplasts. Aust J Plant Physiol 14:21-28.

Chow WS, Qian L, Goodchild DJ, Anderson JM (1988). Photosynthetic acclimation of *Alocasia macrorrhiza* (L.) G Don to growth irradiance: structure, function and composition of chloroplasts. Aust J Plant Physiol 15:107-122.

Chow WS, Telfer A, Chapman DJ, Barber J (1981). State 1 - State 2 transition in leaves and its association with ATP-induced chlorophyll fluorescence quenching. Biochim Biophys Acta 638:60-68.

Davies EC, Chow WS, LeFay JM, Jordan BR (1986). Acclimation of tomato leaves to changes in light intensity: effects on the function of the thylakoid membrane. J Exp Bot 37:211-220.

Douce R, Block MA, Dorne A-J, Joyard J (1984). The plastid envelope membranes: their structure, composition, and role in chloroplast biogenesis. Subcell Biochem 10:1-84.

Ellis RJ (1981). Chloroplast proteins: synthesis, transport and assembly. Ann Rev Plant Physiol 32:111-137.

Evans JR (1987a). The dependence of quantum yield on wavelength and growth irradiance. Aust J Plant Physiol 14:69-79.

Evans JR (1987b). The relationship between electron transport components and photosynthetic capacity in pea leaves grown at different irradiances. Aust J Plant Physiol 14:157-170.

Ghirardi ML, Melis A (1983). Localization of photosynthetic electron transport components in mesophyll and bundle sheath chloroplasts of *Zea mays*. Arch Biochem Biophys 224:19-28.

Goodchild DJ, Anderson JM, Andersson B (1985a). Immunocytochemical localization of the cytochrome b/f-complex of chloroplast thylakoid membranes. Cell Biol Int Rep 9:715-721.

Goodchild DJ, Andersson B, Anderson JM (1985b). Immunocytochemical localization of polypeptides associated with the oxygen evolving system of photosynthesis. Eur J Cell Biol 36:294-298.

Goodchild DJ, Björkman O, Pyliotis NA (1972). Chloroplast ultrastructure, leaf anatomy and soluble protein in rainforest species. Carnegie Inst Wash Year Book 71:102-107.

Gounaris K, Brain ARR, Quinn PJ, Williams WP (1984). Structural reorganization of chloroplast thylakoid membranes in response to heat-stress. Biochim Biophys Acta 766:198-208.

Greenberg BM, Gaba V, Mattoo AK, Edelman M (1987). Identification of a primary *in vivo* degradation product of the rapidly-turning-over 32 kd protein of photosystem II. EMBO J 6:2865-2869.

Haehnel W (1984). Photosynthetic electron transport in higher plants. Ann Rev Plant Physiol 35:659-693.

Horton P (1985). Interactions between electron transfer and carbon assimilation. In Barber J, Baker NR (eds) "Topics in photosynthesis, Vol 6: Photosynthetic mechanisms and the environment," Amsterdam: Elsevier, pp 135-187.

Horton P (1987). Interplay between environmental and metabolic factors in the regulation of electron transport in higher plants. In Biggens J (ed) "Progress in Photosynthesis Research Vol II," Dordrecht: Martinus Nijhoff, pp 681-688.

Larsson UK, Andersson B (1985). Different degrees of phosphorylation and lateral mobility of two polypeptides belonging to the light-harvesting complex of Photosystem II. Biochim Biophys Acta 809:396-402.

Mattoo AK, Edelman M (1987). Intramembrane translocation and post-translational palmitoylation of the chloroplast 32-kDa herbicide-binding protein. Proc Natl Acad Sci USA 84:1497-1501.

Miller KR, Staehelin LA (1976). Analysis of the thylakoid outer surface. Coupling factor is limited to unstacked membrane regions. J Cell Biol 68:30-47.

Murphy DJ (1982). The importance of non-planar bilayer region in photosynthetic membranes and their stabilization by galactolipids. FEBS Lett 150:19-26.

Murphy DJ (1986). The molecular organization of the photosynthetic membranes of higher plants. Biochim Biophys Acta 864:33-94.

Paolillo DJ (1970). The three-dimensional arrangement of intergranal lamellae in chloroplasts. J Cell Sci 6:243-255.

Pearcy RW (1987). Photosynthetic responses of tropical forest trees. In Dooley J, Osmond CB, Wongkaew WC (eds) "Ecophysiology of Tropical Plants: Biotrop Special Publication no 24," Bogor: Biotrop, pp in press.

Rogers S, Wells R, Rechsteiner M (1986). Amino acid sequences common to rapidly degraded proteins: the PEST hypothesis. Science 234:364-368.

Sane PV, Goodchild DJ, Park RB (1970). Characterization of chloroplast photosystems I and II separated by a non-detergent method. Biochim Biophys Acta 216:162-178.

Schmidt GW, Mishkind ML (1986). The transport of proteins into chloroplasts. Ann Rev Biochem 55:879-912.

Shaw PJ, Henwood JA (1985). Immuno-gold localization of cytochrome f, light-harvesting complex, ATP synthase and ribulose 1,5-bisphosphate carboxylase/oxygenase. Planta 165:333-339.

Staehelin LA, Artzen CJ (1983). Regulation of chloroplast membrane function: protein phosphorylation changes the spatial organization of membrane components. J Cell Biol 97:1327-1337.

Sundby C, Andersson B (1985). Temperature-induced reversible migration along the thylakoid membrane of photosystem II regulates its association with LHCII. FEBS Lett 191:24-28.

Thielen APGM, Van Gorkom HJ (1981). Quantum efficiency and antenna size of photosystems IIα, IIβ and I in tobacco chloroplasts. Biochim Biophys Acta 635:111-120.

Thomson WW, Moeller CH (1983). Effects of Tween-20, polyoxyethylene sorbitan monolaurate on the ultrastructure and organization of chloroplast membranes. Protoplasma 114:173-178.

Vallon O, Høyer-Hansen G, Simpson DJ (1987). Photosystem II and cytochrome b-559 in the stroma lamellae of barley chloroplasts. Carlsberg Res Commun 52:405-421.

Vallon O, Wollman FA, Olive J (1985). Distribution of intrinsic and extrinsic subunits of the PSII protein complex between appressed and non-appressed regions of the thylakoid membrane: an immunocytochemical study. FEBS Lett 183:245-250.

Vallon O, Wollman FA, Olive J (1986). Lateral distribution of the main protein complexes of the photosynthetic apparatus in *Chlamydomonas reinhardtii* and in spinach: an immunocytochemical study using intact thylakoid membranes and a PSII-enriched membrane preparation. Photobiochem Photobiophys 12:203-220.

Webber AN, Platt-Aloia KA, Heath RL, Thomson WW (1988). The marginal regions of thylakoid membranes: A partial characterization by polyoxyethylene sorbitan monolaurate (Tween 20) solubilization of spinach thylakoids. Physiol Plant 72:288-297.

Weier TE (1961). The ultramicro structure of starch-free chloroplasts of fully expanded leaves of *Nicotiana rustica*. Am J Bot 48:615-630.

Weier TE, Thomson WW (1962). The grana of starch-free chloroplasts of *Nicotiana rustica*. J Cell Biol 13:89-108.

Whitmarsh J (1986). Mobile electron carriers in thylakoids. In Staehelin LA, Arntzen CJ (eds) "Encyclopedia of Plant Physiology, New Series Vol 19, Photosynthesis III," Berlin: Springer-Verlag, pp 508-527.

Wild A, Höpfner M, Rühle W, Richter M (1986). Changes in the stoichiometry of photosystem II components as an adaptive response to high-light and low-light conditions during growth. Z Naturforsch 41c:597-603.

Witt HT (1971). Coupling of quanta, electrons, fields, ions and phosphorylation in the functional membrane of photosynthesis. Results by pulse spectroscopic methods. Q Rev Biophys 4:365-477.

Photosynthesis, pages 183–205

THE ALLOCATION OF PROTEIN NITROGEN IN THE PHOTOSYNTHETIC APPARATUS: COSTS, CONSEQUENCES, AND CONTROL

John R. Evans and Jeffrey R. Seemann

Plant Environmental Biology Group, Research School of Biological Sciences, Australian National University, P.O. Box 475, Canberra, ACT 2601 Australia (JRE) and; Department of Biochemistry, University of Nevada, Reno, Nevada 89557 USA (JRS)

INTRODUCTION

Nitrogen is an essential component of all proteins. Because photosynthesis is the major physiological process in the leaves of plants, the nitrogen invested in Calvin cycle and thylakoid proteins represent the majority of leaf nitrogen (Fig. 1 and Chapin *et al.*, 1987; Evans, 1988b; Stocking and Ongun, 1962). The photosynthetic capacity of a leaf is thus strongly related to its nitrogen content, although the gain of this relationship may differ between species. In both natural and agricultural environments, nitrogen availability is often limiting the rate of leaf area development and hence the capacity for photosynthesis, plant growth, and reproductive output (Field and Mooney, 1986). For this reason, there have been many studies of nitrogen distribution in the photosynthetic apparatus and its effect on photosynthetic performance, from the ecological to molecular level.

Nitrogen associated with proteins of the photosynthetic apparatus is found in one of two major pools of nitrogen (Fig. 1). The first pool encompasses proteins involved in the light reactions of photosynthesis, and represents about one-quarter of leaf protein nitrogen in a C_3 sun leaf. This pool thus includes thylakoid membrane-bound proteins associated with light harvesting, electron transport and photophosphorylation. The second pool encompasses proteins associated with the photosynthetic carbon reduction cycle, or Calvin cycle, and also represents about one-quarter of leaf protein nitrogen. This pool thus includes soluble proteins involved in CO_2 assimilation, photorespiration, RuBP regeneration, and starch and sucrose synthesis. This chapter explores how protein nitrogen is allocated between thylakoid and CO_2 assimilation functions, the nitrogen costs of each of these two major nitrogen pools in the leaf, how

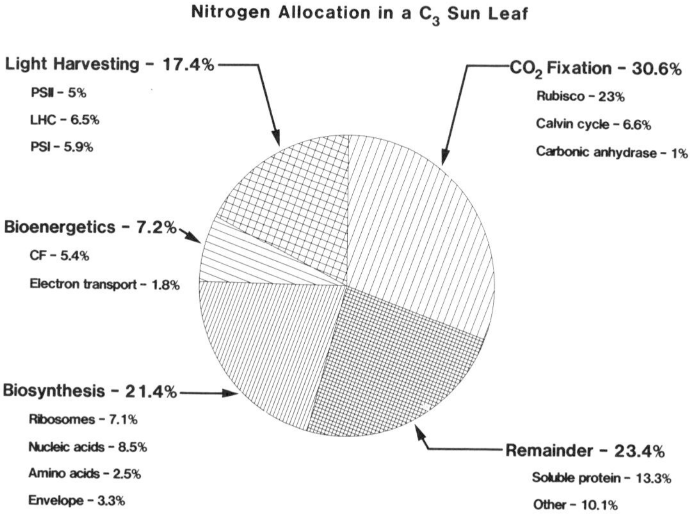

Figure 1. Allocation of protein nitrogen in a leaf of a typical C_3 sun plant. Light harvesting plus bioenergetics represents thylakoid-associated photosynthetic functions; CO_2 fixation represents all soluble protein-associated photosynthetic functions.

genetic and environmental factors may affect this allocation, and the impact of this allocation on photosynthesis and nitrogen use efficiency.

THYLAKOID PROTEIN NITROGEN

Light Harvesting

There are five major protein complexes in the chloroplast thylakoid membranes (Table I). The most prominent of these is the light-harvesting chlorophyll *a/b* protein complex (LHC), which contains at least half of the chlorophyll in the membranes. The reaction centers of photosystems I and II are the other two major chlorophyll containing complexes. The last two complexes, the cytochrome *b/f* complex and the coupling factor (CF), are involved in electron transport and photophosphorylation, respectively.

The complexes occur in different domains of the thylakoid membranes. The first complex that was found to be asymmetrically distributed in the membrane was the coupling factor (Miller and Staehelin, 1976). It is exclusively located in stroma-exposed membranes.

Table I. Components of the thylakoid membranes, their molecular weight (kD) and pigment content.

Complex/Components	PAGE MW	kD	mol N mol^{-1} Chl
Photosystem II			
~52 Chl *a*	47,43	56,51.8	
D2	34	39.5	
D1,HB	32	38.9	
Cyt b_{559}	9,4	9.4,4.4	
	22,7,6.5,5.5,51		
OEC, 2 copies of each	33,23,16,10	26.7	
TOTAL		380	83.5
Photosystem I - LHCI			
~40 Chl *a*	70,70	83.2,82.5	
	18,16		
SUB-TOTAL		200	
LHCI, 8 Chl *a*+*b*, 18 copies	20-25	22	
TOTAL		396	37.0
LHCII			
11 Chl *a*+*b*	25-28	24.5	
TOTAL		24.5	25.5
Cyt *b/f*			
cyt *f*	34	31.8	
cyt b_{563}, 2 copies	23	23	
Rieske Fe-S	20		
2 copies	17	17	
	5		
SUB-TOTAL		137	
Ferredoxin NADP Reductase		37	
SUB-TOTAL		37	
Coupling Factor		500	
TOTAL (*f*:FNR:CF=1:1:1.2)		774	8.85 mol N mmol^{-1} *f*

Subsequently, photosystem I has also been found, through the use of immunogold localization, to be exclusively located in stroma-exposed membranes (Vallon *et al.*, 1986; Anderson and Goodchild, 1987). Photosystem II and much of the light-harvesting chlorophyll *a/b* complex are found in the region of membrane appression which form the grana stacks (Goodchild *et al.*, 1985; Vallon *et al.*, 1986). The cytochrome *b/f* complex has been found in both regions (Allred and Staehelin, 1985; Shaw and Henwood, 1985). The lateral separation of photosystem II from photosystem I means that there can be no transfer of energy between them. The relative stoichiometry of the complexes depends upon growth conditions (see below). Determination of molecular weights from polyacrylamide gel electrophoresis frequently differ from that determined from the amino acid sequence. Consequently, the amino acid sequence has been used whenever possible to estimate the nitrogen costs. Obtaining the stoichiometries of pigment to protein is difficult, because the pigment-protein complex is not a covalent linkage. The values presented in Table I must thus be considered only as a best estimate and contain a considerable element of uncertainty.

The composition of the photosystem II complex (Table I) is taken from Ljungberg (1987) and contains more polypeptides than is usual. Two copies of each of the oxygen-evolving complex (OEC) polypeptides are assumed per reaction center. The chlorophyll content is derived from the chlorophyll distribution in green gels. The photosystem II complex contains about 126 Chl per 1000 total Chl, estimated from discontinuous SDS-PAGE (Leong and Anderson, 1984a) and 2.44 atrazine binding sites (Evans, 1987a), suggesting ~52 Chl per reaction center. Ghirardi *et al.* (1986) calculated that the core of photosystem II contained ~50 Chl in a Chl *b*-less barley mutant, while Thielen and van Gorkom (1981) obtained ~40 with tobacco aurea mutants. The nitrogen cost of photosystem II can be calculated (Table I), given that 52 Chl are complexed in 380 kD of protein: 83.5 mol N mol^{-1} Chl. This is higher than an estimate by Picaud *et al.* (1982) of 49.4 mol N mol^{-1} Chl, but when allowance was made for the probable contamination by LHC, their value increases to 70 mol N mol^{-1} Chl.

Estimation of the nitrogen cost of the photosystem I complex (Table I) is more speculative. Benghis and Nelson (1975) obtained 55 mol N mol^{-1} Chl for a photosystem I preparation while Picaud *et al.* (1982) found only 36 mol N mol^{-1} Chl. Since it is easy to lose pigment, the lower value has been favored. From PAGE, about 32% of the chlorophyll is associated with photosystem I bands, and given ~1.74 P700 per 1000 Chl, this means ~184 Chl per photosystem I. Benghis and Nelson (1975) found that ~20 Chl *a* were associated with each of the two 70 kD polypeptides, which leaves 144 Chl associated with the peripheral antenna and LHC associated with photosystem I. By assuming 8 Chl per LHCI monomer, and 18 copies, a nitrogen cost for photosystem I of 37 mol N mol^{-1} Chl is obtained. This is approximately that determined by Picaud *et al.* (1982). There is a marked similarity in the amino acid sequence between LHCII and LHCI, but LHCI has less Chl *b*. The necessity of constraining LHCI to 8 Chl per monomer suggests that it is 20% more costly than LHCII for complexing pigments.

The number of chlorophylls associated with each monomer of the major pigment-protein complex, LHCII, is still uncertain. The proteins are coded by a nuclear multigene family (Dunsmuir *et al.*, 1983) and different types are observed in the thylakoids. They differ in their apparent molecular weights and are differentially expressed during acclimation to the irradiance during growth (Larsson *et al.*, 1987). Mullet (1983) found 11 Chl per monomer, while 13 can be calculated from the data of Picaud *et al.* (1982) for a monomer of 25.7 kD.

The remainder of the thylakoid nitrogen is mainly associated with the cytochrome *b/f* complex, ferredoxin NADP reductase (FNR), and coupling factor (CF). All of these are known to vary with acclimation to the irradiance during growth (Terashima and Inoue, 1985). For peas, the ratio of CF activity to cytochrome *f* content was unaltered across growth irradiance (Evans, 1987a), whereas it increased with increasing irradiance in spinach (Terashima and Evans, 1988). If a 1:1:1.2 ratio is assumed for *f*:FNR:CF, the cost is approximately 8.85 mol N mol^{-1} cytochrome *f*.

To illustrate the nitrogen partitioning in the thylakoids, examples are given for pea and spinach leaves that were grown under high or low irradiance (Table II). When grown under low irradiance, the Chl *a/b* ratio declines due to an increase in the proportion of Chl associated with LHC at the expense, primarily, of photosystem II. This reduces the N-cost of light harvesting about 5%. The greater saving overall is due to the marked reduction in the content of the cytochrome *f* and CF complexes. The reasonable agreement between the calculated and measured N-costs means that if assumptions for one complex are subsequently found to need correcting, an equivalent change in another fraction will be necessary.

Absorption Spectra

The absolute absorption spectra for the three main Chl-protein complexes are shown in Fig. 2A. The complexes of the two photosystems contain predominantly Chl *a* and β-carotene, whereas LHC contains both

Table II. Nitrogen cost of thylakoids from leaves grown in low or high irradiance.

	Chl *a/b*	Proportion of Chl PSII	LHC	PSI	Light-harvesting a	Cyt *f* b	Bio-energetics c	Total d	e
N-cost		83.5	25.5	37.0	mol N/ mol Chl	8.85/*f*			mol N/ mol Chl
Spinacia									
High	2.80	.14	.56	.30	37.1	2.63	23.3	64.4	64±3
Low	2.42	.10	.62	.28	34.5	1.58	14.0	52.5	53±9
Pisum									
High	2.86	.13	.57	.30	36.5	1.89	16.7	57.2	58±2
Low	2.50	.10	.61	.28	34.6	1.05	09.3	47.9	44±1

a) N-cost of PSII, LHC and PSI proteins.
b) bioenergetics cost, 8.85 mol N mol^{-1} *f*, with the cyt *f* content mmol *f* mol^{-1} Chl.
c) N-cost of bioenergetics.
d) N-cost of pigment-protein complexes, plus bioenergetics plus 4 N Chl^{-1}.
e) observed N-cost (Evans, 1987a; Terashima and Evans, 1988).

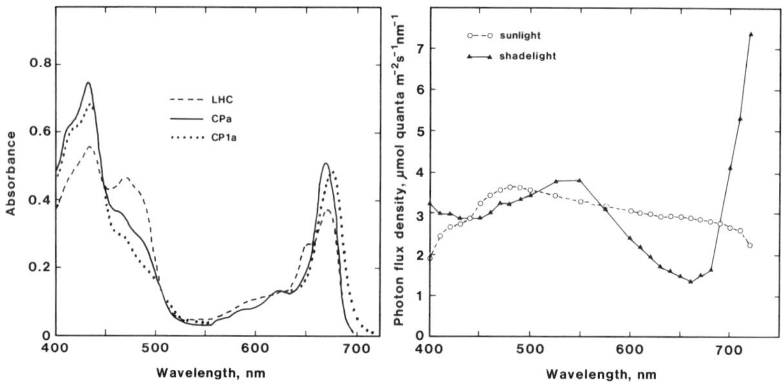

Figure 2. A) Absorption spectra for the three major chlorophyll-protein complexes. The spectra represent the *in vivo* absorption for 10 μM Chl solutions in a 1 cm cuvette (Evans and Anderson, 1987). Dashed line, LHC monomer; solid line, photosystem II complex; dotted line, photosystem I complex. B) Spectral distribution of sunlight (Lemon, 1963) and shade light at ground level in a tropical rainforest (Björkman and Ludlow, 1972). Both spectra give an irradiance of 1 mmol quanta m^{-2} s^{-1} between 400 and 720nm.

Chl *a* and *b* as well as 27 xanthophylls per 100 Chl (Siefermann-Harms, 1985). The Chl *b* and xanthophylls result in greater absorption between 460 and 500nm. The mean molar extinction coefficient from 400 to 720nm is similar for all three complexes, averaging 2230±40 m^2 mol^{-1} Chl (Evans and Anderson, 1987). Given the spectra for sunlight and shade light (Fig. 2B), it is possible to calculate the relative efficiencies of the three complexes when mixed in the appropriate ratio in solution (Table III). The enrichment of far-red light in the shade light versus the sunlight spectrum marginally increased the efficiency of the photosystem I complex. For both sun- and shade-light, LHC is slightly more efficient due to the relative advantage around 480nm. However, in all cases, the differences are small. The effect of path-lengthening due to the scattering of light within the leaf marginally favors photosystem II in the leaf (Table III, right column), but the clumping of pigment, which causes

Table III. Relative efficiencies of the three major chlorophyll-protein complexes. Efficiency was calculated as the proportion of the PAR absorbed divided by the proportion of total chlorophyll in the complex.

Spectrum	PSI	PSII	LHC	Leaf/chloroplast[1]
Sunlight	0.992	0.979	1.020	1.008
Shadelight	1.010	0.971	1.012	1.016

[1]Leaf/chloroplast represents the ratio of light absorbed by PSII/PSI in the leaf divided by the same ratio in the chloroplast (Evans, 1986a).

changes in the opposite direction and which are more difficult to quantify, would probably negate these differences. The proportion of light reaching each photosystem under natural conditions, whether it be sunlight, shade light, or sunfleck is thus virtually proportional to the fraction of chlorophyll associated with that photosystem and is nearly independent of the relative abundance of the pigment complexes (Evans, 1986a, 1988a).

From the comparison of the nitrogen costs, acclimation to low irradiance reduced the protein cost of pigment complexing by about 5%, but there would be little difference in the proportion of light absorbed. The performance of Chl *a/b*-containing complexes (Table II) compares favorably with Chl *a/c* and phycobilin complexes. The three complexes have N-costs of 35, 30 and 66-180 mol N mol^{-1} pigment (Raven, 1984), respectively. Their mean molar extinction coefficients are 2230, 2660 and 1320 m^2 mol^{-1} pigment respectively, so the phycobilins are even more expensive in terms of light capture. Thus while some green light is not absorbed by the Chl complexes in leaves, improving light absorption around 550nm by utilizing for example, phycoerythrin, would require a prohibitive amount of extra nitrogen. This is not the case for underwater environments where the red and violet ends of the spectrum are rapidly reduced with depth.

Electron Transport Capacity

The rate of electron transport or photophosphorylation per unit of Chl depends on the irradiance during growth (Boardman, 1977). The rates of oxygen evolution per unit Chl for high and (low)-light grown leaves were 81 (32) *Pisum*, 102 (54) *Spinacia*, 113 (26) *Cucumis* and 94 (57) *Phaseolus* (mmol O_2 mol^{-1} Chl s^{-1}). While there are increases in the number of photosystem II reaction centers for leaves grown under high irradiance, the more dramatic changes are in the cytochrome *f* and CF complexes. The rate of oxygen evolution is approximately proportional to the cytochrome *f* content (Fig. 3), making cytochrome *f* a good basis for expression of the N-cost of electron transport/photophosphorylation. The nitrogen involved in the non-pigment complexes of about 9 mol N mol^{-1} cytochrome *f* is small relative to the nitrogen cost of the proteins complexing the pigments (~40 mol N mol^{-1} Chl). Consequently, doubling the cytochrome *f* content from 1 to 2 mmol cyt *f* mol^{-1} Chl only increases the nitrogen cost of the thylakoids by 18%, but is associated with a doubling of the electron transport capacity.

CO_2 ASSIMILATION

Calvin Cycle Enzymes

The photosynthetic carbon reduction cycle consists of 12 enzymes, but it is dominated by the enzyme that catalyzes the fixation of CO_2, RuBP carboxylase (Rubisco). The properties of most of the enzymes have been determined (Table IV). Because of the reticulate nature of the cycle, not all steps proceed at the same rate. Three of the reactions proceed at less than their potential maximal rates. Rubisco has a low affinity for CO_2 so that at 300 μbar p(CO_2), the rate of carboxylation is only one third of the potential Vmax. Since Rubisco also catalyzes the oxygenation of RuBP, the actual rate of RuBP consumption at 300 μbar p(CO_2) is about 56% of Vmax. Triose-P-isomerase and Ru5P kinase operate at approximately 20 and 33% of Vmax, respectively, due to low levels of substrate *in vivo* (Robinson and Walker, 1981). When the observed activities are compared with the expected activities, there is close agreement for all the major enzymes with the exceptions of an excess of PGA kinase and a deficiency

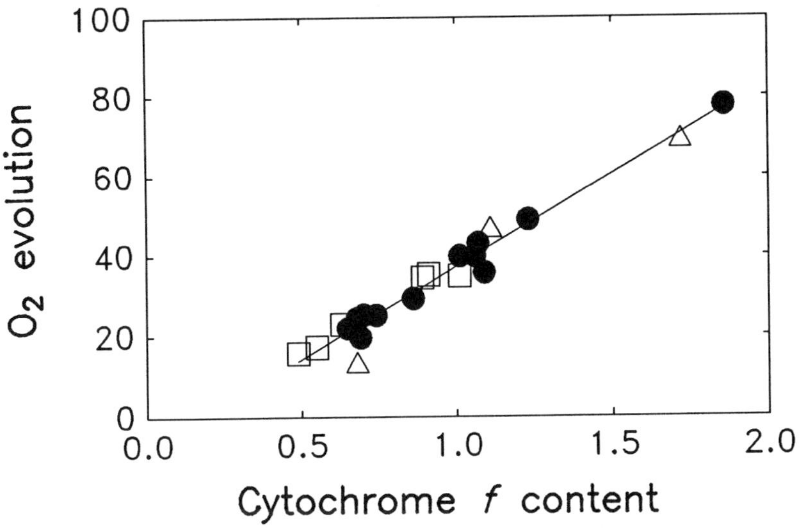

Figure 3. The relationship between the light and CO_2 saturated rate of oxygen evolution (μmol m^{-2} s^{-1}) and the cytochrome *f* content (μmol m^{-2}). Circles, *Spinacia* (Terashima and Evans, 1988); squares, *Pisum* (Evans, 1987a); triangles, *Atriplex patula* (Björkman *et al.*, 1972). The rate of CO_2 assimilation was divided by 0.56 to convert it to the rate of oxygen evolution. See also Holloway *et al.* (1983).

of SBPase. Woodrow (1982) found that SBPase constituted about 0.6% of the soluble protein, but with the low specific activity, this would be insufficient to maintain the observed rates of RuBP regeneration. The specific activity is probably an underestimate such that the SBPase in reality would constitute less than 2% of the leaf nitrogen.

The protein cost was calculated from the rate and specific activity using both the required and observed rates, and were virtually identical (64 mol N mol^{-1} Chl). The cost is similar to the protein cost of the thylakoids acclimated to high irradiance (Table II).

Rubisco/Mesophyll Conductance

Strong relationships between Rubisco and the rate of CO_2 assimilation have been demonstrated for several species (Fig. 4A,B,C), particularly when the *in vivo* activity of Rubisco is determined by measurement of the initial slope of the photosynthesis vs intercellular CO_2 partial pressure response (mesophyll conductance). In some species, the relationship between the *in vivo* rate and Rubisco activity appears to decline at high

Table IV. Calvin cycle enzymes, their properties and relative amounts.

Enzyme	MW 1	Vmax 2	F 3	Vreq 4	O/R 5	Sp.Act 6	Amount 7	 8	%N 9
Rubisco	550	250	1	241a	1.04	3.2	4520	4690	20.4
PGA kinase	47	1250	2	270	4.63	900	18	83	0.36
GAP DHD	140	375	2	270	1.39	130	125	174	0.76
Triose P Isom.	53	1500	.83	560b	2.68	750e	45	121	0.53
F1,6 aldolase	150	75d	.5	67.5	1.11	44	92	102	0.44
F1,6 Pase	160	37.5	.5	67.5	0.56	110	37	20	0.16
Transketolase	140	75d	1	135	0.56	150	54	30	0.23
Aldolase	142	75d	.33	44.6	1.68	25	107	180	0.78
SBPase	73	6.3	.33	44.6	0.14	5	535	76	2.32
R5P epimerase	46	375d	.67	90.5	4.14	1200e	4.5	19	0.08
R5P isomerase	54	750	.33	44.6	16.8	2170	1.2	20	0.09
Ru5P kinase	46	625	1	409c	1.53	320	77	118	0.51
TOTAL							5616	5633	26.7

a) 56% of Vmax - assuming 300 μbar $p(CO_2)$, (pc - G)/(pc + Kc (1 + 0/Ko)) * (pc + 7G/3)/(pc - G) * 4.5/4, pc=300, G=40 μbar, Ko=369 mbar, O=210 mbar (Caemmerer and Farquhar, 1981).
b) 20% of Vmax
c) 33% of Vmax
d) could be underestimates
e) assumed

1) kD.
2) mmol mol^{-1} Chl s^{-1}, observed Vmax (data of Robinson and Walker (1981) and Kelly *et al.* (1976).
3) factor relative to RuBP turnover.4) mmol mol^{-1} Chl s^{-1}, rate required to regenerate 135 mmol RuBP mol^{-1} Chl s^{-1}.
5) observed Vmax / required Vmax.
6) specific activity, μmol mg^{-1} min^{-1} (data of Robinson and Walker, 1981; Kelly *et al.*, 1976; Evans and Seemann, 1984).
7) required amount, g mol^{-1} Chl, col 4/col 6 x 60.
8) observed amount, g mol^{-1} Chl, col 2/col 6 x 60.
9) % of total nitrogen, sun plants on average have 3.8 mmol Chl mol^{-1} N (Evans, 1988b).

Rubisco contents (Fig 4A). This decline in the efficiency of Rubisco use has been attributed to a decline in the $p(CO_2)$ at the site of carboxylation

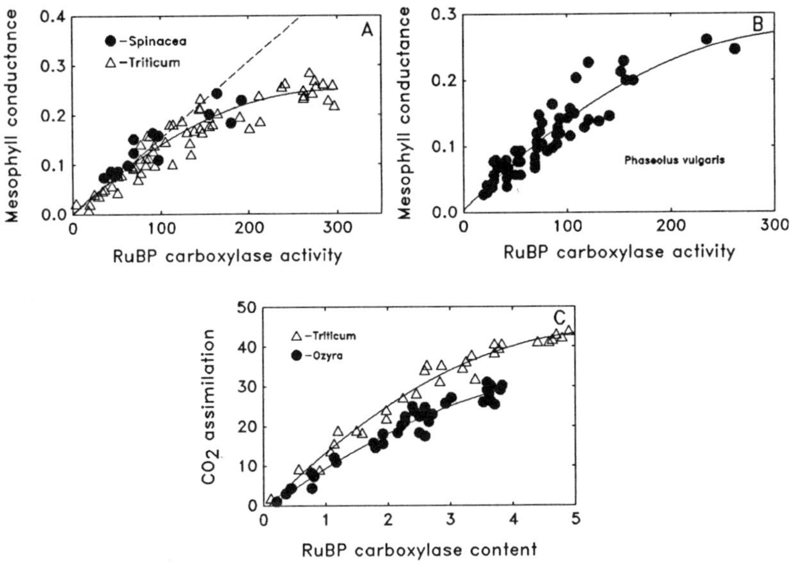

Figure 4. A) The relationship between the mesophyll conductance (the slope of the curve relating the rate of CO_2 assimilation to intercellular $p(CO_2)$ near the CO_2 compensation point) ($mol\ m^{-2}\ s^{-1}\ bar^{-1}$) and the Rubisco activity ($\mu mol\ m^{-2}\ s^{-1}$). Triangles, *Triticum* (Evans, 1983); circles, *Spinacia* (Evans and Terashima, 1988). B) *Phaseolus* (Caemmerer and Farquhar, 1981). C) The rate of CO_2 assimilation versus Rubisco content ($g\ m^{-2}$). Triangles, *Triticum*; circles, *Oryza* (Makino *et al.*, 1988). The rice photosynthesis data has been reduced by 10% to account for the consistently higher intercellular $p(CO_2)$ than was observed for the wheat.

relative to the intercellular $p(CO_2)$ calculated from the gas exchange. For Fig. 4C, the relationship is between the rate of CO_2 assimilation with normal ambient $p(CO_2)$, which therefore also includes variation due to stomatal conductance and Rubisco content. For both the wheat and rice, the curvature was mainly due to declining intercellular $p(CO_2)$ at higher Rubisco contents. The difference between the curves for the wheat and rice is due to different specific activities of the Rubisco enzyme from the two species (Makino *et al.*, 1988). To achieve rates of CO_2 assimilation similar to wheat, rice partitions relatively more of the total leaf nitrogen into Rubisco and operates with a higher intercellular $p(CO_2)$. Variation in Rubisco specific activity has also been seen in other species comparisons (Seemann *et al.*, 1981, 1982, 1984, 1987). Differences in the specific activity of Rubisco between C_3 and C_4 species (Seemann *et al.*, 1984) is also an important component in the significantly higher nitrogen use efficiency of photosynthesis in C_4 than C_3 species (Brown, 1978; Sage *et al.*, 1987).

ADAPTATION AND ACCLIMATION TO IRRADIANCE

Plant species are often genetically predisposed (adapted) for growth over a specific range of irradiance. These so-called sun and shade plants may also possess the capacity to respond to differences in irradiance within the irradiance range which they are adapted to grow (acclimation). Both adaptation and acclimation to different irradiances involve numerous changes in the morphology, physiology and biochemistry of the plant (for reviews see Anderson, 1986; Björkman, 1981). Changes in the form and functioning of the photosynthetic apparatus are part of the long term adjustment to growth at different irradiances. Acclimation of the photosynthetic apparatus involves changes in the levels of carbon reduction cycle enzymes, electron transport components, and pigment-protein complexes associated with light harvesting (Besford, 1986; Chow and Anderson, 1987a, b; Davies *et al.*, 1986a, b; Evans, 1987a; Ferrar and Osmond, 1986; Leong and Anderson, 1984a, b; Terashima and Evans, 1988;). This acclimation is often characterized by a redistribution of resources among these components of the photosynthetic apparatus (Evans, 1988a).

Light Harvesting

Perhaps the most notable characteristic of leaves growing in low irradiance is the relatively lower chlorophyll *a/b* ratio (Egle, 1960). With the advent of non-denaturing systems for separating the pigment-protein complexes (Anderson *et al.*, 1978), it became possible to examine how the composition of the thylakoids changed. While the morphology of chloroplasts in leaves acclimated to low irradiance changes, it never reaches the extremes evident in shade plants (Anderson *et al.*, 1973; Chow *et al.*, 1988). However, the pigment-protein complexes show similar relative abundances in shade plants and plants acclimated to low irradiance.

Chlorophyll *b* and xanthophylls are only complexed by the light-harvesting complexes, which contain more than half of the chlorophyll in the membranes. The lowering of the Chl *a/b* ratio is caused by a relative increase in the LHC content at the expense of the photosystem II complex and, to a lesser extent, the photosystem I complex (Table II). The reduction in the pigment content in the photosystem II complex is accompanied by an equivalent decline in the herbicide binding sites, indicating that there are fewer photosystem II reaction centers in thylakoids isolated from leaves grown under low irradiance. Consequently, the antenna size of photosystem II increases from 200 to 250 Chl as relatively more LHC becomes associated with each reaction center. In contrast, the reduction in pigment associated with photosystem

I is not accompanied by any changes in the P700 content (Chow and Hope, 1987; Evans, 1987a; Terashima and Evans, 1988). It presumably arises from a decline in the LHCI content at low irradiance. The antenna of photosystem I must therefore be supplemented by relatively more LHCII to maintain the balance in pigment distribution between the two photosystems. The balance is inferred from the fact that the quantum yield of photosynthesis is independent of the irradiance during growth and the pigment composition of the leaves (Björkman *et al.*, 1972; Evans, 1987b).

Leaf Absorptance

The amount of light absorbed by pigments in leaves is far greater than if the pigment were in free solution due to path lengthening. Light passing through the leaf is rapidly scattered, particularly by the spongy mesophyll cells (Terashima and Saeki, 1983). The absorptance of white light by leaves with increasing Chl content is shown in Fig. 5. Even with severe nitrogen deficiency, spinach and wheat leaves absorbed around 75% of the white light. With abundant nitrogen, leaves generally absorb

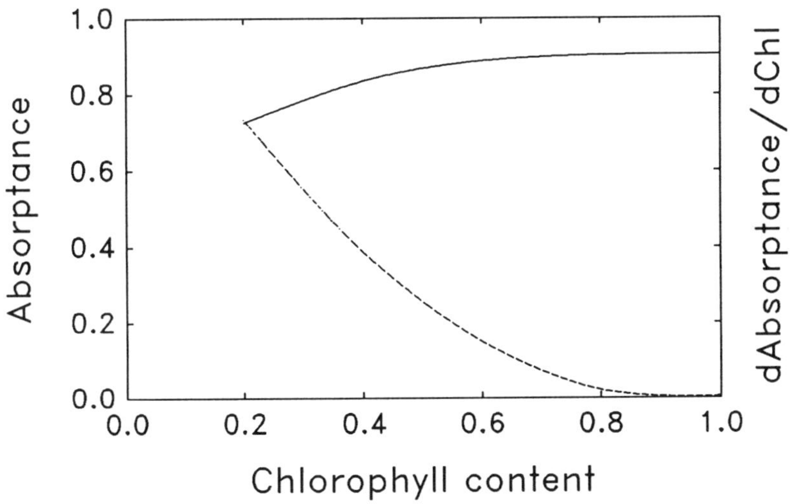

Figure 5. The relationship between leaf chlorophyll content (mmol m^{-2}) and absorptance of white light. The dashed line represents the incremental change in leaf absorptance for a change in chlorophyll content.

85-90% of the light and leaves of shade-species can even approach 95%. The increase in leaf absorptance for each increment in Chl is also shown in Fig. 5. While the gains are very small, they are directly reflected in the apparent quantum yield. Consequently, for plants growing at low irradiance, the small increases are of considerable benefit. Much larger benefits are seen at high irradiance because the electron transport capacity of the leaf is directly proportional to the Chl content (Evans and Terashima, 1987). The return on each additional increase in electron transport capacity decreases only slightly at high irradiance (Fig. 6).

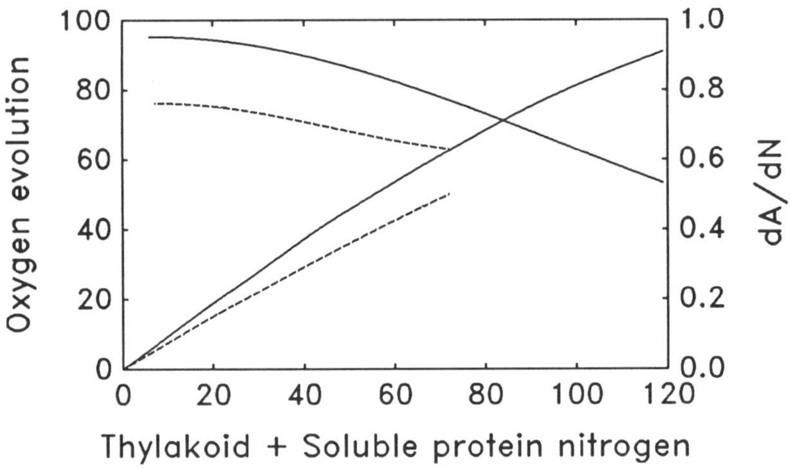

Figure 6. The rate of oxygen evolution (μmol m^{-2} s^{-1}) at 2000 μmol quanta m^{-2} s^{-1} as a function of increasing nitrogen in the thylakoids and soluble protein (mmol m^{-2}). The two curves represent leaves acclimated to high (solid lines) or low (dashed lines) irradiance. Also shown are the incremental gains in the rate of oxygen evolution for each change in nitrogen content (mmol O_2 mol^{-1} N s^{-1}). Data for *Spinacia* from Terashima and Evans (1988).

Electron Transport

While acclimation to lower irradiance reduced the protein cost of light absorption (Table II), it was accompanied by a lower electron transport capacity per unit of chlorophyll. That is, the protein cost of electron transport is greater for leaves acclimated to low irradiance (Fig. 6; Evans, 1987a). The curvature in the response arises from the constraint that the rate of oxygen evolution is measured at an irradiance of 2000 μmol quanta m^{-2} s^{-1}. The higher the nitrogen and hence chlorophyll content, the higher the irradiance required to reach the same point on the irradiance response curve (Evans and Terashima, 1987). The curvature is not great, so the gain in the rate of photosynthesis for each increment in nitrogen drops slowly.

There is a trade-off between performance at low versus high irradiance, such that only one can be maximized. When the relative performance of a leaf acclimated to low irradiance is compared to a leaf acclimated to high irradiance, both leaves containing equal Chl contents, the trade-off is clearly apparent (Fig. 7). The irradiance at which the cross-over occurs depends on the basis for comparison. If the electron transport capacity per unit of thylakoid nitrogen is compared, the cross-over occurs around 220 μmol quanta m^{-2} s^{-1}. However, to utilize the extra electron transport capacity, greater amounts of the Calvin cycle enzymes are needed. When these are included, the cross-over does not occur until nearly 700 μmol quanta m^{-2} s^{-1}. Generally leaves that develop under low irradiance have less nitrogen for the same Chl content. Since there is always a fraction of leaf nitrogen that is 'structural', or not directly involved in photosynthetic capacity, and which is not greatly reduced under low irradiance, when the performance is compared on the basis of whole leaf nitrogen, the cross-over occurs around 400 μmol quanta m^{-2} s^{-1}.

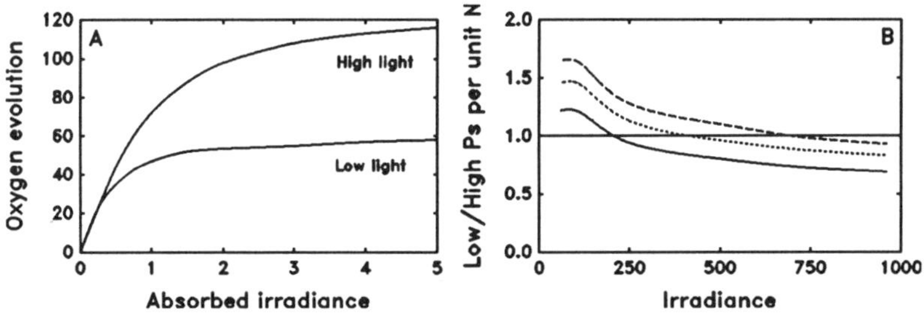

Figure 7. A) The rate of oxygen evolution (μmol O_2 mol^{-1} Chl s^{-1}) as a function of absorbed irradiance (mol quanta mol^{-1} Chl s^{-1}) for leaves grown under high or low irradiance. B) The relative performance of photosynthesis per unit nitrogen in the low versus the high light leaf a function of irradiance (μmol quanta m^{-2} s^{-1}). Solid line, on the basis of thylakoid nitrogen; dashed line, on the basis of thylakoid and soluble protein nitrogen; dotted line, on the basis of total leaf nitrogen. Calculated from Evans and Terashima (1987) and Terashima and Evans (1988).

Rubisco

Changes in the activity of Rubisco are well correlated with changes in photosynthetic capacity which occur in response to the irradiance during growth (Björkman, 1981). There are strong linear relationships between the nitrogen content of the leaf and the Rubisco content in many species (Evans, 1988b). There is usually a non-zero intercept such that Rubisco is not present below a certain minimum leaf nitrogen content (usually determined by extrapolation of data from senescing leaves.). The consequence of the non-zero intercept is that the proportion of leaf nitrogen present in Rubisco increases with increasing nitrogen content per unit leaf area. It is important to distinguish between this and the difference between species in the proportion of nitrogen present in Rubisco for a given leaf nitrogen content. For example, the relationship between Rubisco and leaf nitrogen was independent of the irradiance during growth in *Phaseolus* and *Alocasia*, but the shade plant *Alocasia* had a lower proportion of leaf nitrogen in Rubisco (10% versus 19% for *Phaseolus* with 80 mmol N m^{-2}; Seemann *et al.*, 1987). Generally, shade plants have a lower commitment to Calvin cycle enzymes and electron transport capacity to enable a greater fraction of protein to be committed to the pigment-protein complexes. The electron transport capacity and Rubisco activity are generally closely coupled so that if acclimation reduces the electron transport capacity, less protein is found in the Calvin cycle enzymes (Caemmerer and Farquhar, 1981, 1984; Makino *et al.*, 1983; Evans and Terashima, 1988).

In the short term, Rubisco activity is regulated in response to irradiance. The activity of Rubisco increases with irradiance (Caemmerer and Edmondson, 1986; Kobza and Seemann, 1988; Mott *et al.*, 1984) and this is reflected in the mesophyll conductance (Evans, 1986b). This light-dependent regulation potentially involves changes in the extent of CO_2/Mg^{2+} carbamylation of Rubisco, the level of the tight-binding inhibitor, 2-carboxyarabinitol 1-phosphate, and the degree of noncatalytic binding of RuBP to the enzyme (Brooks and Portis, 1988; Kobza and Seemann, 1988). Light acclimation appears not to involve changes in the light response of Rubisco regulation (J.R. Seemann, unpublished). The light response of Rubisco activation and inhibitor metabolism were found to be quantitatively the same in the sun species *Glycine max* and the shade species *Alocasia macrorrhiza*, despite the fact that photosynthesis was light saturated in *Alocasia macrorrhiza* at a substantially lower PFD than *Glycine max*. An increase in the ratio of Rubisco capacity to electron transport capacity has been found in several other species following acclimation to lower irradiance (Evans, 1988a). The nitrogen use efficiency of photosynthesis is also reduced in *Alocasia* relative to sun species by the apparently lower specific activity of Rubisco *in vivo* and an apparent RuBP regeneration limitation on photosynthesis. All of these factors contribute to the low nitrogen use efficiency of photosynthesis in *Alocasia macrorrhiza*, which is typical of shade species in general (Evans, 1988b). This extreme inefficiency of N use in shade plants seems to support the suggestion by Evans (1988b) that nitrogen may not be a limiting factor for growth in low light environments, as it often is in high light ones (Field and Mooney, 1986). However, both sun and shade species possess some capacity to conserve the rate of photosynthesis per unit Rubisco following acclimation.

LEAF NITROGEN BUDGETS

It is now possible to construct a general picture of the nitrogen-partitioning within the leaf (Fig. 8). For leaves that develop under high irradiance, approximately 50% of the total nitrogen is present as soluble protein. The thylakoids account for approximately 25% and nucleic acids about 8% of total nitrogen. At low irradiance, the leaves generally have lower nitrogen contents but proportion relatively more towards the thylakoids (~30%). The costs involved in leaf structure and nucleic acids are approximately the same, so that the proportion of total leaf nitrogen for the latter increases to 15%.

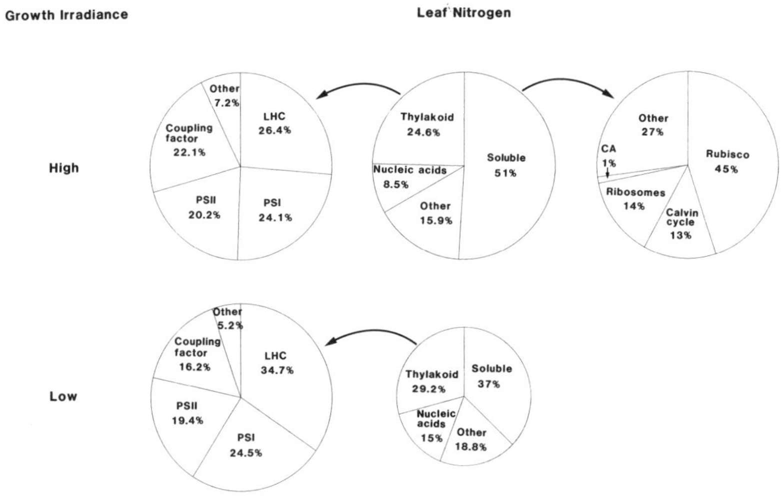

Figure 8. Leaf nitrogen budgets. The two circles in the center represent the nitrogen contents of leaves grown under high or low irradiance, containing 107 and 60 mmol N m^{-2}, respectively. On the left-hand side are represented the nitrogen-containing components of the thylakoids, and on the right-hand side, the soluble protein components.

The changes which occur in the thylakoid fraction are detailed on the left hand side of Fig. 8. At high irradiance, 70% of the thylakoid nitrogen is associated with the pigment-protein complexes. This increases to nearly 80% at low irradiance due to the increase in LHC at the expense of the CF, cytochrome *f* and photosystem II complexes. Despite the 44% reduction in nitrogen per unit leaf area, the altered partitioning results in only a 17% reduction in Chl content. Although many species increase their nitrogen partitioning towards thylakoids when grown at low irradiance, there are notable exceptions; e.g. *Pisum* (Evans, 1987a) and *Spinacia* (Terashima and Evans, 1988).

The soluble protein fractions are shown on the right hand side of Fig. 8. Approximately 45% of the soluble protein is Rubisco while the other Calvin cycle enzymes represent only 13%. Ribosomal protein also represents about 12% (Dean and Leech, 1982; Kirk and Tilney Bassett, 1978) with carbonic anhydrase and Rubisco activase about 3%. In young pea leaves, about 5% of the soluble protein is in the large subunit binding protein involved in the assembly of Rubisco (Hemmingren and Ellis, 1986). The changes that occur with acclimation to low irradiance vary with species, and thus only one circle is shown. The proportion of soluble protein that Rubisco represented was independent of growth irradiance in *Spinacia* (Terashima and Evans, 1988) and *Phaseolus* (Seemann *et al.*, 1987), but declined with decreasing irradiance in *Atriplex patula* (Medina, 1971), *Solidago* (Björkman, 1968), and *Alocasia* (Seemann *et al.*, 1987).

REFERENCES

Allred DR, Staehelin LA (1986). Spatial organization of the cytochrome *b-f* complex within chloroplast thylakoid membranes. Biochim Biophys Acta 849:94-103.

Anderson JM (1986). Photoregulation of the composition, function and structure of thylakoid membranes. Annu Rev Plant Physiol 37:93-136.

Anderson JM, Goodchild DJ (1987). Lateral heterogeneity of thylakoid complexes and the transverse arrangement of the light-harvesting chlorophyll *a/b* protein of photosystem II of thylakoid membranes. Chemica Scripta 27B:181-188.

Anderson JM, Goodchild DJ, Boardman NK (1973). Composition of the photosystems and chloroplast structure in extreme shade plants. Biochim Biophys Acta 325:573-585.

Anderson JM, Waldron JC, Thorne SE (1978). Chlorophyll-protein complexes of spinach and barley thylakoids. Spectral characterization of six complexes resolved by an improved electrophoretic procedure. FEBS Lett 92:227-233.

Benghis C, Nelson N (1975). Purification and properties of the photosystem I reaction center from chloroplasts. J Biol Chem 250:2783-2788.

Besford RT (1986). Changes in some Calvin cycle enzymes of the tomato during acclimation to irradiance. J Exp Bot 37:200-210.

Björkman O (1968). Further studies on differentiation of photosynthetic properties in sun and shade ecotypes of *Solidago virgaurea.* Physiol Plant 21:84-99.

Björkman O (1981). Responses to different quantum flux densities. In Lange OL, Nobel PS, Osmond CB, Zeigler H (eds): "Encyclopedia of Plant Physiology, New Series, Vol 12A," Berlin: Springer-Verlag, pp 57-107.

Björkman O, Boardman NK, Anderson JM, Thorne SW, Goodchild DJ, Pyliotis NK (1972). Effect of light intensity during growth of *Atriplex patula* on the capacity of photosynthetic reactions, chloroplast components and structure. Carnegie Inst Wash Yrbk 71:115-135.

Björkman O, Ludlow MM (1972). Characterization of the light climate on the floor of a Queensland rainforest. Carnegie Inst Wash Yrbk 71:85-94.

Boardman NK (1977). Comparative photosynthesis of sun and shade plants. Annu Rev Plant Physiol 28:355-377.

Brooks A, Portis AR (1988). Protein-bound ribulose bisphosphate correlates with deactivation of ribulose bisphosphate carboxylase in leaves. Plant Physiol 87:244-249.

Brown RH (1978). A difference in N use efficiency in C_3 and C_4 plants and its implications in adaptation and evolution. Crop Sci 18:93-98.

Caemmerer S von, Edmondson DL (1986). Relationship between steady-state gas exchange, *in vivo* ribulose bisphosphate carboxylase activity and some carbon reduction cycle intermediates in *Raphanus sativum.* Aust J Plant Physiol 13:669-688.

Caemmerer S von, Farquhar GD (1981). Some relationships between the biochemistry of photosynthesis and the gas exchange of leaves. Planta 153:376-387.

Caemmerer S von, Farquhar GD (1984). Effect of partial defoliation, changes of irradiance during growth, short-term water-stress and growth at enhanced $p(CO_2)$ on the photosynthetic capacity of leaves of *Phaseolus vulgaris* L. Planta 160:320-329.

Chapin FS, Bloom AJ, Field CB, Waring RH (1987). Plant responses to multiple environmental factors. BioScience 37:49-57.

Chow WS, Anderson JM (1987a). Photosynthetic responses of *Pisum sativum* to an increase in irradiance during growth. I. Photosynthetic activities. Aust J Plant Physiol 14:1-8.

Chow WS, Anderson JM (1987b). Photosynthetic responses of *Pisum sativum* to an increase in irradiance during growth. II Thylakoid membrane components. Aust J Plant Physiol 14:9-19.

Chow WS, Hope AB (1987). The stoichiometries of supramolecular complexes in thylakoid membranes from spinach chloroplasts. Aust J Plant Physiol 14:21-28.

Chow WS, Qian L, Goodchild DJ Anderson JM (1988). Photosynthetic acclimation of *Alocasia macrorrhiza* L. to growth irradiance: structure, function and composition of chloroplasts. In Caemmerer S von, Evans

JR, Adams WW III (eds): "Ecology of Photosynthesis in Sun and Shade," Melbourne: CSIRO, pp 107-122.

Davies EC, WS Chow, LeFay JM, BR Jordan (1986a). Acclimation of tomato leaves to changes in light intensity; effects on the function of the thylakoid membrane. J Exp Bot 37:211-220.

Davies EC, Chow WS, Jordan BR (1986b). A study of factors which regulate the membrane appression of lettuce thylakoids in relation to irradiance. Photosynth Res 9:359-370.

Dean C, Leech RM (1982). Genome expression during normal leaf development. 1. Cellular and chloroplast numbers and DNA, RNA and protein levels in tissues of different ages within a seven-day-old wheat leaf. Plant Physiol 69:904-910.

Dunsmuir P, Smith SM, Bedbrook J (1983). The major chlorophyll *a/b* binding protein of *Petunia* is composed of several polypeptides encoded by a number of distinct genes. J Mol Appl Genet 2:285-300.

Egle K (1960). Menge und verhaltais der pimente. In Ruhland W (ed): "Encyclopedia of Plant Physiology," Springer-Verlag, 5:452-496.

Evans JR (1983). Photosynthesis and nitrogen partitioning in leaves of *Triticum aestivum* and related species. Ph.D thesis, Australian National University.

Evans JR (1986a). A quantitative analysis of light distribution between the two photosystems, considering variation in both the relative amounts of the chlorophyll-protein complexes and the spectral quality of light. Photobiochem Photobiophys 10:135-147.

Evans JR (1986b). The relationship between carbon-dioxide-limited photosynthetic rate and ribulose 1,5-bisphosphate carboxylase content in two nuclear-cytoplasm substitution lines of wheat, and the coordination of ribulose-bisphosphate carboxylation and electron-transport capacities. Planta 167:351-358.

Evans JR (1987a). The relationship between electron transport components and photosynthetic capacity in pea leaves grown at different irradiances. Aust J Plant Physiol 14:157-170.

Evans JR (1987b). The dependence of quantum yield on wavelength and growth irradiance. Aust J Plant Physiol 14:69-79.

Evans JR (1988a). Acclimation by the thylakoid membranes to growth irradiance and the partitioning of nitrogen between soluble and thylakoid proteins. In Caemmerer S von, Evans JR, Adams WW III (eds): "Ecology of Photosynthesis in Sun and Shade," Melbourne: CSIRO, pp 93-106.

Evans JR (1988b). Photosynthesis and nitrogen relationships in leaves of C_3 plants. Oecologia, in press.

Evans JR, Anderson JM (1987). Absolute absorption spectra for the five major chlorophyll-protein complexes and their 77K fluorescence excitation spectra. Biochim Biophys Acta 892:759-765.

Evans JR, Seemann JR (1984). Differences between wheat genotypes in specific activity of ribulose 1,5-bisphosphate carboxylase and the relationship to photosynthesis. Plant Physiol 74:759-765.

Evans JR, Terashima I (1987). Effects of nitrogen nutrition on electron transport components and photosynthesis in spinach. Aust J Plant Physiol 14:59-68.

Evans JR, Terashima I (1988). Photosynthetic characteristics of spinach leaves grown with different nitrogen treatments. Plant Cell Physiol 29:157-165.

Ferrar PJ, Osmond CB (1986). Nitrogen supply as a factor influencing photoinhibition and photosynthetic acclimation after transfer of shade grown *Solanum dulcamara* to high light. Planta 168:563-570.

Field C, HA Mooney (1986). The photosynthesis-nitrogen relationship in wild plants. In Givnish TJ (ed): "On the Economy of Plant Form and Function," London: Cambridge University Press, pp 25-55.

Ghirardi ML, McCauley SW, Melis A (1986). Photochemical apparatus organization in the thylakoid membrane of *Hordeum vulgare* wild type and chlorophyll *b*-less chlorina f2 mutant. Biochim Biophys Acta 851:331-339.

Goodchild DJ, Andersson B, Anderson JM (1985). Immunocytochemical localization of polypeptides associated with the oxygen evolving system of photosynthesis. Eur J Cell Biol 36:294-298.

Hemmingren SM, Ellis RJ (1986). Purification and properties of ribulosebisphosphate carboxylase large subunit binding protein. Plant Physiol 80:269-276.

Holloway PJ, Maclean DJ, Scott KJ (1983). Rate-limiting steps of electron transport in chloroplasts during ontogeny and senescence of barley. Plant Physiol 72:795-801.

Kelly GJ, Latzko E, Gibbs M (1976). Regulatory aspects of photosynthetic carbon metabolism. Annu Rev Plant Physiol 27:181-205.

Kirk JTO, Tilney Bassett RAE (1978). The plastids, their chemistry, structure, growth and inheritance. Elsevier/Nth Holland Biomedical Press, 2nd Ed., Amsterdam.

Kobza J, Seemann JR (1988). Mechanisms for light-dependent regulation of ribulose 1,5-bisphosphate carboxylase activity and photosynthesis in intact leaves. Proc Natl Acad Sci USA 85:3815-3819.

Larsson UK, Anderson JM, Andersson B (1987). Variations in the relative content of the peripheral and inner light-harvesting chlorophyll *a/b*-protein complex (LHCII) subpopulations during thylakoid light adaptation and development. Biochim Biophys Acta 894:69-75.

Lemon E (1963). Energy and water balance of plant communities. In Evans LT (ed): "Environmental Control of Plant Growth," New York: Academic Press, pp 55-77.

Leong T-Y, Anderson JM (1984a). Adaptation of the thylakoid membranes of pea chloroplasts to light intensities. I. Study on the distribution of chlorophyll-protein complexes. Photosyn Res 5:105-115.

Leong T-Y, Anderson JM (1984b). Adaptation of the thylakoid membranes of pea chloroplasts to light intensities. II. Regulation of electron transport capacities, electron carriers, coupling factor (CF_1) activity and rates of photosynthesis. Photosynth Res 5:117-128.

Ljungberg U (1986). Identification and investigation of photosystem II polypeptides. Ph.D thesis, Lund University, Lund, Sweden.

Makino A, Mae T, Ohira K (1983). Photosynthesis and ribulose 1,5-bisphosphate carboxylase in rice leaves. Plant Physiol 73:1002-1007.

Makino A, Mae T, Ohira K (1988). Differences between wheat and rice in the enzymic properties of ribulose-1,5-bisphosphate carboxylase/oxygenase and the relationship to photosynthetic gas exchange. Planta 174:30-38.

Medina E (1971). Effect of nitrogen supply and light intensity during growth on the photosynthetic capacity and carboxydismutase activity of leaves of *Atriplex patula* ssp. hastata. Carnegie Inst Wash Year Book 70:551-559.

Miller KR, Staehelin LA (1976). Analysis of the thylakoid outer surface. Coupling factor is limited to unstacked membrane regions. J Cell Biol 68:30-47.

Mott KA, Jensen RG, O'Leary JW, Berry JA (1984). Photosynthesis and ribulose 1,5-bisphosphate concentration in intact leaves of *Xanthium strumarium* L. Plant Physiol 76:968-971.

Mullet JE (1983). The amino acid sequence of the polypeptide segment which regulates membrane adhesion (grana stacking) in chloroplasts. J Biol Chem 258:9941-9948.

Picaud A, Acker S, Duranton J (1982). A single step separation of PSI, PS2 and chlorophyll antenna particles from spinach chloroplasts. Photosynth Res 3:203-213.

Raven JA (1984). A cost benefit analysis of photon absorption by photosynthetic unicells. New Phytol 98:593-625.

Robinson SP, Walker DA (1981). Photosynthetic carbon reduction cycle (Calvin cycle) (including fluxes and regulation). In Stumpf PK, Conn EE (eds): "The Biochemistry of Plants. A Comprehensive Treatise, Vol. 8 (Photosynthesis)," London: Academic Press, pp 193-226.

Sage RF, Pearcy RW, Seemann JR (1987). The nitrogen use efficiency of C_3 and C_4 plants. III. Leaf nitrogen effects on the activity of carboxylating enzymes in *Chenopodium album* (L.) and *Amaranthus retroflexus* (L.). Plant Physiol 85:355-359.

Seemann JR, Badger MR, Berry JA (1984). Variations in specific activity of ribulose-1,5-bisphosphate carboxylase between species utilizing different photosynthetic pathways. Plant Physiol 74:791-794.

Seemann JR, Berry JA (1982). Interspecific differences in the kinetic properties of RuBPCase protein. Carnegie Inst Wash Year Book 81:78-83.

Seemann JR, Sharkey TD, Wang J, Osmond CB (1987). Environmental effects on photosynthesis, nitrogen-use-efficiency, and metabolite pools in leaves of sun and shade plants. Plant Physiol 84:796-802.

Seemann JR, Tepperman JM, Berry JA (1981). The relationship between photosynthetic performance and the levels and kinetic properties of RuBP carboxylase-oxygenase from desert winter annuals. Carnegie Inst Wash Year Book 80:67-72.

Shaw PJ, Henwood JA (1985). Immuno-gold localization of cytochrome *f*, light-harvesting complex, ATP synthase and ribulose 1,5-bisphosphate carboxylase/oxygenase. Planta 165:333-339.

Siefermann-Harms D (1985). Carotenoids in photosynthesis. I. Location in photosynthetic membranes and light-harvesting function. Biochim Biophys Acta 811:325-355.

Stocking CR, Ongun A (1962). The intracellular distribution of some metallic elements in leaves. Am J Bot 49:284-289.

Terashima I, JR Evans (1988). Effects of light and nitrogen nutrition on the organization of the photosynthetic apparatus in spinach. Plant Cell Physiol 29:143-155.

Terashima I, Saeki T (1983). Light environment within a leaf. I. Optical properties of paradermal sections of *Camellia* leaves with special reference to differences in the optical properties of palisade and spongy tissues. Plant Cell Physiol 24:1493-1501.

Terashima I, Inoue Y (1985). Palisade tissue chloroplasts and spongy tissue chloroplasts in spinach: biochemical and ultrastructural differences. Plant Cell Physiol 26:63-75.

Thielen APGM, Van Gorkom NJ (1981). Quantum efficiency and antenna size of photosystem IIa, IIb and I in tobacco chloroplasts. Biochim Biophys Acta 635:111-120.

Vallon O, Wollman FA, Olive J (1986) Lateral distribution of the main protein complexes of the photosynthetic apparatus in *Chlamydomonas reinhardtii* and in spinach: an immunocytochemical study using intact thylakoid membranes and a PSII enriched membrane preparation. Photobiochem Photobiophys 12:203-220.

Woodrow IE (1982). Sedoheptulose-1,7-bisphosphatase from wheat chloroplasts. In Wood WA (ed): "Methods in Enzymology, Vol 90," Academic Press, pp 392-396.

Photosynthesis, pages 207–226

PRODUCTIVE STRUCTURE OF A LEAF

Ichiro Terashima

Plant Environmental Biology Group, Research School of Biological Sciences, The Australian National University, P.O. Box 475, Canberra, A.C.T. 2601, Australia. **Present address**: Department of Botany, Faculty of Science, University of Tokyo, Hongo, Bunkyo-ku, Tokyo 113, Japan

INTRODUCTION

In this chapter, I describe some structural aspects of leaf photosynthesis. The present approach is similar to that used in studying "productive structure" of canopies (c.f. Shul'gin *et al.*, 1978). Since the canopy photosynthesis can be considered as an integration of the photosynthesis performed by each leaf in the canopy (Boysen-Jensen, 1932; Monsi and Saeki, 1953), environmental factors such as light and CO_2, and photosynthetic capacity of the leaves in the different positions in the canopy have been studied as fundamentals. Analogously, in this chapter, I describe light and CO_2 environments and the sun-shade adaptational gradient of photosynthetic capacity of chloroplasts in a single leaf. The consequences of these factors in leaf photosynthesis are discussed.

I. LIGHT ENVIRONMENT WITHIN A LEAF

Dorsiventral (bifacial) leaves receive most light on the adaxial (upper) surface. Thus in this section, the light environment within the typical dorsiventral leaf which is illuminated from the upper side is considered. More general treatments of optical properties of leaves and leaf tissues are found in recent reviews (Gates, 1980; Fukshansky, 1981; Seyfried and Fukshansky, 1983; Hodáňová, 1985; Vogelman, 1986).

In relation to photosynthesis, absorption of light by the mesophyll is most important. The mesophyll in typical dorsiventral leaves is composed of palisade and spongy tissues. These tissues in *Camellia japonica* L. leaves showed different optical properties (Terashima and Saeki, 1983). Figure 1 shows attenuance of light by paradermal sections. The change in slope of lines means that, on a chlorophyll basis, spongy tissue attenuates more light than palisade tissue. The difference in slope

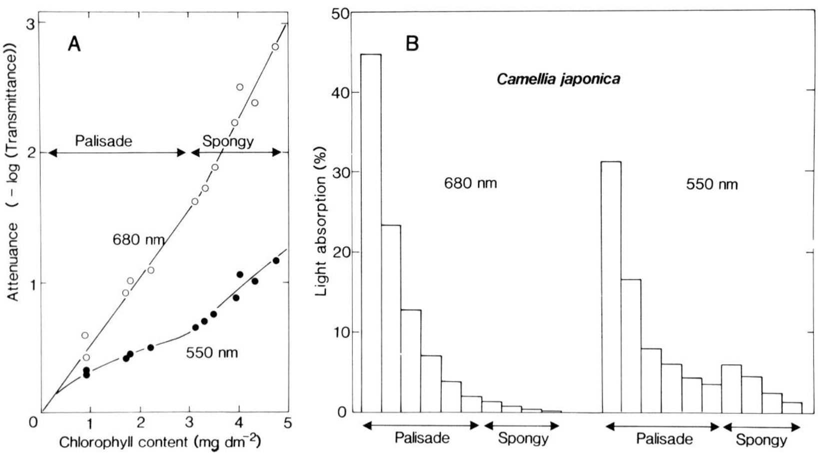

Figure 1. Attenuation of light by paradermal sections (A) and the proportion of light absorption in mesophyll layers (B) of a *Camellia japonica* L. leaf. Various thicknesses of abaxial parts were cut off from the leaf segments using a sliding microtome. Attenuance of adaxial sections was measured with a Shimadzu MPS-5000 spectrophotometer using a collimated measuring beam. Subsequently, chlorophyll contents of the sections were measured. The sections which contained less than 3 mg chlorophyll dm^{-2} had the upper epidermis and palisade tissue, while the thicker sections contained spongy tissue as well as the upper epidermis and the palisade tissue. The proportion was calculated using the data of transmittance and reflectance measured in paradermal sections. Each layer has the same amount of chlorophyll (0.5 mg dm^{-2}). Redrawn from Terashima and Saeki (1983, 1985).

decreased considerably after the infiltration of the air spaces with an oil having a comparable refractive index to that of cells. Thus, the greater attenuation on a chlorophyll basis in spongy tissue is largely due to the greater scattering and increased pathlength in spongy tissue. Chloroplasts are usually appressed to the cell walls. Since most cell walls of palisade tissue cells are parallel to the direction of incident light, the effective area for light interception by chloroplasts is much smaller in palisade tissue than in spongy tissue. Thus, the difference in attenuation on a chlorophyll basis may be partly attributed to the difference in the orientation of chloroplasts. Reflectance was also larger in the spongy tissue than in the palisade tissue (Terashima and Saeki, 1983). Greater scattering in the spongy tissue has been also shown in cotyledons of *Cucurbita pepo* L. using a micro optical fibre method (Vogelman *et al.*, 1988).

Using the data of transmittance (e.g. Fig. 1A) and reflectance, the gradient of light absorption in the *C. japonica* leaf was calculated (Fig.

1B, Terashima and Saeki, 1985). In this approach, multiple reflection was taken into account (Seyfried and Fukshansky, 1983). The gradient is expressed by the proportion of light absorption in 10 mesophyll layers each of which has 0.5 mg chlorophyll dm^{-2} (~ 0.056 mmol chlorophyll m^{-2}). The majority of light is absorbed in a few adaxial layers and light absorption in abaxial layers is small. However, it should be noted that at 550 nm, light absorptions in the 7th and 8th layers (top part of the spongy tissue) are larger than those in the 5th and 6th layers (bottom part of the palisade tissue). This is because, on a unit chlorophyll basis, spongy tissue attenuates more light than palisade tissue especially at wavelengths of weak absorption such as 550 nm (Fig. 1A).

The results in Fig. 1B suggest that the gradient of light absorption is moderated by having the two optically different tissues. The existence of the spongy tissue also seems to contribute to minimizing the loss of photons which enter into the leaf.

Light gradients can be measured more directly using fine optical fibres inserted into plant materials. The refinement of this technique has enabled the measurement of the light gradient in thin leaves (Vogelman, 1986; Vogelman *et al.*, 1988; Bornmann and Vogelman, 1988). The extensive use of this technique in studies on intra-leaf light environment in relation to photosynthesis is awaited.

The other problem of the light environment within a leaf is chloroplast movement. In opaque leaves of land plants, this problem has been approached by monitoring changes of transmittance (Inoue and Shibata, 1973, 1974; Gabryś and Walczak, 1980; Walczak and Gabryś, 1980; Hirata *et al.*, 1983; Chlobowski, 1986). However, the modes of chloroplast movement such as photodinesis (cyclosis), redistribution, and orientational movement (Haupt, 1986) cannot be distinguished by this method. Given the intensive scattering of light in spongy tissue, chloroplast movement in spongy tissue would not change the transmittance markedly. Thus, the transmittance change may be mainly attributed to chloroplast movement in palisade tissue where light is more collimated (Terashima and Saeki, 1983; Vogelman *et al.*, 1988). If this hypothesis is true, the absorbance change just below the palisade tissue can be marked even when the transmittance change by the whole leaf is small.

It has been shown that leaf transmittance changes with irradiance (Fig. 2). At high irradiance, transmittance increases. Thus, at high irradiance, chloroplast movement may also contribute to the homogenization of light environment. The micro-fibre technique can also be applied in this kind of study.

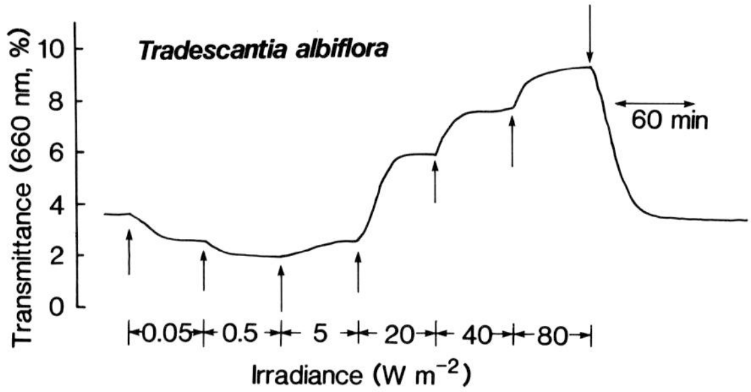

Figure 2. Transmittance change in a *Tradescantia albiflora* Kunth leaf. Transmittance of modulated light at 660 nm was measured under the various irradiances of white light. Note relatively slow responses. Since the scattering is not intense at the wavelengths of strong absorption such as 660 nm, the change of transmittance is more marked than that at the wavelengths of weak absorption or that for the overall photosynthetically active radiation. Redrawn from Gabryś and Walczak (1980).

II. INTRA-LEAF SUN/SHADE ADAPTATION OF CHLOROPLASTS

Although the differentiation of mesophyll into two tissues and the chloroplast movement contribute to moderating the intra-leaf gradient of light, the light gradient is still steep. The extent of the light gradient in some leaves can be equivalent to that found through a closed forest canopy. Therefore, adaptation of chloroplasts to the local light environment in a single leaf is expected.

Recently, the intra-leaf differentiation of chloroplasts has been shown in many species in various ways such as examination of chloroplast anatomy (Ballantine and Forde, 1970; Skene, 1974; Maróti and Gabor, 1976; Fischer and Evert, 1982; Terashima and Inoue, 1985a,b; Terashima and Evans, 1988) and the comparison of fluorescence induction between the two leaf sides (Schreiber *et al.*, 1977; Kulandaivelu *et al.*, 1983). We performed detailed biochemical studied with thylakoids isolated from paradermal sections (Terashima and Inoue, 1984, 1985a,b). The differentiation of chloroplasts occurs in essentially the same way as the so-called sun/shade adaptation (Boardman, 1977; Björkman, 1981; Anderson, 1986; Anderson and Osmond, 1987). On a chlorophyll basis, the photosynthetic rate at light saturation is higher in chloroplasts isolated from the adaxial side than those from the abaxial side (Fig. 3). Grana in chloroplasts in the adaxial part are smaller and have fewer thylakoids than grana in chloroplasts in the abaxial part. Since the intra-leaf sun/shade adaptation of chloroplasts has already been introduced as an example of so-called sun/shade adaptation syndrome (Anderson, 1986; Anderson

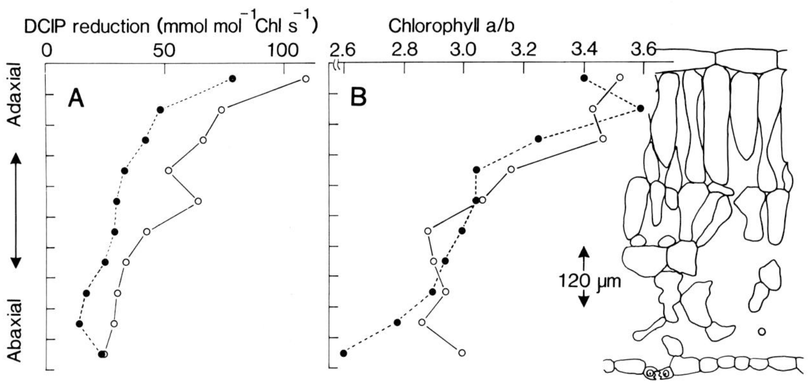

Figure 3. Light-saturated rate of whole chain electron transport (H_2O → dichlorophenol-indophenol, measured at pH 8.0 in the presence of 20 mM methylamine) (A), and chlorophyll a/b ratio (B) in thylakoids isolated from ten serial paradermal sections of *Spinacia oleracea* L. leaf segments. Redrawn from Terashima and Inoue (1985b, •), and Terashima and Takenaka (1986, o).

and Osmond, 1987), I shall only discuss some ambiguous aspects of intra-leaf sun/shade differentiation of chloroplasts.

It is still unclear whether there is a gradient of chloroplast properties in a single cell, although the light gradient in one cell layer is large enough to induce a change. Skene (1974) counted the number of thylakoids per granum of three largest grana in each chloroplast at various positions in leaves of *Malus malus* Britt. (apple) and concluded that the gradient is continuous even in single cells. Based on the biochemical analyses of 10 paradermal sections from *Spinacia oleracea* L. leaf segments (Fig. 3), Terashima and Inoue (1985b) also claimed a continuous gradient across the leaf.

However, when the number of thylakoids per granum for all grana in each chloroplast was counted in *S. oleracea* L. and *Glycine max* (L.) Merrill leaves, the mean thylakoid number for a chloroplast was similar within a single cell but markedly different among cell layers (Fig. 4). Given that morphology of grana is a good indicator for the sun/shade adaptation, the results in Fig. 4 would indicate that sun/shade adaptation of chloroplasts occurs *mainly* at the cellular level (Terashima *et al*., 1986). This may be either due to cellular control of chloroplast development, and/or rapid chloroplast circulation to average light environment.

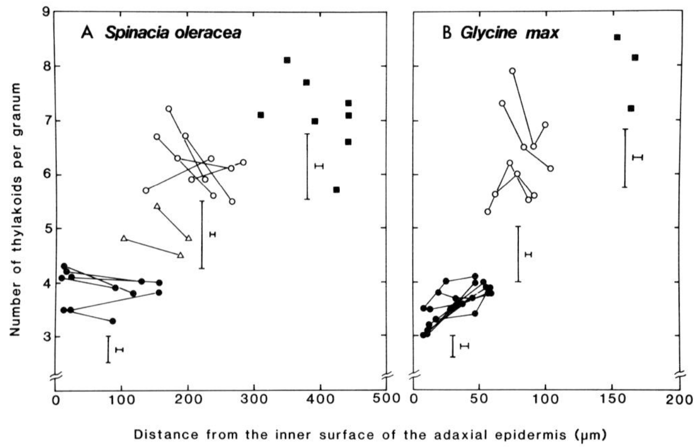

Figure 4. Number of thylakoids per granum (TN) of chloroplasts in a *Spinacia oleracea* L. leaf (A) and a *Glycine max* (L.) Merrill leaf (B). Each symbol represents mean TN for 3 or 4 chloroplasts at similar positions. Each chloroplast section had ca. 30 grana. Each line joining the symbols represents the individual cells. The different symbols represent different cell layers: •, 1st cell layer of palisade tissue; o, 2nd cell layer of palisade tissue; ■, spongy tissue; and Δ, intermediate cells in *S. oleracea*. Vertical bars indicate mean standard deviations of TN for each chloroplast. Redrawn from Terashima *et al.* (1986).

The most serious suspicion is, however, that the chloroplasts that have similar grana morphology, may have different photosynthetic activities (W.S. Chow, personal communication). In our biochemical studies with 10 serial sections of *S. oleracea* leaf segments, the top two fractions were probably from the first cell layer of palisade tissue despite the irregular border between the first and the second cell layers. Between these two sections, the marked decrease in light-saturated electron transport rate contrasted with somewhat constant chlorophyll a/b ratio. Since chlorophyll a/b ratio is an indicator for the relative amount of light harvesting chlorophyll a/b protein complex which plays an important role in grana stacking, constant chlorophyll a/b ratio may be related to the similar grana morphology. Thus, data in Fig. 4 may indicate that the suspicion is justified and the intra-cellular gradient should be re-examined using more direct techniques. The possibility of the rapid circulation of chloroplasts should also be examined.

If the formation of the intra-leaf gradient in chloroplast properties is regulated by the intra-leaf local light environment, artificial change of the intra-leaf light gradient should change the chloroplast gradient. When the first leaves of *G. max* were illuminated from the abaxial side for 8 days during the period of leaf expansion, the gradient in thylakoid number was completely inverted (Terashima *et al.*, 1986). The shape of the cells was also slightly changed. When illumination from the abaxial side was commenced after full expansion, only a slight inversion of the number of thylakoids per granum was observed. These results suggest that the formation of the intra-leaf gradient of chloroplast properties is most affected by the intra-leaf light environment during expansion. However, this may depend on the longevity of the leaf. Chloroplasts of long-lived leaves might adapt to changes in light even after the full expansion, because the turnover or maintenance of chloroplasts may be affected by the light environment.

In spinach leaves grown under 15% full sunlight, the number of thylakoids per granum was around 7 throughout the leaf (Terashima and Evans, 1988). Similar results were obtained in some *Pisum sativum* L. varieties grown at ~50 μmol quanta m^{-2} s^{-1} (Schwarz, 1988). Moreover, Schreiber *et al.* (1977) observed that fluorescence inductions in leaves of sun plants grown under shade were similar between the two leaf surfaces. Therefore, it seems likely that chloroplasts in sun plant species cannot respond to very low light. Interestingly, in leaves of a shade tolerant species, *Alocasia macrorrhiza* (L.) G. Don., the thylakoid number was markedly larger in spongy tissue chloroplasts than in palisade tissue chloroplasts, even in the leaves grown at 10 μmol quanta m^{-2} s^{-1} (Chow *et al.*, 1988).

Some reports claimed that the photosynthetic properties of palisade tissue and spongy tissue chloroplasts are similar (Mokronosov *et al.*, 1973; Outlaw *et al.*, 1976; Seeni *et al.*, 1983; Outlaw, 1987). As discussed above, this could be due to low light conditions during growth. Other factors such as upright leaf orientation, large transmittance due to the low pigment content, and photoinhibition also affect the results.

III. CONSEQUENCES OF GRADIENTS OF LIGHT AND CHLOROPLAST PROPERTIES IN LEAF PHOTOSYNTHESIS

The light response curve of photosynthesis may be considered to be composed of four parts (Leverenz, 1987); (1) A part below the inflexion of the Kok effect, (2) a linear part where the quantum yield of photosynthesis is measured and where the slope is proportional to absorptance (quantum yield region), (3) a non-linear part above the initial slope but below the light saturation (transitional region) and (4) the light saturated region.

Light response curves of dorsiventral leaves are different depending on whether the leaves are illuminated from the adaxial (upper) side or the abaxial (lower) side (Moss, 1964; Malkina, 1976; Oya and Laisk, 1976; Syvertsen and Cunningham, 1970; Tichá *et al.*, 1985; Terashima, 1986; Pospíšilová and Solárová, 1987; Adams *et al.*, 1988). In the quantum yield region and the light saturation region, the curves are similar. In typical dorsiventral leaves, the reflectance of the abaxial side is larger than that of the adaxial side by a few percent. Therefore, the total leaf absorptance, which usually ranges from 70 to 95% for photosynthetically active radiation, is not greatly affected by the direction of illumination. It has been shown that the sun/shade acclimation does not alter quantum yield (Björkman, 1981; Evans, 1987). These two factors explain the similar curves in the quantum yield region. If there is no decline of photosynthetic rate at excessive irradiances, the maximum rate is also independent of the direction of illumination.

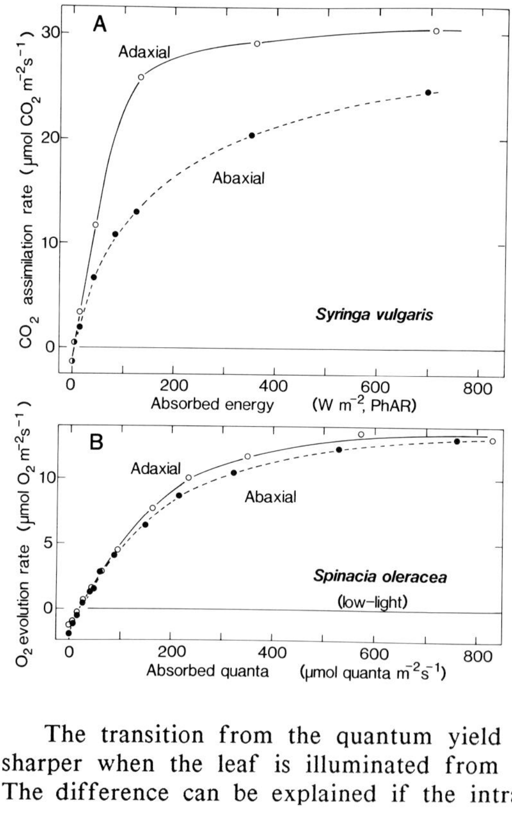

Figure 5. Light response curves for leaf photosynthesis obtained by the unilateral illumination from the adaxial or the abaxial side. A, curves for *Syringa vulgaris* L. leaf measured with CO_2 exchange system. (B) curves for a *Spinacia oleracea* L. leaf grown under low light measured with a leaf disc oxygen electrode system. (A) redrawn from Oya and Laisk (1976). B, Terashima, unpublished.

The transition from the quantum yield region to light saturation is sharper when the leaf is illuminated from the adaxial side (Fig. 5-A). The difference can be explained if the intra-leaf gradients of light and properties of chloroplasts are taken into account (Oya and Laisk, 1976; Terashima and Saeki, 1985; Terashima, 1986; Leverenz, 1987). When the leaf is illuminated from the adaxial side, the gradient of light absorption is still quite steep in spite of the moderating effects mentioned in the first section. This gradient may be considerably counterbalanced by the gradient of light-saturated photosynthetic rate in chloroplasts. The sharpest light response curve of leaf photosynthesis occurs when the photosynthesis by each chloroplast saturates at the same incident irradiance. When the leaf is illuminated from the abaxial side, the intra-leaf gradient of light absorption is very steep, because the light attenuation per unit chlorophyll in the spongy tissue is greater than that in the palisade tissue. Thus, little light reaches the palisade tissue. Moreover, the gradient of the maximal photosynthetic rate is inverse to the light gradient: photosynthesis in the spongy tissue saturates at low incident irradiance, while the saturation of photosynthesis in palisade tissue occurs only at very high incident irradiance. Therefore, the transition from the quantum yield region to the light-saturated region occurs more gradually.

As mentioned in the previous section, chloroplasts of sun plant cannot acclimate to the extreme shade. In shade-grown *S. oleracea* leaves which have nil intra-leaf gradient of chloroplast properties, light response curves differed only slightly depending on the direction of illumination (Fig. 6B).

In leaves with more isobilateral symmetry, the differences in light response curves depending on the direction of unilateral illumination are very small or nil (*Saccharum officinarum* L., Moss, 1964; *Zea mays* L., Moss, 1964; Vâclavík, 1984; *Oryza sativa* L., Tanaka and Matsushima, 1970; *Picea sitchensis* (Bong.) Carr., Leverenz and Jarvis, 1979; *Perezia nana*, Syvertsen and Cunningham, 1970; *Eucalyptus pauciflora* Sieb. ex Spreng., Wong *et al.*, 1985). Moreover, in *Z. mays, O. sativa, P. sitchensis,* and *E. pauciflora*, it has been shown that bilateral illumination result in the sharper light response curves than unilateral illumination. Since leaves of these species receive light on both surfaces under natural conditions, the intra-leaf photosynthetic system is organized so as to receive light bilaterally (Kirschbaum, 1987).

Leverenz (1987) analysed the shape of photosynthetic light response curves in shade-grown conifer needles with varied chlorophyll content. The gradient of chloroplast properties might be small, because the plants were grown under shaded conditions. The sharpest light response curve was obtained in the leaves with the lowest amount of chlorophyll. With increasing chlorophyll content, the sharpness decreased indicating that the large light gradient without corresponding acclimation of chloroplasts reduces the sharpness of light response curves. The very sharp light response curves in needles with low chlorophyll content led Leverenz (1987) to conclude that the light response curves for an individual chloroplast is also very sharp. This is based on the assumption that the sharpness of the light-response curve for the leaf cannot be sharper than that for an individual chloroplast. However, if the chloroplast movements adjust the intra-leaf light environment so as to sharpen the light response curve of whole leaf photosynthesis, this assumption is not absolutely correct. The contribution of the chloroplast movement to the shape of light response curves should be sought.

IV. CO_2 ENVIRONMENT IN THE LEAF

When net photosynthesis occurs, the intercellular spaces are likely to have a CO_2 partial pressure somewhat lower than the ambient atmosphere. In conventional gas exchange studies, CO_2 partial pressure in the intercellular spaces (p_i) is the most important parameter since it indicates what effects the stomata have on leaf photosynthesis (Farquhar and Sharkey, 1982; Sharkey, 1985). This parameter has been also used to assess the effects of stresses on the photosynthetic capacity of the mesophyll. Depression of the relationship between the CO_2 assimilation rate (A) and p_i [$A(p_i)$ relationship] after stress treatment has been attributed to non-stomatal inhibition of photosynthesis.

Intercellular CO_2 partial pressure (p_i) is usually calculated as

$$p_i = p_a - 1.6\ A/g,$$

where p_a, A, and g are CO_2 partial pressure in the ambient air, net CO_2 assimilation rate, and leaf conductance to the diffusion of H_2O, respectively. Recently, a refined version of this equation which takes account of mass-flow effects has been also used (for a review, see Farquhar and Sharkey, 1982). These calculations are based on the assumption that photosynthesis and transpiration are homogeneous over the leaf area under consideration. This assumption appears to be relevant when most of the stomata are open. In some thin amphistomatous leaves, the accuracy of the calculation of p_i has also been experimentally proved.

In these experiments, the amphistomatous leaf was clamped in a double sided assimilation chamber. Letting the leaf photosynthesize only from one surface, the equilibrated CO_2 partial pressure of another side was measured. When the ambient CO_2 partial pressure for the photosynthesizing side was around 330 μbar, the estimated p_i at the stomatal cavities on the photosynthesizing side was from 3 to 70 μbar higher than p_i measured on another side (Sharkey *et al.*, 1982; Mott and O'Leary, 1984; Parkhurst *et al.*, 1988). The value is affected by the thickness of the leaves, porosity and tortuosity of intercellular air spaces, and photosynthetic capacity (see Parkhurst, 1977). Very large differences in p_i (~70 μbar) have been measured in thick and compact leaves of *Eucalyptus pauciflora* Sieb. ex Spreng. (Parkhurst *et al.*, 1987). In situ, the difference in p_i in such amphistomatous leaves must be much smaller because stomata on both sides can function. The distance between the neighbouring stomata is comparable to the thickness of the thin leaves (~100 μm). Thus, the drop of p_i in lateral directions may be negligible. Therefore, in ordinary thin leaves, when the most stomata are open, p_i can be assumed to be relatively homogeneous throughout the mesophyll. This may be also relevant to thin hypostomatous leaves. However, in leaves with decreased stomatal conductance, it is dangerous to assume the homogeneity of p_i and the relevancy of the calculation of p_i.

Although it is not widely established, apertures of individual stomata seem to vary considerably (Ishihara *et al.*, 1971; Laisk *et al.*, 1980; Laisk, 1983; Omasa *et al.*, 1985; Kappen *et al.*, 1987). Moreover, when the stomatal conductance decreases, the number of completely closed stomata increases (Laisk *et al.*, 1980; Terashima *et al.*, 1988; Raschke and Patzke, personal communication). In such cases, spacing among the open stomata becomes wider and therefore, the supply of CO_2 by lateral diffusion in intercellular air spaces becomes important.

Bundle sheath extensions are sheets of cells extended from bundle sheath cells to the epidermes (Wylie, 1952). Since bundle sheath extensions lack intercellular air spaces, they obstruct the lateral diffusion of gases in the mesophyll. Many dicot leaves have netted (reticulate) venation. In some leaves, even the very small veinlets have bundle sheath extensions so that the mesophyll is separated into many small compartments. The leaves with small bundle sheath extension compartments such as *Helianthus annuus* L., *Vitis vinifera* L. and *Xanthium strumarium* L. are referred to as heterobaric leaves (Neger, 1918; see Fig. 6A).

In a small bundle sheath extension compartment, the diameter of which is not more than several hundred μm, p_i may be assumed to be homogeneous. However, p_i can be different among the compartments depending on the number and aperture of open stomata. When the compartment has no open stomata, and the photosynthetic capacity of mesophyll is not impaired, p_i may approach the CO_2 compensation point.

Veinlets in leaves of some species such as *Spinacia oleracea* L. and *Vicia faba* L. lack bundle sheath extensions. Such leaves are called homobaric leaves (Fig. 6B). In these leaves, lateral movement of gases in the intercellular spaces of mesophyll is relatively free. Accordingly, within a small area, p_i may be homogeneous irrespective of the stomatal

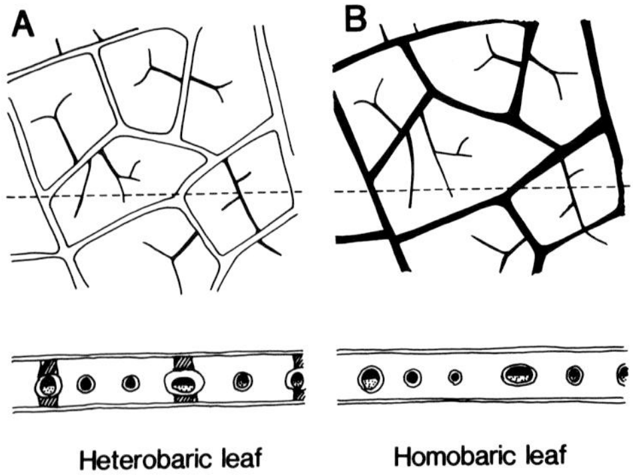

Figure 6. Diagrams showing compartmentation of mesophyll in heterobaric leaves (A) and homobaric leaves (B). Double lines, veins with bundle sheath extensions; solid lines, veins without bundle sheath extensions. In a transverse diagram, bundle sheath extensions are hatched.

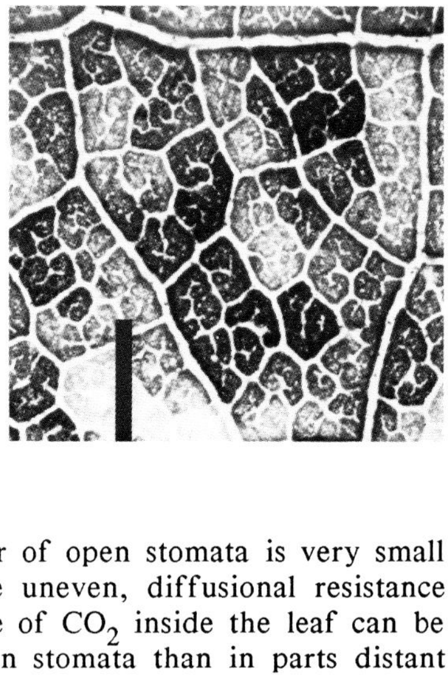

Figure 7. Distribution of photosynthetic starch formation in a *Helianthus annuus* L. leaf in the presence of abscisic acid (ABA). ABA was applied by sandwiching leaf discs between the pieces of filter paper wetted with 3 x 10^{-5} ABA solution for 30 min in the dark. The leaf disc was illuminated at 1,200 μmol quanta m^{-2} s^{-1} for 2 hr in a chamber of a leaf disc oxygen electrode system at ambient CO_2 concentration of 4 mbar. The leaf disc was boiled in methanol, then stained with an iodine-potassium iodide solution. A scale bar indicates 1 mm. Note the bundle sheath extension compartments (Terashima *et al.*, 1988).

conductance. However, when the number of open stomata is very small and the distribution of open stomata are uneven, diffusional resistance cannot be neglected. The partial pressure of CO_2 inside the leaf can be considerably higher in parts near the open stomata than in parts distant from the open stomata.

Abscisic acid (ABA) reduces stomatal conductance. When *Helianthus annuus* L. leaves were illuminated with ABA present in the transpirational stream, photosynthetic starch accumulation detected by starch-iodine staining, was non-uniform and the density of staining was different among the respective bundle sheath extension compartments. In some compartments, no starch was detected (c.f. Fig. 7). The distribution of appreciably open stomata, examined using the silicone rubber impression technique, gave an identical pattern to that of starch formation. The measurements of photosynthetic oxygen evolution in leaves with or without epidermes at various ambient CO_2 partial pressures, revealed that the photosynthetic capacity of mesophyll is not impaired by ABA. Thus, the non-uniform photosynthesis at the ambient p_a level is exclusively due to the non-uniform p_i caused by stomatal/structural factors (Terashima *et al.*, 1988).

Using an autoradiography technique, Downton *et al.* (1988) found the similar non-uniform photosynthesis in *Vitis vinifera* L. leaves after the application of ABA. In *Xanthium strumarium* L. leaves, non-uniform photosynthesis induced by ABA has been revealed by a fluorescence imaging technique (P. Daley, K. Raschke, T. Ball and J. Berry; personal communication). Leaves of *Vicia faba* L. which are homobaric, showed the decreased starch formation in the presence of ABA, but the distribution of starch was relatively homogeneous (Terashima *et al.*, 1988).

It is obvious that the calculation of p_i is erroneous in leaves treated with ABA. When photosynthesis is non-uniform due to the stomatal/structural factors, calculated p_i is always higher than the real p_i (for a theoretical consideration, see Terashima *et al.*, 1988). Moreover, the equations used for the calculation of p_i usually lack the term for cuticular transpiration. So-called "cuticle" resistance may be very large to CO_2, but relatively small to H_2O. The cuticle layer is the only barrier to water in the outer cell walls of epidermes, while, for CO_2, not only the cuticle layer but the epidermal cells are barriers. In some species, the cuticles of some parts of the cell walls near the stomatal pores are fairly permeable to H_2O. This is responsible for "peristomatal" transpiration (Maier-Maercker, 1979; Lösch and Tenhunen, 1981; Appleby and Davies, 1983; Mansfield, 1985; Meidner, 1985). Therefore, when the stomatal conductance to H_2O is low and, thus, the contribution of cuticular conductance (H_2O) to leaf conductance (H_2O) is relatively large, p_i tends to be overestimated (see Kirschbaum and Pearcy, 1988).

Non-uniform photosynthesis due to stomatal/structural factors, (and accompanying low conductance) result in the false depression of $A(p_i)$ relationships. The false depression is most marked in heterobaric leaves with small bundle sheath extension compartments. These erroneous effects are smaller in homobaric leaves (Raschke and Hedrich, 1985; Terashima *et al.*, 1988).

Many stress factors are known to reduce stomatal conductance. When the mode of stomatal closure is similar to that induced by ABA, patchy photosynthesis and transpiration will occur. In some species, effects of water stress on leaf photosynthesis seem to be mediated by ABA (Bunce,

1987). Accordingly, the non-uniform photosynthesis probably occurs under water stress. Non-uniform starch accumulation was actually observed in *H. annuus* leaves photosynthesizing at high vapour pressure deficit (J. Virgona; personal communication).

Another line of evidence, which supports the occurrence of non-uniform photosynthesis, has been obtained using a leaf disc oxygen electrode system operated at very high CO_2 partial pressure (p_a). Under some "acute" stress conditions, depression of the capacity of mesophyll photosynthesis assessed with the leaf disc oxygen electrode systems at high p_a was smaller than the depression of $A(p_i)$ relationships obtained with conventional gas exchange techniques (Ben *et al.*, 1987; Kaiser, 1987; S.P. Robinson, personal communication; L. Flanagan, personal communication). However, in a case of chronic water stress in *Gossypium hirsutum* L., both assessments agreed reasonably (Björkman, this volume).

The underlying mechanism for the non-uniform stomatal behaviour is unknown. The sensitivity of individual stomata may be intrinsically different. The meaning of the existence of bundle sheath extensions is not clear. In terms of water use efficiency, the non-uniform photosynthesis caused by the bundle sheath extensions is not advantageous. One suggested function of bundle sheath extensions is a pathway for water from veins to the epidermis (Wylie, 1943; Canny, 1986). If this is relevant, water status of mesophyll can be buffered. Bundle sheath extensions may also act as an effective light path (Schanderl and Kaempfert, 1933) to moderate the intra-leaf light gradient.

CONCLUSION

In this chapter, I have tried to describe the intra-leaf light gradient and CO_2 distribution in relation to the "productive structure" of a leaf. I have also tried to point out problems in this research field. So far, a lot of hypotheses or speculations have been proposed concerning functional meaning of structure. These speculations should be proved or disproved based on sound experiments using the new techniques. Thorough understanding of leaf function cannot be achieved without progress in this research field.

I thank Prof. C.B. Osmond and Drs. J.R. Evans and W.S. Chow for helpful comments, and Ms. M. Whittaker for technical assistance.

REFERENCES

Adams WW III, Terashima I, Brugnoli E, Demmig B (1988). Comparisons of photosynthesis and photoinhibition in the CAM vine *Hoya australis* and several C_3 vines growing on the coast of eastern Australia. Plant Cell Environ 11:173-181.

Anderson JM (1986). Photoregulation of the composition, function, and structure of thylakoid membranes. Annu Rev Plant Physiol 37:93-136.

Anderson JM, Osmond CB (1987). Shade-sun responses: compromises between acclimation and photoinhibition. In Kyle DJ, Osmond CB, Arntzen CJ (eds): "Photoinhibition," Amsterdam: Elsevier, pp 1-38.

Appleby RF, Davies WJ (1983). The structure and orientation of guard cells in plants showing stomatal responses to changing vapour pressure differences. Ann Bot 52:459-468.

Ballantine JEM, Forde BJ (1970). The effect of light intensity and temperature on plant growth and chloroplast ultrastructure in soybean. Amer J Bot 57:1150-1159.

Ben GY, Osmond CB, Sharkey TD (1987). Comparisons of photosynthetic responses to *Xanthium strumarium* and *Helianthus annuus* to chronic and acute water stress in sun and shade. Plant Physiol 84:476-481.

Björkman O (1981). Responses to different quantum flux densities. In Lange OK, Nobel PS, Osmond CB, Ziegler H (eds): "Encyclopedia of Plant Physiology, New Series, Vol 12A," Berlin: Springer, pp 57-107.

Boardman NK (1977). Comparative photosynthesis of sun and shade plants. Ann Rev Plant Physiol 28:355-377.

Bornmann JF, Vogelman TC (1988). Penetration of blue and UV radiation measured by fiber optics in spruce and fir needles. Physiol Plant 72:699-705.

Boysen-Jensen P (1932). "Die Stoffproduktion der Pflanzen." Jena: G. Fischer, p 108.

Bunce J (1987). Species-specific responses to water stress of gas exchange parameters mimicked by applied abscisic acid. Can J Bot 65:103-106.

Caemmerer S von, Farquhar GD (1981). Some relationships between biochemistry and the gas exchange of leaves. Planta 153:376-387.

Canny MJ (1986). Water pathways in wheat leaves. III. The passage of mestome sheath and the function of the suberized lamellae. Physiol Plant 66:637-647.

Chlobowski A (1986). Photometric method in chloroplast movement investigation. Photosynthetica 20:486-493.

Chow WS, Quian L, Goodchild DJ, Anderson JM (1988). Photosynthetic acclimation of *Alocasia macrorrhiza* to growth irradiance: structure, function and composition of chloroplasts. Aust J Plant Physiol 15:107-122.

Downton WJS, Loveys BR, Grant WJR (1988). Stomatal closure fully accounts for the inhibition of photosynthesis by abscisic acid. New Phytol 108:263-266.

Evans JR (1987). The dependence of quantum yield on wavelength and growth irradiance. Aust J Plant Physiol 14:67-79.

Farquhar GD, Caemmerer S von (1982). Modelling of photosynthesis response to environmental conditions. In Lange OL, Nobel PS, Osmond CB, Ziegler H (eds): "Encyclopedia of Plant Physiology, New Series, Vol 12A," Berlin: Springer, pp 549-587.

Fischer DG, Evert RF (1982). Studies on the leaf of *Amaranthus retroflexus* (Amaranthaceae): Chloroplast polymorphism. Bot Gaz 143:146-155.

Fukshansky L (1981). Optical properties of plants. In Smith H (ed): "Plant and Daylight Spectrum," London: Academic Press, pp 21-40.

Gabryś H, Walczak T (1980). Photometric study of chloroplast phototranslocations in leaves of land plants. Acta Physiol Plant 2:281-290.

Gates DM (1980). "Biophysical Ecology." New York: Springer, p 611.

Haupt W (1986). Photomovement. In Kendrick RE, Kronenberg GHM (eds): "Photomorphogenesis in Plants," Dordrecht: Martinus Nijhoff, pp 415-441.

Hirata M, Ishii R, Kumura A, Murata Y (1983). Photoinhibition of photosynthesis in soybean leaves. III. Leaf transmittance changes in response to incident light intensity. Japan J Crop Sci 52:430-434.

Hodáňová D (1985). Leaf optical properties. In Šesták Z (ed): "Photosynthesis during Leaf Development," Dordrecht: Dr. W Junk, pp 107-127.

Inoue Y, Shibata K (1973). Light induced chloroplast rearrangements and their action spectra as measured by absorption spectrophotometry. Planta 114:341-358.

Inoue Y, Shibata K (1974). Comparative examination of terrestrial plant leaves in terms of light-induced absorbance changes due to chloroplast rearrangements. Plant Cell Physiol 15:717-721.

Ishihara K, Nishihara T, Ogura T (1971). The relationship between environmental factors and behaviour of stomata in aperture. Proc Japan Soc Crop Sci 40:491-496.

Kaiser WM (1987). Non-stomatal, primary dehydration effects on photosynthesis: Possible mechanism for reversible and irreversible damage. Current Topics Plant Biochem Physiol 6:119-133.

Kappen L, Andresen G, Lösch R (1987). In situ observation of stomatal movements. J Exp Bot 38:126-141.

Kirschbaum MUF (1987). A model of photosynthesis for isobilateral leaves incorporating biochemical constraints and the light gradient within the leaf. In Biggins J (ed): "Progress in Photosynthesis Research Vol IV," Dordrecht: Martinus Nijhoff, pp IV 5:261-264.

Kirschbaum MUF, Pearcy RW (1988). Gas exchange analysis of the relative importance of stomatal and biochemical factors in photosynthetic induction in *Alocasia macrorrhiza*. Plant Physiol 86:782-785.

Kulandaivelu GA, Nooruden AM, Sampath PS, Pyriyanan S, Raman K (1983). Assessment of the photosynthetic electron transport properties of upper and lower leaf sides in vivo by fluorometric method. Photosynthetic 17:206-209.

Laisk A (1983). Calculation of leaf photosynthetic parameters considering the statistical distribution of stomatal apertures. J Exp Bot 34:1627-1635.

Laisk A, Oja V, Kull K (1980). Statistical distribution of stomata aperture of *Vicia faba* and *Hordeum vulgare* and Spannungsphase of stomatal opening. J Exp Bot 31:49-58.

Leverenz JW (1987). Chlorophyll content and the light response curve of shade-adapted conifer needles. Physiol Plant 71:20-29.

Leverenz JW, Jarvis PG (1979). Photosynthesis in Sitka spruce. VIII. The effects of light flux density and direction on the rate of net photosynthesis and the stomatal conductance of needles. J Appl Ecol 16:919-932.

Lösch R, Tenhunen JD (1981). Stomatal responses to humidity. In Jarvis PG, Mansfield TA (eds): "Stomatal Physiology," Cambridge: Cambridge Univ. Press, pp 137-161.

Maier-Mercker U (1979). Peristomatal transpiration and stomatal movement: A controversial view. I. Additional proof of peristomatal transpiration by hygrophotography and a comprehensive discussion in the light of recent results. Z. Pflanzenphysiol 91:25-43.

Malkina IS (1976). On the degree of palisade parenchyma participation in photosynthesis of *Acer platanoides* and *Betula pendula* leaves. Lesovendenie 1976: 51-57. (referred in Tichá *et al.* (1985)).

Mansfield TA (1985). Porosity at a price: The control of stomatal conductance in relation to photosynthesis. In Barber J, Baker NR (eds): "Photosynthetic Mechanisms and the Environment," Amsterdam: Elsevier, pp 419-452.

Maróti I, Gabor G (1976). Thylakoid aggregation and pigment ratios in the spongy and palisade parenchymas. Acta Biol Szeged (Hungary) 22:15-27.

Meidner H (1986). Cuticular conductance and the humidity response of stomata. J Exp Bot 37:517-525.

Mokronosov AT, Bagautdinova RI, Bubnova EA, Kobeleva IV (1973). Photosynthetic metabolism in palisade and spongy tissues of the leaf. Fiziol Rast 20:1191-1197. (Sov Plant Physiol 20:1013-1018).

Monsi M, Saeki T (1953). Über den Lichtfaktor in den Pflanzengesellschaften und seine Bedeutung für die Stoffproduktion. Japan J Bot 14:22-52.

Moss DN (1964). Optimum lighting of leaves. Crop Sci 4:131-136.

Moss RA, Loomis WE (1952). Absorption spectra of leaves. I. The visible spectrum. Plant Physiol 27:370-391.

Mott KA, O'Leary JW (1984). Stomatal behavior and CO_2 exchange characteristics in amphistomatous leaves. Plant Physiol 74:4-51.

Neger FW (1918). Die Wegsamkeit den Laubblätter für Gase. Flora 111:152-161.

Omasa K, Hashimoto Y, Kramer PJ, Strain BR, Aiga I, Kondo J (1985). Direct observation of reversible and irreversible stomatal response of attached sunflower leaves to SO_2. Plant Physiol 79:153-158.

Outlaw WH Jr (1987). A mini review: Comparative biochemistry of photosynthesis in palisade cells, spongy cells, and guard cells of C_3 leaves. In Biggins J (ed): "Progress in Photosynthesis Research Vol IV," Dordrecht: Martinus Nijhoff, pp IV 5:265-272.

Outlaw WH Jr, Schmuck CL, Tolbert NE (1976). Photosynthetic carbon metabolism in the palisade parenchyma and spongy parenchyma of *Vicia faba* L. Plant Physiol 58:186-189.

Oya V, Laisk A (1976). Adaptation of the photosynthetic apparatus to the profile of light in the leaf. Fiziol Rast 23:445-451. (Sov Plant Physiol 23:381-386.)

Parkhurst DF (1977). A three-dimensional model for CO_2 uptake by continuously distributed mesophyll in leaves. J Theor Biol 67:471-488.

Parkhurst DF, Wong SC, Farquhar GD, Cowan IR (1987). Gradients of intercellular CO_2 levels across the leaf mesophyll. Plant Physiol 86:1032-1037.

Pospíšilová J, Solárová J (1987). Adaptations and acclimations of dorsiventral leaves to irradiance: Epidermal diffusive conductances and net photosynthetic rate. Photosynthetica 21:349-356.

Raschke K, Hedrich R (1985). Simultaneous and independent effects of abscisic acid and on stomata and photosynthetic apparatus in whole leaves. Planta 163:105-118.

Schanderl H, Kaempfert W (1933). Über die Strahlungsdurchlässigkeit von Blättern und Blattgeweben. Planta 18:700-750.

Schreiber U, Fink R, Vidaver W (1977). Fluorescence induction in whole leaves: Differentiation between the two leaf sides and adaptation to different light regime. Planta 133:121-129.

Schwarz HP (1988). Comparative ultrastructural properties of palisade tissue and spongy tissue chloroplasts from normal and mutant leaves of *Pisum sativum* L. Plant Cell Environ 11:99-109.

Seeni S, Mariappan T, Gopalan G, Gnanam A (1983). Mechanical separation of palisade and spongy parenchyma cells from the leaves of mesomorphic dicotyledon for photosynthetic studies. Planta 157:105-110.

Seyfried M, Fukshansky L (1983). Light gradients in plant tissue. Appl Opt 22:1402-1408.

Sharkey TD (1985). Photosynthesis in intact leaves of C_3 plants: Physics, physiology and rate limitations. Bot Rev 51:53-105.

Sharkey TD, Imai K, Farquhar GD, Cowan IR (1982). A direct confirmation of the standard method of estimating intercellular partial pressure of CO_2. Plant Physiol 69:657-659.

Shul'gin IA, Murei IA, Nichiporovich AA (1978). Structuro-functional organization of the leaf as an integral photosynthesizing system. Fiziol Rast 25:76-85. (Sov Plant Physiol 25:55-62.)

Skene DS (1974). Chloroplast structure in mature apple leaves grown under different levels of illumination and their response to changed illumination. Proc R Soc London B 186:75-78.

Syvertsen JP, Cunningham GL (1979). The effects of irradiating adaxial or abaxial leaf surface on the rate of net photosynthesis of *Perezia nana* and *Helianthus annuus*. Photosynthetica 13:287-293.

Tanaka T, Matsushima S (1970). Analysis of yield-determining process and its application to yield prediction and culture improvement of low land rice. XCIV. Relation between the light intensity on both sides and the amount of carbon assimilation in each side of a single leaf-blade. Proc Crop Sci Soc Japan 39:325-329.

Terashima I (1986). Dorsiventrality in photosynthetic light response curves of a leaf. J Exp Bot 37:399-405.

Terashima I, Evans JR (1988). Effects of light and nitrogen nutrition on the organization of the photosynthetic apparatus in spinach. Plant Cell Physiol 29:143-155.

Terashima I, Inoue Y (1984). Comparative photosynthetic properties of palisade tissue chloroplasts and spongy tissue chloroplasts of *Camellia japonica* L.: Functional adjustment of the photosynthetic apparatus to light environment within a leaf. Plant Cell Physiol 25:555-563.

Terashima I, Inoue Y (1985a). Palisade tissue chloroplasts and spongy tissue chloroplasts in spinach: Biochemical and ultrastructural differences. Plant Cell Physiol 26:63-75.

Terashima I, Inoue Y (1985b). Vertical gradient in photosynthetic properties of spinach chloroplasts dependent on intra-leaf light environment. Plant Cell Physiol 26:781-785.

Terashima I, Saeki T (1983). Light environment within a leaf. I. Optical properties of paradermal sections of *Camellia* leaves with special reference to difference in the optical properties of palisade and spongy tissues. Plant Cell Physiol 24:1493-1501.

Terashima I, Saeki T (1985). A new model for leaf photosynthesis incorporating the gradients of light environment and of photosynthetic properties of chloroplasts within a leaf. Ann Bot 56:489-499.

Terashima I, Sakaguchi S, Hara N (1986). Intra-leaf and intracellular gradients in chloroplast ultrastructure of dorsiventral leaves illuminated from the adaxial or abaxial side during their development. Plant Cell Physiol 27:1023-1031.

Terashima I, Takenaka A (1986). Organization of photosynthetic system of dorsiventral leaves as adapted to the irradiation from the adaxial side. In Marcelle R, Clijsters H, VanPouke M (eds): "Biological Control of Photosynthesis," The Hague: Martinus Nijhoff/Dr W Junk, pp 219-230.

Terashima I, Wong SC, Osmond CB, Farquhar GD (1988). Characterisation of non-uniform photosynthesis induced by abscisic acid in leaves having different mesophyll anatomies. Plant Cell Physiol 29:385-394.

Tichá I, Catsky J, Hodáňová D, Pospíšilová J, Kaše M, Šesták Z (1985). Gas exchange and dry matter accumulation during leaf development. In Šesták Z (ed): "Photosynthesis During Leaf Development," Dordrecht: Dr W Junk, pp 157-216.

Václavík J (1984). Photosynthetic CO_2 uptake by *Zea mays* leaves as influenced by unilateral irradiation of adaxial and abaxial leaf surfaces. Biol Plant 26:206-214.

Vogelman TC (1986). Light within the plant. In Kendrick RE, Kronenberg GHM (eds): "Photomorphogenesis in Plants," Dordrecht: Martinus Nijhoff, pp 307-337.

Vogelman TC, Knapp AK, McClean TM, Smith WK (1988). Measurement of light within thin plant tissues with fibre optic microprobes. Physiol Plant 72:623-630.

Walczak T, Gabryś H (1980). New type of photometer for measurements of transmission changes corresponding to chloroplast movement in leaves. Photosynthetica 14:65-72.

Wong SC, Cowan IR, Farquhar GD (1985). Leaf conductance in relation to rate of CO_2 assimilation. II. Effects of short-term exposure to different photon flux densities. Plant Physiol 78:826-829.

Wylie RB (1943). The role of the epidermis in foliar organization and its relations to the minor venation. Amer J Bot 30:273-280.

Wylie RB (1952). The bundle sheath extension in leaves of dicotyledons. Amer J Bot 39:645-651.

Photosynthesis, pages 227–241

CANOPY PHOTOSYNTHESIS

John M. Norman and Wayne Polley

Department of Soil Science, University of Wisconsin-Madison, Madison, WI 53706 (J.M.N.) and USDA-ARS, Temple, TX 76502 (W.P.)

INTRODUCTION

The photosynthetic productivity of an individual plant is the consequence of a great many processes, perhaps too many to be reasonably accommodated in an integrated description. However, one can define some limited scope and work consistently with as many processes as possible. The purview of this chapter is to use measurements on the plant organ spatial scale and approximately minute-to-hour time scale, and infer functioning on the community scale over time scales of days to months. A few definitions may be useful. An organ can be defined as "a part composed of specialized tissues and adapted to perform specific functions." An organism may be described as "an individual plant having diverse organs that function together as a whole to maintain life." A community can be thought of as a group of interacting organisms. To summarize, in this chapter characteristics at the leaf spatial scale (organs) are taken as given from appropriate measurements, and characteristics on the canopy spatial scale (community) are obtained from the appropriate integrals over time and space.

One question that immediately arises is how information about organs can be integrated to infer functioning of a community of organisms, namely a canopy? An approach can be divided into three parts: 1) defining characteristics of individual organs, 2) describing the arrangement of and connections among organs, and 3) predicting the environment immediately surrounding the various organs. Thus this chapter will emphasize the physical and physiological aspects of understanding canopy photosynthesis in terms of leaf characteristics. This is almost a "bootstrapping" problem because the photosynthetic rate of a leaf depends on its immediate environment; however, its immediate environment is modified by the presence of that leaf and its neighbors. The extent of this modification depends on environmental conditions. This problem of predicting canopy photosynthesis is much simpler if the microenvironmental modifications caused by the plant are ignored. But much can be learned about limitations to plant growth by studying the soil-plant-atmosphere system carefully.

This chapter describes a methodology for combining leaf physiological characteristics, canopy architecture, and microenvironment to obtain estimates of canopy photosynthesis from an integrative model based on reasonably obtainable measurements. The plant-environment model (entitled Cupid) that forms the basis for this approach to predicting canopy photosynthesis has been described for water balance problems (Norman, 1979; Norman and Campbell, 1983). Here the extension of Cupid to consider leaf photosynthesis and its coupling to stomata is discussed.

ORGAN CHARACTERISTICS

An estimation of canopy photosynthesis with a plant-environment model, such as Cupid, requires a mathematical formulation of the dependence of the appropriate leaf physiological processes on environment, time, position, etc. For such an integrative model to be practical, the leaf physiological characteristics must be relatively straightforward to measure. Predictions of canopy photosynthesis require leaf stomatal conductance, photosynthetic rate, and respiration rate as a function of age and environmental factors. The main physical attributes of leaves that are essential for predicting photosynthesis are size and shape, and absorptance to photosynthetically active radiation (PAR); absorptance to solar radiation is essential for leaf temperature considerations.

The stomatal conductance (g_s) and assimilation rate (A) of *Andropogon gerardii*, a C4 warm season grass from the Konza Prairie reserve near Manhattan, Kansas, was measured on intact leaves in the natural environment using a portable gas exchange system (LI-6200 from LI-COR, Inc., Lincoln, NE, USA). The dependence of assimilation rate on PAR (E) and internal carbon dioxide concentration (C_i) is shown in Fig. 1. The data in Fig. 1 were obtained with a closed cuvette system where light was controlled by placing neutral density filters between the chamber and the sun, and the variable carbon dioxide concentrations were the result of a photosynthesizing leaf drawing down the carbon dioxide within the closed system. The data in Fig. 1 could be obtained in 1 to 2 hours. The smooth curves in Fig. 1 represent a fit to the photosynthesis model of Farquhar and von Caemmerer (1982). Since the Farquhar and von Caemmerer (1982) model is for a C3 leaf, and no C4 models are yet available, the use of their equations simply represents an empirical fit.

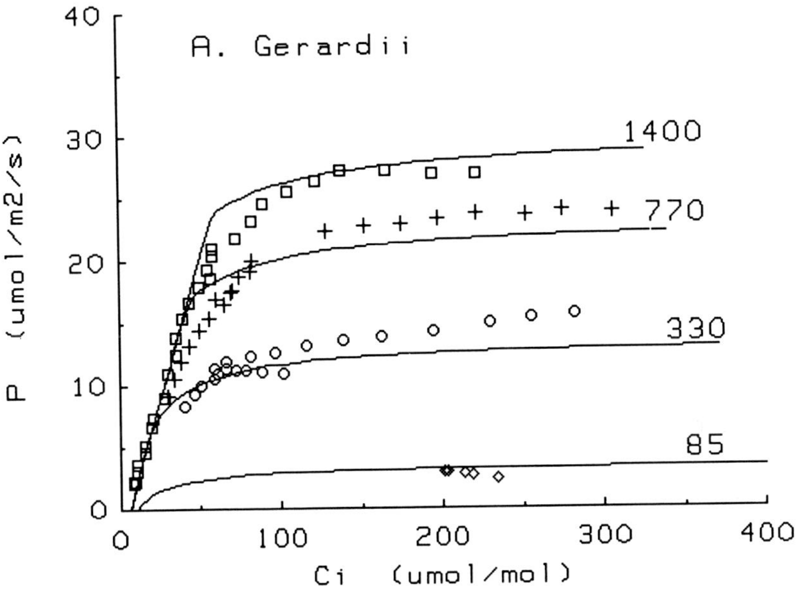

Figure 1. Leaf photosynthetic rate as a function of internal CO_2 concentration and absorbed photosynthetically active radiation (μmol q $m^{-2}s^{-1}$) for *Andropogon gerardii* measured on intact leaves in the field at a leaf temperature of 38°C. The solid curves are an empirical fit to the model of Farquhar and von Caemmerer (1982).

Four factors in their model were changed to produce reasonable fits with data and to be consistent with differences between C3 and C4 photosynthesis: 1) The carbon dioxide compensation point in the absence of day respiration is set to 2 μmol/ mol; 2) the photochemical efficiency is set to 0.062 mol CO_2/mol q (Ehleringer and Pearcy, 1983); 3) the coefficient that determines the curvature in the relation between PAR and the rate of electron transfer (2.1 based on incident light in Farquhar *et al.*, 1980) is changed to 3.3 based on absorbed quanta; and 4) the temperature dependence of the maximum rate of electron transfer is given by Schoolfield *et al.* (1981). For the curves in Fig. 1, the maximum velocity of carboxylation (Vc_{max}) is 624 μmol CO_2 $m^{-2}s^{-1}$ and the maximum rate of electron transfer (J_{max}) is 153 μmol q $m^{-2}s^{-1}$. The leaf photosynthetic rate appears to be unaffected by leaf water potential for potentials greater than -1 MPa; the photosynthetic rate then decreases approximately linearly to zero at -3 MPa for slowly imposed stress.

The stomatal conductance (g_s) to water vapor can be calculated from the photosynthetic rate by assuming C_i to be some fraction of the ambient air concentration (C_a) by using

$$r_s = \frac{C_a - C_i}{1.6\ P} - \frac{1.3}{1.6}\ r_a = \frac{1}{g_s} \tag{1}$$

where r_a is the leaf boundary layer resistance for both sides of a leaf. In the event that stomata respond directly to atmospheric humidity independent of photosynthesis, then g_s from Eq. (1) can be modified by an atmospheric humidity function (Norman, 1985). An alternative method is to use the stomatal index developed by Ball *et al.* (1986):

$$\text{Index} = \frac{A\ H_s}{C_s} \tag{2}$$

where H_s is the relative humidity at the leaf surface and C_s is the carbon dioxide concentration at the leaf surface. Fig. 2 is a sketch of the meaning of surface humidity and carbon dioxide concentration. The stomatal conductance for *A. gerardii* is plotted as a function of stomatal index in Fig. 3 and an approximate relationship is:

$$g_s = 3.6 * \text{Index} + 0.03\ . \tag{3}$$

These two methods both produce very reasonable relationships for the dependence of stomatal conductance on light, temperature, vapor pressure, ambient carbon dioxide concentrations above 250 μmol/mol, and slowly imposed water stress.

The accumulation of photosynthate over a day depends not only on the daytime photosynthesis but also on the nighttime dark respiration rate. This dark respiration rate depends mainly on temperature (Fig. 4), but also on the light environment of the leaf (Anderson and Osmond, 1987).

The physical characteristics of the leaf that are most important are leaf size, shape, and radiative properties. The leaf size and shape affect the boundary layer resistance (r_a) and thus heat and mass exchange between the leaf and the immediately surrounding air (Grace, 1981). The leaf absorptance in the PAR (400-700 nm) is important for estimating the

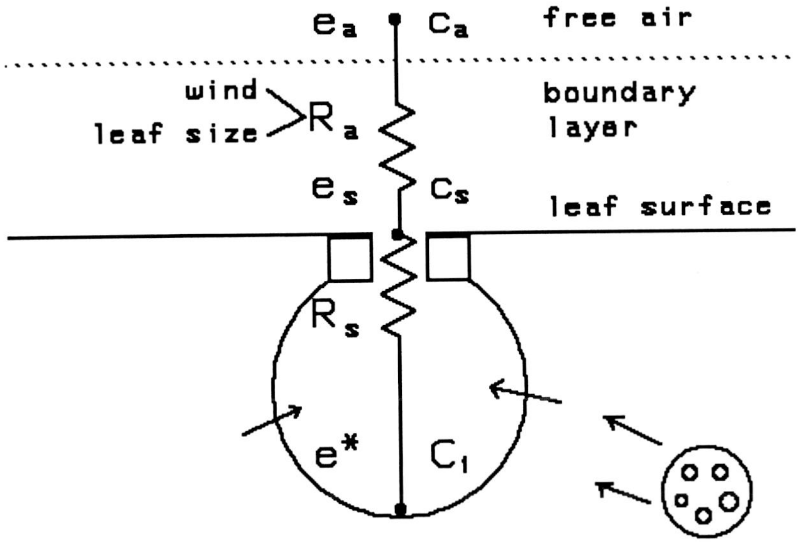

Figure 2. Schematic representation of leaf surface CO_2 concentration (C_s) and vapor pressure (e_s) which is calculated from boundary layer resistance (R_a) and stomatal resistance (R_s).

energy available for photosynthesis, and the absorptance in the solar wavelength band (400-2500 nm) is important for determining the energy balance of the leaf and ultimately leaf temperature. The PAR absorptance of *A. gerardii* is 0.78 and the solar absorptance 0.56; the wavelength dependence of the PAR absorption is shown in Fig. 5 as measured by an LI-1800 spectroradiometer and integrating sphere (LI-COR, Inc., Lincoln, NE, USA).

ORGAN ARRANGEMENT AND CONNECTIONS

The architecture of a plant canopy is remarkably elegant, being composed of leaves of varying shapes, sizes, and orientations that are connected to stems with petioles that may not only support the leaf and connect it to the rest of the plant, but may also contain a pulvinus to orient that leaf in response to environmental conditions.

With grasses, the leaves may be attached to the stem with a sheath and the long thin blades might have surface conformations that direct intercepted rain water toward the stem or ripples and surface

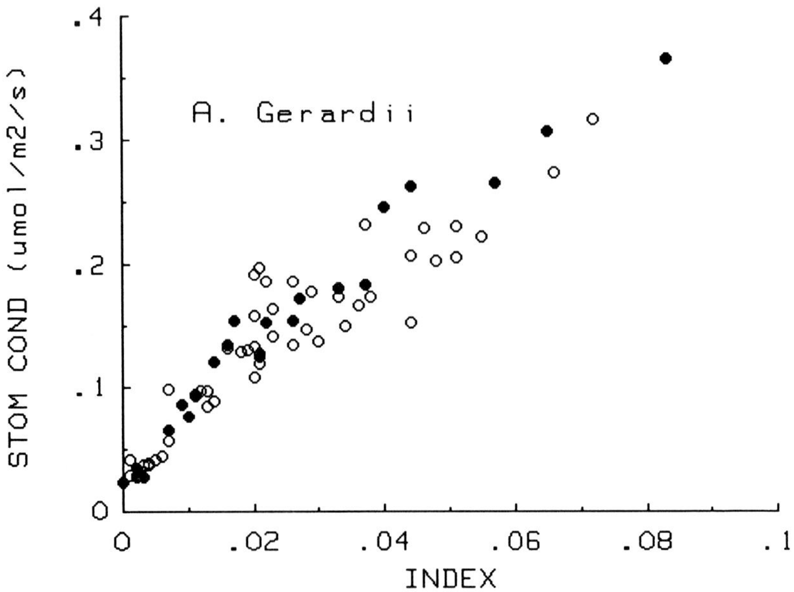

Figure 3. Relation between stomatal conductance and the index defined by Eq. (3).

irregularities that affect radiation interception. Unfortunately to use an integrative model, the canopy architecture must be described by mathematical equations, and these equations are not particularly faithful to the more subtle and elegant aspects of canopy structure. The single most important architectural feature is the leaf area index (LAI-silhouette leaf area per unit ground area). Usually these leaves are distributed vertically according to measured leaf area index profiles or simplified equations (Norman, 1979) and assumed to be randomly positioned in the horizontal (Monsi and Saeki, 1953). Quasi-random formulas have been developed (Nilson, 1971; Whitfield and Conner, 1980), but most canopy representations remain dependent on randomness assumptions at some level. The inclination distribution of leaves to the horizontal and orientation about the points of the compass can be handled with more generality in the mathematical equations than observations of these characteristics merit. Usually canopies are assumed to have leaves symmetrically distributed about the azimuth so the inclination angle distribution is sufficient to describe leaf angles. deWit (1965) proposed a useful system for characterizing canopy leaf angle distributions, but usually mean leaf inclination angle is sufficient for photosynthesis and

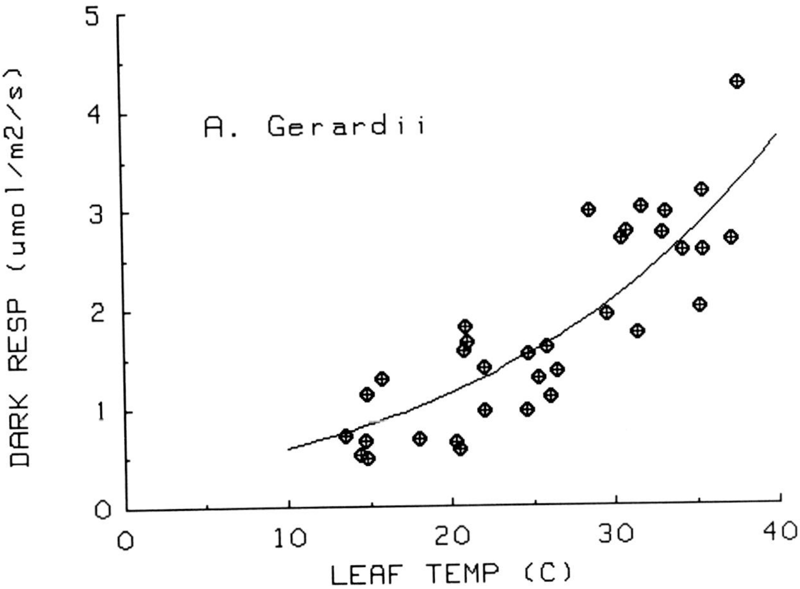

Figure 4. Dark respiration as a function of leaf temperature measured on intact leaves in the field and exponential curve fit to the data.

transpiration estimates (Norman, 1980). Table I contains a list of some canopy types.

The implications for light interception and photosynthesis of these various canopy architectures are important to consideration of productivity. Exposed horizontal-leaf canopies tend to have a low leaf area index (less than 2) with high light-saturation values of leaf photosynthesis, whereas vertical leaf canopies seem able to support much higher leaf area indices (3-7). These vertical leaf canopies can use light very efficiently even though leaves light saturate at less than half full sunlight, because the oblique angles between most leaves and the sun mean that few leaves receive light at more than half full sunlight. Understory canopies, such as at the bottom of forests, usually have horizontal leaves because they absorb the most diffuse light for a given amount of leaf area. Light interception, canopy architecture, and photosynthesis have been treated with considerable mathematical detail for random canopies (deWit, 1965; Idso and deWit, 1971).

The movement of mass through a plant is constrained by the connections between organs. The movement of water and nutrients from

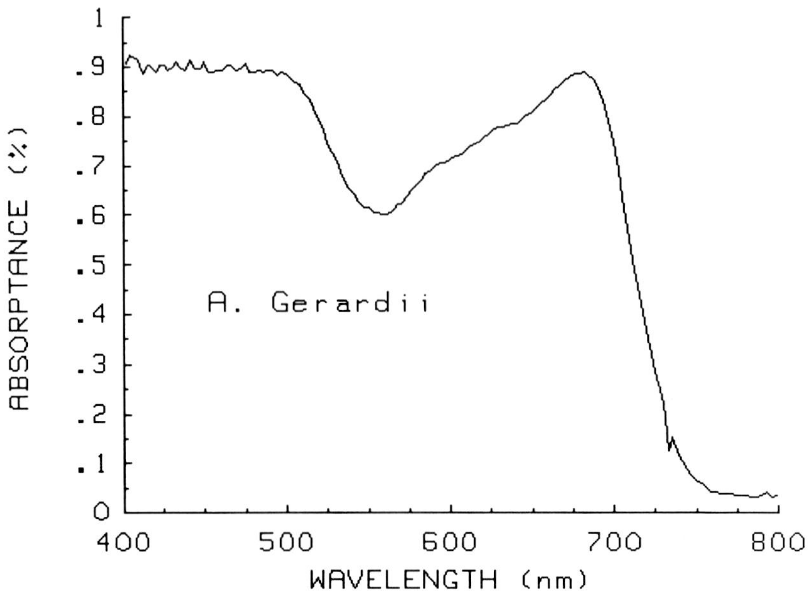

Figure 5. Absorptance of photosynthetically active radiation as a function of wavelength measured on intact leaves in the field.

the soil to functioning organs, and the distribution of synthesized products throughout the plant depend on the vascular system. Of particular importance to plant environment relations is the conduction of water from the soil to the leaves. The continuous stream of water from the root surface to the substomatal cavity of the leaf results in a connection between the water status of the leaf, the water status at the root surface, and the rate of water flow through the plant (transpiration rate). At the present time, our limited knowledge of plant vascular systems prohibits a quantitative description of water flow in plants at the level of individual xylem vessels. However, a useful and practical approach to predicting the limiting effect of water on photosynthesis can be obtained by a simple Ohm's law analogy:

$$T = \frac{P_{soil} - P_{leaf}}{R_{plant+} R_{soil}} \tilde{=} K_{plant}(P_{soil} - P_{leaf})$$

Table I. Some leaf angle distributions that are used to represent plant canopies.

Name	Characteristics	Mean leaf inclination angle
planophile	mainly horizontal	20°
uniform	equal leaf area at all angles	45°
spherical	distributed as the surface area on a sphere	57°
crcctophilc	mainly vertical	70°
plagiophile	mainly at a single leaf angle	--

where T is the plant transpiration rate, P_{soil} and P_{leaf} are the soil and leaf water potentials, respectively, and R_{soil} and R_{plant} are the resistances to water flow in the soil and plant, respectively (Kramer, 1985). Soil resistances rarely are limiting so often the denominator is just R_{plant} and $K_{plant} = 1/R_{plant}$ is the plant hydraulic conductance. This plant hydraulic conductance usually is assumed to be relatively constant so the regulation of transpiration is simulated by the dependence of stomatal conductance on leaf water potential (P_{leaf}). This scenario seems reasonable if we accept the appropriateness of the electrical analog for water flow in the plant. Unfortunately, this analog may not always be appropriate. Gollan *et al.* (1986) have shown that stomatal conductance may be more closely related to the water content immediately surrounding the roots than to the leaf water potential, for slowly imposed water stress. Most likely both leaf water potential and rhizosphere water content may be important.

ORGAN ENVIRONMENT

The exchange of gases at the organ level clearly depends strongly on environmental conditions immediately surrounding those organs. For example, the transpiration rate of a leaf, which can affect leaf water potential and thus stomata and photosynthesis, depends on air temperature, air relative humidity, wind speed, and incident solar radiation. These same four environmental factors affect leaf temperature through the leaf energy budget. Consequently, the leaf environment affects photosynthesis indirectly through impact on stomatal aperture, or directly through impact on leaf temperature or light. Obtaining the dependence of various plant physiological processes on these environmental factors is difficult enough; however, understanding canopy photosynthesis is yet one step more difficult. The presence of the plant itself alters the environment immediately surrounding the organs of that plant. Furthermore the plant experiences gradients of environmental conditions over its extent, and the well adapted plant alters the physiological characteristics of various organs to make best use of the environment that the plant itself has influenced. Thus the environmental conditions measured at a weather station outside of the plant canopy are not directly usable in estimating the photosynthetic rate of a canopy. The approach that we use here is to solve simultaneously for both the canopy microenvironment and the functioning state of the leaves at various levels in the canopy by using meteorological conditions at some level above the canopy, soil conditions at a level just below the root zone, canopy architecture, and physiological and physical characteristics of leaves (Goudriaan, 1977; Norman and Campbell, 1983). This method accommodates gradients of state variables (such as temperature, humidity, radiation, etc.) and fluxes (transpiration, photosynthesis, etc.) throughout the canopy and thus can provide a reasonably general solution to this complex problem.

SYNTHESIS

The soil-plant-atmosphere model, Cupid, represents an attempt to combine organ characteristics, organ arrangement, and organ microenvironment together using fundamental principles to produce hourly, daily, or monthly estimates of canopy photosynthesis from a synthesis of more detailed information.

A comparison of predictions from the Cupid model with direct measurements of canopy photosynthesis is possible by using data from the First ISLSCP Field Experiment (FIFE) conducted near the Konza Prairie Research Natural Area, located just south of Manhattan, Kansas (ISLSCP is an acronym for the International Satellite Land Surface Climatology Program). Three species of native prairie grasses dominate the landscape: *A. gerardii* (Big blue stem), *Sorghastrum nutans* (Indian grass), and *Panicum virgatum* (Switch grass). The leaf measurements presented on *A. gerardii* in this chapter were also made on the other two species. A mean of the characteristics for the three species was used in the simulations reported here.

The direct measurements of canopy carbon dioxide fluxes were made by two methods: 1) Eddy correlation above the grass canopy, and 2) a closed, transparent chamber that was placed over a grass canopy. The eddy correlation method requires rapid response measurements of vertical velocity and carbon dioxide concentration that can be made a few meters above a grass canopy (Verma *et al.*, 1986). Thus, this method provides a flux of carbon dioxide that is averaged over tens to hundreds of square meters. The closed, transparent chamber, which encloses an area of grass canopy about two square meters in size, seals the canopy off from the atmosphere so that the carbon dioxide flux can be estimated from a change in carbon dioxide concentration within the chamber (Field *et al.*, 1989). The eddy correlation measurements were averaged over 30 minutes, whereas the chamber measurements were averaged over only a few minutes.

Both of these direct measurement methods provide a measurement of the sum of canopy and soil carbon dioxide fluxes. Thus an estimate of soil carbon dioxide flux is necessary for a comparison of measured and modeled canopy fluxes. Unfortunately, the measurement of soil carbon dioxide flux below a grass canopy is extremely difficult and no direct measurements were available on the day of interest. Nighttime eddy correlation carbon dioxide flux measurements suggest a soil carbon dioxide flux of 9 μmol $m^{-2}s^{-1}$; therefore the predicted canopy fluxes were adjusted by 9 μmol $m^{-2}s^{-1}$ for the comparison with measured fluxes.

The results of this comparison are shown in Fig. 6, with ambient environmental conditions in Table II. The model predictions fall between the results of the two measurement methods. The closed chamber is likely to underestimate canopy carbon dioxide fluxes because the two layers of plastic only transmit about 85% of the photosynthetically active radiation. Thus increasing the closed chamber estimates by 15% would seem appropriate; this was not done to the results in Fig. 6. Overall, the model predictions compare reasonably well with direct measurements.

The methodology presented in this chapter is generally applicable to many kinds of plant canopies, provides reasonable estimates of canopy photosynthesis from a knowledge of leaf characteristics, and is capable of predicting many other quantities such as net radiation, sensible heat, evaporation, transpiration, and soil heat and water fluxes. Table III contains a comparison of measured and modeled fluxes of net radiation, soil heat conduction, evapotranspiration, and sensible heat for the same site and time as the photosynthesis estimates in Fig. 6. The plant-

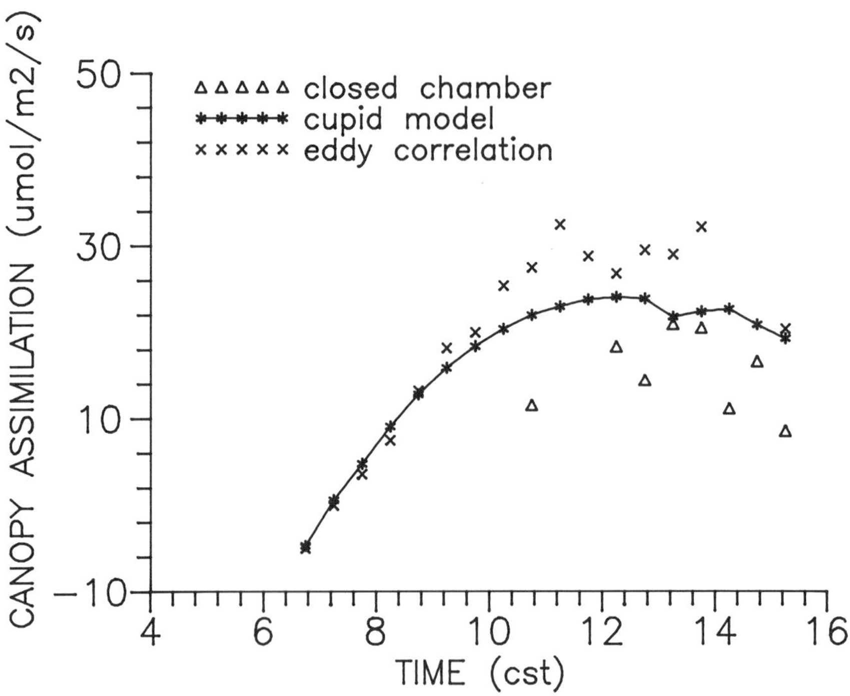

Figure 6. Comparison of carbon dioxide fluxes from a native grassland by two direct-measurement methods with predictions from the model Cupid. Predicted fluxes are adjusted by a 9 μmol $m^{-2}s^{-1}$ soil flux to be comparable to the measured fluxes. Environmental data appropriate for this simulation are in Table 2.

environment model, Cupid, also can be used to address many kinds of problems in plant-environment relations. For example, Cupid has been used to assess a detailed water balance of an irrigated crop (Norman and Campbell, 1982), to study the interaction between canopy microenvironment and Banks grass mite population dynamics (Toole *et al.*, 1984), and to link canopy characteristics to remotely-sensed radiation (Norman *et al.*, 1985). Plant productivity can be simulated by combining photosynthesis estimates with maintenance respiration (Amthor, 1984) and growth respiration (Vertregt and Penning deVries, 1987) concepts; this has been done for corn (Norman and Arkebauer, 1989). The generality of this soil-plant-atmosphere system approach makes it both challenging and useful.

Table II. Ambient environmental conditions at 2 m above the ground for a canopy 0.4 m tall[1].

Time (CST)	T_{air} (C)	e_{air} (kPa)	Wind (ms^{-1})	Solar (Wm^{-2})
0645	23.7	2.79	2.9	66
0715	24.4	2.87	3.3	138
0745	25.2	2.93	4.1	208
0815	25.8	2.93	4.6	299
0845	26.5	2.94	4.9	396
0915	27.2	2.94	5.0	492
0945	27.8	2.96	5.1	588
1015	28.3	2.98	5.2	678
1045	29.0	3.03	4.8	749
1115	29.5	3.06	4.7	813
1145	30.0	3.07	4.6	859
1215	30.3	3.08	3.9	917
1245	30.9	3.14	4.2	950
1315	31.1	3.12	4.9	1009
1345	31.2	3.10	4.8	943
1415	31.3	3.04	4.7	786
1445	31.4	3.04	5.2	967
1515	31.5	3.02	5.3	871

[1] Leaf area index = 3.0; ambient environmental conditions include air temperature (T_{air}), air vapor pressure (e_{air}), wind speed, and solar radiation. These data go with Fig. 6.

Table III. Comparison of energy fluxes from a native grassland by direct measurement with predictions from the model Cupid[1].

Time	Measured fluxes (Wm^{-2})				Predicted fluxes (Wm^{-2})			
(CST)	Rn	E	H	G	Rn	E	H	G
0645	32	37	2	-3	31	41	14	-13
0745	118	113	3	8	124	97	30	10
0845	262	192	48	22	256	214	37	18
0945	402	274	88	41	398	338	43	29
1045	518	349	106	63	522	424	58	46
1145	627	413	129	84	608	458	88	70
1245	670	456	122	93	679	464	131	87
1345	624	426	110	88	672	469	146	69
1445	555	405	75	74	678	468	149	67

[1] Rn = net radiation; E = evapotranspiration; H = sensible heat; G = soil conduction heat flux.

ACKNOWLEDGMENTS

The authors thank NASA as the principal source of support for this work through their general funding of FIFE (First ISLSCP Field Experiment) and specifically grant no. NAGW-1099. We also thank Dr. Piers Sellers of the University of Maryland for sharing his closed chamber data and Dr. S.B. Verma of the University of Nebraska for sharing his eddy correlation data.

REFERENCES

Amthor JS (1984). The role of maintenance respiration in plant growth. Plant Cell Environ 7:561-569.

Anderson JM, Osmond CB (1987). Shade-sun responses: compromises between acclimation and photoinhibition. In Kyle DJ, Osmond CB, Arntzen CJ (eds): "Topics in Photosynthesis: Photoinhibition, Vol 9," Amsterdam: Elsevier, pp 1-38.

Ball JT, Woodrow IE, Berry JA (1986). A model predicting stomatal conductance and its contribution to the control of photosynthesis under different environmental conditions. In Biggins J (ed): "Progress in Photosynthesis Research, Vol 4," Dordrecht: Martinus-Nijhoff,pp 221-224.

Ehleringer J, Pearcy, RW (1983). Variation in quantum yield for CO_2 uptake among C3 and C4 plants. Plant Physiol 73:555-559.

Farquhar GD, von Caemmerer S (1982). Modelling of photosynthetic response to environmental conditions. In Lange, OL, Nobel PS, Osmond CB, Ziegler H (eds): "Physiological Plant Ecology, II: Water Relations and Carbon Assimilation," New York: Springer-Verlag, pp 549-587.

Farquhar GD, von Caemmerer S, Berry JA (1980). A biochemical model of photosynthetic CO_2 assimilation of C3 species. Planta 149:78-90.

Field CB, Ball JT, Berry JA (1989). Photosynthesis: Principles and field techniques. In Pearcy RW, Ehleringer JR, Mooney HA, Rundel PW (eds): "Plant Physiological Ecology: Field Methods and Instrumentation," Chapman & Hall, in press.

Grace J (1981). Some effects of wind on plants. In Grade J, Ford ED, Jarvis PG (eds): "Plants and Their Atmospheric Environment," London: Blackwell Sci. Publ, pp 31-56.

Gollan J, Passioura JB, Munns R (1986). Soil water status affects the stomatal conductance of fully turgid wheat and sunflower leaves. Aust J Plant Physiol 13:459-464.

Goudriaan J (1977). "Crop Micrometeorology: A Simulation Study." Wageningen: PUDOC, 249 pp.

Idso SB, deWit CT (1970). Light relations in plant canopies. Appl Optics 9:177-184.

Kramer PJ (1983). "Water Relations of Plants." New York: Academic Press, 489 pp.

Monsi M, Saeki T (1953). Über den Lichtenfaktor in den Pflanzengessellschafte und seine Bedeuting für die Stoffproduktion. Jap J Bot 14:22-52.

Nilson T (1971). A theoretical analysis of the frequency of gaps in plant stands. Agric Meteorol 8:25-38.

Norman JM (1979). Modeling the complete crop canopy. In Barfield BJ, Gerber J (eds): "Modification of the Aerial Environment of Crops," St. Joseph, Michigan: Am Soc Agric Engr, pp 249-277.

Norman JM (1980). Interfacing leaf and canopy light interception models. In Hesketh JD, Jones JW (eds): "Predicting Photosynthate Production and Use for Ecosystem Models," Boca Raton, Florida: CRC Press, pp 49-67.

Norman JM (1986). Instrumentation use in a comprehensive description of plant-environment interactions. In Gensler W (ed) "Advanced Agricultural Instrumentation," Dordrecht: Martinus-Nijhoff, pp 249-307.

Norman JM, Campbell GS (1983). Application of a plant-environment model to problems in irrigation. In Hillel DI (ed): "Advances in Irrigation," New York: Academic Press, pp 155-188.

Norman JM, Arkebauer TJ (1989). Predicting canopy light-use efficiency from leaf characteristics. In Ritchie JT, Hanks J (eds): "Modeling Plant and Soil Systems," Madison, Wisconsin: Am Soc Agron, in press.

Norman JM, Welles JM, Walter EA (1985). Contrasts among bidirectional reflectance of leaves, canopies and soils. Trans Geosci and Remote Sensing, GE-23:695-704.

Schoolfield RM, Sharp PJH, Magnuson CE (1981). Non-linear regression of biological temperature-dependent rate models based on absolute reaction-rate theory. J Theor Biol 88:719-731.

Toole, JL, Norman JM, Holtzer T, Perring T (1984). Simulating Banks grass mite population dynamics as a subsystem of a corn canopy-microenvironment model. Environ Entomol 13:329-337.

Verma SB, Baldocchi DD, Anderson DE, Matt DR, Clement RJ (1986). Eddy fluxes of CO_2, water vapor and sensible heat over a deciduous forest. Boundary-Layer Meteorol 36:71-91.

Vertregt N, Penning deVries, FWT (1987). A rapid method for determining the efficiency of biosynthesis of plant biomass. J Theor Biol 128:109-119.

deWit CT (1965). Photosynthesis of leaf canopies. Agric Res Rep no 663, Wageningen: Central Agric Publ Doc, 57 pp.

Whitfield DM, Conner DJ (1980). Penetration of photosynthetically active radiation into tobacco crops. Aust J Plant Physiol 7:449-461.

Photosynthesis, pages 243–255

THE "CO_2 CONCENTRATING MECHANISM" OF CYANOBACTERIA: PHYSIOLOGICAL, MOLECULAR AND THEORETICAL STUDIES.

Aaron Kaplan[1], Devorah Friedberg, Rakefet Schwarz, Rina Ariel, Julie Seijffers and Leonora Reinhold.

Life Science Institute, The Hebrew University of Jerusalem, 91904 Jerusalem, Israel.

In 1976 a group at the Carnegie Institution, headed by Dr. J.A. Berry, demonstrated that a syndrome of biochemical and physiological changes is involved in the adaptation of green alga and cyanobacteria from high ambient CO_2 (1-5% CO_2 in air) to a low (air) level of CO_2 (Badger *et al.*, 1977; Badger *et al.*, 1978; Berry *et al.*, 1976). This syndrome leads to a large increase in the activity of a mechanism for concentrating inorganic carbon (Ci) at the carboxylating site (Abe *et al.*, 1987; Badger *et al.*, 1980; Kaplan *et al.*, 1980; Miller and Colman, 1980). Hence, it enables the cells to grow in the presence of the air level of CO_2, conditions under which the concentration of CO_2 in the medium, at equilibrium with air, is some twenty times lower than the Km(CO_2) of the cyanobacterial Rubisco (Badger, 1980).

In this communication we shall briefly describe some of our recent physiological and molecular studies on the "CO_2 concentrating mechanism" of cyanobacteria. We shall also present some of the main conclusions drawn from our quantitative model for inorganic carbon (Ci) fluxes in cyanobacteria.

STUDIES WITH THE AID OF HIGH CO_2 REQUIRING MUTANTS.

Very little information is as yet available on the molecular mechanism(s) of Ci uptake or of the adaptation to the air level of CO_2. An analysis of the polypeptide pattern of a membrane preparation, as observed following SDS-PAGE, has indicated a large accumulation of a

1 Address for correspondence: Prof. A. Kaplan, Department of Botany, The Hebrew University of Jerusalem 91904 Jerusalem, Israel, Bitnet: AARONKA@HBUNOS.

42kDa polypeptide in a cytoplasmic membrane-enriched fraction during adaptation to low CO_2 (Omata and Ogawa, 1986). The role, if any, of this polypeptide in Ci transport and accumulation is not yet understood but it might be elucidated with the aid of a mutant which does not accumulate this protein. Should the protein indeed have a role in adaptation to low CO_2 or in Ci uptake and accumulation, such a mutant might not be able to adapt to or grow at the air level of CO_2. We have recently reported on the physiological characters of a mutant of *Synechococcus* PCC7942 (E_1) which requires a high concentration of CO_2 for growth (Kaplan *et al.*, 1987; Marcus *et al.*, 1986). The E_1 mutant however, is capable of accumulating the 42kDa polypeptide even when exposed to a CO_2 concentration as high as 0.3% (Omata *et al.*, 1987).

We have isolated another high CO_2-demanding mutant, O_{221}, which does not accumulate the 42kDa polypeptide (Schwarz *et al.*, 1988). Its photosynthetic performance, apparent photosynthetic affinity for extracellular Ci, and Vmax (Fig. 1), resemble those of the E_1 mutant (Marcus *et al.*, 1986). Uptake of the different Ci species was slightly reduced (Fig. 2), compared to the wild type (WT), but this difference was too small to account for the very large difference in the apparent

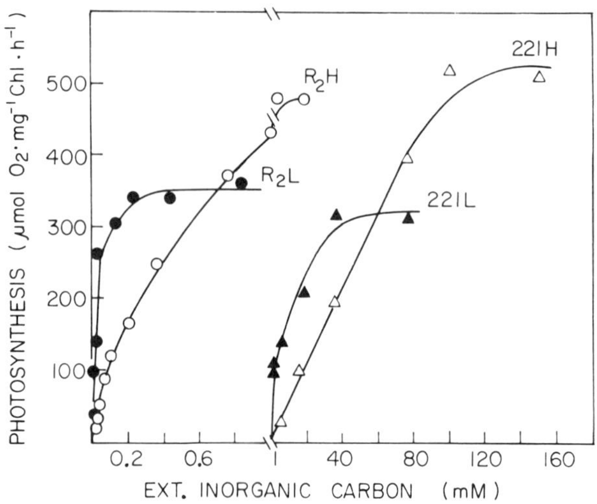

Figure 1. The rate of photosynthetic O_2 evolution as a function of the extracellular inorganic carbon (Ci) concentration in wild type *Synechococcus* PCC7942 and the mutant O_{221}. The cells were grown in the presence of 5% CO_2 in air. Cells were either kept under the high CO_2 (H) or exposed to a 1:1 mixture of air and CO_2-free air (L), for 15 hours before the experiments. The experiments were conducted at pH 8.0 in the O_2 electrode chamber, 30C. Note the different scales for the Ci concentration in the case of the mutant.

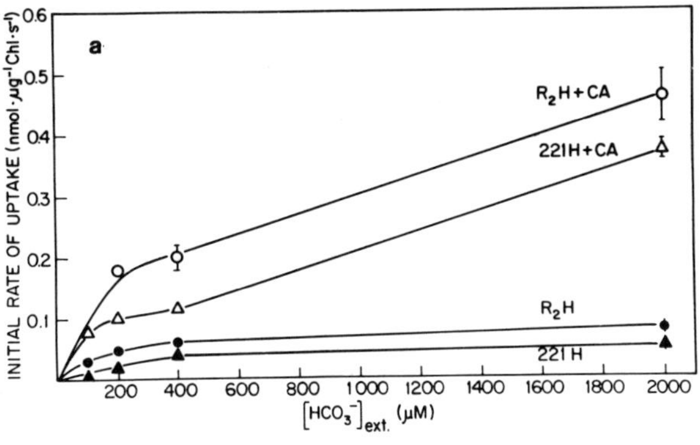

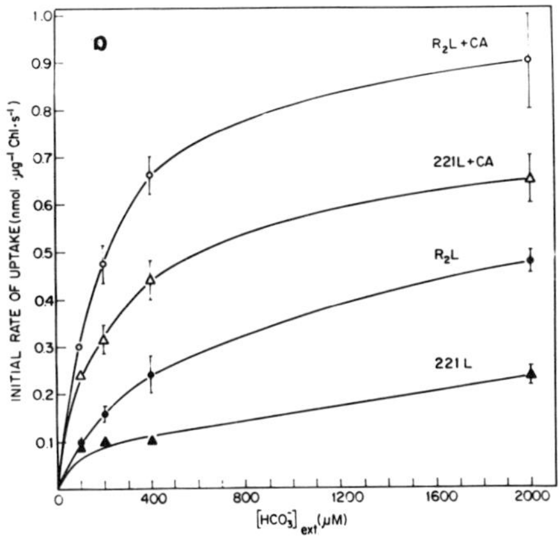

Figure 2. The initial rate of Ci uptake as a function of the nature and concentration of the Ci species supplied. High (a) and low CO_2-exposed (b) cells of the wild type (R2) and the mutant O_{221} (221) were provided with HCO_3- (a&b) in the presence or absence of carbonic anhydrase (CA), or with CO_2 (c) for 5 sec.

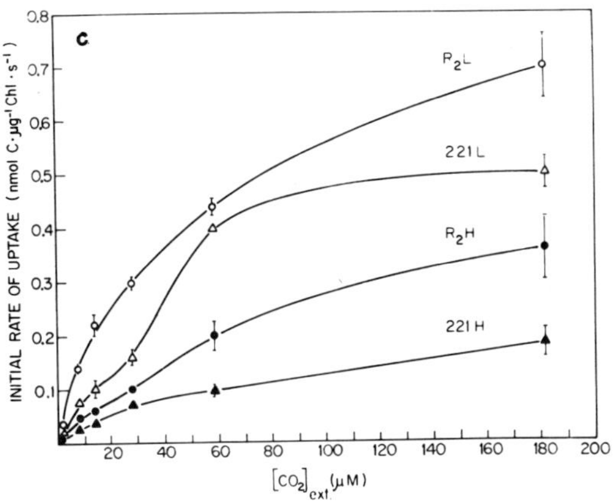

photosynthetic affinity for extracellular Ci (and growth) as between the WT and O_{221}. The differing kinetics for CO_2 and HCO_3- and further the effect of carbonic anhydrase on HCO_3 uptake (Fig. 2) might suggest that CO_2 is the principal Ci species taken up by *Synechococcus*. However, a firm conclusion as to the relative roles of each Ci species would have to rest on detailed analysis of the uptake parameters for each species and the competition between them during uptake, taking into account the relative abundance of each Ci species under the specific growth and experimental conditions (see Kaplan *et al.*, 1988; Schwarz *et al.*,1988).

The curve relating the rate of photosynthesis to intracellular Ci concentration in the mutant lies well below that depicted for the WT under both low (Fig. 3) and high CO_2 (not shown) conditions. Thus, the mutant appears to be defective in its ability to utilize the intracellular Ci

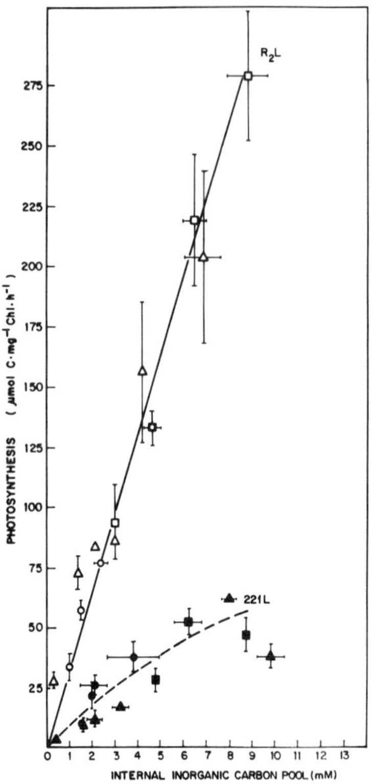

Figure 3. The rate of acid stable ^{14}C accumulation (photosynthetic products) as a function of the intracellular ^{14}Ci concentration. The cells were provided with different concentrations of HCO_3- (open symbols), HCO_3- plus carbonic anhydrase (CA,) or CO_2 (closed symbols) for 5 sec. Data (presented as the mean +- SD, n=5) are from the same experiments where the uptake of CO_2 and HCO_3- were determined (Fig 2).

pool photosynthetically. This lack could stem from a defective Rubisco with reduced affinity, i.e. a much higher $Km(CO_2)$. However, we could not detect large differences in the kinetic parameters of Rubisco as between the mutant and the WT (not shown). The data presented in Fig. 3 could contribute to the inability of the mutant to grow under limiting CO_2 levels. Since the mutant accumulated Ci internally almost as efficiently as did the WT, we must conclude that the 42kDa polypeptide does not participate directly in Ci uptake. This polypeptide might nevertheless have some role in the low CO_2 syndrome.

Synechococcus PCC7942 is a recombinant-proficient host capable of inserting homologous DNA into its genome by single or double crossing over (Williams and Szalay, 1983). We have constructed a genomic DNA library from this strain in pBRK322, a modified pBR322 plasmid into which we have inserted the NPTII gene which codes for kanamycin resistance. The high CO_2-demanding E_1 and O_{221} mutants were transformed with the genomic library. Transformed cells capable of growing under the air level of CO_2, in the presence of kanamycin, were isolated and characterized. Their physiological and biochemical characters, including the accumulation of the 42kDa polypeptide were similar to those in the WT. The plasmid pE12 which complemented both E_1 and O_{221} was rescued from the genome and partly characterized (Friedberg *et al.*, 1987). Comparison of the restriction map of the pE12 with that of the structural gene which codes for the 42kDa polypeptide (kindly provided by Dr. J. Pierce and Dr. T. Omata) clearly indicated that the pE12 fragment does not contain the structural gene for the 42kDa polypeptide. We have subcloned a BstXI/BstEII fragment (about 1.7kb) from the pE12, which complemented the O_{221} mutant but not the E_1 mutant.

Insertion of the NPTII gene, which codes for kanamycin resistance, into specific sites of a DNA fragment, has been used as a powerful means for the inactivation of various genes. Transformation of the WT with a construct containing the NPTII gene, inserted in the BstEII site turned out to be lethal. The cells were unable to grow in the presence of kanamycin under high or low-CO_2 conditions. On the other hand growth under high (but not low) CO_2 conditions, in the presence of kanamycin, was obtained following the insertion of the NPTII gene at the BstXI site. These data strongly support our previous conclusion that the BstEII/BstXI clone contain DNA sequences essential for growth under low CO_2. We are currently sequencing the cloned DNA fragment in order to define the mutation in O_{221}. This should clarify the nature of this fragment which is involved in the ability to grow under low CO_2 and in the accumulation of the 42kDa polypeptide. The data (Friedberg *et al.* in preparation) bear on the possibility that this fragment might be involved in the organization of the carboxysomes.

ROLE OF CARBOXYSOMES.

The possibility that the cyanobacterial Rubisco gene itself might be involved in the ability to grow at the air level of CO_2 should be considered, particularly in view of the recent data of Pierce *et al.* (Pierce *et al.*, 1988). These workers replaced the structural gene of Rubisco in *Synechocystis* 6803 with that from the photosynthetic bacterium *R. rubrum.* The resulting organism, named cyanorubrum, demanded high CO_2 for growth, was also very sensitive to oxygen, and lacked visible carboxysomes.

Very little information is as yet available on the structural organization and the functional significance of the carboxysomes in cyanobacterial photosynthesis (Codd and Marsden, 1984). The number of carboxysomes within the cells increases during adaptation to low CO_2 (Turpin *et al.*, 1984). The structure of the carboxysomes from *Thiobacillus* has recently been partly clarified (Holthuijzen *et al.*, 1986). They are surrounded by a protein layer which does not appear to contain any lipids. The Rubisco molecules are concentrically arranged. SDS-PAGE gels of the carboxysomes proteins (Codd and Marsden, 1984, Volokita unpublished) indicate the presence of about 8 different polypeptides but apart from the Rubisco, none of these polypeptides appears to be an enzyme of the PCR cycle. The completion of the PCR cycle might therefore require large fluxes of metabolites to and from the carboxysomes and it is not known to what extent these fluxes rate limit the maximal photosynthetic rate of cyanobacteria. It is recognized that back-diffusion of CO_2 from the accumulated Ci pool could result in a prohibitively high energy cost of the CO_2 concentrating mechanism (Raven and Lucas, 1985) and it has therefore been assumed that the plasmalemma of these organisms must have very low permeability for CO_2 (Badger *et al.*, 1985; Marcus *et al.*, 1986). Assigning such a high resistance to CO_2 diffusion to the plasmalemma poses at least two serious problems. First, it runs counter to the known PCO_2 of lipid bilayers. Second, assuming that permeability to O_2 is of the same order as that to CO_2, such a diffusion barrier would seriously hamper the escape of photosynthetically evolved O_2 from the cell. Badger *et al.* (1985) have concluded that the partial pressure of O_2 would build up to about 2 atm O_2 in *Synechococcus* at Ci concentrations just saturating for photosynthesis.

We have recently proposed (Reinhold *et al.*, 1987; Reinhold *et al.*, submitted) two models for Ci fluxes and photosynthesis in cyanobacteria which obviate the need for assigning a high resistance to CO_2 to the plasmalemma. The models assume that the plasmalemma in cyanobacteria does not differ in this respect from the normal boundary membranes encountered in other organisms. The first model locates a barrier to CO_2

diffusion at the surface of the carboxysome. The second requires no such diffusion barrier in the cell, since CA is placed centrally in the interior of the carboxysome and much of the CO_2 formed at this site is fixed as it diffuses outwards towards the carboxysome surface, past Rubisco sites located along the diffusion path (see schematic drawing in Fig. 4).

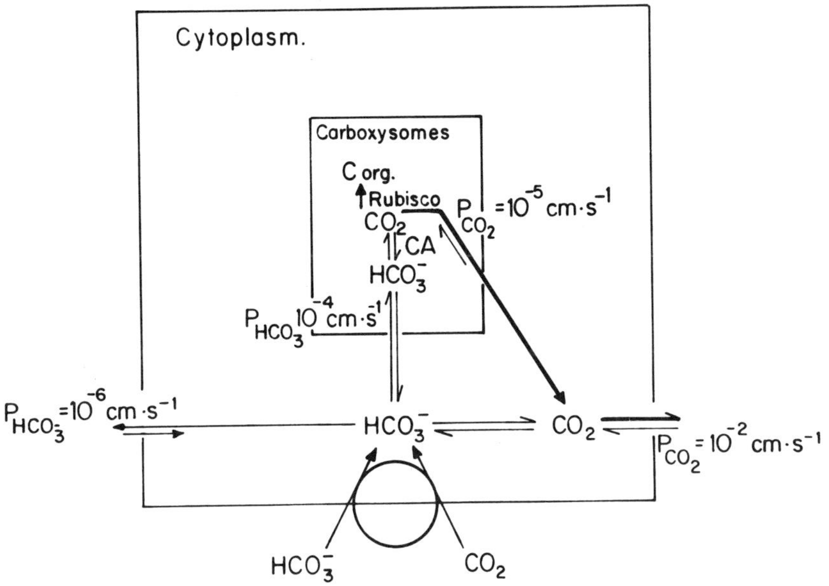

Figure 4. Schematic representation of proposed model for Ci fluxes and photosynthesis in a cyanobacterial cell, indicating the permeability coefficients assigned to the various membranes. The carboxysome compartment represents the total combined volume of the carboxysomes. The outer layers of the cell envelope are not shown.

Based on our previous findings (Schwarz *et al.*, 1988; Volokita *et al.*, 1984) the model postulates that the transport mechanisms take up both HCO_3- ions and CO_2 from the medium but that the Ci species released into the cell interior is HCO_3- in both cases (Abe *et al.*, 1987; Schwarz *et al.*, 1988; Volokita *et al.*, 1984). It further proposes that conversion of HCO_3- to CO_2 in the cytoplasm is uncatalyzed and thus relatively slow. The accumulated HCO_3- ions penetrate into the carboxysomes where carbonic anhydrase is present; CO_2 is rapidly generated in the immediate vicinity of Rubisco and part of it is fixed. The CO_2 concentration in the carboxysomes will be much higher than that in the external medium; some of the CO_2 will leak from the carboxysomes into the cytoplasm and may diffuse into the medium.

The predicted relation between CO_2 fixation and external Ci concentration (based on the calculated carboxysomal CO_2 concentration) is shown in Fig. 5. The kinetic parameters for photosynthesis and Ci uptake, per *Anabaena* cell has been deduced from the results of Kaplan *et al.* (1980) and Volokita *et al.* (1984). The curve does not depart markedly from a rectangular hyperbola in spite of the curvilinear

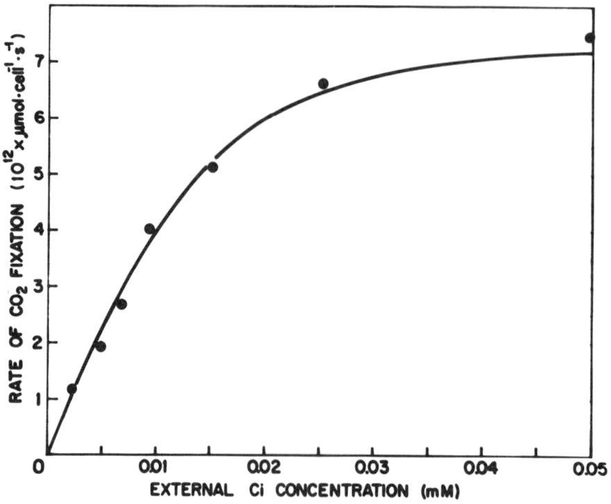

Figure 5. The predicted curve for photosynthesis by an *Anabaena* cell as a function of external Ci concentration, based on the carboxysomal CO_2 concentration predicted by the model.

relations shown for carboxysomal CO_2 concentration (Reinhold *et al.*, 1987), and is a remarkably good fit through the points obtained experimentally (Kaplan *et al.*, 1980). In order to fit experimental points obtained for other species certain parameters may need adjustment, e.g. the kinetic parameters of the Ci transport process or the value for carboxysome surface area (Reinhold *et al.*, submitted). Figure 6 shows the dependence of photosynthesis on the degree of catalysis by CA, at 3 external Ci concentrations, (the external concentration allowing half maximal rate of photosynthesis at this pH is 0.01mM Ci). At saturating external Ci concentration (0.1 mM) CA concentration such as to accelerate the rate of inter-conversion of Ci species by a factor as low as x10 is sufficient to allow photosynthesis to proceed at almost 90 per cent of maximum rate. This degree of catalysis, though very low, is nevertheless essential - when the factor is x1 (i.e. the uncatalyzed rate) the rate of photosynthesis drops to about 20 per cent maximum rate. As the external Ci concentration is reduced the amount of CA required to support photosynthesis is progressively raised (Fig. 6).

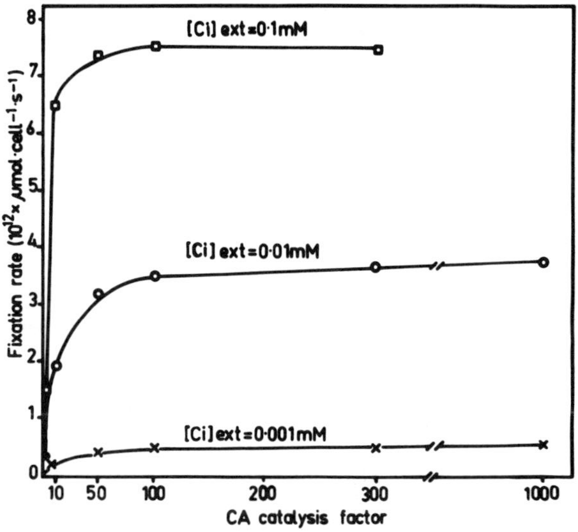

Figure 6. Rate of CO_2 fixation by an Anabaena cell as a function of carbonic anhydrase (CA) activity in the carboxysome, at three external Ci concentrations. The CA catalysis factor gives the speed of interconversion of Ci species relative to the uncatalyzed rate.

HOW MUCH CO_2 LEAKS BACK INTO THE MEDIUM?

The HCO_3- content of the cytoplasm falls sharply with increase in carboxysomal CA factor up to about CA x100 (Fig. 7A). Presumably this fall in steady state HCO_3- concentration reflects the fact that, with increase in CA factor, the rate of HCO_3- consumption in the carboxysomes increases, whereas the rate of HCO_3- entry into the cytoplasm via the uptake mechanism remains constant. The rather surprising finding that the CO_2 leak at any one external Ci concentration is greatest when no CA is present in the carboxysomes, and falls sharply as the CA factor is increased to x100 (Fig. 7B) is probably explained by the parallel changes in cytoplasmic HCO_3- concentration (Fig. 7A). The direct source of the CO_2 leak to the medium is the cytoplasmic CO_2 pool. Below a CA factor of x50 nearly all the cytoplasmic CO_2 originates from cytoplasmic HCO_3- by uncatalyzed conversion (see Fig. 7B). CO_2 leaking out of the carboxysomes becomes a significant fraction of total cytoplasmic CO_2 at a carboxysomal CA factor of about x50, and forms the greater proportion of cytoplasmic CO_2 at factors above x100 (Fig. 7B). The ratio fixed CO_2/leaked CO_2 (Fig. 8) is thus a function both of the CA factor and of the external Ci concentration as well as of other factors such as carboxysome surface area (not shown but see Reinhold *et al.*, submitted). The leak of CO_2 predicted by the model is considerable - the ratio CO_2 leaked to CO_2 fixed is approximately one to four at external Ci concentration equal to the Km value, rising to about one to two at Ci concentrations just saturating photosynthesis when the CA factor is x1000, but it is not in marked conflict with that deduced by Badger *et al.* (1985) for *Synechococcus*. In other species this ratio

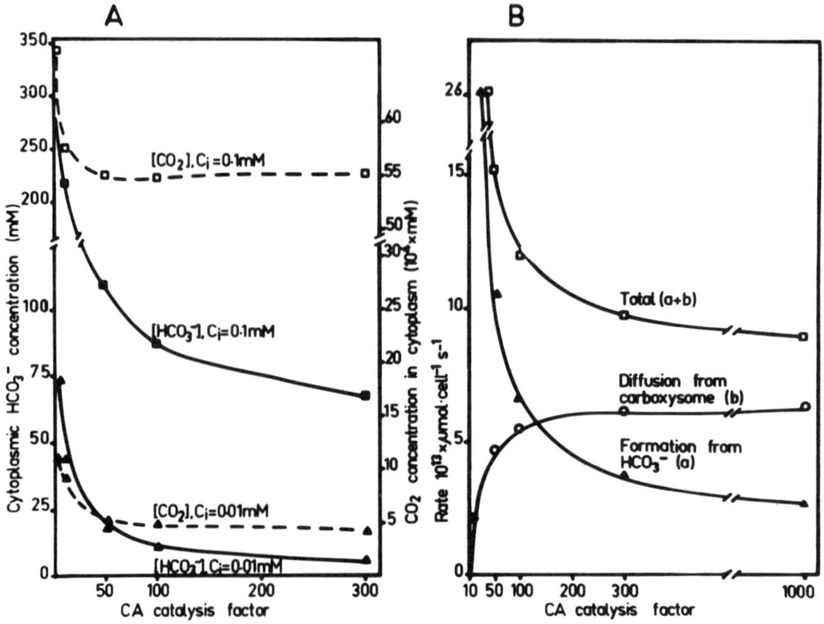

Figure 7. A) Cytoplasmic HCO_3- and CO_2 concentrations as a function of carbonic anhydrase activity in the carboxysome, at two external Ci concentrations as indicated on the curves. B) The components of the cytoplasmic CO_2 pool: Rate of CO_2 formation from cytoplasmic HCO_3-, or rate of arrival in the cytoplasm by diffusion from the carboxysomes, as a function of carbonic anhydrase activity in the carboxysome. External Ci concentration 0.01 mM.

may vary owing to varying values for the different parameters for Ci transport. Modifying some of the values taken for the various parameters, e.g. decreasing the value for total surface area of the carboxysomes, would increase the proportion of the CO_2 fixed (not shown). At saturating external Ci concentrations back-diffusion of CO_2 from the accumulated Ci pool might constitute a waste of energy serious enough to detract substantially from the usefulness to the organism of the CO_2-concentrating mechanism (Raven and Lucas, 1985).

The leak our model predicts would be an overestimation if there were a steep gradient in CO_2 concentration within the carboxysome, such that the CO_2 concentration at the carboxylation site(s) was considerably higher than that at the boundary of the carboxysome. Such a gradient might well arise if CA were placed within the carboxysome, and if a high proportion of the CO_2 formed at this site were fixed at Rubisco sites

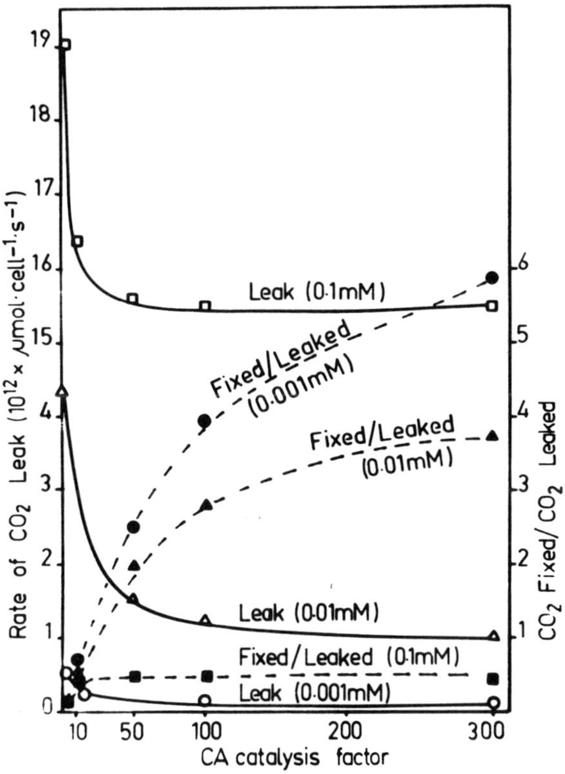

Figure 8. Rate of leakage of CO_2 (solid lines) and the ratio CO_2 fixed/CO_2 leaked (dashed lines) as functions of carbonic anhydrase (CA) activity in the carboxysome. The curves have been calculated for 3 external Ci concentrations as indicated in the Figure.

along the diffusion path as CO_2 diffused outwards towards the carboxysome surface (see Reinhold *et al.*, 1987).

THE EFFECT OF VARYING THE Km(CO_2) FOR RUBISCO.

Figure 9 shows that as the Km is varied from 0.01 to 0.5 mM, the rate of CO_2 fixation falls. The CO_2 concentration in the carboxysome rises accordingly, and the leak of CO_2 from the carboxysome and subsequently from the cytoplasm rises in consequence. Partly owing to the damping effect of the rise in CO_2 concentration, CO_2 fixation rate only falls by about 20 per cent over the range from Km=0.1 to Km=0.5. Thus, if the standard Km(CO_2) chosen for working out the model, 0.25mM, is inaccurate, the impact on the model would be minor. The advantage to be gained, from the point of view of CO_2 fixation rate, by decreasing the Km for CO_2 fixation, is very modest. This prediction may have serious implications for attempt to produce and isolate a cyanobacterial mutant of Rubisco, which exhibits an "improved" CO_2/O_2 specificity, in an organism where normal carboxysomes are present.

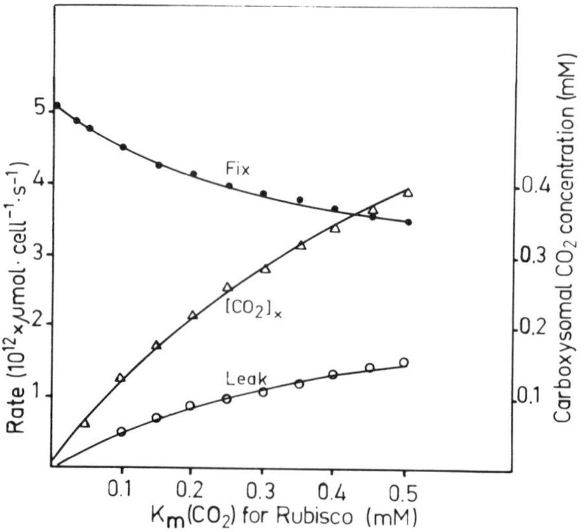

Figure 9. The rates of CO_2 fixation and leakage respectively, and the carboxysomal CO_2 concentration, as functions of the Km(CO_2) taken for Rubisco. External Ci 0.01 mM.

REFERENCES.

Abe T, Tsuzuki M, Miyachi S (1987). Plant Cell Physiol 28:867-874.

Badger MR (1980). Arch Biochem Biophys 201:247-254.

Badger MR, Kaplan A, Berry JA (1977). Carnegie Inst YB76:362366.

Badger MR, Kaplan A, Berry JA (1978). Carnegie Inst YB77:251-261.

Badger MR, Kaplan A, Berry JA (1980). Plant Physiol 66:407413.

Badger MR, Bassett M, Comins HN (1985). Plant Physiol 77:465-471.

Berry J, Boynton J, Kaplan A, Badger M (1976). Carnegie Inst. YB75:423-432.

Codd GA, Marsden WJN (1984). Biol Rev 59:389-422.

Friedberg D, Schwarz R, Kaplan A (1987). In Papageorgiou GC (ed): "EMBO workshop on oxygenic and unoxygenic electron transport system in cyanobacteria." Cape Sounion Greece, pp 34.

Holthuijzen YA, van Breeman JFL, Kuenen JG, Konings WN (1986). Arch Microbiol 144:398-404.

Kaplan A, Badger MR, Berry JA (1980). Planta 149:219-226.

Kaplan A, Zenvirth D, Marcus Y, Omata T, Ogawa T (1987). Plant Physiol 84:210-213.

Kaplan A, Lerner M, Scherer S (1988). Plant Physiol, in press.

Marcus Y, Schwarz R, Friedberg D, Kaplan A (1986). Plant Physiol 82:610-612.

Miller AG, Colman B (1980). J Bacteriol 143:1253-1259.

Omata T, Ogawa T (1986). Plant Physiol 80:525-530.

Omata T, Ogawa T, Marcus Y, Friedberg D, Kaplan A (1987). Plant Physiol 83:892-894.

Pierce J, Carlson TJ, Williams JGK (1988). PNAS (in press).

Raven JA, Lucas WJ (1985). In Lucas WJ, JA Berry (eds): "Carbon Uptake by Aquatic Photosynthetic Organisms." ASPP pp 305-324.

Reinhold L, Zviman M, Kaplan A (1987). In Biggins J (ed): "Progress in Photosynthesis Research." Vol IV Martinus Nijhoff Pub Dordrecht, pp 6.289-6.296.

Reinhold L, Zviman M, Kaplan A (1989). Plant Physiol (submitted).

Schwarz R, Friedberg D, Kaplan A (1988). Plant Physiol 88:284-288.

Turpin DH, Miller AG, Canvin DT (1984). J Phycol 20:249253.

Volokita M, Zenvirth D, Kaplan A, Reinhold L (1984). Plant Physiol 76:599-602.

Williams JGK, Szalay AA (1983). Gene 24:34-51.

Photosynthesis, pages 257–268

LIGHT-REGULATION AND THE PEA RBCS-3A GENE

Cris Kuhlemeier and Nam-Hai Chua

Laboratory of Plant Molecular Biology, The Rockefeller University, 1230 York Avenue, New York, NY 10021-6399

Light is of critical importance to the plant as an energy source to drive photosynthesis. In addition, plants monitor the quality and the quantity of the incident light to regulate their growth, differentiation, and metabolism. It is hardly an exaggeration to state that light controls almost every process in plants. For example, the rate and direction of growth, leaf movement, stomatal opening, germination, chloroplast development, and flowering can all be influenced by light (Shropshire and Mohr, 1983). Lettuce seeds will only germinate if they are irradiated with light. A red flash is sufficient to trigger the response but a far red pulse immediately following will counteract the red light treatment and no germination will occur. Shade-adapted plants such as *Oxalis oregana* respond to high light intensities by closing their leaflets, a reaction that takes place within minutes (Briggs and Iino, 1983). Such an enormous diversity in the plant's response to light raises at least three questions: [1] how is the light signal perceived; [2] how is it transmitted; [3] how is the specific response elicited? To address these questions we have chosen to study a relatively simple light response, namely light-regulated expression of genes coding for the chloroplast proteins ribulose-1,5-bisphosphate carboxylase (rbcS) and the chlorophyll *a/b* binding protein (cab or lhcp). These genes rapidly respond to light by a 20- to 50- fold increase in the steady state levels of their mRNAs. The structure of their genes has been studied in great detail and the availability of techniques for *in vitro* mutagenesis and gene transfer allows an analysis of the relationships between gene structure and gene regulation.

The purpose of this contribution is to sketch the latest developments in the molecular characterization of gene expression and to relate this information to the total chain of events that is required to transform the light signal into an on/off command for the transcriptional machinery. Several related reviews have recently appeared (Briggs *et al.*, 1985; Briggs and Iino, 1983; Fluhr *et al.*, 1986a; Kuhlemeier *et al.*, 1987a; Lagarias, 1985; Lagarias *et al.*, 1987; Nagy *et al.*, 1988a; Roux, 1984; Schäfer and Briggs, 1986; Tobin and Silverthorne, 1985). Here we will focus on the pea rbcS gene, a gene which has been under intense investigation in our laboratory.

THE DIVERSITY OF LIGHT-REGULATED GENE EXPRESSION

Both rbcS and cab genes are induced by light. However, in pea, the cab genes are turned on by a fluence of light which is several orders of magnitude lower than what is needed for turning on the rbcS genes (Kaufman *et al.*, 1984). The induction of the chalcone synthase gene in tobacco is mediated by light in the UV-B part of the spectrum. On the other hand, in *Antirrhinum majus* the chalcone synthase gene is induced by visible light (Kaulen *et al.*, 1986). Other genes, coding for phytochrome or protochlorophyllide reductase are expressed in dark-grown seedlings, but repressed when the plants are exposed to light (Apel, 1981; Colbert *et al.*, 1985; Otto *et al.*, 1984). Kaufman *et al.* 1986, have studied the kinetics of induction of a number of light-responsive genes in pea and demonstrated clear-cut differences in the rate of induction, the escape time (that is the time after a red flash that a far red irradiation can still effectively reverse induction) and the duration of the mRNA increase (Kaufman *et al.*, 1984; Kaufman *et al.*, 1985; Kaufman *et al.*, 1986). Thus, different genes may respond very differently to the same light stimulus.

In pea, there are five genes which code for the small subunit of Rubisco (Fluhr *et al.*, 1986b, and references therein). Since all five members code for identical mature polypeptides, the significance of the presence of multiple genes, if any, must lie in the regulation of their expression. Clearly, there are quantitative differences, transcripts from the most highly expressed genes, rbcS-3A and 3C being at least ten times more abundant than those from the low-expressing rbcS-E9 and 3.6 genes (Fluhr *et al.*, 1986b; Kuhlemeier *et al.*, 1988b). In addition, there are subtle qualitative differences: the lowly expressed genes are under-represented in seeds and petals. Moreover, in etiolated seedlings, but not in mature green dark-adapted plants, the low-expressing genes show a longer lag time before their transcripts start to accumulate. Within the tomato gene family, the qualitative differences between rbcS family members are even more striking (Sugita and Gruissem, 1987). The wavelength requirements of the rbcS genes in pea depend on the developmental stage of the plant. In etiolated seedlings, a red light pulse is sufficient for induction and the effect is far-red reversible. In contrast, red light is not very effective in mature plants and blue light is needed to turn on the gene (Fluhr and Chua, 1986). Thus, within a gene family and even for a single gene, the mode of expression changes depending on developmental stage and environmental conditions.

TRANSGENIC PLANTS AND LIGHT-REGULATED GENE EXPRESSION

The current approach for studying the relation between gene structure and gene regulation in higher plants is to introduce wild-type genes and various mutated versions thereof into plant cells by *Agrobacterium*-mediated gene transfer. Intact plants can be generated which carry one or a few copies of the transgene in all their cells and which transmit the acquired trait in a strictly Mendelian fashion (see Kuhlemeier *et al.*, 1988b and references therein). We want to emphasize two important aspects of this transgenic plant system.

First, transgenic plants can be produced and maintained in large numbers. In plants such as tobacco, somatic cells can efficiently regenerate into complete organisms, which obviates the need for tedious germ-line transformation. And whereas lines of transgenic flies or mice need to be propagated continuously, transgenic plants can be stored as seeds which greatly simplifies the task of maintenance. This means that it is feasible to analyze large numbers of mutant constructs in transgenic plants with relatively little effort.

Second, recent work in mammalian systems has confirmed that cell culture systems may behave very differently from whole organisms. For instance, in the upstream region of the mouse alpha-fetoprotein gene, three redundant enhancer elements have been defined in cell-culture expression systems. When the same elements were studied in transgenic mice, a very different picture emerged with both the temporal and the tissue-specific patterns of expression requiring combinations of these elements (Hammer *et al.*, 1987). In plants, the totipotency of cells, which is such an asset for DNA transfer techniques, may complicate the use of cell cultures. For instance, Hirth and co-workers (Fleck *et al.*, 1979; Vernet *et al.*, 1982) showed that the rbcS genes are highly expressed in the leaves of *Nicotiana sylvestris*. When protoplasts were prepared from these leaves, expression was immediately shut off. This problem and the implications for the early transformation experiments involving the expression of light-regulated genes in non-morphogenic calli have been extensively discussed (Fluhr *et al.*, 1986a). In other cases, such as the elicitor/UV-induced expression of defense genes in parsley cells or somatic embryogenesis in carrot cells, cell cultures may more truly reflect the reality of the intact plant.

A complication in the analysis of transgenic organisms is that the level of expression and sometimes also the qualitative pattern of expression is influenced by the site of insertion into the genome. In two cases expression variations as large as 50- to 200-fold could be observed with the same construct, when the levels of RNA were assayed in individual transgenic plants (Jones *et al.*, 1985; Nagy *et al.*, 1985). In our recent experiments with the pea rbcS-3A gene, however, we have observed far less variability: 90 percent of the plants vary no more than 5-fold in absolute RNA levels (Kuhlemeier *et al.*, 1987b ;Kuhlemeier *et al.*, 1988b). This decreased heterogeneity in expression levels may be gene-specific or it may be correlated with the use of new vectors with better selectable markers. It should be noted that in plants there is as yet no unambiguous evidence that the observed variability is due to position effects. Alternatively, the difference could be influenced by other parameters, such as [1] small mutations in the transgene which escape detection by Southern analysis; [2] the plants may not have been harvested under identical conditions; [3] mutations elsewhere in the genome induced by the tissue culture procedures (somaclonal variation); [4] the transgenic plant may in fact be a chimera in which only some of the cells contain the transgene. The rigorous test for true positional variation is that an aberrantly behaving transgene must give rise to the normal scale of expression levels, when reintroduced into plants (Levis *et al.*, 1985). Such analysis has not yet been performed in plants.

CIS-ACTING ELEMENTS

We will focus here on the pea rbcS-3A gene, as to our knowledge, this is the gene studied in most detail (Fluhr and Chua, 1986; Fluhr *et al.*, 1986a; Fluhr *et al.*, 1986b; Green *et al.*, 1987; Kuhlemeier *et al.*, 1987b; Kuhlemeier *et al.*, 1988a; Kuhlemeier *et al.*, 1988b). Other light-regulated genes for which considerable data exist, are the pea rbcS-3.6 (Simpson *et al.*, 1986; Timko *et al.*, 1985), 3C (Fluhr and Chua, 1986; Fluhr *et al.*, 1986b), and E9 (Fluhr *et al.*, 1986a; Morelli *et al.*, 1985; Nagy *et al.*, 1985) genes, a *Nicotiana plumbaginafolia* rbcS gene (Poulsen *et al.*, 1986; Poulsen and Chua, 1988), a pea cab gene (Simpson *et al.*, 1986a; Simpson *et al.*, 1986b), a *N. plumbaginafolia* cab gene (Castresana *et al.*, 1988), a wheat cab (Nagy *et al.*, 1986; Nagy *et al.*, 1987; Nagy *et al.*, 1988b) gene and an *Antirrhinum majus* chalcone synthase gene (Kaulen *et al.*, 1986).

The pea rbcS-3A gene with 410 bp of 5' flanking sequence gives correct and high level expression in transgenic petunia and tobacco, that is, expression is induced by white or blue light in mature green plants and by red light in etiolated seedlings (Fluhr and Chua, 1986; Fluhr *et al.*, 1986a; Fluhr *et al.*, 1986b). In addition the expression is tissue-specific in the same fashion as in pea. The absolute mRNA level is at least 30% of the level in pea.

The upstream region from -330 to -50 bp contains all the information needed for light-responsive and cell-specific expression. In either orientation, this DNA fragment confers these properties upon a constitutive reporter gene consisting of the CaMV 35S TATA box and the bacterial chloramphenicol acetyltransferase (CAT) coding sequence. Therefore the rbcS-3A upstream region acts as a transcriptional enhancer (Aoyagi *et al.*, 1988; Fluhr *et al.*, 1986a).

We want to emphasize that although the upstream region is sufficient for regulated expression, additional regulatory transcriptional elements are located further downstream. Experiments involving transformed calli support the contention that in both the pea rbcS-E9 (Morelli *et al.*, 1985) and rbcS-3.6 (Timko *et al.*, 1985) genes, a sequence in the vicinity of the TATA box plays a role in light-regulation. Transgenic plant data have recently demonstrated a similar function for the pea rbcS-3A TATA box region (C. K. and G. Strittmatter, unpublished).

Sequence inspection of the upstream region revealed several short stretches of perfect homology among all pea rbcS genes. The most striking ones, box II and box III in the -150 bp region have strong homology with mammalian viral enhancer elements. Further upstream, around -220 bp are two sequences, box II* and III*, with homology to box II and box III. A 58 bp segment containing boxes II and III decreases the expression of a 35S-CAT gene in the dark when inserted between the 35S enhancer and the 35S TATA box. Therefore, box II and box III can act as negative transcriptional elements when placed in the proper context (Kuhlemeier *et al.*, 1987b). This result strongly suggests that also in their normal environment in the rbcS-3A gene, they can act as negative regulatory elements to decrease transcription in the dark.

This immediately raises the question where the positive elements can be located. The interesting answer to this question is that negative and positive elements are located very closely together and in fact overlap at least partially. When the gene is deleted from the 5' side to -166, expression is still fully normal. Interruptions of either box II or box III in this -166 mutant abolish expression (Kuhlemeier *et al.*, 1987b; Kuhlemeier *et al.*, 1988a). Thus, box II and box III, not only have a negative function, but also are required for enhancement of transcription. Interestingly, when box II or box III are mutated in the intact gene, with 410 bp of upstream sequence, no effect can be measured and expression appears to be completely normal. This indicates that there is a redundancy of cis-acting elements, prime candidates for such additional elements being the boxes II* and III* in the -220 bp region (Kuhlemeier *et al.*, 1987a).

When it is stated that the -166 mutant is fully expressed and that there is functional redundancy in the enhancer, such a statement requires an important qualification. Namely that the -166 mutant is completely active only under certain conditions. We have always been careful to isolate RNA only from fully expanded leaves in the middle of the transgenic tobacco plant, and in those leaves the sequences upstream and downstream of -166 always appeared equivalent and redundant. In contrast, when 7-day-old etiolated seedlings are irradiated with bright white light, the -166 mutant has a strongly reduced activity relative to the full-length gene (Kuhlemeier *et al.*, 1988a). Similarly, in the immature top leaves of a 20-25 leaf greenhouse-grown plant, expression of the -166 mutant is also reduced. This indicates that the redundancy is not absolute, but varies with the circumstances. To be able to study gene regulation under different experimental conditions and throughout the life cycle of the organism, is one of the most attractive features of the transgenic plant system.

TRANS-ACTING FACTORS

The simplest hypothesis is that the DNA sequences identified in the upstream region act as binding sites for proteins that have a regulatory role in the transcription of rbcS-3A. Indeed, nuclear extracts prepared from light-grown peas contain protein(s) that specifically bind to boxes II, III, II* and III* (Green *et al.*, 1987) and two boxes, II** and III**, located around -370 (P. Green *et al.*, unpublished). Competition assays strongly suggest that the same protein or protein complex (named GT-1) binds to each of these six boxes. Interestingly, extracts from dark-adapted plants also contain a very similar binding activity. This indicates that light regulation of rbcS transcription is not controlled at the level of GT-1 binding. In general, GT-1 binding *in vitro* correlates well with transcriptional activity *in vivo*, that is, all upstream segments that are active in transgenic plants, are also capable of binding GT-1 in the test tube. However, not every GT-1 binding fragment is active *in vivo.* These results suggest that GT-1 is a positive transcription factor, which is necessary, but not sufficient for turning on rbcS-3A. Purification of GT-1 to homogeneity, functional tests for its activity, *in vivo* binding studies and characterization of additional factors will be required to elucidate the exact role of GT-1 in rbcS regulation.

LIGHT RECEPTION AND SIGNAL TRANSDUCTION

The first step in the chain of events leading to any light-regulated response is the perception of the incident radiation by a photoreceptor. Action spectra have provided evidence for the existence of at least three photoreceptors in higher plants: phytochrome, which absorbs in the red, and a minimum of two photoreceptors that monitor the blue-UV region of the spectrum (Assmann *et al.*, 1985; Briggs *et al.*, 1985; Briggs and Iino, 1983; Iino *et al.*, 1985; Lagarias, 1985; Lagarias *et al.*, 1987; Schäfer and Briggs, 1986). Whereas the latter two photoreceptors are poorly defined in molecular terms, a wealth of data is available on the structure and function of phytochrome. In oats the pigment can be isolated from dark-grown seedlings as a 124 kDa protein covalently linked to a linear tetrapyrrole (Lagarias, 1985; Lagarias and Mercurio, 1985; Lagarias *et al.*, 1987). Irradiation with red light induces a conformational change which shifts the absorption maxima towards longer wavelengths (Lagarias and Mercurio, 1985; Wong *et al.*, 1986); a subsequent irradiation with far red light reverses the effect. In the near future DNA sequence analysis and epitope mapping are likely to give detailed insights into the structure-function relation of this photoreceptor (Hershey *et al.*, 1984; Shimazaki and Pratt, 1985; Tokuhisa *et al.*, 1985).

The conformational change in the photoreceptor phytochrome is the signal for the cell to turn on specific genes and turn others off. Phytochrome-mediated regulation of gene expression has mainly been studied in etiolated plants. One of the reasons for this is that regulation in green plants may be more complex, as was clearly demonstrated for the pea rbcS-3A and 3C genes (Fluhr and Chua, 1986; Fluhr *et al.*, 1986b) and even more strikingly for the wheat cab gene (Nagy *et al.*, 1988b). Even in etiolated seedlings, changes other than just phytochrome photoconversion occur upon illumination. First, when etiolated seedlings are irradiated, the steady-state mRNA levels and the amount of phytochrome protein decrease to less than 5% of the initial level (Colbert *et al.*, 1985). Second, in green plants a major form of phytochrome is detectable that is immunologically distinct from the "etiolated" species (Abe *et al.*, 1985; Otto *et al.*, 1984; Tokuhisa *et al.*, 1985).

It is important to realize that light regulation of transcription has very different physiological effects in dark-grown seedlings compared to mature green plants grown under a normal light/dark cycle. In etiolated seedlings, light-induced stimulation of rbcS transcription will result in increased production of rbcS protein and hence allow for increased carbon assimilation. In mature green plants the diurnal changes in the abundance of the rbcS mRNAs may be dramatic, but the level of the very stable rbcS protein will remain virtually constant. Therefore light regulation of rbcS transcription in mature plants is more likely to be related to the general energy economy of the cell than to control of photosynthetic carbon fixation.

There is a scarcity of data on the components of the signal transduction chain between the phytochrome molecule and the ultimate response. A very interesting observation concerning the phytochrome molecule itself has recently been made (Wong *et al.*, 1986). It appears that a protein kinase activity is associated with phytochrome, implying that phosphorylation plays a role in signal transduction.

In two cases concerning lower plants, there is strong evidence for the involvement of Ca^{2+} as a second messenger. In the green alga *Mougeotia*, the chloroplast rotates in response to light, a classical phytochrome-mediated response. This light-response can be mimicked by the application of Ca^{2+} in combination with the Ca^{2+}-specific ionophore A23187. Conversely, calmodulin inhibitors prevent the rotation even in the presence of stimulating light (Serlin and Roux, 1984). The other example is the phytochrome-mediated spore germination in the fern *Onoclea sensibilis* (Wayne and Hepler, 1984). Here, Ca^{2+} in combination with A23187, can substitute for light to initiate the germination process. Both the fact that Ca^{2+} inhibitors can prevent a response, and the finding that Ca^{2+} can substitute for light is a strong indication (but not unequivocal proof) for the direct involvement of Ca^{2+} in the signal transduction pathway. In maize, the phosphorylation of specific proteins is promoted by Ca^{2+} and calmodulin (Veluthambi and Poovaiah, 1984a; Veluthambi and Poovaiah, 1984b). Such experiments are taken to indicate in plants, as in animal cells, that protein kinases play a role in stimulus-response coupling. Although this seems entirely plausible, direct proof that such phosphorylation is indeed an immediate result of phytochrome photoconversion is lacking. Neither is there any evidence that phosphorylation causes gene activation.

CONCLUSIONS

The pea rbcS-3A gene provides an attractive model system for the study of light-regulated phenomena in higher plants. This gene responds to light by a 20- to 50-fold increase in steady state mRNA. Depending on the developmental stage, the expression is mediated either by phytochrome or by a blue photoreceptor in conjunction with phytochrome. Members of the rbcS-gene family have different levels of expression and induction kinetics. Thus, a variety of specific light responses can be studied in this relatively simple system.

At present, we have gained a fairly precise understanding of the regulatory DNA-sequences in the rbcS-3A gene that are involved in light-regulation. We are beginning to understand how these cis-elements interact with specific DNA binding proteins. In the future, we not only want to extend these studies, but also, to characterize the molecules that link the trans-acting factors with the primary photoreceptors. Elucidation of the complete transduction chains is needed to be able to explain the diversity and specificity in light regulation.

Acknowledgements: Work performed in the authors' laboratory was supported by a grant from Monsanto Company. We thank Wendy Roine for skillfully assembling the manuscript.

REFERENCES

Abe H, Yamamoto KT, Nagatani A, Furuya M (1985). Characterization of green tissue-specific phytochrome isolated immunochemically from pea seedlings. Plant Cell Physiol 26:1387-1399.

Aoyagi K, Kuhlemeier C, Chua N-H (1988). The pea rbcS-3A enhancer-like element directs cell-specific expression in transgenic tobacco. Molec Gen Genet, 213:179-185.

Apel K (1981). The protochlorophyllide holochrome of barley (*Hordeum vulgare* L). Phytochrome-induced decrease of translatable mRNA coding for the NADPH: protochlorophyllide oxidore-ductase. Eur J Biochem 120:89-93.

Assmann SM, Simoncini L, Schroeder JI (1985). Blue light activates electrogenic ion pumping in guard cell protoplasts of *Vicia faba.* Nature 318:285-287.

Briggs WR, Mandoli DF, Shinkle JR, Kaufman LS, Watson JC,Thompson WF (1985). Phytochrome regulation of plant development at the whole plant, physiological and molecular levels. In Colombetti G, Song P-S (eds): Sensory perception and transduction in aneural organisms. Plenum Publishing Corp.

Briggs WR, Iino M (1983). Blue-light absorbing photoreceptors in plants. Phil Trans R Soc Lond B303:347-359.

Castresana C, Garcia-Luque I, Alonso E, Malik VS, Cashmore AR (1988). Both positive and negative regulatory elements mediate expression of a photoregulated cab gene from *Nicotiana plumbaginifolia.* EMBO J 7:1929-1936.

Colbert JT, Hershey HP, Quail PH (1985). Phytochrome regulation of phytochrome mRNA abundance. Plant Mol Biol 5:91-101.

Fleck J, Durr A, Lett MC, Hirth L (1979). Changes in protein synthesis during the initial stage of life of tobacco protoplasts. Planta 145:279-285.

Fluhr R, Chua N-H (1986). Developmental regulation of two genes encoding ribulose-bisphosphate carboxylase small subunit in pea and transgenic petunia plants: phytochrome responses and blue-light induction. Proc Natl Acad Sci USA 83:2358-2362.

Fluhr R, Kuhlemeier C, Nagy F, Chua N-H (1986a). Organ-specific and light-induced expression of plant genes. Science 232:1106-1112.

Fluhr R, Moses P, Morelli G, Coruzzi G, Chua N-H (1986b). Expression dynamics of the pea rbcS multigene family and organ distribution of the transcripts. EMBO J 5:2063-2071.

Green PJ, Kay SA, Chua N-H (1987). Sequence-specific interactions of a pea nuclear factors with light-responsive elements of the rbcS-3A gene. EMBO J 6:2543-2549.

Hammer RE, Krumlauf R, Camper SA, Brinster R, Tilghman S (1987). Diversity of Alpha-fetoprotein gene expression in mice is generated by a combination of separate enhancer elements. Science 235:53-58.

Hershey HP, Colbert JT, Lissemore JL, Barker RF, Quail PH (1984). Molecular cloning of cDNA for Avena phytochrome. Proc Natl Acad Sci USA 81:2332-2336.

Iino M, Ogawa T, Zeiger E (1985). Kinetic properties of the blue-light response of stomata. Proc Natl Acad Sci USA 82:8019-8023.

Jones JDG, Dunsmuir P, Bedbrook J (1985). High level expression of introduced chimaeric genes in regenerated transformed plants. EMBO J 4:2411-2418.

Kaufman LS, Thompson WF, Briggs WR (1984). Different red light requirements for phytochrome-induced accumulation of cab RNA and rbcS RNA. Science 226:1447-1449.

Kaufman LS, Briggs WR, Thompson WF (1985). Phytochrome control of specific mRNA levels in developing pea buds. Plant Physiol 78:388-393.

Kaufman LS, Roberts LL, Briggs WR, Kaufman LS (1986). Phytochrome control of specific mRNA levels in developing pea buds: kinetics of accumulation, reciprocity, and escape kinetics of the low fluence response. Plant Physiol 81:1033-1038.

Kaulen H, Schell J, Kreuzaler F (1986). Light-induced expression of the chimeric chalcone synthase-NPTII gene in tobacco cells. EMBO J 5:1-8.

Kuhlemeier C, Cuozzo M, Green PJ, Goyvaerts E, Ward K, Chua N-H (1988a). Localization and conditional redundancy of regulatory elements in rbcS-3A, a pea gene coding for the small subunit of ribulose bisphosphate carboxylase. Proc Natl Acad Sci USA 85:4662-4666

Kuhlemeier C, Fluhr R, Chua N-H (1988b). Upstream sequences determine the difference in transcript abundance of pea rbcS genes. Molec Gen Genet 212:405-411.

Kuhlemeier C, Green PJ, Chua N-H. (1987a). Regulation of gene expression in higher plants. Ann Rev Plant Physiol 38:221-257.

Kuhlemeier C, Fluhr R, Green PJ, Chua N-H (1987b). Sequences in the pea rbcS-3A gene have strong homology to mammalian enhancers, but function as negative regulatory elements. Genes Devel 1:247-255.

Lagarias JC (1985). Progress in the molecular analysis of phytochrome. Photochem Photobiol 42:811-820.

Lagarias JC, Mercurio FM (1985). Structure function studies on phytochrome. Identification of light induced conformational changes in 124 kDa Avena phytochrome *in vitro*. J Biol Chem 260:2415-2423.

Lagarias JC, Wong Y-S, Berkelman TR, Kidd DC, McMichael RW (1987). Structure-function analysis of Avena phytochrome. In: Phytochrome and Photoregulation in Plants. Academic Press, New York.

Levis R, Hazelrigg RT, Rubin G (1985). Effects of genomic position on the expression of transduced copies of the white gene of Drosophila. Science 229:558-561.

Morelli G, Nagy F, Fraley RT, Rogers SG, Chua N-H (1985). A short conserved sequence is involved in the light-inducibility of a gene encoding ribulose 1,5-bisphosphate carboxylase small subunit of pea. Nature 315:200-204.

Nagy F, Boutry M, Hsu M-Y, Wong M, Chua N-H (1987). The 5' proximal region of the wheat cab gene contains a 268 bp enhancer-like sequence for phytochrome response. EMBO J 6:2537-2542.

Nagy F, Kay SA, Boutry MA, Hsu M-Y, Chua N-H (1986). Phytochrome-controlled expression of a wheat Cab gene in transgenic tobacco seedlings. EMBO J 5:1119-1124.

Nagy F, Kay SA, Chua N-H (1988a). Gene regulation by phytochrome. Trends Genet 4:37-42

Nagy F, Kay SA, Chua N-H (1988b). A circadian clock regulates transcription of the wheat Cab-1 gene. Genes Devel 2:376-382.

Nagy F, Morelli G, Fraley RT, Rogers SG, Chua N-H (1985). Photoregulated expression of a pea rbcS gene in leaves of transgenic plants. EMBO J 4:3063-3068.

Otto V, Schäfer E, Nagatani A, Yamamoto KT, Furuya M (1984). Phytochrome control of its own synthesis in Pisum sativum. Plant and Cell Physiol 25:1579-1584.

Poulsen C, Fluhr R, Kauffman JM, Boutry M, Chua N-H (1986). Characterization of an rbcS gene from *Nicotiana plumbaginifolia* and expression of an rbcS-CAT chimeric gene in homologous and heterologous nuclear background. Molec Gen Genet 205:193-200.

Poulsen C, Chua N-H (1988). Dissection of 5' upstream sequences for selective expression of the *Nicotiana plumbaginifolia* rbcS-8B gene. Molec Gen Genet, 214:16-23.

Roux SJ (1984). Ca^{2+} and phytochrome action in plants. Bioscience 34:25-29.

Schäfer E, Briggs W (1986). Photomorphogenesis from signal perception to gene expression. Photobiochem Photobiophys 12:305-320.

Serlin BS, Roux SJ (1984). Modulation of chloroplast movement in the green alga *Mougeotia* by the Ca^{2+} ionophore A23187 and by calmodulin antagonists. Proc Natl Acad Sci USA 81:6368-6372.

Shimazaki Y, Pratt LH (1985). Immunochemical detection with rabbit polyclonal and mouse monoclonal antibodies of different pools of phytochrome from etiolated and green Avena shoots. Planta 164:333-344.

Shropshire W, Mohr H (1983). Encyclopedia of Plant Physiology: Photomorphogenesis. New Series, Vol 16, Springer Verlag, Berlin.

Simpson J, Schell J, Van Montagu M, Herrera-Estrella L (1986a). The light-inducible and tissue specific expression of a pea LHCP gene involves an upstream element combining enhancer- and silencer-like properties. Nature 323:551-553.

Simpson J, Van Montagu M, Herrera-Estrella L (1986b). Photosynthesis-associated gene families: differences in response to tissue-specific and environmental factors. Science 233:34-38.

Sugita M, Gruissem W (1987). Developmental, organ-specific and light-dependent expression of the tomato ribulose-1,5-bisphosphate carboxylase small subunit gene family. Proc Natl Acad Sci USA 84:7104-7108.

Timko MP, Kausch AP, Castresana C, Fassler J, Herrera-Estrella L, Van den Broeck G, Van Montagu M, Schell J, Cashmore AR (1985). Light regulation of plant gene expression by an upstream enhancer-like element. Nature 318:579-582.

Tobin EM, Silverthorne J (1985). Light regulation of gene expression in higher plants. Ann Rev Plant Physiol 36:569-93.

Tokuhisa JG, Daniels SM, Quail PH (1985). Phytochrome in green tissue: spectral and immunochemical evidence for two distinct molecular species of phytochrome in light-grown *Avena sativa* L Planta 164:321-332.

Veluthambi K, Poovaiah BW (1984a). Calcium and calmodulin-regulated phosphorylation of soluble and membrane proteins from corn coleoptiles. Plant Physiol 76:359-365.

Veluthambi K, Poovaiah BW (1984b). Calcium-promoted protein phosphorylation in plants. Science 223:167-169.

Vernet T, Fleck J, Durr A, Fritsch C, Pinck M, Hirth L (1982). Expression of the gene coding for the small subunit of ribulose bisphosphate carboxylase during differentiation of tobacco plant protoplasts. Eur J Biochem 126:489-495.

Wayne R, Hepler PK (1984). The role of calcium ions in phytochrome-mediated germination of spores of *Onoclea sensibilis* L. Planta 160:12-20.

Wong Y-S, Cheng H-C, Walsh A, Lagarias JC (1986). Phosphorylation of Avena phytochrome *in vitro* as a probe of light-induced conformational changes. J Biol Chem 261:12089-12097.

Photosynthesis, pages 269–284

CONTROL OF CHLOROPLAST GENE EXPRESSION DURING PLANT DEVELOPMENT AND ENVIRONMENTAL CHANGES

Wilhelm Gruissem, Xing-wang Deng and John Tonkyn

Department of Botany, University of California, Berkeley, California 94720

INTRODUCTION

Chloroplasts as the organellar site of photosynthetic CO_2 fixation, ATP synthesis, and O_2 evolution are unique to algae and plants. Based on their central importance in photosynthesis as well as their developmental and differentiation capacities during plant development, plastids have become an important model system to study the regulation of photosynthetic gene expression and the interaction of nuclear and organelle genomes. In developing plants, chloroplasts are derived from precursor plastids termed proplastids, or from etioplasts in seedlings which are grown in the dark (Kirk and Tilney-Basset, 1978). Both proplastids and etioplasts are low in pigmentation and lack a functional thylakoid membrane system, but differ significantly in their ultrastructure. While proplastids are small, unstructured, and contain few internal membranes, etioplasts can reach the size of the mature chloroplast. In addition, etioplasts synthesize an internal structure, called the prolamellar body, which is thought to consist of mainly protochlorophyllide, NADPH protochlorophyllide reductase, and membranes (Boardman *et al.*, 1978; Castelfranco and Beale, 1983).

During leaf formation, the biogenesis of proplastids and etioplasts into photosynthetically active chloroplasts is one of the most significant morphogenetic changes. The principal effector in the development of a functional chloroplast is light (Tobin and Silverthorne, 1985; Harpster and Apel, 1985; Jenkins *et al.*, 1984). Phytochrome, protochlorophyllide and blue-light receptors mediate the signal transduction to elicit a cascade of molecular responses (Mohr, 1984; Kasemir, 1983; Briggs and Iino, 1983). Proplastids and chloroplasts can also differentiate into specialized plastid types that assume other functions in non-photosynthetic organs of higher plants, such as amyloplasts in roots and tubers, or chromoplasts in many flowers and fruits (Kirk and Tilney-Basset, 1978). The diverse functions of chloroplasts and non-photosynthetic plastids require the products of several hundred genes, of which only about 120 are located in the chloroplast genome. All other chloroplast proteins are expressed from nuclear genes. Thus, the temporal and spatial expression of plastid genes during development and differentiation of chloroplasts and other plastid types is not only regulated in different plant organs, but must also be coordinated with the expression of nuclear genes for photosynthetic proteins. In this review we will discuss the recent progress in our understanding of the developmental regulation of plastid gene expression and how environmental factors can affect this control at the molecular level.

MORPHOGENETIC AND MOLECULAR CHANGES DURING CHLOROPLAST DEVELOPMENT

The initial phase of chloroplast biogenesis is characterized by the rapid synthesis and assembly of the photosynthetic thylakoid membrane system. The most visible changes are the rapid accumulation of chlorophylls, which associate with the light-harvesting chlorophyll *a/b* and reaction center proteins of photosystems I and II. Together with the two photosystems, the cytochrome b_6/f complex and ATP synthase subunit polypeptides are synthesized and assembled into the functional thylakoid membrane. Each of these complexes consists of up to 20 polypeptides (Anderson, 1984; Barber, 1987). Thus, many previously undetectable photosynthetic proteins appear during the greening process. After illumination, mRNAs from several plastid and nuclear genes that encode these different photosynthetic proteins accumulate during chloroplast development. However, the extent to which these mRNAs increase in the light is variable and may reflect species-specific regulatory mechanisms (Bedbrook *et al.*, 1978; Reisfeld *et al.*, 1978; Smith and Ellis, 1981; Nelson *et al.*, 1984; Rodermel and Bogorad, 1985; Kreuz *et al.*, 1986; Deng and Gruissem, 1987; Mullet and Klein, 1987). For example, it has been shown that most plastid mRNAs for photosynthetic proteins are also present in etioplasts of dark-grown monocotyledonous and dicotyledonous plant species (Herrmann *et al.*, 1985; Klein and Mullet, 1986; Deng and Gruissem, 1988), and may even be detectable in cells containing mostly proplastids (Barkan, 1988). While the level of some mRNAs can increase drastically after illumination, other mRNAs show little or no fluctuations. Examples for this differential accumulation pattern of RNA populations are the mRNAs for the 32-kd quinone-binding protein of photosystem II (*psb*A gene) and the large subunit of ribulose-1,5-bisphosphate carboxylase (*rbc*L gene). A rapid and high accumulation of *psb*A mRNA has been consistently measured after illumination of several plant species, while *rbc*L mRNA levels do not change significantly (Deng and Gruissem, 1987).

Superimposed on changes in plastid and nuclear mRNA levels for photosynthetic proteins during leaf development, the synthesis of photosynthetic proteins is also regulated at the translational and post-translational level. Several experiments have shown that light plays an important role in this control, since certain chloroplast proteins are not detectable in the dark, although their mRNAs have already accumulated to significant levels (Herrmann *et al.*, 1985; Inamine *et al.*, 1985; Fromm *et al.*, 1985; Berry *et al.*, 1988; Klein and Mullet, 1987; Klein *et al.*, 1988). Overall, chloroplast development involves a hierarchy of control steps which regulate the expression of nuclear and plastid genes at the transcriptional, post-transcriptional and translational level.

MORPHOGENETIC ADAPTATIONS OF CHLOROPLASTS TO DIFFERENT LIGHT QUALITIES

Once the chloroplast is fully developed and has established its photosynthetic capacity, it becomes subject to different environmental and physiological changes. For example, there are significant differences in the ultrastructure of chloroplasts from plants adapted to sun and shade, and in plants growing in different light intensities and qualities (Anderson *et al.*, 1973; Melis and Harvey, 1981; Lichtenthaler *et al.*, 1982; Senger, 1980; Glick *et al.*, 1985). Shade plants and plants growing in low light have a higher thylakoid-membrane to stromal-volume ratio compared to plants growing in high light, but similar ratio changes can also occur in plants growing in similar light intensities, but different light qualities (Eskin *et al.*, 1985; Glick *et al.*, 1986). It has been proposed that such changes involve both synthetic and degradative processes that modulate the relative concentration of PS I and PS II, electron transport complexes, and ATP synthase (Melis, 1984). Thus, plants respond to light in significant adaptations of the composition and structural reorganization of the components of thylakoid membranes. This adaptability of chloroplasts to different environmental light conditions allows higher plants to succeed in a variety of habitats (Boardman, 1977; Melis and Harvey, 1981; Wilhelm and Wild, 1984).

DIFFERENTIATION OF NON-PHOTOSYNTHETIC PLASTIDS

In addition to the complex regulatory circuits that are operational in the coordinated expression of nuclear and plastid genes for correct temporal and spatial chloroplast functions, the differentiation of chloroplasts into other plastid types must also be controlled at the molecular level in higher plants. In non-photosynthetic plant organs proplastids differentiate into other specialized plastid types such as amyloplasts, which are starch-containing plastids in roots (Kirk and Tilney-Basset, 1978). The mRNAs for photosynthetic proteins are present at much lower levels in these organs (Deng and Gruissem, 1988).

Photosynthetically active chloroplasts can also differentiate into plastid types of other distinct functions, for example, the carotenoid-synthesizing chromoplast in certain flower petals and fruits. During chromoplast differentiation, the thylakoid membrane system disintegrates, and photosynthetic mRNAs and their proteins disappear (Piechulla *et al.*, 1987). However, the differentiation of specialized plastid types is not a terminal process, since they maintain their DNA and the competence to revert to chloroplasts. This is most convincingly demonstrated by the conversion of chromoplasts into chloroplasts in greening orange pericarp (Mayfield and Huff, 1986) and by the regeneration of plants from single tomato root protoplasts (Norton and Boll, 1954). Thus, this differentiation capacity implies that all plastids must maintain at least basal levels of transcription and translation activities, since, besides genes for photosynthetic proteins, there are approximately sixty genes on the plastid genome which encode products essential for transcription and translation (Shinozaki *et al.*, 1986; Ohyama *et al.*, 1986).

Most of the ultrastructural, physiological, and molecular changes that occur during chloroplast development in photosynthetically active chloroplasts and during plastid differentiation are now well defined. However, the mechanisms that control the expression of nuclear and plastid genes during these events are less well understood, since the effects of developmental changes, light, or other environmental signals on the regulation of gene expression are extremely complex. In the following we will primarily discuss the advances that have been made in our perception of regulatory mechanisms that operate at the level of gene expression in plastids during plant development. We will especially emphasize those controls that account for the changes in gene expression which occur during chloroplast development and plastid differentiation, as a result of light quality differences, or in response to inhibition of carotenoid biosynthesis. As part of this discussion, data will be presented in support of the conclusion that plastids in higher plants have developed post-transcriptional and translational mechanisms to regulate the expression of genes for photosynthetic proteins.

DEVELOPMENTAL REGULATION OF PLASTID mRNA LEVELS

Developing and differentiating plastids have in common that significant adjustments in the levels of mRNAs for photosynthetic proteins can be detected. For example, mRNAs from *rbc*L and *psb*A accumulate approximately 350- and 140-fold, respectively, in cotyledons of dark-grown spinach seedlings as compared to spinach roots (Table I). However, while the relative mRNA levels for *rbc*L, *psa*A and *atp*BE differ by less than twofold in cotyledons of dark-grown spinach seedlings and photosynthetically active leaves, the *psb*A mRNA level increase another 6- to 7-fold after illumination and during leaf development. Similar observations have been reported for etioplast-to-chloroplast transitions (Rodermel and Bogorad, 1985; Deng and Gruissem, 1987; Mullet and Klein, 1987; Herrmann *et al.*, 1985). The adjustments of mRNA populations during etioplast formation in dark-grown spinach cotyledons consequently are not tightly correlated with the development of photosynthetic functions, but are clearly regulated by the developmental program of the plant (Deng and Gruissem, 1988). Illumination of dark-grown cotyledons and chloroplast development in leaves can further induce changes in the levels of individual plastid mRNAs. Accumulation of plastid mRNAs for photosynthetic proteins in dark-grown spinach cotyledons was first reported by Herrmann *et al.* (Herrmann *et al.*, 1985). The authors concluded that the transcription of plastid photosynthetic genes is not inactivated in this tissue, a concept which was not undisputed at this time (Rodermel and Bogorad, 1985). While most plastid mRNAs show differential increases during etioplast formation and chloroplast development, mRNA levels decrease in differentiating chromoplasts and are extremely low in root amyloplasts (Table I). These differences correlate with the function of these plastid types: amyloplasts and chromoplasts are specialized plastids which store large concentrations of starch and carotenoids, respectively, but lack a functional thylakoid membrane system.

Table I. Changes in relative plastid mRNA levels during plant development (% of leaf; *= *atp*A gene probe).

		16 S *rrn*	*psb*A	*psa*A	*rbc*L	*atp*B/E
Spinach	Amyloplast (Root)	1.18	0.11	0.47	0.26	1.75
	Etioplasts (Cotyledon)	78.8	15.7	77.2	91.1	64.9
	Chloroplast (Leaf)	Control = 100				
Tomato	Chloroplast (Fruit)	44.1	69.9	20.8	24.3	22.7*
	Chromoplast (Fruit)	16.2	20.4	——	12.4	11.4*

The changes in photosynthetic capacity and photosynthetic protein pattern associated with chloroplast/chromoplast differentiation has been extensively studied during fruit ripening in tomato. Chloroplasts in green tomato pericarp tissue are structurally and functionally equivalent to leaf chloroplasts. However, determinations of chlorophyll *a/b* ratios and variable to non-variable fluorescence yields (F_v/F_0) indicate a smaller antenna size in tomato pericarp chloroplasts. The photosynthetic electron transport capacity (on an area basis) and the amount of chlorophyll per protein are lower in green fruit pericarp tissue, suggesting that their photosynthetic efficiency in general may be reduced. This is also consistent with the result that ribulose-1,5-bisphosphate carboxylase activity is approximately 40% of the activity measured in leaves (Piechulla *et al.*, 1987). Despite differences in photosynthetic efficiency, leaf and pericarp chloroplasts have a similar pattern of mRNAs and proteins for photosynthetic components (Piechulla *et al.*, 1985, 1986, 1987). Immunological studies have demonstrated that the differential decrease of mRNAs for photosynthetic proteins during chromoplast differentiation is, for the most part, correlated with the disappearance of these proteins. However, in some cases, e.g., the 32-kd QB-binding protein of photosystem II, the mRNA is still present at relatively high levels in fruit chromoplasts (see Table I), but the protein is undetectable (Piechulla *et al.*, 1987).

TRANSCRIPTIONAL AND POST-TRANSCRIPTIONAL CONTROL OF PLASTID mRNA LEVELS DURING PLANT DEVELOPMENT

Taken together, the development of photosynthetically active chloroplasts from precursor plastids and the differentiation of specialized plastid types are associated with the differential accumulation and decline of mRNA levels for photosynthetic proteins. Most plastid mRNAs are also present at lower levels in tissues and organs, which do not acquire or have lost their photosynthetic competence. However, the development of photosynthetic competence is generally characterized by an increase in the levels of mRNAs for photosynthetic proteins. Earlier studies have suggested that such adjustments could be controlled at the DNA level by changing the transcription rate of the respective genes after illumination (Rodermel and Bogorad, 1985; Sasaki *et al.*, 1984; Link, 1984). Similar transcriptional regulatory mechanisms could also be operational during chromoplast differentiation or in specialized plastid types where mRNA levels decline. The observed differences in steady-state mRNA levels could also be a result of changes in mRNA turnover rates and thus the stability of individual plastid mRNAs during plant development. Both alternatives have recently been investigated in barley, spinach and tomato using a plastid run-on transcription assay (Deng *et al.*, 1987; Mullet and

Klein, 1987; Gruissem *et al.*, 1988). In this assay isolated intact plastids are lysed and briefly pulsed with radioactively labeled nucleotide triphosphates to label nascent mRNA molecules. The labeled mRNAs are subsequently hybridized to DNA fragments specific for photosynthetic genes. The amount of hybridizable material reflects the relative transcription activity of the respective genes or polycistronic transcription units. Such analyses of plastid run-on transcripts and steady-state mRNA levels from different developmental stages of spinach and barley leaves have demonstrated that transcription activity and RNA accumulation act independently for most plastid genes (Deng and Gruissem, 1987; Klein and Mullet, 1986). Most changes in the relative transcriptional activities of genes for photosynthetic proteins are small and are not sufficient to explain the differential accumulation of their mRNAs. The results therefore suggest that adjustments in plastid RNA populations can occur by post-transcriptional mechanisms, and that changes in mRNA stability are important determinants in the control of gene expression during chloroplast development (Gruissem *et al.*, 1988; Stern and Gruissem, 1987).

The detection of transcription activity, together with noticeable levels of mRNAs for photosynthetic proteins (Table I) and polysomes in root amyloplasts, indicate that the expression of the plastid genome is constitutive in all plant tissues and organs (Deng and Gruissem, 1988). Constitutive plastid gene transcription would be required for at least the genes in the chloroplast genome that encode genetic functions. This concept would be consistent with the totipotency of most higher-plant cells and the differentiation capacity of plastids during plant development. Since the plastid genome encodes approximately 60 genes for transcription and translation functions besides the genes for photosynthetic proteins, the organelle must maintain at least basal levels of transcription and translation activities to ensure the expression of these genes. In the absence of significant transcriptional regulation of individual plastid genes, therefore, it is most likely that all plastid genes, including those for photosynthetic proteins, are transcribed in plastids and plant organs that lack photosynthetic competence. Transcription run-on experiments with amyloplasts isolated from dark-grown spinach seedlings and chromoplasts from tomato fruits have demonstrated that genes such as *rbc*L, *psb*A and *psb*B are in fact transcribed in these non-photosynthetic organelles (Deng and Gruissem, 1988; Gruissem *et al.*, 1987). These results in general support a model of constitutive plastid gene transcription in all plant organs. Thus, regulation of plastid gene expression during plant development involves a significant post-transcriptional control of mRNA populations, most likely by alterations in the stability of individual transcripts.

LIGHT-QUALITY CONTROL OF mRNA LEVELS FOR PHOTOSYNTHETIC PROTEINS

Photosynthetically active chloroplasts adapt to different environmental light qualities by modulation of thylakoid membrane complexes. For example, the chloroplasts from plants grown in PS II-sensitizing ("yellow") light are enriched in PS I complexes, while plants grown in light preferentially absorbed by PS I ("red" light) are enriched in PS II complexes (Eskin *et al.*, 1985; Melis, 1984). These structural reorganizations of the components of the thylakoid membrane are in response to the imbalance in light absorption between PS II and PS I and the subsequent imbalance in electron flow between the two photosystems. The light-quality response is different from the response observed by variation in the photon flux density that causes changes in the antenna size of PS II and PS I (Lichtenthaler *et al.*, 1982; Boardman, 1977; Leong and Anderson, 1984; Leong and Anderson, 1983), and in the concentration of components of the electron-transport chain between PS II and PS I.

The ultrastructural and biochemical changes that are caused by yellow and red light can be quantitated at the level of the PS II/PS I reaction center ratio. When pea plants are grown in light primarily absorbed by PS I, the PS II/PS I reaction center ratio is approximately twofold greater than that in plants grown in PS II-sensitizing, yellow light (Glick *et al.*, 1986). These changes are reflected in differences at the plastid mRNA levels for PS II and PS I proteins. For example, in plants grown under yellow PS II-sensitizing light there is a significantly (approx. 5-fold) higher level of the mRNA for the PS I P700 chlorophyll *a*-apoprotein as compared to plants grown under red light. In contrast, the ratio of the mRNA for the 51-kd chlorophyll *a*-apoprotein to that of the PS I P700 chlorophyll *a*-apoprotein increases approximately 2.6-fold in plants grown under red light, which is consistent with the increased PS II/PS I reaction center ratio in these plants (Glick *et al.*, 1986). Thus, the differential reaction center concentrations in the thylakoid membranes of red and yellow light-grown plants is accompanied by adjustments of the mRNA levels for PS II and PS I proteins. This result suggests that light quality can also affect the synthesis or stability of specific plastid mRNAs.

To address the extent of transcriptional or post-transcriptional control of mRNA levels in plants grown under different light qualities, spinach seedlings were germinated in either red or yellow light. The differential accumulation of mRNAs for PS II and PS I proteins in spinach cotyledons is comparable to or even more pronounced than that in pea leaves (Table II), indicating that this is not an organ-specific property. The observed differences in the transcript levels cannot be attributed to changes in plastid DNA content that may have been a result of plants grown in red or yellow light. Hybridization of DNA blots of restriction enzyme digests of total DNA from spinach cotyledons and pea leaves shows no significant difference in plastid DNA levels in these plants. The analysis of general transcription activities in chloroplasts from spinach cotyledons also shows that they are similar in plants grown in white, yellow or red light. This suggests that the differences observed in the chloroplast mRNA populations are most likely the result of post-transcriptional changes in

Table II. Changes in relative plastid mRNA levels induced by different light qualities.

Plastid RNA was isolated from 15 d old spinach plants grown in yellow, red and white light at 50 $\mu Em^{-2}s^{-1}$ in a 16 hr light / 8 hr dark cycle. RNAs were quantitated by Northern hybridizations with specific DNA probes.

Light	16S *rrn*	*psa*A	*psb*A	*psb*B	*rbc*L	*atp*BE
White	Control = 1					
Yellow	1	1.01	0.94	0.90	1.09	1.12
Red	1	0.06	0.79	0.49	0.44	0.56

the accumulation of transcripts. Evidence that the stability of mRNAs for PS II and PS I proteins rather than their synthesis is affected by different light qualities was obtained from transcription run-on experiments with isolated spinach cotyledon chloroplasts. Quantitations of run-on transcripts from genes encoding PS II and PS I proteins in plants grown in red and yellow light confirm that the relative transcription activities are similar and do not differ significantly from those in chloroplasts of plants grown in white light (Deng *et al.*, manuscript in preparation). Thus, as in specialized plastid types and developing chloroplasts, the light-quality mediated changes are regulated by post-transcriptional mechanisms that affect the stability of certain mRNAs for PS II- and PS I-specific proteins. The molecular mechanism by which light-quality signals are identified in the cell and transmitted to the chloroplast to adjust the levels of specific plastid mRNAs is presently unknown. It is unlikely that phytochrome-mediated processes are involved, since under the experimental conditions both yellow and red light sources emit light of sufficient intensity at 660 nm to affect phytochrome activation.

DIURNAL FLUCTUATIONS OF PHOTOSYNTHETIC mRNAs

While the biochemical, ultrastructural, and molecular changes in plastids during chloroplast development and light adaptation have been well studied, there is little information on the regulation of genes for photosynthetic proteins during the diurnal cycle. Photosynthetic activity in plants is subject to a diurnal cycle, and it is now well established that the activity of several enzymes in photosynthate metabolism is regulated during the diurnal cycle (Laval-Martin and Farineau, 1977; Servaites *et al.*, 1984; Stitt *et al.*, 1984). The level of the nuclear encoded mRNAs for the light-harvesting chlorophyll *a/b*-proteins also cycle together with nuclear and plastid mRNAs of other photosynthetic proteins, as has been reported for pea, maize, and tomato, but the extent to which individual mRNA levels fluctuate are gene specific (Kloppstech, 1985; Piechulla and Gruissem, 1987).

The diurnal fluctuations in mRNA levels for photosynthetic proteins is not specific to leaves, but the regulatory mechanisms are also operational in other plant organs. For example, an extensive survey of plastid and nuclear mRNAs for photosynthetic proteins, tubulin, and the β-subunit of the mitochondrial ATPase in developing tomato fruits has shown that their levels also change during the diurnal cycle (Piechulla and Gruissem, 1987). When Northern blots with fruit RNA from 4 hr intervals over a 38 hr period were hybridized with gene probes specific for the large and small subunit of ribulose-1,5-bisphosphate carboxylase, fluctuations of *rbc*L and *rbc*S mRNA levels are detectable over a 24 hr period. These mRNAs reach their lowest levels in the afternoon, after which they rapidly increase over the following 4 hrs, and then start to decline again to their lowest levels in the next 20 hrs. However, *rbc*S and *rbc*L mRNAs are detectable throughout the entire diurnal cycle. This pattern in is in contrast to the nuclear encoded mRNA for the light-harvesting chlorophyll *a/b*-binding proteins (*cab*), which shows significant changes over a 24 hr period. The *cab* mRNA levels are highest at noon, after which they decline to undetectable levels over the next 16 hrs, followed by a rapid accumulation to their high levels at noon. Analysis of *cab* proteins using LHCII-specific antibodies, however, does not reveal significant changes in their levels during the diurnal cycle. This result suggests that the proteins are relatively stable, and that their accumulation is uncoupled from the relative mRNA abundance and regulated by post-translational mechanisms.

Although plastid mRNAs for the reaction center proteins of photosystem I (*psa*A) and II (*psb*A , *psb*B) also fluctuate in a pattern similar to those for the *rbc*S, *rbc*L and *cab* mRNAs, their differences in mRNA levels throughout the diurnal cycle are less pronounced. The mRNA levels for the non-photosynthetic proteins are also affected by the diurnal cycle, but their relative changes are smaller compared to the significant fluctuations of the *cab* mRNA (Piechulla and Gruissem, 1987). These results suggest that light and/or diurnal cycles also control the mRNA levels for photosynthesis-specific genes in other organs of higher plants besides leaves. The similar mRNA fluctuation pattern for nuclear and plastid genes indicates that a general control mechanism may be responsible for regulating the expression of genes in different compartments. It still remains to be established at which regulatory levels this control operates, but based on other results it seems likely that transcriptional changes may be the principal effector for *rbc*S and *cab* mRNA fluctuations, while post-transcriptional changes in RNA stability may primarily affect the levels of plastid mRNAs.

EFFECT OF CAROTENOID BIOSYNTHESIS ON TRANSCRIPTION AND STABILITY OF mRNAs FOR PHOTOSYNTHETIC PROTEINS

Besides a developmental program and the modulation by light, plastid gene expression and chloroplast development are also controlled by other regulatory mechanisms. Under normal growth conditions, the assembly of the photosynthetic apparatus appears to follow a coordinated sequence of events, which requires the interaction of plastid and nuclear genomes. However, mutations in pigment biosynthesis have a significant effect on the correct assembly of thylakoid membranes. Carotenoid-deficient mutants have been shown to have altered chloroplast membrane structures and to be deficient in chloroplast membrane proteins (Smillie *et al.*, 1977), which in at least some cases results from a block in the biosynthesis of certain thylakoid membrane proteins or their rapid turnover (Weidner *et al.*, 1976; Harpster *et al.*, 1984; Mayfield and Taylor, 1984). The lack of carotenoids in mutants which block this pathway also reduces the levels of certain nuclear encoded mRNAs for photosynthetic proteins, with the most pronounced effect on the levels of the mRNAs for the light-harvesting chlorophyll *a*-binding proteins (Scolnik *et al.*, 1987; Thelander *et al.*, 1986). While there is now evidence that the decline of the specific nuclear mRNAs may result from a decrease in transcription activity of the corresponding genes (Guiliano and Scolnik, 1988), very little is known how a block in carotenoid biosynthesis affects the level of plastid mRNAs or the transcription of plastid genes.

These questions were addressed by treatment of fully established hydroponically grown spinach plants with norflurazon, a herbicide which blocks the synthesis of carotenoids, resulting in the photooxidation of chlorophyll and a white leaf phenotype (Bartels and McCullough, 1972). Thus, following the addition of norflurazon to the hydroponic solution, newly developing leaves are white, while the established leaves retain chlorophyll, presumably due to the stability of carotenoids present in the photosynthetically active chloroplasts. Quantitation of plastid RNA levels from young and mature leaves of norflurazon-treated plants and comparison with the levels of RNAs in control leaves (Table III) shows that the herbicide has little effect on the established, green leaves, but that RNA levels are significantly affected in young white leaves. While levels are reduced for most mRNAs coding for photosynthetic proteins, there is, strikingly, no significant difference in the level of ribosomal RNA and the mRNA for the D1 protein of photosystem II (*psb*A).

The decline in mRNA populations for the other photosynthetic proteins in the carotenoid-deficient young leaves could be a result of the

Table III. Changes in spinach plastid mRNA levels in young and mature leaves after treatment with norflurazon.

Spinach plants were grown hydroponically in 0.5 X Hoagland's solution for three weeks with white light (200 $\mu Em^{-2}s^{-1}$) , followed by treatment with norflurazon (3.5 mg/l) for 8 additional days. RNA was isolated from young and mature leaves of norflurazon-treated (N) and control (C) plants using standard procedures, and quantitated by Northern hybridizations with specific DNA probes.

Leaf	16S *rrn*	*psa*A	*psb*A	*psb*B	*rbc*L	*atp*B/E
N - Young	1	0.24	1.18	0.35	0.43	0.38
C - Young	Control = 1					
N - Mature	1.45	0.89	3.28	1.32	0.80	1.47
C - Mature	1.40	1.68	2.59	1.36	0.87	1.36

general decrease in transcription activity, which is approximately 60% reduced relative to the untreated control leaves. In mature control and herbicide-treated leaves, the general plastid transcription activity is similar, and there are no significant differences in the levels of most RNAs except for the mRNA coding for the photosystem I reaction center protein (*psa*A), which is reduced approximately two-fold in the treated plants. Thus, the high levels of ribosomal RNA and *psb*A mRNA in treated young leaves, and the decrease of *psa*A mRNA in treated mature leaves indicate that their abundance is regulated by post-transcriptional mechanisms, which are affected differently by norflurazone. This conclusion is also supported by transcription run-on assays using plastids from treated and control young and mature spinach leaves, which show that the relative transcription activity of most plastid transcription units at a genomic resolution (Figure 1) and specifically of the genes presented in Table III (not shown) do not change considerably. Taken together, lack of carotenoid biosynthesis has a pronounced effect on the transcription activity in plastids of young leaves and the stability of different mRNAs. Although it is most likely that transcription activity and mRNA accumulation in plastids are both under the control of nuclear genes, at present it is not known how the regulation of these functions is related to a block in carotenoid biosynthesis, and how the information flow for such a regulatory decision is established and coordinated between the nuclear and organelle genomes.

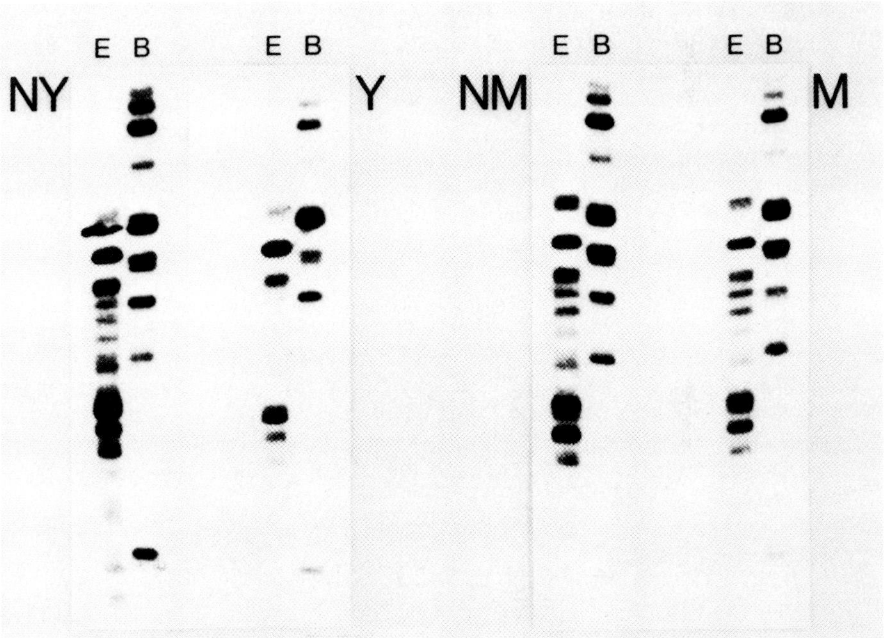

Figure 1. Relative transcription activities at genomic resolution of plastids from norflurazone-treated and control plants.

Spinach plants were grown and treated as described in the legend of Table III. Plastids from treated (N) and control young (Y) and mature (M) leaves were isolated and subjected to run-on assays (Deng *et al.*, 1987). Run-on transcripts were isolated and hybridized to total spinach chloroplast DNA digested with EcoRI (E) and BamHI (B).

CONCLUSIONS

The coordinated assembly of photosynthetic protein complexes during chloroplast development, together with the effect of environmental factors on chloroplast structure and activity, as wells as the differentiation of non-photosynthetic plastids with specific functions, reveal a complex array of regulatory processes that control the expression of plastid genes at the transcriptional, post-transcriptional, and translational level. The data presently available indicate that the plastid genome is constitutively transcribed in all organs of the plant, and that mRNAs for photosynthetic proteins can be detected even in non-photosynthetic plastids. Thus, transcriptional regulation of plastid genes appears to be of minor importance and may be restricted to general changes in transcription activity without significantly affecting the relative transcription activities of individual plastid transcription units. Changes in plastid RNA populations during plant development therefore must be controlled by adjustments in the stability of certain mRNAs and rRNA. Such post-transcriptional control would provide a mechanism to regulate the relative levels of ribosomal RNAs and mRNAs for specific plastid proteins independently of the transcriptional state of their genes. Similar adjustments in mRNA stabilities could also be operational in plants that experience changing environments. Superimposed on the fluctuating mRNA levels are translational or post-translational controls that regulate the synthesis or turnover of photosynthetic proteins independently of the mRNA levels. Translational and post-translational control in the chloroplast appear to have evolved as regulatory mechanisms that allow the plant to uncouple the synthesis of photosynthetic proteins from the constitutive transcription of their genes, thus significantly increasing the fidelity in correct temporal and spatial assembly of electron transfer complexes. Much exciting work still lies ahead to determine the molecular mechanisms of transcriptional, post-transcriptional, and translational control and the nuclear and plastid factors involved in these processes.

ACKNOWLEDGEMENTS

The work in our laboratory is supported by grants from the National Institute of Health and the Department of Energy to W.G. X.W.D. is the recipient of a University of California fellowship.

REFERENCES

Anderson JM (1984). In Sybesma C (ed): "Advances in Photosynthesis Research," Nijhoff/Dr W Junk, The Hague 3:1-10.

Anderson JM, Goodchild DJ, Boardman NK (1973). Biochim Biophys Acta 325:573-585.

Barber J (1987). TIBS 12:321-326.
Barkan A (1988). Mol Gen Genet, manuscript submitted.
Bartels PG, McCullough C (1972). Biochem Biophys Res Comm 48:16-22.
Bedbrook JR, Link G, Coen DM, Bogorad L, Rich A (1978). Proc Natl Acad Sci USA 75:3060-3064.
Berry JO, Carr JP, Klessig DF (1988). Proc Natl Acad Sci USA 85:4190-4194.
Boardman NK (1977). Ann Rev Plant Physiol 28:355-377.
Boardman NK, Anderson JM, Goodchild DJ (1978). Curr Top Bioenerg 8:35-109.
Briggs WR, Iino M (1983). Phil Trans R Soc London Ser B 303:347-359.
Castelfranco PA, Beale SJ (1983). Ann Rev Plant Physiol 34:241-278.
Deng XW, Gruissem W (1987). Cell 49:379-387.
Deng XW, Gruissem W (1988). EMBO J 7: in press.
Deng XW, Stern DB, Tonkyn JC, Gruissem W (1987). J Biol Chem 262:9641-9648.
Eskin K, Duysen M, Dybas L. McCarthy S (1985). Plant Physiol 77:29-34.
Fromm H, Devic M, Fluhr R, Edelman M (1985). EMBO J 4:291-295.
Giuliano G, Scolnik PA (1988). Plant Physiol 86:7-9.
Glick RE, McCauley SW, Gruissem W, Melis A (1986). Proc Natl Acad Sci USA 83:4287-4291.
Glick RE, McCauley SW, Melis A (1985). Planta 164:487-494.
Gruissem W, Barkan A, Deng XW, Stern DB (1988). Trends in Genetics 4: in press.
Gruissem W, Callan K, Lynch J, *et al.* (1987). In Nevins DJ, Jones RA (eds): "Tomato Biotechnology," Alan R Liss Inc, New York, pp. 239-249.
Harpster M, Apel K (1985). Physiol Plant 64:147-152.
Harpster MH, Mayfield SP, Taylor WC (1984). Plant Mol Biol 3:59-71.
Herrmann RG, Westhoff P, Alt J, Tittgen J, Nelson N (1985). In van Vloten-Doting L, Groot GSP, Hall TC (eds): "Molecular Form and Function of the Plant Genome," Plenum Press, Amsterdam, pp. 233-256.
Inamine G, Nash B, Weisbach H, Brot N (1985). Proc Natl Acad Sci USA 82:5690-5694.
Jenkins GI, Gallagher TF, Hartley MR, Bennet J, Ellis RJ (1984). In Sybesma C (ed): "Advances in Photosynthesis Research," Nijhoff/Dr W Junk, The Hague, 4:863-872.
Kasemir H (1983). Photochem Photobiol 37:701-708.
Kirk JTO, Tilney-Basset RAE (1978). The Plastids. Elsevier/North Holland Biomedical Press, Amsterdam
Klein RR, Mullet JE (1986). J Biol Chem 261:11138-11145.
Klein RR, Mullet JE (1987). J Biol Chem 262:4341-4348.
Klein RR, Mason HS, Mullet JE (1988). J Cell Biol 106:289-301.

Kloppstech K (1985). Planta 165:502-506.
Kreuz K, Dehesh K, Apel K (1986). Eur J Biochem 159:459-467.
Laval-Martin D, Farineau J (1977). Plant Physiol 60:872-876.
Leong TY, Anderson JM (1983). Biochim Biophys Acta 723:391-399.
Leong TY, Anderson JM (1984). Photosynth Res 5:105-115.
Lichtenthaler HK, Kuhn G, Prenzel U, Buschmann C, Meier D (1982). Z Naturforsch Teil C 37:464-475.
Link G (1984). Plant Mol Biol 3:243-248.
Mayfield SP, Huff A (1986). Plant Physiol 81:30-35.
Mayfield SP, Taylor WC (1984). Eur J Biochem 144:79-84.
Melis A (1984). J Cell Biol 24:271-285.
Melis A, Harvey GW (1981). Biochim Biophys Acta 637:138-145.
Mohr H (1984). Chloroplast Biogenesis. In Baker NR, Barber J (eds): Elsevier/North Holland Biomedical Press, Amsterdam, 304-347.
Mullet JE, Klein RR (1987). EMBO J 6:1571-1579.
Nelson T, Harpster MH, Mayfield SP, Taylor WC (1984). J Cell Biol 98:558-564.
Norton JP, Boll WG (1954). Science 119:220-221.
Ohyama K, Fukuzawa H, Kohchi T. *et al.* (1986). Nature 322:572-574.
Piechulla B, Chonoles-Imlay KR, Gruissem W (1985). Plant Mol Biol 5:373-384.
Piechulla B, Glick RE, Bahl H, Melis A, Gruissem W (1987). Plant Physiol 84:911-917.
Piechulla B, Gruissem W (1987). EMBO J 6:3593-3599.
Piechulla B, Pichersky E, Cashmore AR, Gruissem W (1986). Plant Mol Biol 7:367-376.
Reisfeld A, Gressel J, Jakob KM, Edelman M (1978). Photochem Photobiol 27:161-165.
Rodermel SR, Bogorad L (1985). J Cell Biol 100:463-476.
Sasaki Y, Tomoda Y, Kamikubo T (1984). FEBS Lett 173:31-34.
Scolnik PA, Hinton P, Greenblatt IM, Giuliano G, Delanoy MR, Spector DL, Pollock D (1987). Planta 171:1-18.
Senger H (1980). The Blue Light Syndrome. Springer Verlag, Berlin.
Servaites JC, Torisky RS, Chao SF (1984). Plant Sci Lett 35:115-121.
Shinozaki K, Ohme M, Tanaka M, *et al.* (1986). EMBO J 5:243-249.
Smillie RM, Nielson NC, Henningsen KW, von Wettstein D (1977). Aust J Plant Physiol 4:415-438.
Smith SM, Ellis RJ (1981). J Mol Appl Genet 1:127-137.
Stern DB, Gruissem W (1987). Cell 51:1145-1157.
Stitt M, Herzog B, Heldt HW (1984). Plant Physiol 75:548-553.
Thelander M, Narita JO, Gruissem W (1986). Curr Top Plant Biochem Physiol 5:128-141.
Tobin EM, Silverthorne J (1985). Ann Rev Plant Physiol 36:569-593.
Weidner M, Steinbiss H, Kremer BP (1976). Planta 131:262-270.
Wilhelm C, Wild A (1984). J Plant Physiol 115:115-135.

Photosynthesis, pages 285–298

ILLUMINATION OF EIGHT-DAY-OLD DARK-GROWN BARLEY SEEDLINGS ACTIVATES CHLOROPLAST PROTEIN SYNTHESIS; EVIDENCE FOR REGULATION OF TRANSLATION INITIATION

Patricia E. Gamble, Robert R. Klein and John E. Mullet

Department of Biochemistry & Biophysics, Texas A&M University, College Station, Texas 77843-2128

SUMMARY

Plastid protein synthesis increased when 8-day-old dark-grown barley seedlings were illuminated for 4 hr. This treatment did not cause a change in rRNA, *rbcL*, *psbA* or *psaA-psaB* RNA abundance. However, plastid polysome content increased and *rbcL* and *psbD-psbC* transcripts shifted into larger polysomes indicating initiation of translation was stimulated by light. Plastid protein synthesis was increased to a similar extent by 4 hr of continuous red or white light (soluble proteins 4-fold; membrane proteins 7-fold). A pulse of red light followed by 4 hr of darkness increased both membrane and soluble protein synthesis 1.6-fold; far-red light given immediately after the red light pulse partly attenuated the increase in protein synthesis. These results indicate that one photoreceptor controlling the reactivation of chloroplast protein synthesis in 8-day-old dark-grown barley is phytochrome.

INTRODUCTION

Cells of monocot leaves are produced primarily by a basal intercalary meristem. In a growing barley leaf, dividing cells occupy a 0.5 to 1 cm region at the leaf base. Immediately adjacent to the basal meristem is a 2 to 3 cm zone of cell elongation and older, more mature leaf cells are located toward the leaf apex. Cell development along the developing leaf blade is paralleled by chloroplast development (see Mullet, 1988). Therefore, monocot leaves provide a useful system to study the sequence of events which occur during leaf, cell, and chloroplast development (cf. Boffey and Leech, 1982; Dean and Leech, 1982, Robertson and Laetsch, 1974; Apel, 1979; Klein and Mullet, 1987; Mackender, 1978; Martineau and Taylor, 1985; Rodermel and Bogorad, 1985).

In barley, cells of the leaf basal meristem contain 10 to 15 undifferentiated proplastids which are 0.5 to 2 μm in diameter (Robertson and Laetsch, 1974). These non-photosynthetic organelles contain only a few ribosomes and exhibit low transcription activity (Baumgartner *et al.*, 1988). On the other hand, the proplastids of the basal meristem are actively dividing and synthesizing plastid DNA. When cell division ceases, cells are displaced toward the leaf apex due to cell enlargement and growth of additional cells produced by the basal meristem. Plastid replication and DNA synthesis continue after cell division stops until mesophyll cells contain an average of 60 plastids each with approximately 200 copies of DNA. Plastid transcription activity increases 10-fold per plastid (30-fold per cell) in elongating leaf cells (Baumgartner *et al.*, 1988) and plastid RNA and proteins accumulate during this phase of chloroplast development. Interestingly, plastid number, DNA, and transcription increase to a similar extent during this early phase of chloroplast development in dark-grown and illuminated seedlings.

We have previously examined light-induced changes in plastid gene expression in the apical portion of 4.5-day-old barley leaves (Klein and Mullet, 1986). We found that incorporation of radiolabel into a limited number of chloroplast-encoded membrane proteins was activated within 15 min of illumination. This induction was selective since dark-grown plastids accumulated most of the soluble and membrane polypeptides found in light-grown chloroplasts. Polypeptides showing light-induced accumulation included the plastid-encoded chlorophyll *a* apoproteins of PSI and PSII (gene products of *psaA*, *psaB*, *psbA*, *psbB*, *psbC*). Accumulation of the plastid-encoded chlorophyll *a* apoproteins is regulated post-transcriptionally (Klein and Mullet, 1986; Klein *et al.*, 1988a; Kreuz *et al.*, 1986), and the accumulation of these apoproteins requires chlorophyll *a* biosynthesis (Klein *et al.*, 1988b). Accumulation of the nuclear-encoded chlorophyll *a/b* apoproteins also requires chlorophyll biosynthesis, but is controlled to a large extent at the level of transcription by a red light photoreceptor, phytochrome (Simpson *et al.*, 1986; Nagy *et al.*, 1986). Illumination, therefore, is required for the formation of photosynthetically active chloroplasts; once chloroplast development is completed, plastid transcription activity, mRNA levels, translation activity, ribosome, and DNA levels decline (Klein and Mullet, 1987; Baumgartner *et al.*, 1988).

If 4.5-day-old dark-grown barley seedlings are not illuminated and are kept in darkness for additional time, transcription and translation activity of plastids in the leaf apex gradually decline (Klein and Mullet, 1987; Mullet and Klein, 1987). However, in contrast to illuminated plants, plastid ribosome and DNA copy number do not decrease, and RNA levels decline to a smaller extent in the older dark-grown plants compared to illuminated plants. These properties suggests that plastids of older dark-grown leaf sections retain a latent capacity for plastid transcription and translation activity.

In the present study, we examined plastid protein synthesis and mRNA levels during the illumination of 8-day-old dark-grown barley leaves. The present results confirm that the illumination of older dark-grown barley leaves reactivates the synthesis of plastid-encoded polypeptides (both membrane and soluble) and that the reactivation occurs without a change in mRNA abundance. Further, the examination of polysome profiles revealed that the light-activated synthesis of several plastid-encoded polypeptides occurs at the level of translation initiation. Finally, we examined the nature of the photoreceptor governing the activation of plastid protein synthesis in older dark-grown barley.

MATERIALS AND METHODS

Plant growth, tissue selection, illumination conditions

Barley (*Hordeum vulgare* L. var. Morex) seedlings were grown as previously described (Klein and Mullet, 1987). Seeds were germinated and grown for 8 days in a dark chamber located in a light-tight room. At this stage of development, primary leaves were 15 cm in length (Baumgartner *et al.*, 1988). Plastids and RNAs were isolated from the apical 4 cm of the primary leaf. Eight-day-old seedlings were kept in darkness or illuminated with white light, red light, or far-red light as previously described (Klein *et al.*, 1988b). Duration of illumination and fluence rates for individual experiments are given in the text and figure legends. All manipulations of dark-grown plants were performed in either complete darkness or in light provided by a dim, green safe-light which did not photoconvert measurable amounts of protochlorophyllide.

RNA, protein, and translation assays

RNA was extracted from plastids or tissue as previously described (Mullet *et al.*, 1985; Klein *et al.*, 1988b). Northern blots were performed by separating RNA on 0.8% agarose gels containing 6% formaldehyde. RNA was transferred to Gene Screen nylon membranes and the blots probed, washed and autoradiographed as described (Gamble *et al.*, 1988).

Conditions for ATP-driven protein synthesis by isolated intact plastids were as previously described (Klein and Mullet, 1986). Plastids (1.375×10^7) were added to an optimized protein synthesis mixture (75 μl final volume) and incubated for 10 min. In some experiments, unlabeled methionine (8.5 mM final concentration) was added to block further incorporation of ^{35}S-methionine and to allow chain elongation of incomplete nascent polypeptides (the chase period was 10 min). Following the labeling period, plastids were fractionated into membrane and soluble polypeptides. Measurements of trichloroacetic acid-insoluble radioactivity were obtained and samples were subsequently electrophoresed and autoradiographed (Klein and Mullet, 1986). Total chloroplast polysomes were isolated from intact plastids and separated on sucrose density gradients as described previously (Klein *et al.*, 1988a). Messenger RNA from sucrose density gradient fractions was isolated by phenol extraction as previously described (Mullet *et al.*, 1985; Klein *et al.*, 1988a).

RESULTS

We have previously found that plastid protein synthesis activity declines when 4.5-day-old dark-grown seedlings are kept in darkness for an additional 84 hr (total of 8 days dark) (Fig. 1, Klein and Mullet, 1987). In the present study, when 8-day-old dark-grown barley seedlings were illuminated with white light (350 $\mu E/m^2/s$), plastid translation activity increased within 4 hr after the onset of illumination. Translation activity continued to increase for 14 hr then declined to lower levels after 24 to 48 hr of continuous illumination with white light (Fig. 1). The light-induced activation and eventual decline in protein synthesis was observed for both membrane and soluble polypeptides.

The light-induced increase in plastid translation activity might result from increased synthesis of a few proteins or from an overall increase in plastid protein synthesis. To address this question, radiolabeled plastid membrane and soluble proteins were separated by SDS-PAGE and autoradiographed (Fig. 2). When compared to dark-grown plastids, little change in radiolabel incorporation into membrane and soluble proteins was observed after 1 hr of illumination (lanes 1 vs 2; 4 vs 5). However, by 4 hr after the onset of illumination, radiolabel incorporation into nearly all plastid proteins increased (lanes 3 and 6).

The increase in ^{35}S-methionine incorporation into protein could result from increased plastid transcript levels. To check this possibility, RNA levels of several genes were measured in plastids of 8-day-old dark-grown seedlings and plastids isolated from seedlings illuminated with white light for 1 or 4 hr (Fig. 3). This analysis showed that plastid ribosomal RNA as well as *rbcL* and *psbA* mRNA abundance were not significantly increased by 4 hr of illumination. Although not shown in Figure 3, *psaA-psaB* transcript levels were also not altered by 4 hr of illumination. During the same period of illumination, however, the synthesis of polypeptides encoded by these mRNAs increased several-fold (see Fig. 2).

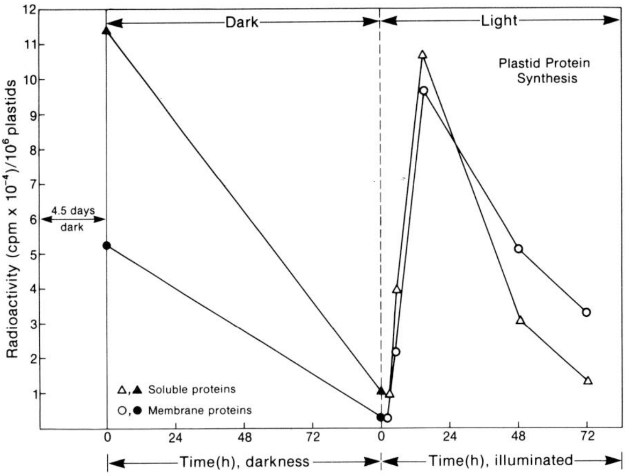

Figure 1. Protein synthesis in isolated plastids. Plastids were isolated from 4.5-day-old dark-grown (closed symbols) barley seedlings or 8 day-old dark-grown seedlings exposed to continuous illumination (open symbols) for up to 72 hr and ^{35}S-methionine incorporation into soluble (Δ, ▲) and membrane (o, •) proteins during a 10 min pulse-labeling assay determined.

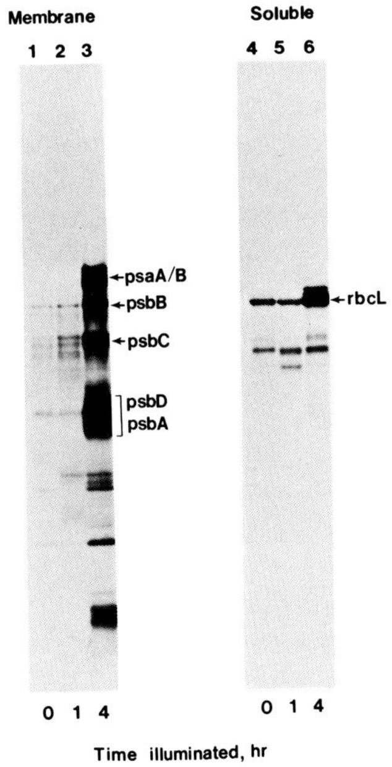

Figure 2. Accumulation of radiolabeled membrane and soluble polypeptides by plastids isolated from dark-grown and illuminated barley seedlings. Plastids were isolated from 8-day-old dark-grown seedlings (lanes 1 and 4) and from 8-day-old dark-grown seedlings transferred to the light for 1 (lanes 2 and 5) or 4 hr (lanes 3 and 6). Intact plastids were incubated for 10 min in the presence of ^{35}S-methionine followed by fractionation into membrane and soluble polypeptides. The samples were subsequently electrophoresed on an SDS-PAGE gel containing 4M urea; membrane proteins from 8 x 10^5 plastids were loaded in lanes 1-3; soluble proteins from 3.2 x 10^5 plastids were loaded in lanes 4-6.

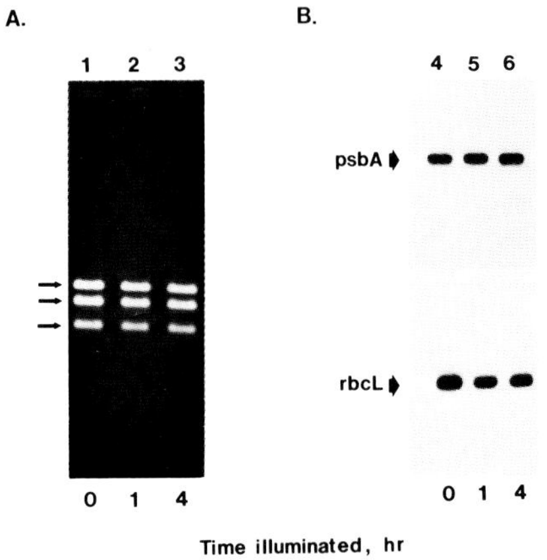

Figure 3. RNA levels of plastids from dark-grown and illuminated seedlings. Nucleic acid was isolated from plastids of 8-day-old dark-grown seedlings (lanes 1 and 4) and from seedlings transferred to the light for 1 (lanes 2 and 5) or 4 hr (lanes 3 and 6). (A) Nucleic acid, isolated from an equal number of plastids, was loaded on a methyl mercury gel and separated RNA detected by ethidium bromide staining. Arrows next to lane 1 mark the position of rRNA. (B) Northern blot analysis of *psbA* and *rbcL* mRNA. Total nucleic acid was separated on 6% formaldehyde gels, transferred to nylon membranes, and hybridized with nick-translated DNA from *psbA* or *rbcL*.

The increase in plastid protein synthesis upon illumination could be due to activation of translation initiation, elongation, and/or termination. The arrest of translation at each of these points can result in a unique polysome profile that permits the site of translational regulation to be determined (Klein *et al.*, 1988a). For instance, if translation was inhibited at initiation, then a significant portion of the transcripts would be restricted to nonpolysomal fractions. Upon illumination, the block in translation initiation would be reduced and transcripts would be recruited into polysomal fractions. To test whether translation is blocked in the dark at the point of initiation, polysomes from plastids of dark-grown plants and plants illuminated for 4 hr were separated on sucrose density gradients, polysome profiles compared (Fig. 4A, B), and the distribution of RNA in the polysome fractions determined by Northern analysis (Fig. 4; C-F). This analysis demonstrated that illumination of dark-grown plants resulted in a significant increase in plastid polysome content (polysome to total ribosome ratio (P/T ratio) increased from 0.15 to 0.36) and the ratio of large polysomes (mRNA associated with >3 ribosomes) to small polysomes also increased (Fig. 4A, B).

Associated with this increase in polysome content was a decrease in free 30S and 50S ribosomal subunits indicating that translationally inactive ribosomal subunits were recruited into polysomes upon illumination (compare Fig. 4A vs 4B). Examination of the distribution of *rbcL* and *psbD-psbC* transcripts across the polysome profile indicated that illumination shifted these mRNAs into larger polysomes (Fig. 4C-F, fractions 4 and 5). In particular, a significant portion of *rbcL* mRNA was recruited from nonpolysomal fractions (1 and 2) into polysomes (fractions 3-5) upon illumination while a smaller shift was observed for

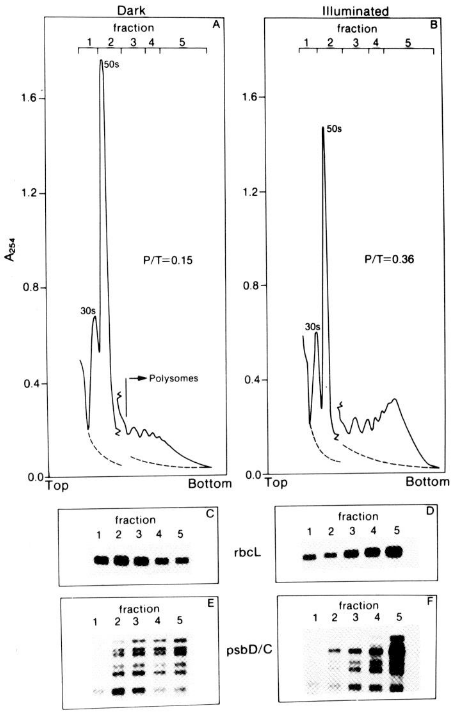

Figure 4A, B. Sucrose density gradient profiles of ribosomal material of plastids isolated from dark-grown and illuminated barley seedlings. Intact plastids were isolated from 7.5-day-old dark-grown seedlings (A) and from seedlings illuminated for 4 hr prior to harvest (B). Plastids were solubilized followed by separation on 15-60% (w/v) sucrose gradients. Gradients were scanned at 254 nm and P/T ratios calculated. Gradients of ribosomal material were fractionated into 5 fractions as indicated. A 2-fold decrease in full scale absorbance is indicated by the break in the absorbance tracing of each profile. Dashed lines represent base lines. **Figure 4C-F.** Northern blot analyses of sucrose density gradient profiles of ribosomal material from plastids of 7.5-day-old dark-grown seedlings (C and E) and dark-grown seedlings transferred to the light for 4 hr (D and F). Nucleic acid was isolated from sucrose density gradient fractions by phenol extraction. Total nucleic acid from each fraction was separated on 6% formaldehyde gels, transferred to nylon membranes, and hybridized with nick-translated DNA probes from *rbcL* (C and D) and *psbD-psbC* (E and F).

psbD-psbC mRNA. Therefore, these results suggest that illumination of 8-day-old dark-grown barley activates protein synthesis, in part, by stimulating the initiation of translation.

Higher plants contain at least three photoreceptors which could participate in the activation of plastid protein synthesis observed here; a blue light photoreceptor and two red light receptors, phytochrome and protochlorophyllide-holochrome. As a first step toward understanding the involvement of these photoreceptors in this process, the effect of a variety of light regimes on plastid translation activity was assessed (Table I). After 4 hr of continuous white light, incorporation of ^{35}S-methionine into soluble plastid proteins increased 3- to 4-fold, while a 7-fold increase by membrane proteins was observed. Continuous red light was

Table I. Influence of Various Light Treatments on Plastid Protein Synthesis

	Radiolabel Incorporation --cpm/1 x 10^6 plastids--	
Treatment[1]	Membrane	Soluble
Dark	8,460	12,050
Continuous white light (4.68×10^6 $\mu E/m^2$)	66,940	41,430
Continuous red light (2.88×10^5 $\mu E/m^2$)	65,960	49,940
Red light pulse (1×10^3 $\mu E/m^2$)	12,800	15,800
Red light pulse followed by far-red light (1×10^3 $\mu E/m^2$ red light; 6×10^3 $\mu E/m^2$ far-red light)	11,200	13,300

[1] 8-day-old dark-grown seedlings were illuminated continuously for 4 hr or illuminated with pulses of red, far-red or red followed by far-red light and then placed in darkness for 4 hr. Red light pulses were 1 min in duration; far-red light pulses were 5 min in duration.

equally effective in stimulating plastid protein synthesis suggesting that this process does not require activation of a blue light receptor. A 1 min pulse of red light followed by 4 hr of darkness led to a small increase (1.6-fold) in both membrane and soluble protein synthesis. If the red light pulse was followed immediately by far-red light, the increase in protein synthesis was partially attenuated.

These results indicate that a red light photoreceptor participates in the activation of protein synthesis in 8-day-old dark-grown barley leaves. Two photoreceptors that absorb red light are phytochrome and the protochlorophyllide-holochrome (for recent references, see Klein *et al.*, 1988b). Given that far-red light partially reversed the stimulation of

^{35}S-methionine incorporation by red light, phytochrome, in part, controls the reactivation of protein synthesis in older barley leaves. To examine further the effect of red and far-red light on the synthesis of individual membrane polypeptides, radiolabeled proteins were separated by SDS-PAGE gels and autoradiographed (Fig. 5). In apical portions of 8-day-old dark-grown barley leaves the synthesis of the *psbD* gene product and several other polypeptides was detectable (Fig. 5A, lane 1).

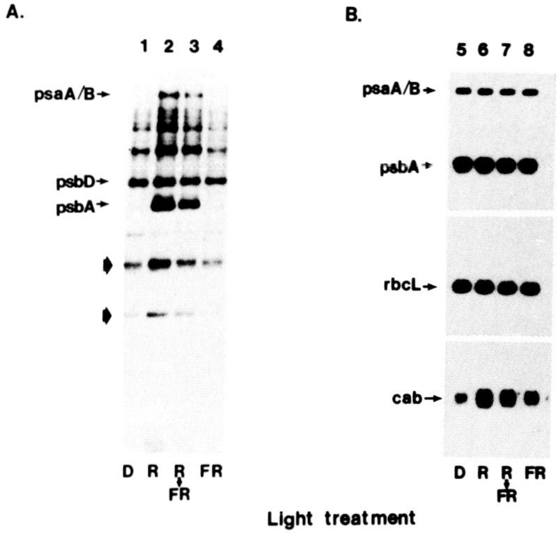

Figure 5. Phytochrome regulation of radiolabeled chlorophyll *a* apoprotein accumulation and mRNA abundance. 8-day-old dark-grown seedlings were illuminated with red light (1 x 10^3 $\mu E/m^2$ over 1 min period). Following irradiation with red light, plants were either returned to darkness for 4 hr or were immediately irradiated with far-red light (6 x 10^3 $\mu E/m^2$ over 5 min period). Red + far-red-treated plants were then returned to darkness for 4 hr. A third set of seedlings were irradiated with far-red light alone (6 x 10^3 $\mu E/m^2$ over 5 min) and placed in darkness for 4 hr. (A) Intact plastids were isolated and labeled for 10 min in the presence of ^{35}S-methionine. Plastids were fractionated and membrane polypeptides from 8 x 10^5 plastids were loaded on an SDS-PAGE gel containing 8M urea. (B) Northern blot analysis of *psaA*/B, *psbA*, *rbcL* and *cab* mRNA. Total nucleic acid was isolated from intact plastids and separated on 6% formaldehyde gels on an equal plastid basis. Following electrophoresis, the nucleic acid was transferred to nylon membranes and hybridized with nick-translated DNA probes from *psaA*/B, *psbA* and *rbcL*. For analysis of *cab* mRNA abundance, nucleic acid was isolated from whole leaf tissue. Total RNA was loaded on 6% formaldehyde gels on an equal RNA basis (approximately 2 μg per lane). After transfer to a nylon membrane the blot was hybridized with a nick-translated cDNA probe for *cab*. Irradiation treatments: Dark, lanes 1 and 5; red light, lanes 2 and 6; red plus far-red light, lanes 3 and 7; far-red light alone, lanes 4 and 8.

In contrast, little radiolabel accumulated in the *psbA* or *psaA-psaB* gene products. When plants received a pulse of red light followed by 4 hr of darkness, radiolabel incorporation into the *psbD* gene product was stimulated to a small extent. In contrast, this treatment increased radiolabel incorporation into the *psbA* and *psaA-psaB* gene products to a much greater extent (Fig. 5A, lane 2). The increase in *psbD* translation induced by the red light pulse could be almost completely reversed by far-red light given immediately after the red light treatment. Far-red light was less effective in attenuating the red light-induced increase in the synthesis of the *psbA* and *psaA-psaB* gene products (Fig. 5A, lane 3). Northern blot analyses indicated that the red light-induced accumulation of the *psaA-psaB*, *psbA* and *rbcL* gene products occurred without a corresponding increase in mRNA levels (Fig. 5B). As a control, we examined the accumulation of *cab* mRNA under our illumination regime. As expected, *cab* mRNA abundance was increased by a red light pulse and the increase was partially attenuated if the red light pulse was followed immediately by far-red illumination (Fig. 5B).

DISCUSSION

Leaf and chloroplast development procede to only a limited extent in dark-grown dicots such as pea. In contrast, in monocots such as barley, primary foliage leaf development is not significantly inhibited in darkness and chloroplast development is arrested only late in development (see Mullet, 1988). In 4.5-day-old dark-grown barley, chloroplast transcription activity is high in the apical 2 to 3 cm of the primary leaf and illumination of these plants for 16 hr causes transcription activity to decline (Mullet and Klein, 1987). Translation of soluble proteins is also high in plastids of 4.5-day-old dark-grown seedlings and illumination causes only a transient increase in soluble protein synthesis followed by a gradual decline as chloroplasts mature (Klein *et al.*, 1988b; Klein and Mullet, 1986). Illumination of 4.5-day-old dark-grown barley also results in a dramatic but selective increase in radiolabel accumulation in the plastid encoded chlorophyll *a* apoproteins (Klein and Mullet, 1986). The increase is dependent on chlorophyll biosynthesis and occurs in the absence of changes in chlorophyll-apoprotein mRNA levels (Klein *et al.*, 1988b). These results indicate that plastid transcription and translation activity are nearly fully expressed in 4.5-day-old dark-grown seedlings and that illumination alters this situation primarily by promoting chlorophyll formation and accumulation of the chlorophyll *a* apoproteins.

If 4.5-day-old dark-grown barley seedlings are kept in darkness for additional time, plastid transcription and translation activity decline (Klein and Mullet, 1987; Mullet and Klein, 1987). The decrease in translation is greater than can be accounted for by decreases in plastid mRNA levels. For example, *rbcL* mRNA levels declined 2-fold whereas translation of the *rbcL* gene product decreased approximately 10-fold when 4.5-day-old barley seedlings were kept in darkness an additional 96 hr. Furthermore, a decrease in the synthesis of all the plastid-encoded proteins was observed suggesting operation of a global control mechanism (Klein and Mullet, 1987). A similar decline in translation of the *rbcL* gene product has been observed in older dark-grown *Amaranthus hypochondriacus* cotyledons (Berry *et al.*, 1985; 1986) and when light-grown plants were transferred to darkness (Berry *et al.*, 1988). The decline in translation in both these instances could not be explained by changes in *rbcL* mRNA levels or translatability. In *Spirodela oligorrhiza*, transfer of light-grown plants to darkness caused decreased radiolabel accumulation into the *rbcL* and *psbA* gene products (Fromm *et al.*, 1985). Here again the decrease in radiolabel incorporation could not be accounted for by changes in RNA levels suggesting that regulation of translation or protein stability was involved.

In this paper, we show that illumination of 8-day-old dark-grown barley seedlings reactivates plastid protein synthesis. The activation occurs within 4 hr after the onset of illumination, reaches a high level after 14 hr and then protein synthesis activity declines to lower levels after 48 to 72 hr of continuous illumination with white light. A similar increase in plastid translation activity was reported when 8-day-old dark-grown wheat seedlings were illuminated (Obokata, 1984). The increase in plastid protein synthesis which occurs when 8-day-old dark-grown barley plants are illuminated for 4 hr is not due to changes in RNA levels (rRNA and *rbcL*, *psbA*, *psaA-psaB* mRNA reported here) and can be explained, in part, by an overall activation of translation initiation. This conclusion is consistent with the light-induced increase in plastid polysome content and a shift of nonpolysome-associated mRNA into large polysomes. At present the mechanism which alters the rate of translation initiation is unknown but it could involve reversible modification of initiation factors as has been observed in other systems (Scorsone *et al.*, 1987).

Protein synthesis in plastids of 8-day-old dark-grown barley is stimulated to approximately the same extent by continuous white or red light. This result indicates that blue light is not required to enhance plastid protein synthesis. A pulse of red light (low fluence range) induced both membrane and soluble protein synthesis activity. Interestingly, the red light pulse treatment enhanced radiolabel incorporation into some membrane proteins to a greater extent than incorporation into soluble proteins and other membrane proteins. Specifically, the synthesis of the *psbA* and *psaA-psaB* gene products was nearly undetectable in dark-grown plants and the red light pulse treatment caused a greater increase in radiolabel incorporation into these chlorophyll-apoproteins than into other polypeptides. We attribute the enhanced light-induced increase in *psbA* and *psaA-psaB* gene product translation to the activity of the photoreceptor, protochlorophyllide. We have recently shown that accumulation of radioactivity into chlorophyll-apoproteins in barley plastids requires chlorophyll *a* which is absent from dark-grown barley (Klein *et al.*, 1988b). The synthesis of other proteins, such as the gene products of *rbcL* or *psbD*, do not exhibit a dependence on chlorophyll formation. Therefore, the differential effect of a red light pulse on the incorporation of radioactivity into membrane proteins such as the *psbA* and *psaA-psaB* gene products is related to photoreduction of protochlorophyllide and subsequent chlorophyll *a* formation which selectively enhances accumulation of radioactivity into these proteins.

Finally, the overall increase in membrane and soluble protein synthesis induced by a pulse of red light could be attenuated if the red light treatment was immediately followed by far-red light. This result suggests that phytochrome plays a role in reactivating plastid protein synthesis in 8-day-old dark-grown barley.

ACKNOWLEDGEMENTS

We thank Sharyll Pressley and Belinda Dowdy for their help in assembling this manuscript. This work was supported by NSF grant No. DCB 86-16156 to JEM.

REFERENCES

Apel K (1979). Phytochrome-induced appearance of mRNA activity for the apoprotein of the light-harvesting chlorophyll *a/b* protein of barley (*Hordeum vulgare*). Eur J Biochem 97:183-188.

Baumgartner BJ, Rapp JC, Mullet JE (1988). Chloroplast transcription and DNA copy number in developing barley leaves. Plant Physiol (in press).

Berry JO, Carr JP, Klessig DF (1988). mRNAs encoding ribulose-1,5-bisphosphate carboxylase remain bound to polysomes but are not translated in Amaranth seedlings transferred to darkness. Proc Natl Acad Sci USA 85:4190-4194.

Berry JO, Nikolau BJ, Carr JP, Klessig DF (1985). Transcriptional and post-transcriptional regulation of ribulose 1,5-bisphosphate carboxylase gene expression in light- and dark-grown Amaranth cotyledons. Molecular and Cellular Biology 5:2238-2246.

Berry JO, Nikolau BJ, Carr JP, Klessig DF (1986). Translational regulation of light-induced ribulose 1,5-bisphosphate carboxylase gene expression in Amaranth. Molecular and Cellular Biology 6:2347-2353.

Boffey SA, Leech RM (1982). Chloroplast DNA levels and the control of chloroplast division in light-grown wheat leaves. Plant Physiol 69:1387-1391.

Dean C, Leech RM (1982). Genome expression during normal leaf development. I. Cellular and chloroplast numbers and DNA, RNA, and protein levels in tissues of different ages within a seven-day-old wheat leaf. Plant Physiol 69:904-910.

Fromm H, Devic M, Fluhr R, Edelman M (1985). Control of *psbA* gene expression: in mature *Spirodela* chloroplasts light regulation of 32-kd protein synthesis is independent of transcript level. EMBO J 4:291-295.

Gamble PE, Sexton TB, Mullet JE (1988). Light-dependent changes in *psbD* and *psbC* transcripts of barley chloroplasts: accumulation of two transcripts maintains *psbD* and *psbC* translation capability in mature chloroplasts. EMBO J 7:1289-1297.

Klein RR, Gamble PE, Mullet JE (1988b). Regulation of chlorophyll *a*-apoprotein accumulation. Plant Physiol 88:1246-1256.

Klein RR, Mason HS, Mullet JE (1988a). Light-regulated translation of chloroplast proteins. Transcripts of *psaA-psaB*, *psbA* and *rbcL* are associated with polysomes in dark-grown and illuminated barley seedlings. Journal of Cell Biology 106:289-301.

Klein RR, Mullet JE (1986). Regulation of chloroplast-encoded chlorophyll-binding protein translation during higher plant chloroplast biogenesis. Journal of Biological Chemistry 261:11138-11145.

Klein RR, Mullet JE (1987). Control of gene expression during higher plant chloroplast biogenesis. Journal of Biological Chemistry 262:4341-4348.

Kreuz K, Dehesh K, Apel K (1986). The light-dependent accumulation of the P700 chlorophyll *a* protein of the photosystem I reaction center in barley. Eur J Biochem 159:459-467.

Mackender RO (1978). Etioplast development in dark-grown leaves of *Zea mays* L. Plant Physiol 62:499-505.

Martineau B, Taylor WC (1985). Photosynthetic gene expression and cellular differentiation in developing maize leaves. Plant Physiol 78:399-404.

Mullet JE (1988). Chloroplast development and gene expression. Annu Rev Plant Physiol Plant Mol Biol 39:475-502.

Mullet JE, Klein RR (1987). Transcription and RNA stability are important determinants of higher plant chloroplast RNA levels. EMBO J 6:1571-1579.

Mullet JE, Orozoco EM, Chua N-H (1985). Multiple transcripts for higher plant *rbcL* and *atpb* genes and localization of the transcription initiation site of the *rbcL* gene. Plant Molecular Biology 4:39-54.

Nagy R, Kay SA, Boutry M, Hsu M-Y, Chua N-H (1986). Phytochrome controlled expression of a wheat *cab* gene in transgenic tobacco seedlings. EMBO J 5:119-1124.

Obokata J (1984). Protein synthesis in isolated plastids during chloroplast development in wheat. Plant & Cell Physiol 25(5):821-830.

Robertson D, Laetsch WM (1974). Structure and function of developing barley plastids. Plant Physiol 54:148-159.

Rodermel SR, Bogorad L (1985). Maize plastid photogenes: mapping and photoregulation of transcript levels during light-induced development. Journal of Cell Biology 100:463-476.

Simpson J, Schell J, Montagu MV, Herrera-Estrella L (1986). Light-inducible and tissue-specific pea LHCP gene expression involves an upstream element combining enhancer- and silence-like properties. Science 323:551-554.

Scorsone KA, Panniers R, Rowlands AG, Henshaw EC (1987). Phosphorylation of eukaryotic initiation factor 2 during physiological stresses which affect protein synthesis. Journal of Biological Chemistry 262:14539-14543.

Photosynthesis, pages 299–309

THE ASSEMBLY OF RIBULOSE BISPHOSPHATE CARBOXYLASE IN *Escherichia coli* AND ISOLATED CHLOROPLASTS

Anthony A. Gatenby, Pierre Goloubinoff, Thomas H. Lubben and George H. Lorimer.

Central Research and Development Department, Experimental Station, E.I. DuPont de Nemours and Co., Wilmington, Delaware 19880-0402.

The initial demonstration of the successful expression of a chloroplast ribulose bisphosphate carboxylase (RubisCO) large subunit gene (*rbcL*) in *Escherichia coli* (Gatenby *et al.*, 1981), has been followed by several reports which have expanded the range of both *rbcL* and small subunit (*rbcS*) genes from plant and prokaryotic sources which can be expressed in *E. coli* (reviewed by Gatenby, 1988; Tabita, 1988). This has enabled experimental systems to be developed that result in the assembly of catalytically active dimeric (Somerville and Somerville, 1984), hexameric (Quivey and Tabita, 1984) and hexadecameric (Gatenby *et al.*, 1985) enzymes. These systems can now be used to study site-specific mutations of the active site (Gutteridge *et al.*, 1984), to examine the assembly pathway (van der Vies *et al.*, 1986), and to examine the formation of hybrid enzymes (Bradley *et al.*, 1986). One of our interests has centered on the mechanism of assembly of the RubisCO enzyme. RubisCO assembly can be examined in a number of ways. In this paper, however, we shall concentrate on experiments that examine RubisCO assembly in *E. coli*, and in isolated chloroplasts following the import of large (L) and small (S) subunits.

SYNTHESIS OF CHLOROPLAST RUBISCO IN *E. coli*

The maize and wheat chloroplast *rbcL* genes were the first plant genes expressed in *E. coli* (Gatenby *et al.*, 1981). The L subunits synthesized had a similar M_r to that observed in chloroplasts, suggesting that translation initiated correctly at the chloroplast ribosome binding site. The production of L subunits in *E. coli* was low if expression relied on transcription from the endogenous chloroplast promoters. Because of the similarities between chloroplast and bacterial translational initiation sites, high levels of expression of *rbcL* can be obtained by simply increasing its rate of transcription. This is conveniently achieved by using a plasmid (pPBI3) encoding a lambda bacteriophage P_L promoter, a transcription antiterminator protein (pN), and the maize *rbcL* gene (Gatenby and Castleton, 1982). Synthesis of L subunits can be induced by thermal denaturation of the *c*I repressor. Under these conditions synthesis of the L subunits continued for more than 120 min and the percentage of L in the cell increased to a maximum of 2% of total *E. coli* protein.

Pulse-chase experiments using *E. coli* cells containing plasmid pPBI3 showed that the maize L subunits were stable for at least 4 h (Gatenby, 1984). The L subunits do not possess enzyme activity, nor will they form a stable enzyme.metal.CO_2.[^{14}C]carboxyarabinitol-P_2 quaternary complex. The protein was also insoluble, whether expressed at high or low levels. The presence of the substrate ribulose-P_2, or the positive effector fructose 1,6-bisphosphate, do not improve solubility. At pH 11-12 L subunits were slowly solubilized, suggesting that aggregate structures may form. It is apparent from these experiments that the maize L subunits are inactive, and become insoluble after, or during, synthesis in *E. coli*.

To synthesize both subunits of higher plant RubisCO in the same cell, a plasmid has been constructed that encodes maize *rbcL* and wheat *rbcS* genes. The wheat *rbcS* gene was modified so that S subunits could be synthesized in *E. coli* without the amino-terminal transit peptide. This was achieved by fusion of an appropriate restriction fragment to the 5' end of *lacZ*, and making use of the *lacZ* promoter and ribosome binding sites for expression of fusion proteins. These modified S subunits contained ten or eleven amino-terminal residues from β-galactosidase (Bradley *et al.*, 1986; Gatenby *et al.*, 1987).

The *rbcS* expression clone was recombined with an *rbcL* expression plasmid to form plasmid pLSS308. In this plasmid the wheat *rbcS* gene is under the control of P_{lac}, and the maize *rbcL* gene is under the control of the thermoinducible bacteriophage P_L promoter (Gatenby *et al.*, 1987). The use of different promoters allows the independent expression of either *rbc* gene. Figure 1 shows the result of an induction experiment in which S subunit synthesis is constitutive because of the presence of isopropyl-β-*D*-thiogalactopyranoside in the media, and the synthesis of the L subunits have been increased by shifting the temperature of the culture to 41°C. When the S subunits were synthesized in the absence of the L subunits, they were soluble but unstable, with a half-life of less than 15 min. They appear to be either monomers or dimers as determined by sucrose gradient centrifugation. When both L and S subunits were synthesized together, both were found predominantly in the insoluble fraction of disrupted cells. They were stable for more than 120

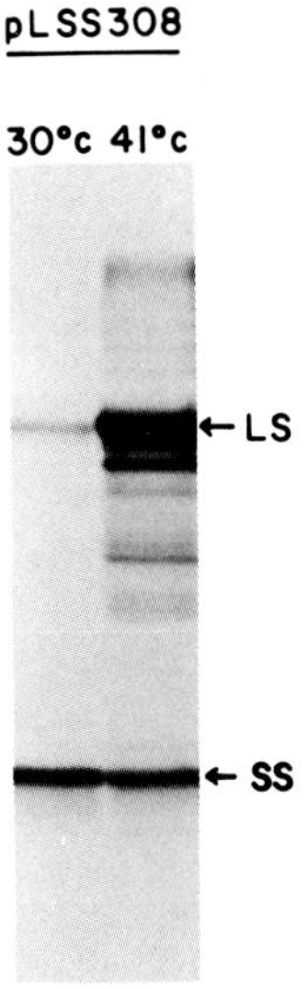

Figure 1. Fluorograph showing the synthesis of RubisCO subunits in *E. coli* following induction of plasmid pLSS308. Cells were labelled with [^{35}S]methionine in the presence of IPTG to induce wheat S subunit (SS) synthesis, and grown at 30°C or 41°C to obtain repression or induction respectively of maize L subunit (LS) synthesis. The radiolabeled polypeptides were incubated with antiserum against wheat RubisCO, before electrophoresis of the products on SDS polyacrylamide gels. Reprinted with permission from Eur. J. Biochem. (Gatenby *et al.*, 1987).

min, suggesting that aggregation of the two subunits may occur leading to stabilization of the S subunits. The two subunits do not assemble together into RubisCO holoenzyme. Similar results were obtained for the expression of *rbcL* and *rbcS* genes from wheat synthesized together in *E. coli* (Bradley *et al.*, 1986).

ASSEMBLY OF PROKARYOTIC RUBISCO IN *E. coli*

The prokaryotic nature of cyanobacteria simplifies the expression of their genes in *E. coli*. Because cyanobacteria do not contain chloroplasts the S subunits are synthesized without a transit peptide, which therefore enables direct expression of the mature S subunits in *E. coli*. In addition, the *rbcL* and *rbcS* genes are physically linked on a contiguous fragment of DNA. Cyanobacterial RubisCO that is enzymatically active can be synthesized and assembled in *E. coli* when the *rbc* genes from *Anabaena* 7120 (Gurevitz *et al.*, 1985), *Anacystis nidulans* (Christeller *et al.*, 1985; Tabita and Small, 1985) or *Synechococcus* PCC6301 (Gatenby *et al.*, 1985) are cloned downstream of suitable promoters. The latter two are different names for the same organism. In each of these examples the *rbcL* and *rbcS* genes were cloned into expression plasmids to make use of plasmid-borne promoters.

The *Synechococcus rbc* genes were transcribed in *E. coli* from a lambda bacteriophage P_L promoter, following temperature induction (Gatenby *et al.*, 1985). Synthesis of the L and S subunits can be detected, together with enzyme activity and [^{14}C] 2-carboxyarabinitol 1,5-

bisphosphate [CABP] binding. In addition, velocity sedimentation on sucrose gradients resolves a peak of enzyme activity that is identical in size to the spinach holoenzyme. These sedimentation studies indicate that L8 cores are formed, but an L8Sn oligomer where n is less than 8 would be difficult to resolve because of the small size of S subunits. Although S subunits must be present to obtain a fully active enzyme (Andrews and Abel, 1981), it is possible that the S binding sites on the L octomers produced in *E. coli* are not fully saturated (Gatenby *et al.*, 1985). First, the specific activity in cell extracts was lower than expected based on the known amount of L8 present (as measured by [^{14}C]CABP binding). Second, the amount of ^{35}S present in the L and S polypeptides of RubisCO isolated from *E. coli* indicates that L8Sn structures are formed where n=2 or 3. This problem of sub-optimal saturation of binding sites can be avoided by using a P_{lac} expression vector to increase the production of S subunits in the cell (van der Vies *et al.*, 1986).

Cloning the cyanobacterial *rbcL* and *rbcS* genes separately results in the absence of enzyme activity, indicating that both L and S subunits are required for normal levels of activity (Christeller *et al.*, 1985; van der Vies *et al.*, 1986). Plasmid pDB50 contains the *rbcS* and *rbcL* genes of *Synechococcus*, and it directs the synthesis of active RubisCO in *E. coli.* In plasmid pDB53 the *rbcS* gene is deleted, and cells containing this plasmid synthesize inactive L subunits. Since this deletion does not result in any change in the amount of soluble L subunits synthesized, it demonstrates that the product of *rbcS* is essential for activity. This was confirmed by infection of *E. coli* cells containing the *rbcS* deletion plasmid (pDB53) with M13 phages encoding the *rbcS* gene. This enabled active RubisCO holoenzyme to be assembled from plasmid- and phage-encoded subunits (van der Vies *et al.*, 1986).

These deletion and infection experiments have provided data on parts of the assembly pathway of RubisCO in *E. coli.* The major soluble RubisCO protein structures formed in cells synthesizing the L subunits alone using pDB53 are oligomers, principally L octomers and L dimers (van der Vies *et al.*, 1986). Therefore, the formation of a *Synechococcus* L8 structure does not require S subunits to be present. The L2 structure may represent an intermediate which then assembles to the L8 form. Demonstration of the assembly of prokaryotic RubisCO in *E. coli*, and the failure to assemble higher-plant RubisCO in bacteria, suggests that the chloroplast binding protein may be required as an auxiliary factor to facilitate assembly of the holoenzyme. The absence of the binding protein in bacteria would therefore prevent assembly of the chloroplast enzyme. Assembly of prokaryotic RubisCO in *E. coli* could indicate that cyanobacteria do not require a binding protein, but it has also been suggested that cyanobacteria do use such a protein for RubisCO assembly and its function is supplied in *E. coli* by a related protein (Gatenby *et al.*,

1988). Evidence for a bacterial protein related to the chloroplast binding protein has been presented by Hemmingsen *et al.* (1988) who found a high level of sequence homology between the *E. coli groEL* protein and the binding protein. The *groEL* protein is involved in bacteriophage morphogenesis and belongs to a class of proteins implicated in the assembly of oligomeric proteins. We have recently obtained data demonstrating that the assembly of cyanobacterial RubisCO in *E. coli* does require the *groEL* protein (Goloubinoff *et al.*, 1989).

ASSEMBLY OF A HYBRID RUBISCO IN *E. coli*

The ability to assemble an active cyanobacterial RubisCO enzyme using an experimental system in which the *rbcL* gene is plasmid encoded, and the *rbcS* gene is expressed from an infecting M13 phage, provides a simple approach to the formation of hybrid enzymes. Appropriate foreign RubisCO genes can be cloned in two different expression vectors and the assembly and activity of the resulting hybrid enzyme examined. For example, hybrid RubisCO assembly has been demonstrated between cyanobacterial L subunits and wheat S subunits (van der Vies *et al.*, 1986).

To form this hybrid enzyme, *E. coli* cells containing the plasmid pDB53 that encodes the *Synechococcus rbcL* gene (Gatenby *et al.*, 1985) were infected with the phage M13-72 that encodes the wheat *rbcS* gene (van der Vies *et al.*, 1986). Both *rbc* genes were transcribed from a *lac* promoter. Following infection and growth of the cultures the disrupted cells were assayed for RubisCO activity. Activity was detectable at a level of 3 nmol of CO_2 fixed per min per mg total protein for the hybrid enzyme, whereas a homologous cyanobacterial enzyme assembled under identical conditions resulted in the fixation of 32 nmol of CO_2 per min per mg total protein (van der Vies *et al.*, 1986). Therefore, the higher plant S subunits can substitute for the cyanobacterial S subunits in assembly of active RubisCO molecules *in vivo*, but the enzyme has a reduced specific activity.

The nature of the hybrid enzyme was examined by sucrose gradient centrifugation and a peak of RubisCO activity was observed. Its position in the gradient suggests that the enzyme has a *Synechococcus* L8 core, to which wheat S subunits are attached. There was non-stoichiometric binding of wheat S subunits to L8, and only trace amounts of S subunits were present in the active fractions. The majority of the S subunits remained at the top of the sucrose gradient, which is a reflection of the less efficient binding of the heterologous S subunits when compared to cyanobacterial S subunits. The low specific activity of the hybrid enzyme is probably due to incomplete saturation of the binding sites on the L8 cores. There may also be a mismatch in the subunit interactions of those subunits that do bind, giving rise to a lower rate of turnover at the active site.

ASSEMBLY OF IMPORTED PROKARYOTIC RUBISCO LARGE SUBUNITS IN CHLOROPLASTS

It has been shown that cyanobacterial RubisCO L subunits can assemble with higher plant S subunits to form an active enzyme when the genes for the two subunits are expressed in *E. coli* (van der Vies *et al.*, 1986), or if the plant S subunits are added to cyanobacterial L8 cores (Andrews and Lorimer, 1985). To determine if hybrid RubisCO molecules can be assembled in isolated chloroplasts, the transit peptide of the S subunit from soybean was fused to the L subunit from *A. nidulans* and the fusion protein imported into isolated pea chloroplasts (Gatenby *et al.*, 1988).

The transit peptide of the imported radiolabelled fusion protein was removed, and when the stromal fraction was analyzed by sucrose gradient centrifugation the cyanobacterial L subunits co-sedimented with RubisCO holoenzyme. The presence of imported L subunits in assembled RubisCO was also apparent when samples of the stroma were electrophoresed on non-denaturing polyacrylamide gels. Imported L subunits were also associated with a large protein complex, as well as with assembled RubisCO (Fig. 2). The protein complex is the binding protein oligomer, and imported L subunits are initially found associated with it. During a 90 min incubation the L subunits associated with the binding protein decreased, and those found in RubisCO increased. This result is similar to data presented by Barraclough and Ellis (1980) in which L subunits synthesized within chloroplasts were initially associated with the binding protein before assembly into holoenzyme. These results indicate that L subunits are assembled into holoenzyme using a similar pathway whether they are homologous pea L subunits synthesized within chloroplasts, or heterologous cyanobacterial L subunits imported into the organelle.

It is important to note that these data do not necessarily demonstrate a precursor-product relationship between L subunits associated with the binding protein oligomer, and those found in RubisCO holoenzyme. The L subunits that assemble into RubisCO could come from sources other than the oligomeric form of the binding protein, perhaps the monomeric form of the binding protein may be involved. The decrease in L subunits associated with the oligomeric binding protein could then occur by alteration of the equilibrium between this complex and the monomeric form.

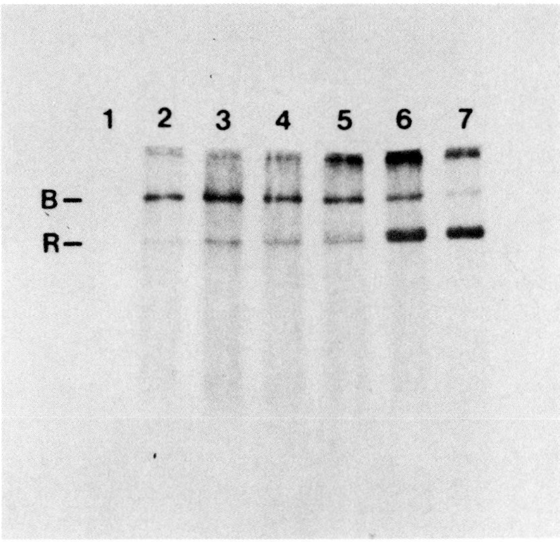

Figure 2. Time-course of *A. nidulans* L subunit import into isolated pea chloroplasts. The L subunit was fused to the soybean S subunit transit peptide to give a protein (TPLS) that was synthesized *in vitro* in the presence of [^{35}S]methionine. The samples were electrophoresed on non-denaturing gels. Imported L associated initially with the binding protein (B), and was then assembled into RubisCO holoenzyme (R). The time points shown in lanes 1-7 are 0, 5, 10, 20, 30, 60, and 90 min. Reprinted with permission from EMBO J. (Gatenby *et al.*, 1988).

THE COMPLEXITY OF RUBISCO ASSEMBLY

In photosynthetic eukaryotes RubisCO is assembled in the chloroplast using imported S subunits, which combine with L subunits synthesized within the organelle (Fig. 3). Prokaryotic cells containing RubisCO do not use an import step, but must still assemble a hexadecameric enzyme from the L and S subunits. An early indication that chloroplast RubisCO assembly might not be a simple process was presented by Barraclough and Ellis (1980), who showed that newly-synthesized L subunits associated with a large protein prior to assembly into RubisCO holoenzyme. This protein became known as the large subunit binding protein, and it contains two polypeptide chains of about 60 kDa, which under certain conditions form an oligomer of about 720 kDa (Hemmingsen and Ellis, 1986; Musgrove *et al.*, 1987). Bloom *et al.* (1983) postulated a complex set of reactions involved in assembly that requires large subunit binding protein, nucleotides, Mg^{2+} and putative intermediates. Imported L and S subunits also associate with the large subunit binding protein in chloroplasts (Gatenby *et al.*, 1988), and the protein has now become known as the RubisCO binding protein.

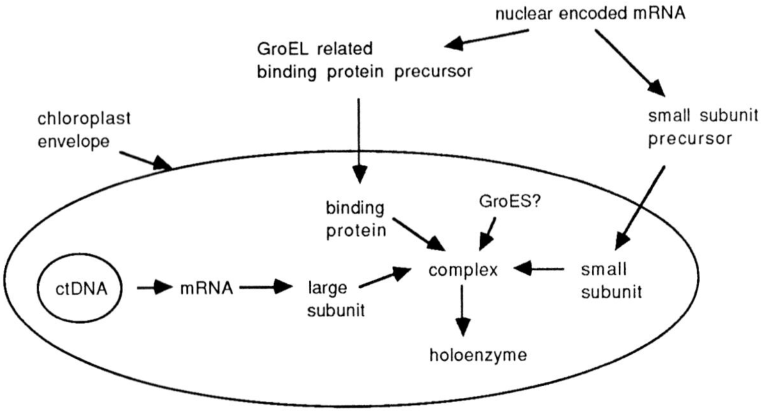

Figure 3. A suggested assembly pathway for higher plant RubisCO. The chloroplast-encoded L subunits are synthesized in the organelle on 70S ribosomes. The nuclear-encoded S subunits are synthesized on 80S cytoplasmic ribosomes as precursor proteins, which are then imported into the chloroplast post-translationally. The transit peptide is removed, presumably before assembly into RubisCO holoenzyme. The assembly complex from which holoenzyme ultimately emerges is incompletely defined. The *groEL*-related RubisCO binding protein is implicated in the reaction, and is itself a nuclear encoded polypeptide that is imported into the chloroplast. Both L and S subunits are known to associate with the binding protein. Our data on RubisCO assembly in *E. coli* has identified a second auxiliary protein (*groES*) required for holoenzyme formation. The presence of a related protein in chloroplasts has not been demonstrated, but we suspect it is required.

Attempts to assemble an active higher plant RubisCO holoenzyme in *E. coli* have been unsuccessful, even though both L and S subunits can be synthesized together (Bradley *et al.*, 1986; Gatenby *et al.*, 1987). In contrast, RubisCO from prokaryotic organisms can readily assemble in *E. coli* (Gatenby *et al.*, 1985). If binding protein is required for higher plant RubisCO assembly, this may explain the inability of the plant enzyme to assemble in *E. coli*, since chloroplast binding protein is not present in the bacterial cell. The experiments to assemble a plant/cyanobacterial hybrid enzyme *in vivo* indicate that there are no intrinsic reasons why wheat S subunits expressed in *E. coli* cannot bind to an L8 structure to form an active site (van der Vies *et al.*, 1986). The observation that cyanobacterial L8 can function as an acceptor for wheat S subunits, but plant L subunits do not, suggests that the principle obstacle for assembly of higher plant RubisCO may be the formation of a suitable L8 core for S subunit binding.

The observation that RubisCO from prokaryotic organisms assembles in *E. coli* suggests that either the bacterial enzyme does not require this auxiliary protein for oligomer formation, or that its function can be supplied by a related *E. coli* protein. The demonstration by Hemmingsen *et al.* (1988) that there is a protein in *E. coli* known as *groEL* that is related to the binding protein, and our observations that *groEL* is involved in cyanobacterial RubisCO assembly (Goloubinoff *et al.*, 1989), show that these auxiliary proteins appear to be required for plant and bacterial RubisCO assembly. It may be possible to assemble plant RubisCO in *E. coli* by co-expressing the chloroplast binding protein, together with the L and S subunits. There may, however, be an additional complication. The *groE* operon contains a second gene (*groES*) that is also involved in RubisCO assembly (Goloubinoff *et al.*, 1989). An equivalent protein may be present in chloroplasts to mediate assembly of RubisCO. If such a component exists, this may also be required for assembly of higher plant RubisCO in *E. coli* cells or *in vitro*. Figure 3 illustrates a suggested overall assembly pathway.

REFERENCES

Andrews TJ, Abel KM (1981). Kinetics and subunit interaction of ribulose bisphosphate carboxylase/oxygenase from the cyanobacterium *Synechococcus* sp. J Biol Chem 256:8445-8451.

Andrews TJ, Lorimer GH (1985). Catalytic properties of a hybrid between cyanobacterial large subunits and higher plant small subunits of ribulose bisphosphate carboxylase-oxygenase. J Biol Chem 260:4632-4636.

Barraclough R, Ellis RJ (1980). Protein synthesis in chloroplasts IX. Assembly of newly-synthesized large subunits into ribulose bisphosphate carboxylase in isolated intact pea chloroplasts. Biochim Biophys Acta 608:19-31.

Bloom MV, Milos P, Roy H (1983). Light-dependent assembly of ribulose-1,5-bisphosphate carboxylase. Proc Natl Acad Sci USA 80:1013-1017.

Bradley D, van der Vies SM, Gatenby AA (1986). Expression of cyanobacterial and higher-plant ribulose 1,5-bisphosphate carboxylase genes in *Escherichia coli*. Phil Trans R Soc London B 313:447-458.

Christeller JT, Terzaghi BE, Hill DF, Laing WA (1985). Activity expressed from cloned *Anacystis nidulans* large and small subunit ribulose bisphosphate carboxylase genes. Plant Mol Biol 5:257-263.

Gatenby AA (1984). The properties of the large subunit of maize ribulose bisphosphate carboxylase/oxygenase synthesized in *Escherichia coli*. Eur J Biochem 144:361-366.

Gatenby AA (1988). Synthesis and assembly of bacterial and higher plant rubisco subunits in *Escherichia coli*. Photosynthesis Res 17:145-157.

Gatenby AA, Castleton JA (1982). Amplification of maize ribulose bisphosphate carboxylase large subunit synthesis in *E. coli* by transcriptional fusion with the lambda *N* operon. Mol Gen Genet 185:424-429.

Gatenby AA, Castleton JA, Saul MW (1981). Expression in *E. coli* of maize and wheat chloroplast genes for large subunit of ribulose bisphosphate carboxylase. Nature 291:117-121.

Gatenby AA, Lubben TH, Ahlquist P, Keegstra K (1988). Imported large subunits of ribulose bisphosphate carboxylase/oxygenase, but not imported β-ATP synthase subunits, are assembled into holoenzyme in isolated chloroplasts. EMBO J 7:1307-1314.

Gatenby AA, van der Vies SM, Bradley D (1985). Assembly in *E. coli* of a functional multi-subunit ribulose bisphosphate carboxylase from a blue-green alga. Nature 314:617-620.

Gatenby AA, van der Vies SM, Rothstein SJ (1987). Co-expression of both the maize large and wheat small subunit genes of ribulose-bisphosphate carboxylase in *Escherichia coli*. Eur J Biochem 168:227-231.

Goloubinoff P, Gatenby AA, Lorimer GH (1989). GroE heat-shock proteins promote assembly of foreign prokaryotic ribulose bisphosphate carboxylase oligomers in *Escherichia coli*. Nature 337:44-47.

Gurevitz M, Somerville CR, McIntosh L (1985). Pathway of assembly of ribulosebisphosphate carboxylase/oxygenase from *Anabaena* 7120 expressed in *Escherichia coli*. Proc Natl Acad Sci USA 82:6546-6550.

Gutteridge S, Sigal I, Thomas B, Arentzen R, Cordova A, Lorimer GH (1984). A site-specific mutation within the active site of ribulose-1,5-bisphosphate carboxylase of *Rhodospirillum rubrum*. EMBO J 3:2737-2743.

Hemmingsen SM, Ellis RJ (1986). Purification and properties of ribulosebisphosphate carboxylase large subunit binding protein. Plant Physiol 80:269-276.

Hemmingsen SM, Woolford C, van der Vies SM, Tilly K, Dennis DT, Georgopoulos CP, Hendrix RW, Ellis RJ (1988). Homologous plant and bacterial proteins chaperone oligomeric protein assembly. Nature 333:330-334.

Musgrove JE, Johnson RA, Ellis RJ (1987). Dissociation of the ribulosebisphosphate-carboxylase large-subunit binding protein into dissimilar subunits. Eur J Biochem 163:529-534.

Quivey RG, Tabita FR (1984). Cloning and expression in *Escherichia coli* of the form II ribulose 1,5-bisphosphate carboxylase/oxygenase gene from *Rhodopseudomonas sphaeroides*. Gene 31:91-101.

Somerville CR, Somerville SC (1984). Cloning and expression of the *Rhodospirillum rubrum* ribulose bisphosphate carboxylase gene in *E. coli*. Molec Gen Genet 193:214-219.

Tabita FR (1988). Molecular and cellular regulation of autotrophic carbon dioxide fixation in microorganisms. Microbiol Rev 52:155-189.

Tabita FR, Small CL (1985). Expression and assembly of active cyanobacterial ribulose-1,5-bisphosphate carboxylase/oxygenase in

Escherichia coli containing stoichiometric amounts of large and small subunits. Proc Natl Acad Sci USA 82:6100-6103.

van der Vies SM, Bradley D, Gatenby AA (1986). Assembly of cyanobacterial and higher plant ribulose bisphosphate carboxylase subunits into functional homologous and heterologous enzyme molecules in *Escherichia coli*. EMBO J 5:2439-2444.

Photosynthesis, pages 311–329

NITROGEN-DEPENDENT BIOGENESIS OF CHLOROPHYLL-PROTEIN COMPLEXES

F. Gerald Plumley, Susan E. Douglas, Alison Branagan Switzer and Gregory W. Schmidt

Department of Botany, University of Georgia, Athens, Georgia 30602 and Atlantic Research Laboratory, Halifax, NS, B3H 3Z1, Canada (SED)

INTRODUCTION

Photosynthetic efficiency and the capacity for nitrogen assimilation are metabolically interdependent. Moreover, N is the mineral nutrient most limiting to plant growth worldwide (Field and Mooney, 1986). Since plants grown with reduced N are chlorotic and photosynthetically sluggish (but accumulate carotenoids and storage reserves of starch and/or lipid within the chloroplast), an obvious effect of N limitation is on chloroplast biogenesis. There has been a rather large amount of data accumulated on the metabolic characteristics of chloroplasts from N-limited plants but very little is known about the underlying reorganization of chloroplast proteins or the regulatory mechanisms involved in adaptation to the availability of this nutrient.

Our studies concern the relationship between N availability, alteration of the photosynthetic apparatus, and the molecular basis for adaptation to nutrient provision. A dramatic effect of N limitation is on the accumulation of nuclear-encoded mRNAs for the apoproteins of light-harvesting chlorophyll complexes. In chloroplasts, N limitation leads to severe repression of organelle protein synthesis but, in this case, the effect is exerted at the translational level. Overall, N-limited cells possess photosynthetic membranes that are altered with regard to antenna functions of reaction centers. In part, this is reflected in altered ratios of antenna pigments in thylakoids as well as photosynthetic protein abundance.

METHODS

Wild type cells of *Chlamydomonas reinhardtii* were grown in a continuous culture system with autotrophic medium containing 150 μM NH_4Cl_2 supplied at a dilution rate of 0.25 per day. Thus, N provisions are precisely controlled and vegetative cells are maintained in a steady state of N limitation. After several weeks of adaptation, aliquots were removed for analyses and a pulse of NH_4Cl_2 to 7.5 mM was administered to the cultures; aliquots were subsequently removed at intervals after the N-pulse. The benefit of this culture system is that it eliminates problems inherent with N remobilization, which leads to senescence, in N-limited higher plants (Gauch, 1972). N-sufficient batch cultures were grown autotrophically with 7.5 mM NH_4Cl_2 and maintained in log phase.

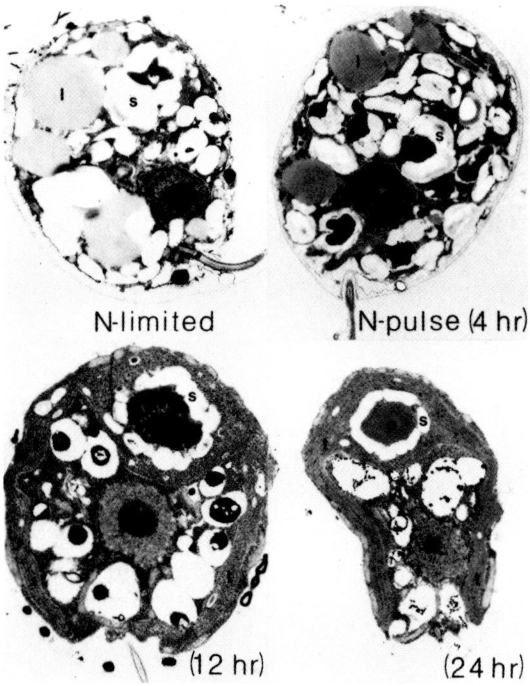

Figure 1. Electron Microscopy of N-Limited Cells and Cells Under-going N-Dependent Re-Greening.

l, lipid deposits;
s, starch granules;
p, pyrenoid;
t, thylakoids

RESULTS

Cell Ultrastructure as Revealed by Electron Microscopy

Consistent with earlier observations of N limitation in plant and algal species, *Chlamydomonas* accumulates large amounts of starch and lipids; both are visible in electron micrographs and appear to be located exclusively within the chloroplast (Fig. 1). Much of the lipid in N-limited cells is triglycerides, a species almost undetectable in N-sufficient cells (D. Wilmont, *et al.*, in prep.). Also evident from electron micrographs is the low level of thylakoid membranes, all unpaired throughout most of their length. Following an N pulse, the starch and lipid reserves are rapidly consumed such that their corresponding plastid inclusions are almost completely destroyed within 24 h. Thylakoid membrane proliferation also occurs rapidly, but after a lag of about 4 h.

Accumulation, Localization and Fluorescence Characteristics of Photosynthetic Pigments

N-limited *Chlamydomonas* cells have only 30% of the normal levels of Chl. Although chlorosis invariably is a symptom of N limitation, it has been equivocal as to whether Chl *a*, Chl *b* or both are reduced (Terashima and Evans, 1988; Hunt *et al.*, 1985). Moreover, N-limited plants accumulate carotenoids excessively (Goodwin, 1980) but neither the relative contribution of various carotenoid species nor their cellular localization have been well documented. To resolve these questions, we compared the whole cell absorption spectra of N-limited and N-sufficient (batch) cells. When expressed on a per cell basis, the reduction in Chl in N-limited cells (Fig. 2, top left) or purified thylakoid membranes therefrom (Fig. 2, bottom left) is apparent. When normalized to an equal Chl basis and the difference spectra plotted, there appears to be an overall reduction in Chl *b* but an enhancement of carotenoids. In contrast, there is reduced absorption by carotenoids in purified thylakoid membrane fractions of N-limited cells.

To quantitate these differences, pigments were resolved by HPLC and the spectra normalized to equal Chl *a* concentrations. The chl *a/b* ratio increases from 2.5 to 3.5 as a result of N limitation in both whole cell and thylakoid extracts; the similarities in the Chl *a/b* ratios of whole cells and thylakoids within treatments demonstrates that all Chl protein complexes were recovered during cell fractionation (Fig. 2, right). Loroxanthin is virtually absent in N-limited cells. This decrease is in contrast to the high levels of violaxanthin and lutein; expressed on a per cell basis, these xanthophylls are present at levels commensurate with those of control cells. The percentage increase in these two xanthophylls is not evident in purified thylakoid membranes (Fig. 2, bottom right). Thus, greater than 75% of the lutein and violaxanthin are localized either in plastoglobuli or chloroplast envelopes of N-limited cells but less than 45% of these pigments are in non-thylakoid fractions of N-sufficient cells.

The low level of thylakoid membranes in N-limited cells (Fig. 1) is consistent with the reduction of Chl normally associated with N limitation. Moreover, the absence of appressed membranes suggests that photosynthetic components that are restricted to nonstacked regions (i.e.,

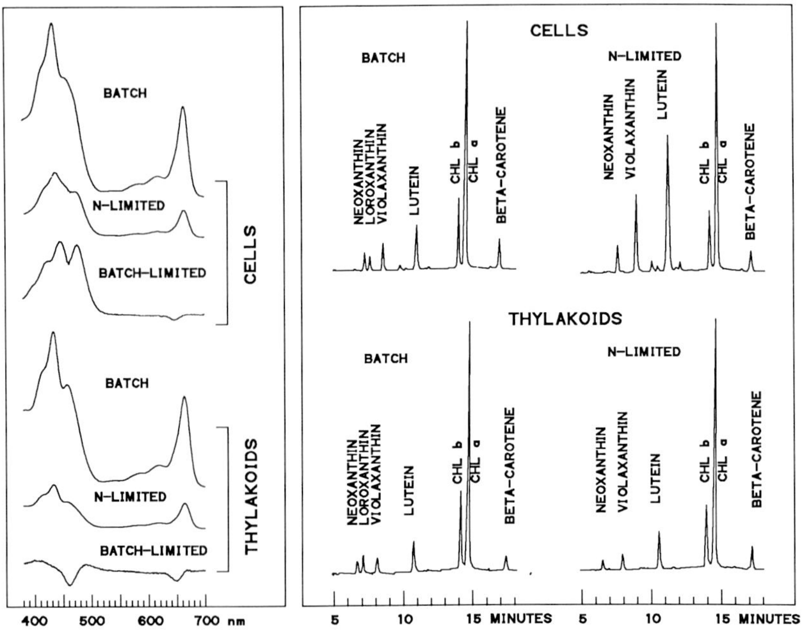

Figure 2. Absorption Spectra and HPLC Analyses of Whole Cells and Purified Thylakoid Membranes. (*Left Panel*) Absorption spectra of equal numbers of cells (upper) or thylakoids from the same number of cells (lower). Difference spectra (Batch - Limited) were calculated after normalizing to equivalent Chl *a* absorption. (*Right Panel*) Pigments in 80% acetone were resolved on a RP-18 column with acetone (76% for 3 min followed by an 8 min ramp to 100%; 1 ml/min). HPLC profiles were normalized to equivalent Chl *a* for whole cell extracts (upper) or thylakoids (lower).

PSI) may be relatively more abundant in N-limited cells. To determine if this is the case, thylakoid membrane fractions were analyzed by non-dissociating LDS-PAGE (Fig. 3). N-limited cells have high levels of the PSI reaction center complex, CCI, when thylakoids are normalized to equivalent Chl levels; in this case, however, the electrophoretic mobility of the complex is significantly faster than that of normal CCI. Preliminary analysis of the pigment composition of the CCI complex from N-limited cells indicates a substantial increase in β-carotene relative to Chl *a*. The antenna of PSI, LHCI, is not observed in the preparations from N-limited cells and the antenna of PSII, LHCII, is severely depleted. The PSII complexes, CCII, which are difficult to isolate on green gels, are virtually absent despite the fact that the N-limited cells are photosynthetically competent; the absence of CCII complexes appears to indicate increased lability in N-limited cells.

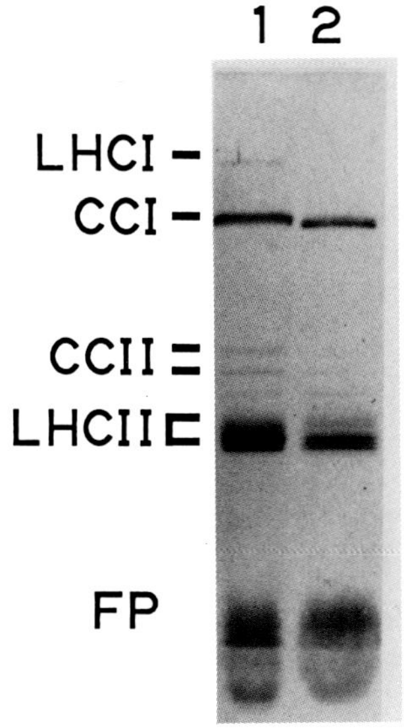

Figure 3. LDS-Polyacrylamide Gel Electrophoresis of Pigment-Protein Complexes from Thylakoids of N-Limited and Batch Culture Cells. Thylakoid membranes were purified by flotation through sucrose step gradients. After solubilization at 0° with 2% lithium dodecyl sulfate in gel sample buffer, samples corresponding to 9 μg of Chl were subjected to electrophoresis in 10-20% polyacrylamide gradient gels at 4°C. Unstained gels were photographed through a blue filter for visualization of the indicated pigment-protein complexes. LHCI and LHCII are light-harvesting complexes for Photosystem I and Photosystem II, respectively. CCI and CCII are the major Chl *a* complexes of the Photosystem I and Photosystem II reaction centers, respectively.

N-limited cells exhibit a unique 77 K fluorescence emission when Chl *a* (440 nm) is excited (Fig. 4). Difference spectra indicate the emission peak is at 676 nm. Since there is no 676 nm signal when Chl *b* (470 nm) or carotenoids (485 nm) are excited (not shown), the anomalous fluorescence emission arises from an altered, perhaps unorganized, pigment bed which contains only Chl *a*.

PHOSPHORYLATION OF THYLAKOID MEMBRANE PROTEINS

Phosphorylation of integral thylakoid membrane proteins has been correlated with modification of their function. Two proteins phosphorylated in *Chlamydomonas* and other plants are the 43 kD CCII apoprotein and the LHCII apoproteins (Ikeuchi *et al.*, 1987a). When N-limited cells were labeled *in vivo*, the recovery of ^{32}P in protein fractions is reduced. The reduction is not due to reduced uptake of P_i (not shown) but may be due to enhanced formation of polyphosphate bodies in N-limited cells. Thus, while the quantitative aspects of these data are equivocal, their qualitative features are not. Each of the LHCII apoproteins are phosphorylated at almost identical rates in N-limited cells while there is preferential phosphorylation of specific apoproteins (i.e., polypeptides 11 and 17; Fig. 5) in batch culture controls. Moreover, the LHCII apoproteins are phosphorylated in a light-dependent fashion in batch culture cells (not shown, Ikeuchi *et al.*, 1987b) but, in N-limited cells, they are phosphorylated independently of illumination conditions (Fig. 5). The CC II apoprotein is not phosphorylated in N-limited cells

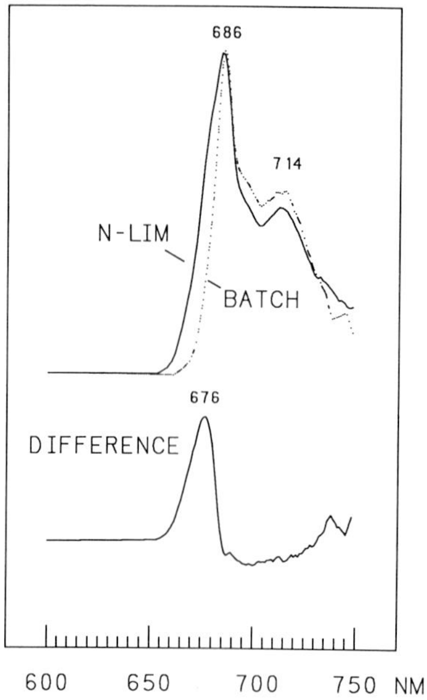

Figure 4. Low Temperature (77 K) Emission Spectra of N-Limited and Batch Culture Cells. Cells at 5 μg Chl/ml in 50% glycerol were frozen in liquid nitrogen. Excitation and emission bandpasses were 4 nm. The spectra were corrected for differential responses of the photomultiplier tube to wavelength and normalized to their maxima at approximately 685 nm.

in either the light or dark. In contrast, two other PSII associated phosphoproteins, B and C (Ikeuchi *et al.*, 1987b), are seen (in longer exposures) to be phosphorylated in the light and dark in N-limited cells.

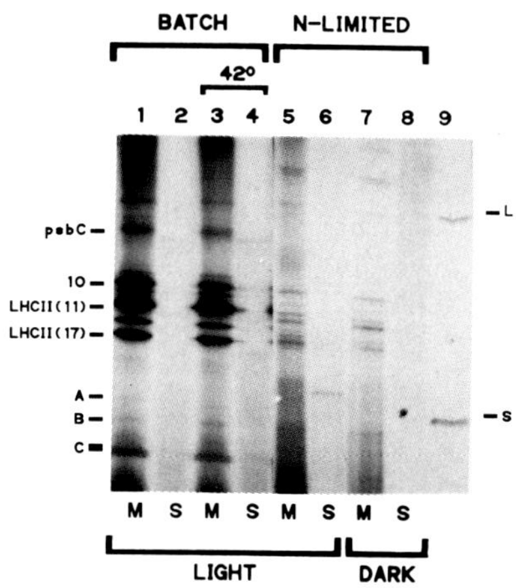

Figure 5. *In Vivo* Phosphorylation of Thylakoid Membrane Proteins. Cells were labeled with $^{32}PO_4$, the integral membrane proteins recovered as NaOH insoluble fractions, resolved by SDS-PAGE and autoradiographed. Batch cells were labeled also after a 30 min shift to 42°C. NaOH-insoluble membrane (M) and soluble (S) fractions are shown. B and C, phosphoproteins associated with Photosystem II; A, associated with Photosystem I; L and S, large and small subunits of RuBPCase; 10, 11 and 17 are individual apoproteins of antenna complexes.

The unique pattern of LHCII polypeptide phosphorylation in N-limited cells could be due to changes in the conformation and/or relative abundance of the apoproteins, elevated levels of Photosystem I reaction centers, or changes in the levels of active protein kinases. The absence of CCII phosphorylation indicates that the modification of Photosystem II reaction center complexes is also affected in N-limited cells, perhaps accounting for the lability of the pigment-protein complexes.

The N-limited Stress Response Is Not The Same as Heat-Shock

Several stress responses in plants are associated with synthesis of a set of proteins frequently referred to as heat-shock proteins (Lindquist, 1986). To ascertain if N limitation induces a response similar to a general stress response (as typified by heat shock), cells were labeled *in vivo* and the radioactive proteins resolved by two-dimensional electrophoresis. At 42°C, *Chlamydomonas* synthesizes a set of high molecular weight proteins whose synthesis is either reduced or absent from both batch culture cells (Douglas *et al.*, 1986) and N-limited cells (Fig. 6). In contrast, N-limited cells synthesize a number of smaller, mostly acidic proteins which are absent from batch culture and heat-shocked cells (Fig. 6).

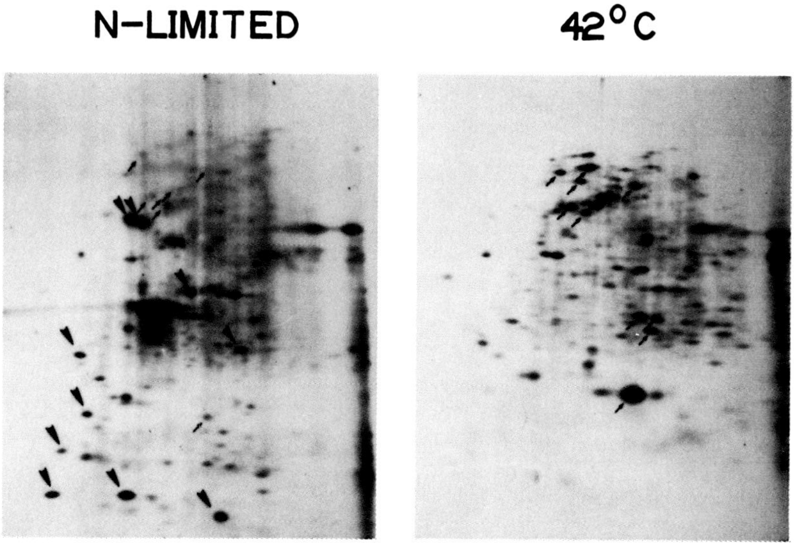

Figure 6. Nitrogen-Limited Cells Do Not Synthesize Heat-Shock Proteins *In Vivo*. Cells were shifted to 42°C for 30 min then labeled with $^{35}SO_4$. Urea solubilized proteins were resolved by isoelectric focusing followed by SDS-PAGE in the second dimension and fluorography.

It is possible that N-limited cells accumulate heat-shock mRNAs but they are not translated at readily detectable rates. The capacity to synthesize heat shock proteins can be developmentally regulated (Bonato *et al.*, 1987). Moreover, synthesis of some stress proteins is regulated posttranscriptionally, possibly at the level of initiation (Kloppstech *et al.*, 1987). Since the rate of protein synthesis is low in N-limited cells (i.e., possibly due to low initiation) and most cells in chemostat cultures are blocked at discrete cell-cycle stages, we translated total RNA in the cell-free wheat germ translation system as an assay for heat-shock transcripts. These assays confirm the absence of heat-shock mRNAs in N-limited cells and also identify transcripts which are relatively more abundant in N-limited cells than batch culture cells (Fig. 7).

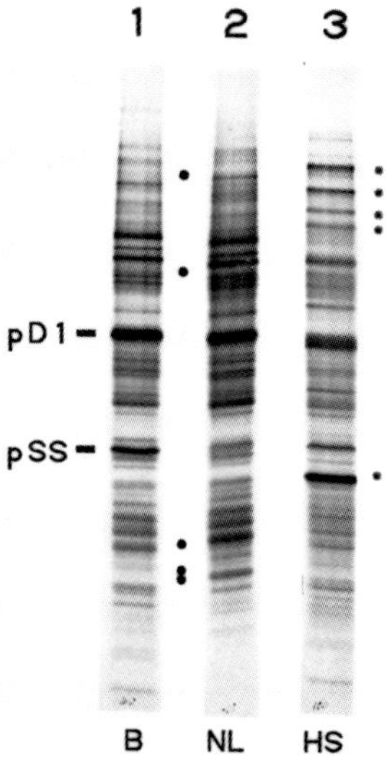

Figure 7. Translatable Heat Shock mRNAs Are Not Accumulated in Nitrogen-Limited *Chlamydomonas*. Total RNA was translated in the wheat germ cell-free translation system. The protein products were resolved by SDS-PAGE and visualized by fluorography. B, NL and HS respectively correspond to RNA preparations from Batch, N-limited and cells subjected to 42°C for 30 min. prior to RNA extraction. pD1 and pSS are the precursors of the chloroplast psbA gene product and the small subunit of RuBPCase, respectively. Dots indicate translation products whose mRNAs are more abundant in N-limited cells than in batch cells; stars indicate the positions of heat shock proteins.

REDUCED AND DIFFERENTIAL ACCUMULATION OF PHOTOSYNTHETIC PROTEINS

Quantitative rocket-immunoelectrophoresis of nine photosynthetic proteins indicate the LHCII and LHCI apoproteins are severely reduced (20% of batch) in N-limited *Chlamydomonas* (Fig. 8). Proteins of Photosystem II, including the psbB, psbC and oec1 gene products, are present at moderate levels (35% of batch) while the β subunit of ATP synthetase and the large and small subunits of RuBPCase are less affected

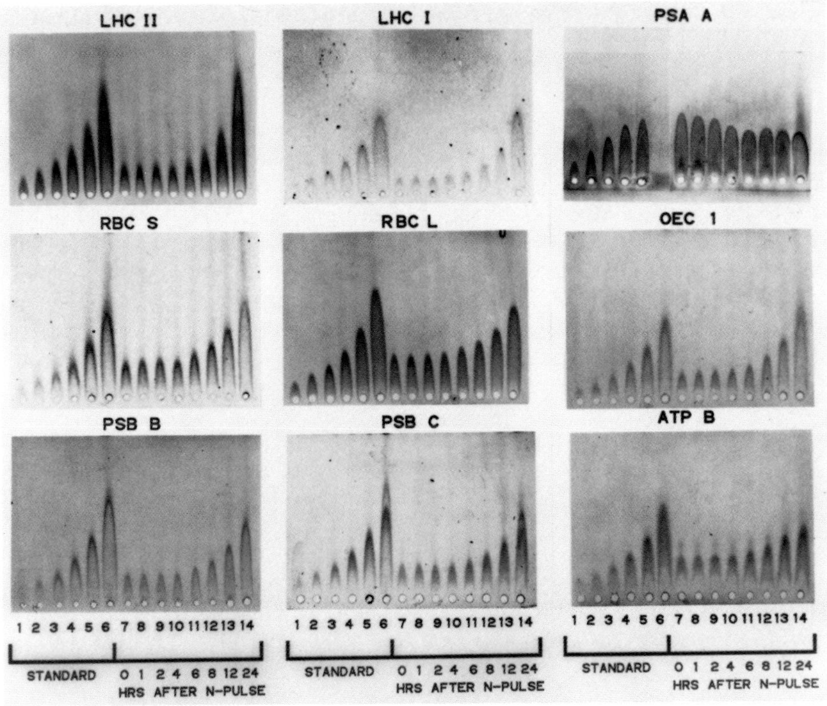

Figure 8. Rocket-Immunoelectrophoresis of Photosynthetic Proteins from N-Limited Cells Undergoing N-Dependent Greening. SDS-solubilized cells were subjected to rocket-immunoelectrophoresis (Plumley and Schmidt, 1983). Batch culture cells (standards) are two-fold dilutions at 3.12 X 10^5 cells (lane 6) through 0.10 X 10^5 cells (lane 1); lane 6 was omitted for PSA A. For N-limited cells, 10 ml of culture was harvested at each time point; cell numbers increase after the N-pulse as shown in Fig. 2. The dilutions required for stainable rockets in the zone of equivalence varied: the number of cells (lane 2) loaded for each antibody were: RBC S, LHCI, LHCII, PSA A and OEC 1, 4.22 X 10^5; RBC L, ATP B, PSB B and PSB C, 2.81 X 10^5. RBC: RuBPCase; PSA: Photosystem I; OEC 1: largest water oxidation complex subunit; PSB B and C: CCII apoproteins of Photosystem II; ATP B: β subunit of ATP synthetase.

(60% of batch). Photosynthetic proteins remain at low levels for 5 to 6 hr following the N-pulse but then accumulation proceeds rapidly, and differentially, with batch culture levels achieved in 24 h. The amounts of Photosystem I reaction center apoprotein are high in N-limited cells and do not change appreciably.

RATES OF PROTEIN SYNTHESIS AND ACCUMULATION OF MACRO-MOLECULES

Protein synthesis rates in N-limited cells are 10% of those of control cells, increase 10 to 20-fold within 30 minutes of ammonium provision and peak 5 hours later (Fig. 9), a time corresponding to the maximum accumulation of high molecular weight RNA. The rise in RNA levels exhibits a lag relative to that of protein synthesis. Chlorophyll accumulation is stimulated by N; maximum accumulation coincides with the period of elevated protein synthesis. The percentage increases in both chlorophyll (3-fold) and total N (2-fold) are modest relative to the effects on protein synthesis. The ratio of Chl *a/b* progressively declines (Fig. 9), reflecting preferential accretion of light-harvesting complexes.

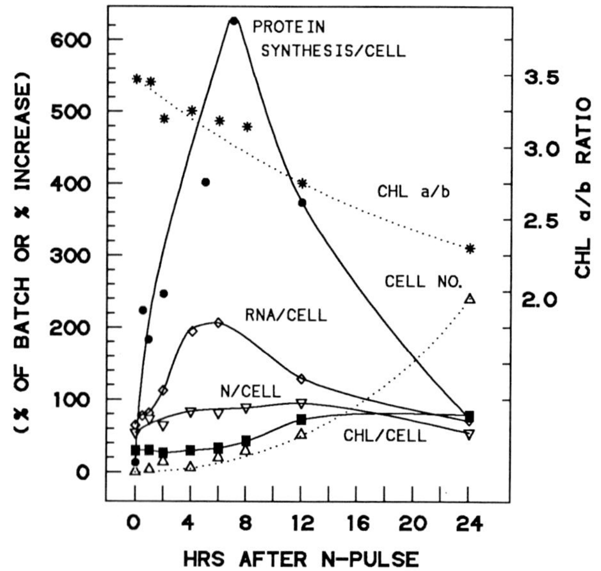

Figure 9. Protein, RNA, Pigment and Cell Nitrogen in N-Limited Cells Responding to Nutrient Repletion. Cell number, protein synthesis ($^{35}SO_4$ incorporated in hot TCA/ether-insoluble material), high mw RNA levels (phenol extracted, LiCl precipitated), particulate N (CHN Analyzer) and Chl *a* and *b* were determined before and after addition of ammonium to N-limited cultures. Data are percentage comparisons to N-replete batch cells.

Thylakoid Proteins are Synthesized at Reduced, but Differential, Rates

The low abundance of thylakoid proteins in N-limited cells could result from a variety of control mechanisms, one of which is rapid turnover. Enhanced degradation of proteins in N-limited *Lemna* has been reported (Ferreira and Davies, 1987). To determine the rates of synthesis of individual photosynthetic proteins in N-limited *Chlamydomonas*, cells were pulse-labeled and the membrane proteins resolved by SDS-PAGE. As indicated above, protein synthesis in N-limited cells is greatly reduced (Fig. 9). By loading equal counts for SDS-PAGE, however, relative rates of protein synthesis can be visualized. The most pronounced affect of N limitation is on the LHCII apoproteins whose synthesis is 40-fold lower than in batch culture cells (Fig. 10). In contrast, the ATP subunits and the psbA gene products appear to be synthesized at comparable *relative* rates in batch culture and N-limited cells (the absolute extent of their

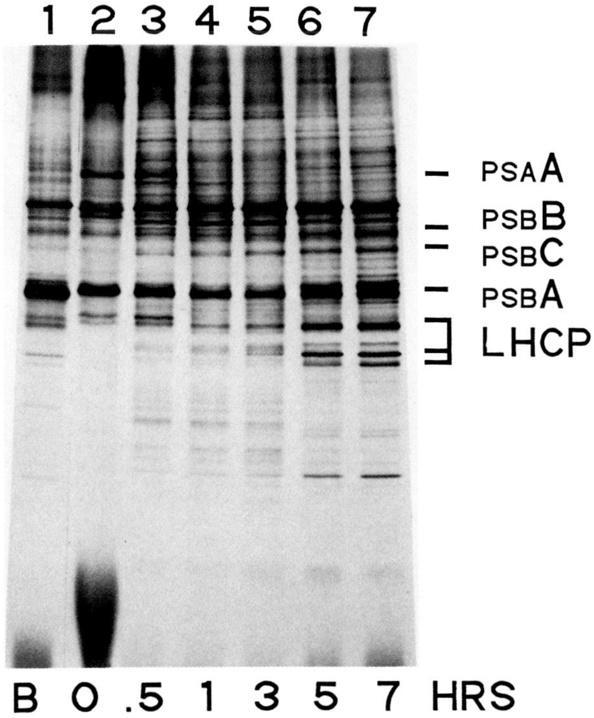

Figure 10. Pulse-Labeling of Proteins in Batch Culture and N-Limited Cells Following Application of a N-Pulse. Cells, concentrated to 10^7 cells/ml, were pulse-labeled 30 min with $^{35}SO_4$. Membrane proteins were partitioned with 0.2 N NaOH, resolved by SDS-PAGE and visualized by fluorography. Equal cpm were loaded for each lane; protein synthesis rates for each time point are as in Fig. 2.

synthesis, however, is reduced 10-15 fold). Moreover, the *relative* synthesis rates of the psaA products are enhanced in N-limited cells (Fig. 10), accounting in part to their preferential accumulation. The identity of these thylakoid proteins has been confirmed by labeling cells in the presence of cycloheximide or chloramphenicol. Finally, turnover of these thylakoid proteins in N-limited cells is comparable to those of batch culture cells during an 11 h chase period (not shown).

When N-limited cells are supplied with N, the rate of protein synthesis increases almost immediately (Fig. 9). Synthesis of several thylakoid proteins, such as the gene products of psbB and psbC, increase rapidly while the relative rate of synthesis of the psaA products decline (Fig. 10). Synthesis of LHCII apoproteins is virtually undetectable for the first few hours following the N-pulse but these proteins are major translation products after a 4 to 5 h lag.

Synthesis of LHCII is Transcriptionally Regulated While Chloroplast Encoded Proteins are Translationally Controlled

The low rates of synthesis of thylakoid membrane proteins could result from blocks in translation and/or transcription. To distinguish between these two possibilities, total RNA was purified and the relative levels of individual transcripts determined by Northern gel analysis (Plumley and Schmidt, submitted) and dot-blot hybridization analyses (Fig. 11). LHCII transcripts are present at very low levels in N-limited cells; one LHCII transcript, cab1, is present at moderate levels while the cab2 transcript is hardly detectable. Transcripts for chloroplast encoded photosynthetic proteins are relatively abundant in N-limited cells. Thus, the synthesis of these proteins is regulated at the translational level. Transcripts for two mitochondrial proteins are more abundant in N-limited cells than in batch culture controls. During N repletion, the LHCII transcripts decrease in abundance for the first few hours after the N-pulse but then accumulate to levels several hundred fold higher than those in batch culture cells (at 6 hr). In contrast, there are moderate increases in other nuclear encoded transcripts (e.g., rbcS), chloroplast transcripts, and mitochondrial transcripts (Figs. 11 and 12).

Localization of Chloroplast Transcripts with Polysomes

The presence of relatively high levels of chloroplast transcripts in N-limited cells (Figs. 11 and 12), in view of the reduced rates of synthesis

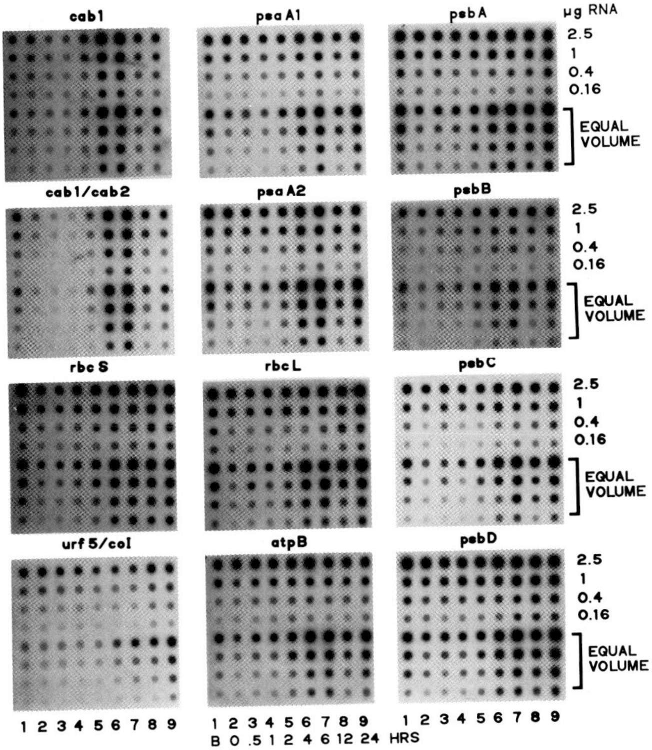

Figure 11. RNA Dot Blot Analyses of Photosynthetic Protein Transcript Abundance as a Function of Nitrogen Provision in *Chlamydomonas*. Total high molecular weight RNA (see Fig. 9) was sequentially diluted, wheat germ tRNA added to yield equivalent RNA concentrations in each sample and then applied to Zeta-Probe membranes. Filters were hybridized with cloned DNA fragments containing sequences encoding LHCII apoproteins (cab1 and cab1/cab2), psbA, the large and small subunits of RuBPCase (rbcL and rbcS), mitochondrial genes encoding cytochrome oxidase subunit I (coI) and an unidentified reading frame (urf5), PSI apoproteins (psaA1 and psaA2), the ß subunit of ATP synthase (atpB) or PSII proteins psbA, psbB, psbC, psbD. (Upper Panel) Dots were loaded with equivalent amounts of RNA as indicated to the right. (Lower Panel) Dots with RNA from an equivalent number of batch culture (lane 1) or N-limited cells (lane 2); the number of cells in lanes 3 through 9 are as indicated in Fig. 9.

of their corresponding proteins (Figs. 9 and 10), indicate a prevalent translational block of chloroplast protein synthesis. Protein synthesis is typically inhibited during various conditions of starvation by the levels of active initiation factors (Ochoa and de Hara, 1979). To determine the

Figure 12. Relative Accumulation of Transcripts in N-Limited Cells and Cells Undergoing N-Dependent Regreening. The relative amount of RNA present at each time point following N replenishment was determined by densitometric scanning of the dots in Fig. 11.

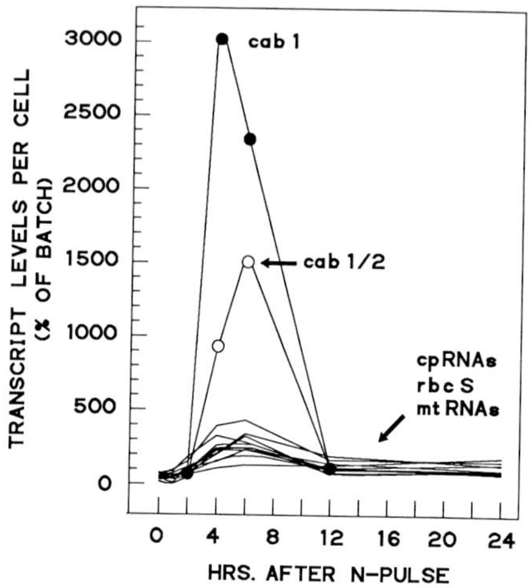

nature of the translational block in N-limited *Chlamydomonas*, polysomes were resolved on sucrose gradients and the location of chloroplast messages determined by hybridization to DNA probes. One technical note is required. The method used for visualizing polysome profiles (Fig. 13) does not allow resolution of mRNA which does not enter the sucrose gradient. Thus, uninitiated transcripts are not detected by this procedure.

A striking feature of the polysome profile of N-limited cells is that most of the rRNA seen on stained gels is in monosomes or larger polysomes (Fig. 13). In batch culture cells, much of the rRNA exists as free subunits or as polysomes containing at least two ribosomes. In contrast to the nuclear rRNA which can be visualized by staining, the less abundant chloroplast rRNA can only be detected by hybridization. Surprisingly, most of the chloroplast rRNA (16S or 5S) is present in ribosome monomers or larger polysomes. Moreover, transcripts encoding psbA and rbcL gene products are predominantly on polysomes containing two or more ribosomes.

We have not analyzed mRNA in low molecular weight complexes but the intensity of the signals in the Northern blots suggests that relatively more chloroplast transcripts are associated with ribosomes in N-limited cells than in controls.

Chloroplast DNA Levels Are Correlated With mRNA Accumulation

Chloroplast DNA levels also are reduced in N-limited cells (Fig. 14). In general, the abundance of cpDNA and chloroplast transcripts are

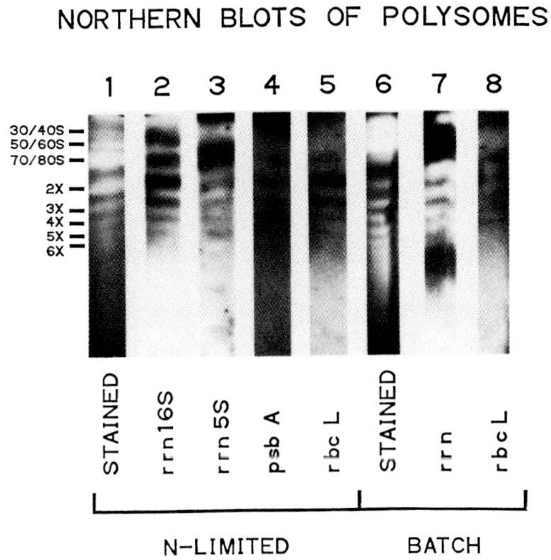

Figure 13. Northern Gel Analysis of Polysome Profiles. Cells were rapidly cooled, concentrated to 3 X 109 cells/ml at 5000g, broken at 4000 psi in a French press and solubilized with 3% (w/v) NP-40 by gentle mixing at 0°C. Lysates were layered on 20-50% sucrose density gradients (containing 4.75% acrylamide, 0.25% bis-acrylamide, 0.025% TEMED, 3.5 μg/ml riboflavin) and centrifuged 14 h at 40,000g. Gradients were polymerized at 4°C by exposure to light (Murthy *et al.*, 1986) and cut longitudinally into 2 mm thick slices which were stained (ethidium bromide; lanes 1, 5) or SDS/NaOH denatured and electrophoretically transferred to Zeta-Probe membrane according to manufacturers directions and hybridized in formamide/SDS/dextran sulfate/NaCl to cloned DNA fragments (lanes 2-5, 7, 8).

highly correlated. Moreover, the initial increase in cpDNA following the administration of an N-pulse occurs with the same kinetics as the increase in chloroplast transcripts. However, cpDNA levels remain at high levels following the period of rapid greening (6 to 12 h after the N-pulse) while most transcripts decline. Mitochondrial DNA abundance is also depressed in N-limited cells. Thus, the levels of mtDNA and mitochondrial transcripts are uncoupled.

DISCUSSION

In this report we have demonstrated that *Chlamydomonas* responds to N limitation in a manner similar to other plant species. N limitation is accompanied by severe chlorosis, but importantly, there is a preferential reduction in Chl *b*. Moreover, two other pigments normally associated with Chl *b* in LHCII complexes, loroxanthin and neoxanthin (Cammarata

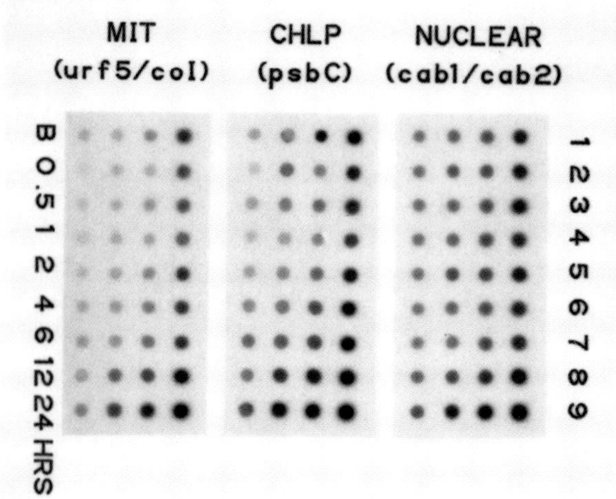

Figure 14. DNA Dot Blot Analysis of N-Limited *Chlamydomonas* Before and During the Period of N-Dependent Greening. Total DNA was recovered as the LiCl-soluble fraction following phenol/chloroform extraction. DNA was serially diluted 2-fold, brought to equal DNA concentration with herring sperm DNA, denatured with NaOH, and diluted with 0.125N NaOH, 0.125X SSPE so that 2 μg total DNA were loaded per dot on GeneScreenPlus membrane. DNA from the same number of cells was loaded for batch culture (lane 1) or N-limited cells (lane 2). DNA from an increasing number of cells was loaded in lanes 3 through 9 following the N-pulse (see Fig. 9 for increase in cell number). Blots were hybridized to cloned DNA fragments as described in Fig. 11.

et al., 1988), are also present at low levels. Two xanthophylls, violaxanthin and lutein, accumulate to normal levels but are not localized in thylakoid membranes. Accompanying this adjustment in pigment composition is a dramatic reorganization in the activity of antenna Chl molecules as reflected in an anomalous fluorescence emission. Surprisingly, this emission is observed only when Chl *a* is excited. We have tentatively associated this peak with Photosystem I complexes. The Photosystem I complex, CCI, has altered electrophoretic mobility during SDS-PAGE and an elevated β-carotene content. We cannot rule out the possibility that a Chl *a* antenna of Photosystem II is responsible for the unusual fluorescence peak; both major complexes of CCII are more labile in N-limited cells than in batch culture control cells.

The accumulation of photosynthetic proteins is reduced in N-limited cells. Importantly, all proteins are not reduced to the same extent. The Chl *a/b* antenna apoproteins, LHCI and LHCII, are the most severely reduced while the Photosystem I apoproteins and proteins of the ATP synthase complex are the most abundant. This reorganization of thylakoid membranes presumably decreases noncyclic electron flow while maintaining higher relative rates of cyclic photophosphorylation. The low levels of photosynthetic proteins results from their low rates of synthesis. Turnover of photosynthetic proteins is not affected by growth under N-limited conditions.

The low rates of protein synthesis in N-limited cells result from both transcriptional and translational control mechanisms. Transcripts encoding LHCII apoproteins are severely depressed in N-limited cells. We hypothesize that the low level of LHCII transcripts results from a transcriptional block. The rapid disappearance of cab transcripts following addition of a N-pulse to N-limited cells suggests that turnover may also be important in controlling the levels of cab transcripts. Moreover, the stability (and/or rate of transcription) of the cab1 transcript is greater than that of cab2 in *Chlamydomonas*. The molecular basis for regulation of the cab genes is currently under investigation.

In contrast to nuclear-encoded photosynthetic proteins, the chloroplast-synthesized proteins appear to be regulated translationally. Moreover, at least two of the chloroplast transcripts are present on large polysomes. Thus, initiation does not seem to limit chloroplast protein synthesis. While it might be assumed that the abundance of charged tRNAs limits the rates of chloroplast protein synthesis, the apparently high synthesis rates of select chloroplast proteins, e.g., those encoded by psaA1/2, psbA, cannot be so easily explained. Instead, it is likely that specific factors required for translation of individual chloroplast proteins are differentially limiting. Such factors, which are nuclear encoded, have been reported for Photosystem II proteins (Jensen *et al.*, 1986). The existence of mRNA-specific translation factors have been implicated also in the arrest of chloroplast protein synthesis in heat-shock treated cells and in cells treated with inhibitors of cytoplasmic protein synthesis (Switzer *et al.*, in preparation). Moreover, in those instances when chloroplast protein synthesis is arrested in etiolated seedlings, at least some of chloroplast transcripts are polysome associated (Klein *et al.*, 1988). Thus, a pervasive mode of translational regulation in chloroplasts is at the level of elongation but the molecular details of how this is achieved remains to be resolved. It will be of interest to determine if nitrogen availability affects differentially the synthesis of mRNA-specific factors and/or factors that facilitate the completion of polypeptide chains.

The effects of nitrogen limitation on chloroplast biogenesis are, in many instances, quite similar to the changes in plastid and nuclear gene expression that have been characterized in the past as light-dependent processes (Mullet, 1988; Kuhlemeier *et al.*, 1987). Indeed, some of the key enzymes involved in nitrogen assimilation have been documented to be light-activated and/or synthesized in a light-dependent manner (Aparicio *et al.*, 1985; Mandersheid and Wild, 1986). It is possible, therefore, that many of the light-dependent changes in transcription and translation that have been studied intensively, are actually superimposed upon nitrogen-mediated regulatory processes. Resolution of this matter will require elucidation of the signal transduction pathways for both the light- and nitrogen-dependent control mechanisms.

ACKNOWLEDGMENTS

We are grateful to Drs. D.L. Herrin, P. Boer and M. Goldschmidt-Clermont for provision of many of the DNA probes used in these experiments. This work was supported by grants from NSF and DOE to GWS.

REFERENCES

Aparicio PJ, Balandin T, Maurino SG, Maldonado JM (1985) Photoregulation of nitrate utilization in green algae and higher plants. Photochem Photobiol 42:765-770.

Bonato MC, Silva AM, Gomes SL, Maia JCC, Juliani MH (1987). Differential expression of heat-shock proteins and spontaneous synthesis of HSP during the life cycle of *Blastocladiella emersonii.* Eur J Biochem 163:211-220.

Cammarata K, Plumley FG, Schmidt GW (1988). Pigment and protein composition of reconstituted light-harvesting complexes. Plant Physiol 86:86S.

Douglas SE, Plumley FG, Schmidt GW (1986). Heat shock arrest of chloroplast protein synthesis. Plant Physiol 80:82S.

Ferreira RB, Davies DD (1987). Protein degradation in *Lemna* with particular reference to ribulose bisphosphate carboxylase. II. The effect of nutrient starvation. Plant Physiol 83:878-883.

Field C, Mooney HA (1986). The photosynthesis-nitrogen relationship in wild plants. In Givinish TJ (ed): "On the Economy of Plant Form and Function," New York: Cambridge University Press, pp 25-55.

Gauch HG (1972). "Inorganic Plant Nutrition." Stroudsburg, PA: Dowden, Hutchinson and Ross, Inc. 488pp.

Goodwin TW (1980). "The Biochemistry of the Carotenoids." New York: Chapman & Hall.

Hunt ER, Weber JA, Gates DM (1985). Effects of nitrate application on *Amaranthus powellii* Wats. I. Changes in photosynthesis, growth rates, and leaf area. Plant Physiol 79:609-613.

Ikeuchi M, Plumley FG, Inoue Y, Schmidt GW (1987a). Phosphorylation of photosystem II components, CP43 apoprotein, D1, D2 and 10 to 11 kilodalton protein in chloroplast thylakoids of higher plants. Plant Physiol 85:638-642.

Ikeuchi M, Plumley FG, Inoue Y, Schmidt GW (1987b). Identification of phosphorylated reaction center polypeptides in thylakoids of *Chlamydomonas reinhardtii* and *Pisum sativum*,. In Biggins J (ed): "Progress in Photosynthesis Research, Vol. II." Netherlands: Martinus Nijhoff Publishers, pp 805-808.

Klein, RR, HS Mason and JE Mullet (1988). Light-regulated translation of chloroplast proteins. I. Transcripts of psaA-psaB, psbA, and rbcL are associated with polysomes in dark-grown and illuminated barley seedlings. J Cell Biol 106:289-301.

Kloppstech K, Lorberboum H, deGroot N, Hochberg AA (1987). Translational control at the level of initiation between mRNAs for pre-existing proteins and a 22-kD a heat-shock protein of *Chlamydomonas* by small cytosolic RNAs. Eur J Biochem 167:501-505.

Kuhlemeier C, Green PJ, Chua N-H (1987). Regulation of gene expression in higher plants. Ann Rev Plant Physiol 38:221-257.

Lindquist S (1986). The heat shock response. Ann Rev Biochem 55:1151-1191.

Manderscheid R, Wild, A (1986). Characterization of glutamine synthetase of roots, etiolated cotyledons and green leaves from *Sinapis alba* (L.). Z Naturforsch 41c:712-716.

Mullet JE (1988). Chloroplast development and gene expression. Ann Rev Plant Physiol Plant Mol Biol 39:475-502.

Murthy MRV, Bharuch AD, Charbonneau R (1986). A novel method for sucrose gradient fractionation of polysomes and mRNA. Nucl Acids Res 14:6337.

Ochoa S, de Haro C (1979). Regulation of protein synthesis in eukaryotes. Ann Rev Biochem 48:549-580.

Plumley GP, Schmidt GW (1983). Rocket and crossed immuno-electrophoresis of proteins solubilized with sodium dodecyl sulfate. Anal Biochem 134:86-95.

Terashima I, Evans JR (1988). Effects of light and nitrogen nutrition on the organization of the photosynthetic apparatus in spinach. Plant Cell Physiol 29:143-155.

Photosynthesis, pages 331–345

THE PLASTID ENVELOPE MEMBRANES AND THEIR CONTRIBUTION TO PLASTID MEMBRANE BIOGENESIS

Jacques Joyard, Maryse A. Block, Claude Alban and Roland Douce

Laboratoire de Physiologie Cellulaire Végétale, UA CNRS n°576, Département de Recherche Fondamentale, Centre d'Etudes Nucléaires et Université Joseph Fourier, 85 X, F-38041 Grenoble-cedex, France.

INTRODUCTION

Chloroplasts are limited by a pair of distinct membranes, known as the envelope, which separate the organelle from the surrounding cytosol. Studies with isolated intact chloroplasts have demonstrated the different permeability properties of each envelope membrane (Heber and Heldt, 1981). The outer envelope membrane is freely permeable to small molecules, due to a pore-forming protein (Flügge and Benz, 1984), whereas the inner envelope membrane is the functional border between the plastid stroma and the cytosol. Specific translocators regulate the flow of metabolites across the inner envelope membrane (Heber and Heldt, 1981). Furthermore, biochemical studies of purified envelope membranes (mostly from spinach and pea) have demonstrated the complexity of their enzymatic equipment which has a major role in chloroplast biogenesis: up to 20 different enzymes involved in the synthesis of thylakoid components such as glycerolipids (galactolipids, phosphatidylglycerol, and probably sulfolipid) and terpenoid compounds (prenylquinones and probably carotenoids) have been demonstrated and/or characterized in envelope membranes (Douce *et al.*, 1984). To understand the role of envelope membranes in plastid biogenesis, it is essential to describe the glycerolipid, pigment, and prenylquinone composition of plastid membranes. In addition, studies on envelope membranes from non-green plastids are also important to determine whether envelope membranes have the same properties in all kind of plastids.

CHEMICAL COMPOSITION OF PLASTID ENVELOPE MEMBRANES

A. Glycerolipids

In contrast to extraplastidial membranes, plastid membranes (envelope and thylakoids) are characterized by a low phospholipid content and by the presence of glycolipids (Douce *et al.*, 1973). The major plastid lipid compounds are galactolipids, which contain one or two galactose molecules attached to the *sn*-3 position of the glycerol backbone, corresponding respectively to monogalactosyldiacylglycerol (or MGDG) and digalactosyldiacylglycerol (or DGDG). Interestingly, these glycerolipids are also the major components in cyanobacteria, which are supposed to have a common ancestor with chloroplasts.

Galactolipids in thylakoid and envelope membranes contain a high amount of polyunsaturated fatty acids: up to 95 % (in some species) of the total fatty acids is linolenic acid. In non-green plastids, 18:3 is still a major component although appreciable amounts of 18:1 and 18:2 are present. Therefore, the most abundant molecular species of MGDG and DGDG have 18:3 at both *sn*-1 and *sn*-2 positions of the glycerol backbone. Some plants, such as pea, having little but 18:3 in MGDG are called "18:3 plants." Other plants, such as spinach, contain important amounts of 16:3 in MGDG, and they are called "16:3 plants" (Heinz, 1977). These two types are also found when non-green plastids are analyzed: for instance, sycamore is a 18:3 plant whereas cauliflower is a typical 16:3 plant. The positional distribution of 16:3 in MGDG is highly specific: this fatty acid is only present at the *sn*-2 position of glycerol, and is almost excluded from *sn*-1 position. Therefore, two major structures are found in galactolipids, one with C18 fatty acids at both *sn* position and one with C18 and C16 fatty acids respectively at *sn*-1 and *sn*-2 position. The first one is typical of "eukaryotic" lipids (such as phosphatidylcholine or PC) and the second one corresponds to a "prokaryotic" structure. These differences are probably due to galactolipid biosynthetic pathways (see below).

Plastid membranes sometimes contain galactolipids with 3 (tri-GDG) and 4 (tetra-GDG) galactoses. They are formed by an enzymatic galactose exchange between MGDG and DGDG, owing to a galactolipid: galactolipid galactosyltransferase (Van Besouw and Wintermans, 1978; Heemskerk and Wintermans, 1987) which catalyzes the following reactions:

$$2\ \text{MGDG} \rightarrow \text{DGDG} + \text{diacylglycerol}$$

$$\text{MGDG} + \text{DGDG} \rightarrow \text{Tri-GDG} + \text{diacylglycerol}$$

$$2\ \text{DGDG} \rightarrow \text{tetra-GDG} + \text{diacylglycerol}$$

We have demonstrated that this enzyme (a) is located on the cytosolic side of the outer envelope membrane, (b) is susceptible to proteolytic digestion by thermolysin, a non-penetrant protease, and (c) is present in all plastids analyzed so far (chloroplasts and non-green plastids). Consequently, in order to obtain a glycerolipid composition which could represent the *in vivo* situation within plastid membranes, the galactosyltransferase should be destroyed prior to fractionation of plastids.

The glycerolipid composition of envelope membranes from chloroplasts and non-green plastids is given in Table I. As expected after thermolysin treatment, no diacylglycerol was detected (whereas it represents about 10 % of the total envelope glycerolipids from non-treated plastids). The most striking feature is that the glycerolipid pattern is almost identical in envelope membranes from chloroplasts, etioplasts, or other non-green plastids. The high amount of MGDG reflects the presence of the inner envelope membrane which has a glycerolipid composition close to that of thylakoids as shown in pea (Cline *et al.*, 1981) or spinach (Block *et al.*, 1983): in the inner envelope membrane and in thylakoids, MGDG is the major component. The presence of PC in envelope membranes reflects the presence of the outer envelope membrane: using phospholipase *C* digestion of intact chloroplasts, we have demonstrated that no PC is present in thylakoids. In contrast, PC represents about 30-35% of the outer envelope glycerolipids, where it is concentrated in the outer leaflet of the membrane (Dorne *et al.*, 1985). The presence of this typical eukaryotic lipid in the outer membrane leaflet facing the cytosol could be due to the functioning of the cytosolic phospholipid transfer protein: for instance, a 35% increase of the total envelope PC content could be obtained after incubation of intact spinach chloroplasts in presence of PC liposomes and purified phospholipid transfer protein (Miquel *et al.*, 1987). The major phospholipid in the inner envelope membrane and in thylakoids is phosphatidylglycerol (PG), which is unique because of a $16{:}1_{trans}$ fatty acid at *sn*-2 position of the glycerol backbone. This phospholipid is therefore different from that found in extra-plastidial membranes. Finally, Table I also confirms the observation that plastid membranes are devoid of phosphatidylethanolamine (PE), which is a major component (together with PC) of mitochondrial, endoplasmic reticulum or other extra-plastidial membranes.

B. Pigments and Prenylquinones

In contrast to thylakoids, envelope membranes from chloroplasts or non-green plastids are yellow, due to the presence of carotenoids and the absence of chlorophyll. Envelope membranes from chloroplasts and from amyloplasts present the same absorption spectrum (Fig. 1), thus

Table 1 - Glycerolipid composition of membranes from mitochondria and plastids isolated from various plants.

Except for mitochondria and envelope membranes from sycamore amyloplasts, all these analyses have been done after thermolysin treatment of intact plastids prior to fractionation. Experiments on pea etioplasts were done in collaboration with Dr. J. Soll (Munich, FRG).

ORGANELLE	MGDG	DGDG	SL	PC	PG	PI	PE	DPG
MITOCHONDRIA								
- cauliflower buds	tr	tr	0	37	2	8	38	13
- mung bean hypocotyls								
total	0	0	0	36	1	2.5	46	14
inner membrane	0	0	0	29	1	2	50	17
outer membrane	0	0	0	68	2	5	24	0
- sycamore cells								
total	0	0	0	43	3	6	35	13
inner membrane	0	0	0	41	2.5	5	37	14.5
outer membrane	0	0	0	54	4.5	11	30	0
PLASTIDS								
- spinach chloroplasts								
thylakoids	52	26	6.5	4.5	9.5	1.5	0	0
inner envelope membrane	49	30	5	6	8	1	0	0
outer envelope membrane	17	29	6	32	10	5	0	0
- pea etioplasts								
prothylakoids	42	35	6	9	5	2	0	0
envelope	34	31	6	17	5	4	0	0
- cauliflower bud non-green plastids								
envelope	31.5	27.5	6	20	9	4.5	1	0
- sycamore cells amyloplasts								
envelope	24	21	4	21.5	9	5.5	0.8	0

Abbreviations: MGDG, monogalactosyldiacylglycerol; DGDG, digalactosyldiacylglycerol; SL, sulfolipid; PC, phosphatidylcholine; PG, phosphatidylglycerol; PI, phosphatidylinositol; PE, phosphatidylethanolamine; DPG, diphosphatidylglycerol.

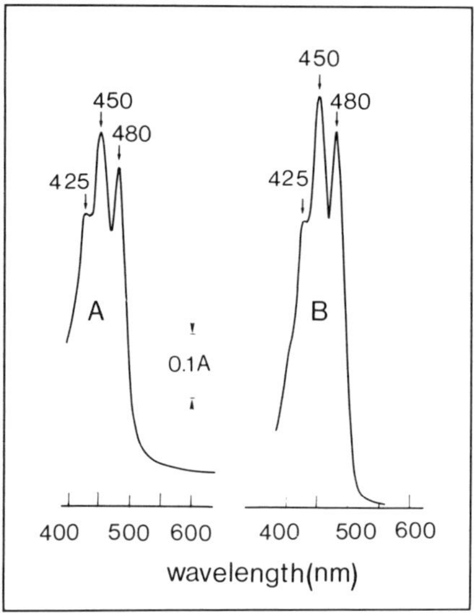

Figure 1. Absorption spectra of chloroform extracts of spinach chloroplast envelope membranes (A) and amyloplast membranes from potato tubers (B). Experimental conditions are described by Neuburger *et al.* (1982).

demonstrating that they have a very similar pigment composition. Envelope membranes from chloroplasts and non-green plastids contain a higher proportion of violaxanthin (even the outer membrane), compared to thylakoids which are rich in β-carotene and xanthophyll (Douce *et al.*, 1984).

Although devoid of the most conspicuous plastid pigment (chlorophyll) some non-green plastids contain chlorophyll precursors such as protochlorophyllide. However, nothing was known about the presence of such pigments in any chloroplast membrane. We have demonstrated that envelope membranes from mature spinach chloroplasts contain low amounts of pigments having the absorption and fluorescence spectroscopic properties, together with the behavior in polar/non-polar solvents of protochlorophyllide and chlorophyllide (Pineau *et al.*, 1986). It is not yet known whether the outer and the inner envelope membranes contain similar or different levels of these pigments. The low temperature fluorescence emission spectrum of total envelope membranes is strikingly different from that obtained with thylakoids (Fig. 2). In addition, upon illumination in presence of NADPH, a decrease of the level of

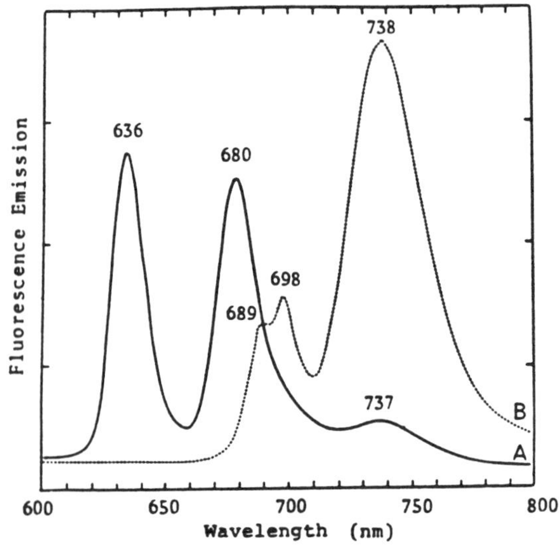

Figure 2. Low temperature (77°K) fluorescence emission spectra of isolated envelope membranes (A) and thylakoids (B) from spinach chloroplasts. Excitation light: 440 nm. Fluorescence emission intensity is expressed in arbitrary units. Experimental conditions are described by Pineau *et al.* (1986).

fluorescence at 636 nm together with a parallel increase of fluorescence level at 680 nm was observed, thus suggesting a possible phototransformation of protochlorophyllide into chlorophyllide (Pineau *et al.*, 1986).

In addition to pigments, envelope membranes from spinach chloroplasts contain prenylquinones as genuine components. Both envelope membranes contain plastoquinone-9, α-tocopherol and phylloquinone (Soll *et al.*, 1985). However, in contrast to thylakoids, the major prenylquinone in envelope membranes is α-tocopherol whereas it is plastoquinone-9 in thylakoids (Soll *et al.*, 1985).

Together, these observations demonstrate that envelope membranes contain qualitatively the same pigments and prenylquinones as thylakoids. However, they are present in different amounts, probably due to their role within envelope membranes. Unfortunately, if the functions of prenylquinones and pigments are well established for thylakoids, nothing is known for their function in envelope membranes. The observation that these compounds are present in envelope membranes from non-green plastids and chloroplasts (Douce *et al.*, 1984; Alban *et al.*, 1988) is most interesting for studies to determine the specific functions of envelope pigments and prenylquinones.

ROLE OF ENVELOPE MEMBRANES IN PLASTID BIOGENESIS

A. Glycerolipid Biosynthesis

The first observation that envelope membranes could be involved in the biosynthesis of plastid components was provided by Douce (1974): isolated envelope membranes purified from spinach chloroplasts were able to catalyze MGDG synthesis from diacylglycerol at high rates. Following the observations of Douce and Guillot-Salomon (1970) who established the presence of the enzymes from the Kornberg-Pricer pathway in chloroplasts and non-green plastids, a considerable body of data has been accumulated which clearly demonstrate that plastids are able to incorporate *sn*-glycerol 3-phosphate into lysophosphatidic acid, phosphatidic acid and diacylglycerol, and then into MGDG (after addition of UDP-galactose). Most of these data concern chloroplasts but we have recently demonstrated that non-green plastids, such as amyloplasts and other starch-containing plastids, behave like chloroplasts as far as glycerolipid biosynthesis by the so-called eukaryotic and prokaryotic pathways are concerned (Alban *et al.*, 1988). Glycerolipid biosynthesis requires the assembly of three parts (Joyard and Douce, 1987): fatty acids, glycerol, and a polar head group (galactose, for galactolipids; sulfoquinovose, for sulfolipid; and phosphorylglycerol, for PG).

Formation of phosphatidic acid

The plastid Kornberg-Pricer pathway is unique in the sense that it is probably the only one to be closely associated with fatty acid synthesis, which is localized within the plastid stroma. All the enzymes of this pathway are closely associated with the envelope membranes (Joyard and Douce, 1977), and more precisely with the inner membrane (Block *et al.*, 1983; Andrews *et al.*, 1985).

The first enzyme of the pathway is a "soluble" enzyme which produces lysophosphatidic acid (lyso-PA):

sn-glycerol 3-phosphate + acyl-ACP $\rightarrow$
1-acyl-*sn*-glycerol 3-phosphate + ACP

In fact, this enzyme is closely associated with the inner envelope membrane and lyso-PA is released directly into the membrane (Joyard and Douce, 1977). This enzyme is most interesting since it is specific for acyl-ACP thioesters (Frentzen *et al.*, 1983). In addition, regardless of the thioester used, a striking specificity for *sn*-1 position of the glycerol is observed (Joyard *et al.*, 1979; Frentzen *et al.*, 1983). Finally, when a mixture of 16:0 and 18:1 thioesters is offered, 18:1 is preferably used (Frentzen *et al.*, 1983). We have found that in non-green plastids from cauliflower buds and amyloplasts from sycamore cells, this enzyme was also soluble (Fig. 3) and presents the same specificities and selectivities as in mature chloroplasts (Alban *et al.*, 1988). This enzyme has been purified from chloroplasts and several isoforms have been demonstrated in chloroplast stroma from several plant species (Bertrams and Heinz, 1981; Nishida *et al.*, 1987). It is likely that the sequence of the gene coding for this protein will be determined soon.

The second enzyme, acyl-ACP:1-acyl-*sn*-glycerol 3-phosphate acyltransferase, catalyses the formation of phosphatidic acid:

1-acyl-*sn*-glycerol 3-phosphate + acyl-ACP $\rightarrow$
1,2-diacyl-*sn*-glycerol 3-phosphate + ACP

This enzyme is firmly bound to the inner envelope membrane from chloroplasts (Joyard and Douce, 1977) and non-green plastids (Fig. 3). Since lyso-PA used for this reaction is esterified at *sn*-1 position, the enzyme will direct fatty acids to the available *sn*-2 position (Joyard *et al.*, 1979; Frentzen *et al.*, 1983). However, the enzyme is highly specific for palmitic acid. Interestingly, the same results were obtained with non-green plastids purified from cauliflower buds and with amyloplasts from sycamore cells (Alban *et al.*, 1988).

Therefore, the two acyltransferases have distinct specificities and selectivities for acylation of *sn*-glycerol 3-phosphate: together, they led to the formation of phosphatidic acid having 18:1 and 16:0 fatty acids respectively at *sn-1* and *sn*-2 positions of the glycerol backbone. This structure is typical for the so-called prokaryotic glycerolipids which are found in 16:3 plants (see above). In contrast, in 18:3 plants, plastid glycerolipids (MGDG) do not contain C16 fatty acids at *sn*-2 position, and therefore probably not derive directly from the envelope Kornberg-Pricer pathway. Their origin is still under investigation.

Formation of diacylglycerol and galactolipids

The phosphatidic acid formed in the inner envelope membrane can be used either for the synthesis of phosphatidylglycerol (Mudd *et al.*, 1987) or for the formation of diacylglycerol (Joyard and Douce, 1977) which is the substrate for MGDG and sulfolipid biosynthesis.

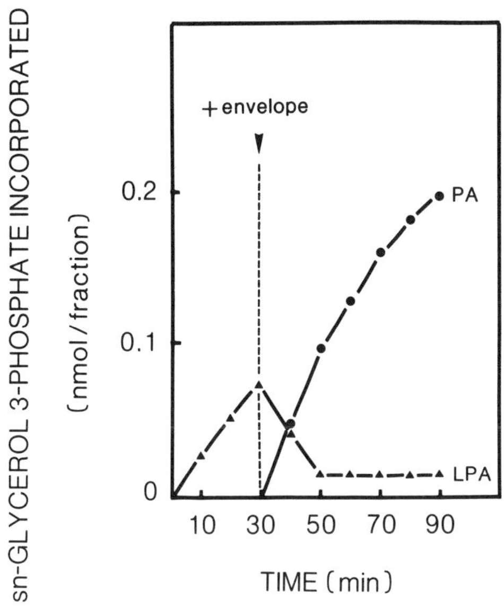

Figure 3. Time course evolution of lysophosphatidic acid (LPA) and phosphatidic acid (PA) biosynthesis by the soluble enzymes and the envelope membranes of sycamore cell amyloplasts. LPA is formed by the soluble enzymes and is used for PA synthesis only when envelope membranes are present. Experimental conditions are described by Alban *et al.* (1988).

Diacylglycerol biosynthesis occurs in the inner envelope membrane owing to a phosphatidate phosphatase (Joyard and Douce, 1977):

1,2-diacyl-*sn*-glycerol 3-phosphate → 1,2-diacyl-*sn*-glycerol + Pi

This enzyme is unique: it is membrane-bound, active at alkaline pH, and highly sensitive to cations (Joyard and Douce, 1979). In addition, 18:3 plants have a rather low phosphatidate phosphatase, in contrast to 16:3 plants (Heinz and Roughan, 1983). It is possible that the level of phosphatidate phosphatase activity in envelope membranes could be responsible for the difference observed between 18:3 and 16:3 plants at the level of MGDG synthesis. We have recently demonstrated that this was also true for non-green plastids (Fig. 4). Chloroplast envelope membranes isolated from 16:3 and 18:3 plants have the same capacity to form MGDG when supplied with diacylglycerol and UDP-galactose (UDP-gal). However, very little MGDG is formed from *sn*-glycerol 3-phosphate in 18:3 plastids (or envelope membranes, see Fig. 4), although

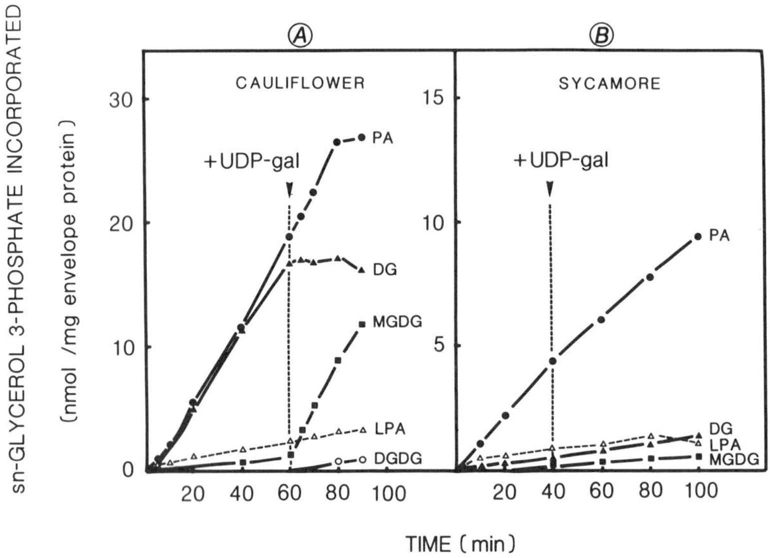

Figure 4. Effect of UDP-gal on the incorporation of *sn*-[^{14}C]-glycerol 3-phosphate into envelope lipids from cauliflower plastids (A) and sycamore amyloplasts (B). The major difference between cauliflower and sycamore is the very low level of diacylglycerol (and therefore of MGDG) formed in sycamore. Experimental conditions are described by Alban *et al.* (1988).

the biosynthesis of phosphatidic acid is very active, thus suggesting that the limiting step is indeed the formation of diacylglycerol (Heinz and Roughan, 1983).

The inner envelope membrane of spinach chloroplasts is characterized by the presence of a UDP-galactose:diacylglycerol galactosyltransferase (or MGDG synthase) which transfers a galactose from a water-soluble donor, UDP-gal, to an hydrophobic acceptor molecule, diacylglycerol, for synthesizing MGDG (Douce, 1974):

UDP-gal + 1,2-diacyl-*sn*-glycerol →
1,2-diacyl-3-O-β-D-galactopyranosyl-*sn*-glycerol + UDP

This enzyme is present in all plastids (chloroplasts and non-green plastids) analyzed so far. We have achieved a partial purification of this enzyme from spinach chloroplasts envelope membranes (Covès *et al.*, 1986). CHAPS was used to solubilize envelope membranes and was rather efficient for MGDG synthase. After solubilization, we used hydroxyapatite chromatography to fractionate envelope proteins. A fraction containing most of the MGDG synthase activity was prepared. It contained less than 5 % of the envelope proteins and was almost entirely devoid of lipids. We have demonstrated a strong lipid requirement for the MGDG synthase. Addition of total envelope lipids stimulated the MGDG synthase activity up to 2 μmol galactose incorporated/h/mg protein, thus corresponding to a 25-30 fold increase of the specific activity, compared to the value in the solubilized envelope. Acidic glycerolipids, and especially PG, were shown to be the best activators of the enzyme: the maximum activity thus obtained was as high as 10 μmol galactose incorporated into MGDG h^{-1} mg^{-1} protein (Fig. 5). The preparation of the delipidated MGDG synthase-enriched fraction, together with the development of optimal assay conditions allowed the determination of kinetic parameters for both UDP-gal and the hydrophobic substrate, diacylglycerol (Covès *et al.*, 1988).

The biosynthesis of DGDG is still a problem. Van Besouw and Wintermans (1978) and Heemskerk and Wintermans (1987) have proposed that this galactolipid could be synthesized mediated by the galactolipid:galactolipid galactosyltransferase (see above).

Formation of other glycerolipids

Sulfolipid is a typical plastid component. This acidic glycolipid is synthesized within intact chloroplasts. We have demonstrated that there is a competition between MGDG and sulfolipid synthesis at the level of diacylglycerol (Joyard *et al.*, 1986), thus suggesting that envelope membranes are involved in sulfolipid biosynthesis. Unfortunately, the natural sulfoquinovosyl donor is still unknown, thus limiting experiments with isolated envelope membranes.

As presented above, chloroplast PG is unique because it contains a $16{:}1_{trans}$ fatty acid at *sn*-2 position. Measurement of PG synthesis in chloroplasts has been unsuccessful for a long time because the optimal experimental conditions for assay in the microsomal fraction were actually inhibitory for chloroplast PG synthesis. Mudd and coworkers (see for instance Mudd *et al.*, 1987) have first demonstrated the ability of intact chloroplasts to synthesize PG and they have also demonstrated that all the enzymes involved in its formation are localized in the inner envelope membrane:

phosphatidic acid + CTP → CDP-diacylglycerol + PPi

CDP-diacylglycerol + *sn*-glycerol 3-phosphate →
3-*sn*-phosphatidyl-1'-*sn*-glycerol 3-phosphate + CMP

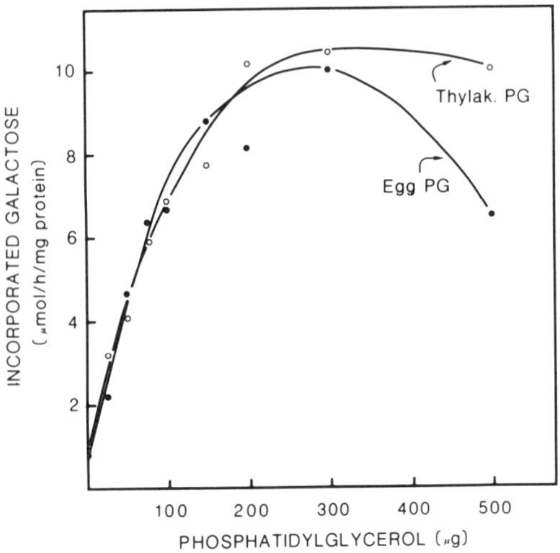

Figure 5. Stimulation of partially purified MGDG synthase activity by PG. Since thylakoid PG contains polyunsaturated fatty acids, whereas egg PG (made by transphosphatidylation of egg PC) contains saturated fatty acids, the saturation level of PG is almost without effect on the stimulation. Experimental conditions are described by Covès *et al.* (1988).

3-*sn*-phosphatidyl-1'-*sn*-glycerol 3-phosphate →
phosphatidylglycerol + Pi

An additional proof for the demonstration of PG formation from phosphatidic acid synthesized via the envelope Kornberg-Pricer pathway was provided by a careful analyses of the nature and the position of the fatty acids on PG.

B. Prenylquinone Biosynthesis

Prenylquinones, and especially plastoquinone-9, are very important compounds for photosynthesis. Since chloroplasts were able to synthesize prenylquinones, we have investigated whether the envelope membranes could be involved in this process.

Tocopherols are synthesized from homogentisic acid and a C20-prenyl unit to form 2-methyl-6-prenyl quinol. By a series of methylations and cyclization, the 2-methyl-6-prenylquinol gives rise successively to 2,3-dimethyl-6-prenylquinol, γ-tocopherol (or γ-tocotrienol) and finally α-tocopherol (α-tocotrienol). Plastoquinone-9 is synthesized from homogentisic acid and solanesyl-pyrophosphate to form 2-methyl-6-solanesyl-quinol, which is methylated and oxidized to form successively plastoquinol-9 and plastoquinone-9. All these enzymes have been localized in the inner envelope membrane of spinach chloroplasts (Soll *et al.*, 1985). Consequently, the presence in envelope membranes of all these enzymes requires the massive transport of prenylquinones from envelope membranes to thylakoids. The mechanism for such a transport is not yet known.

CONCLUSION

Owing to their envelope membranes, chloroplasts contain the complete array for glycerolipid and prenylquinone biosynthesis (Douce and Joyard, 1987). Clearly, envelope membranes are a major structure in plastid biogenesis. The contribution of envelope membranes, and especially the

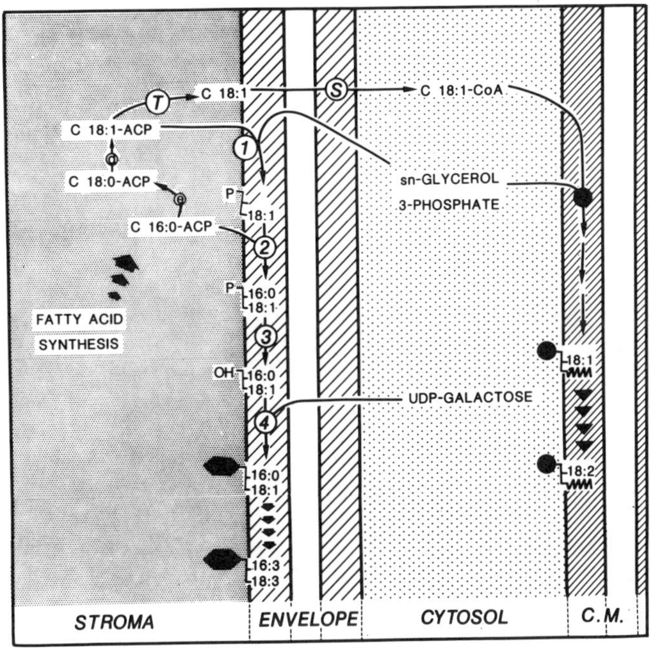

Figure 6. Synthesis of 18:3/16:3 MGDG in envelope membranes. Fatty acid synthesis occurs in the plastid stroma and led to the formation of 16:0-ACP and 18:1-ACP. Acyl-ACP are then used for lysophosphatidic acid, phosphatidic acid, diacylglycerol and MGDG synthesis. MGDG fatty acids are then desaturated.

inner envelope membrane, to the biosynthesis of phosphatidic acid, diacylglycerol and MGDG is well established in plastids (chloroplasts and non-green plastids) from 16:3 plants, as shown in Fig. 6. Unfortunately this is not as clear in 18:3 plants. The contribution of eukaryotic PC to the formation of the C18/C18 backbone is postulated (Roughan and Slack, 1982), but there is a missing link in the pathway: although it is possible to transfer PC to the outer envelope membrane, the existence of a phospholipase *C* type enzyme has not yet been demonstrated in envelope membranes. Such an enzyme would provide diacylglycerol for the MGDG synthase which is active in 16:3 and 18:3 plastids. Another possibility is that C18/C18 diacylglycerol, and not PC, is transfered from the extraplastidal membranes to the outer envelope membrane. This remains to be demonstrated.

Despite this problem, it remains clear that envelope membranes are directly involved in the formation of plastid components, in 16:3 and 18:3 plants, in chloroplasts and in non-green plastids. They are the permanent membrane structure of the plastids, and the presence in all plastids analyzed so far of the enzymes involved in the biosynthesis of plastid compounds such as galactolipids, phosphatidylglycerol or sulfolipid explain the flexibility of the transformation of proplastids into various plastid types.

REFERENCES

Alban C, Joyard J, Douce R (1988). Plant Physiol 88:709-717.

Andrews J, Ohlrogge JB, Keegstra K (1985). Plant Physiol 78:459-465.

Bertrams M, Heinz E (1981). Plant Physiol 68:653-657.

Block MA, Dorne A-J, Joyard J, Douce R (1983). J Biol Chem 258:13280-13286.

Carde JP, Joyard J, Douce R (1982). Biol Cell 44:315-324.

Cline K, Andrews J, Mersey B, Newcomb EH, Keegstra K (1981). Proc Natl Acad Sci USA 78:3595-3599.

Covès J, Block MA, Joyard J, Douce R (1986). FEBS Lett 208:401-406.

Covès J, Joyard J, Douce R (1988). Proc Natl Acad Sci USA 85:4966-4970.

Dorne A-J, Joyard J, Block MA, Douce R (1985). J Cell Biol 100:1690-1697.

Douce R (1974). Science 183:852-853.

Douce R, Guillot-Salomon T (1970): FEBS Lett 11:121-126.

Douce R, Holtz RB, Benson AA (1973). J Biol Chem 248:7215-7222.

Douce R, Block MA, Dorne AJ, Joyard J (1984). In Roodyn DB (ed): "Subcellular Biochemistry, Vol 10," New York: Plenum Press, pp 1-86

Flügge UI, Benz R (1984). FEBS Lett 68:85-89.

Frentzen M, Heinz E, Mc Keon T, Stumpf PK (1983). Eur J Biochem 129:629-639.

Heber U, Heldt HW (1981). Annu Rev Plant Physiol 32:139-168.

Heemskerk JWM, Wintermans JFGM (1987). Physiol Plantarum 70:558-568.

Heinz E (1977). In Tevini M, Lichtenthaler HK (eds): "Lipids and Lipid Polymers," Berlin: Springer, pp 102-120.

Heinz E, Roughan PG (1983). Plant Physiol 72:273-279.

Joyard J, Douce R (1977). Biochim Biophys Acta 486:273-285.

Joyard J, Douce R (1979). FEBS Lett 102:147-150.

Joyard J, Douce R (1987). "The Biochemistry of Plants, Volume 9, Lipids," Stumpf PK (ed): New York: Academic Press, pp 215-274.

Joyard J, Chuzel M, Douce R (1979). In Appelqvist LA, Liljenberg C (eds): "Advances in the Biochemistry and Physiology of Plant Lipids," Amsterdam: Elsevier, pp 181-186.

Joyard J, Blée E, Douce R (1986). Biochim Biophys Acta 879:78-87.

Miquel M, Block MA, Joyard J, Dorne A-J, Dubacq J-P, Kader J-C, Douce R (1987). Biochim Biophys Acta 937:219-228.

Mudd JB, Andrews J, Sparace SA (1987). Methods Enzymol 148:338-345.

Neuburger M, Journet EP, Bligny R, Carde JP, Douce R (1982). Arch Biochem Biophys 217:312-323.

Nishida I, Frentzen M, Ishizaki O, Murata N (1987). Plant Cell Physiol 28:1071-1079.

Pineau B, Dubertret G, Joyard J, Douce R (1986). J Biol Chem 261:9210-9215.

Roughan PG, Slack CR (1982). Annu Rev Plant Physiol 33:97-132.

Soll J, Schultz G, Joyard J, Douce R, Block MA (1985). Arch Biochem Biophys 238:290-299.

Van Besouw A, Wintermans JFGM (1978). Biochim Biophys Acta 529:44-53.

Photosynthesis, pages 347–357

A NEW HYPOTHESIS FOR THE MECHANISM OF PROTEIN TRANSLOCATION INTO CHLOROPLASTS

Kenneth Keegstra

Department of Botany, University of Wisconsin, Madison, Wisconsin 53706

Most chloroplastic proteins are synthesized in the cytosol and transported across two lipid bilayer membranes to arrive at their proper location. Although considerable information is available concerning the topogenic sequences which direct this transport process (Lubben *et al.*, 1988), the mechanism whereby proteins cross the two lipid bilayer membranes of the chloroplastic envelope remains unclear. This manuscript considers two fundamentally different mechanisms to account for protein translocation into chloroplasts and discusses some experimental data about protein transport in terms of these hypotheses.

Before considering transport into chloroplasts, transport across the ER membrane will be briefly reviewed because it is the best characterized example of protein translocation across a biological membrane (Verner and Schatz, 1988). Transport of proteins across the ER membrane normally, but not necessarily, occurs as the proteins are being synthesized (Walter and Lingappa, 1986). Those few instances in which a protein can be transported posttranslationally seem to be limited to special cases where the finished protein does not fold completely (Verner and Schatz, 1988). Regardless of whether posttranslational transport is the exception or the rule, its existence demonstrates that translation and translocation are not obligatorily coupled and that translocation requires a source of energy different from that used for translation (Perara *et al.*, 1986).

The mechanism proposed to account for transport across the ER membrane is that the nascent chain is threaded across the membrane as it comes out of the ribosome (Walter and Lingappa, 1986). In those cases where the protein does not fold it can be threaded across the membrane at a later time. A key aspect of this mechanism is transport of an unfolded protein, with folded proteins being incompetent for transport. It has been suggested, but not proven, that transport occurs via a translocator protein (Singer *et al.*, 1987). The inability to identify translocator proteins has caused some authors to suggest that protein translocation occurs directly through the bilayer (Engelman and Steitz, 1981), possibly with the involvement of lipid rearrangements (Batenburg *et al.*, 1988; Rietveld and de Kruijff, 1986).

Protein translocation into chloroplasts (and mitochondria) differs from transport across the ER membrane in several respects. One is that chloroplastic transport occurs after synthesis of the precursor protein is complete (Schmidt and Mishkind, 1986). Consequently, the precursor forms of chloroplastic proteins have folded into a particular conformation before translocation. At least one precursor has folded into a conformation sufficiently similar to the mature protein that it has enzymatic activity (della-Cioppa *et al.*, 1986). The distinction between cotranslational transport across the ER membrane and posttranslational transport into chloroplasts led to the idea that the mechanisms involved were very different (Wickner and Lodish, 1985). However, the observation that transport across the ER membrane can occur postranslationally has led to the suggestion that a universal mechanism is responsible for transport across all biological membranes (Singer *et al.*, 1987). In order to consider a universal translocation mechanism, it is necessary to postulate that chloroplastic precursors are unfolded before they can be threaded across two bilayer membranes via a mechanism similar to that used for secretory proteins. This unfolding hypothesis has become very popular and is used to interpret most observations regarding protein translocation into both chloroplasts and mitochondria (Verner and Schatz, 1988).

Another difference between translocation across the ER membrane and translocation into chloroplasts is that the former involves transport across a single membrane, whereas the latter requires transport across two membranes. This difference may be more important than the distinction between cotranslational and posttranslational transport. To explain transport across two membranes, it has been postulated that transport occurs at special contact sites where the two membranes are held in close proximity. The exact structure of a contact site is not clear. Some authors have shown schemes that depict the two membrane fusing into one (Douglas *et al.*, 1986). In such schemes, proteins traverse only a single membrane. If this is the case, the fusions must be temporary, because electron microscopy does not support the existence of permanent fusion sites (Carde *et al.*, 1982; Cremers *et al.*, 1988). More likely, a contact site is a location where the two membranes are held in close proximity, probably by the interaction of proteins anchored in each membrane. In this case, translocation is viewed as occurring across the two membranes sequentially, but in one step (Pfanner *et al.*, 1988).

AN ALTERNATIVE HYPOTHESIS

It has been argued that because it is not possible to transport folded proteins across a bilayer membrane, proteins must be unfolded during transport into chloroplasts and mitochondria (Verner and Schatz, 1988). However, another possibility is that the bilayer membrane is temporarily altered to allow a folded protein to pass. Specifically it is possible that lipid molecules are temporarily rearranged from their typical bilayer configuration to some nonbilayer configuration. This model offers a fundamentally different explanation than the protein unfolding model to account for protein translocation into chloroplasts. It is possible to postulate several different types of lipid rearrangements that would suffice. It is also possible to postulate that the precursor protein must undergo some conformational changes before or during translocation. This would be consistent with experimental observations that transport is blocked when a precursor is locked into one conformation (Eilers and Schatz, 1988; della-Cioppa and Kishore, 1988). However, the lipid rearrangement hypothesis places primary emphasis on alterations of membrane structure. This hypothesis can be used to explain many experimental observations regarding precursor structure, the properties of the membranes involved in transport, and the transport process itself. These points are discussed in more detail below.

The Lipid Composition of Chloroplastic Membranes.

The lipid rearrangement hypothesis is attractive because it is consistent with several pieces of information about the translocation system. One such point is that the membranes of chloroplasts are rich in monogalactosyldiacylglycerol (Douce *et al.*, 1983). It is a nonbilayer lipid which normally forms hexagonal H_{II} structures rather than a bilayer (de Kruijff *et al.*, 1984). Indeed, it is rather difficult to cause chloroplastic lipids to form bilayers and special conditions must be used (Sprague and Staehelin, 1984). A similar argument applies to mitochondrial membranes where both phosphatidylethanolamine and cardiolipin are nonbilayer lipids. Given this unusual lipid composition, it is relatively easy to imagine that when the proper stimulus is applied, a localized portion of these lipids could temporarily reorganize from a bilayer configuration to some other arrangement. Such transitions have certainly been shown to occur in model systems (de Kruijff *et al.*, 1984). Local transient rearrangements of this nature would be difficult to detect in biological membranes.

The Properties of Transit Peptides and a Hypothesis for How a Transit Peptide Causes Translocation.

A second relevant point is the membrane activity of chloroplastic transit peptides and mitochondrial presequences (Roise and Schatz, 1988). Several groups have demonstrated that a transit peptide is sufficient to cause the import of a foreign passenger protein (Lubben *et al.*, 1988; Douglas *et al.*, 1986). These observations raise the question of how transit sequences can cause translocation of passenger proteins. Transit peptides do not possess primary sequence similarities that might offer insight into their mechanism of action. Rather, it has been suggested that the important common feature is a particular secondary structure (Verner and Schatz, 1988). For mitochondrial presequences, this secondary structure has been suggested to be amphiphilic, especially an amphiphilic helix (Roise *et al.*, 1988; von Heijne, 1988). This is a particularly attractive hypothesis in that peptides which form amphipathic helices, such as mellitin, are known to cause membrane alterations (Kaiser and Kezdy, 1987). They cause changes in lipid conformation (Batenburg *et al.*, 1987), cause membrane fusion events (Eytan and Almary, 1983) and at higher concentrations lead to cell lysis. Although relatively little work has been reported with chloroplastic transit peptides, several groups have demonstrated that synthetic peptides corresponding to mitochondrial presequences are also membrane active. Presequence peptides alter membrane permeability and lead to uncoupling of mitochondria (Ito *et al.*, 1985; Roise *et al.*, 1986; Furuya *et al.*, 1987). Tamm (1986) demonstrated that a presequence peptide became spontaneously inserted into phospholipid monolayers to form an alpha helix with its axis parallel to the plane of the monolayer.

Although details of the interaction between a transit peptide and the chloroplastic surface are not known, one hypothesis is shown below in Figure 1. It is based in part on the studies with synthetic peptides described above and in part on the evidence that receptor proteins are involved in binding (Lubben *et al.*, 1988). In this scheme, it is postulated that the transit sequence interacts both with a receptor protein and with the lipid bilayer. Binding to the receptor may provide organelle specificity, whereas the interaction with the bilayer may lead to translocation.

The interaction of a membrane-active peptide with a bilayer rich in nonbilayer lipids may trigger a localized and temporary bilayer rearrangement. During this lipid rearrangement, the passenger protein attached to the membrane-active transit peptide would be translocated across the membrane. If this process occurs at a contact site where the two membrane are held in close proximity, then it is possible that lipids from both membranes would be involved in the rearrangement and the protein would be transported across both membranes. It seems unlikely

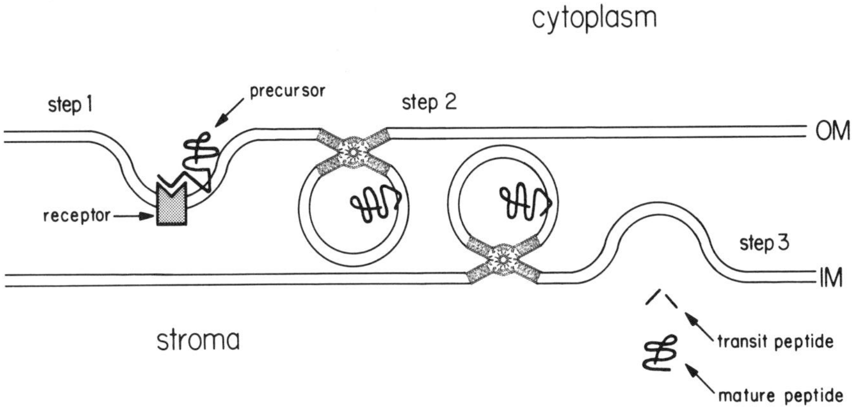

Figure 1. Schematic representation of one particular type of lipid rearrangement that could result in protein translocation. Binding of the precursor (step 1) triggers the formation of a small vesicle (step 2) derived from the outer membrane (OM). The vesicle quickly fuses (step 2) with the inner membrane (IM). In this particular scheme the imported protein is released by proteolytic removal of the transit peptide (step 3). Several variations on this scheme are possible. The scheme is not drawn to scale and the relative dimensions of the lipids, proteins and vesicles are not accurate.

that both formation and disappearance of a lipid rearrangement could occur spontaneously. It is more plausible that proteins of the envelope membranes are necessary to regulate and mediate the process. Indeed, it is well established that the translocation step requires energy that is derived from ATP hydrolysis (Lubben *et al.*, 1988). The precise way in which this energy is used to regulate or mediate translocation is unknown, but this should be a fruitful area of investigation and one that may allow a distinction between the two alternative hypotheses.

In this context, it may be relevant that the envelope membranes have at least two protein kinase activities (Soll, 1988; Soll *et al.*, 1988). It has been suggested that protein phosphorylation may have a role during protein transport. In support of this hypothesis, Hinz and Flugge (1988) have recently demonstrated that a specific envelope membrane protein is phosphorylated during protein translocation. It is possible that reversible phosphorylation of envelope proteins plays a role in membrane rearrangements that allow protein translocation. On the other hand, the protein-unfolding model predicts that ATP is used by an ATP-dependent activity that unfolds the transported protein (Eilers and Schatz, 1988).

The essential feature of the lipid rearrangement hypothesis is that a transit peptide is a membrane-active agent which causes bilayer disruption and lipid rearrangements. If this hypothesis is correct, then not all transit sequences would need to form an amphiphilic helix. Rather an amphiphilic helix would be just one example of the type of peptide that might cause rearrangements. Several variations of this basic idea involving different hypotheses for precursor binding and various types of lipid rearrangements are possible. In all cases, the important feature is the initiation of a lipid rearrangement by the membrane-active peptide and the resulting translocation of the passenger protein across two lipid bilayer membranes without being completely unfolded.

Explanation of Observations on Protein Translocation

The experimental data concerning protein translocation are insufficient to allow rigorous discrimination between the models involving protein unfolding or lipid rearrangement. Many experimental observations can be interpreted in terms of either model. However, some experimental observation are easier to explain in terms of the lipid rearrangement model. One such observation is the translocation of foreign passenger proteins. Several groups have demonstrated that it is possible to cause the import of various passenger proteins into chloroplasts simply by inclusion of an appropriate targeting signal (Lubben *et al.*, 1988). It is possible to envision these various passenger proteins being unfolded and threaded through the membrane, once the process has been initiated by a transit sequence. However, it is easier to imagine that translocation occurs via a lipid rearrangement triggered by a transit peptide and that this rearrangement allows a folded passenger protein to pass.

A second example of an experimental observation that is easier to understand in terms of lipid rearrangement is the import of proteins with stop-transfer sequences. A stretch of hydrophobic amino acids is known to halt the transport of a protein across a single membrane thereby converting the transported protein into a transmembrane protein with the hydrophobic stretch serving as the membrane spanning region (Davis and Model, 1985). Several proteins imported into chloroplasts have such stretches. For example, the chlorophyll *a/b* protein of the thylakoid membrane has at least three membrane-spanning segments (Karlin-Neumann *et al.*, 1985). Yet it is synthesized in the cytoplasm and imported across the two envelope membranes before being inserted into the thylakoid membrane (Cline, 1986). In addition, Lubben *et al.* (1987)

have demonstrated that chimeric precursors containing defined stop-transfer regions obtained from two different proteins, a mouse IgM and the vesicular stomatitis virus G protein, are imported into chloroplasts and do not accumulate in the envelope membranes. It is difficult to understand how proteins containing hydrophobic stretches of amino acids can be transported across two membranes if the proteins are unfolded and threaded directly through the lipid bilayer. It is possible to postulate that transport occurs via a translocator protein. However, these results can be easily understood in terms of a lipid rearrangement, where the protein remains folded and the hydrophobic sequences are not exposed to a hydrophobic environment during transport.

A third example of an observation more easily accommodated by the lipid rearrangement model is that protein import into chloroplasts and mitochondria occurs posttranslationally. It is reasonable to assume that precursors are folded into a specific conformation after synthesis, but before transport. For example, the precursor to the chloroplastic enzyme, EPSP synthase, has enzymatic activity (della-Cioppa, 1986). Thus it seems likely that this precursor is folded into a conformation similar or identical to the mature protein. The protein-unfolding model requires that precursor proteins be in an unfolded state to be threaded through a membrane. Thus, they must be unfolded before or during transport to achieve a transport-competent state. Because this unfolding is energetically unfavorable, it has been postulated that an ATP-dependent unfoldase is involved in protein import (Eilers and Schatz, 1988). Although it is possible that such an activity exists, at the present time it has not been identified in any transport system. On the other hand, a lipid rearrangement model does not require that the folded precursor be completely unfolded before or during import.

The examples described above illustrate some experimental observations that are more easily understood in terms of the lipid rearrangement model for protein import. However many other observations can be interpreted equally well within the context of either model.

A SPECIFIC HYPOTHESIS FOR ONE TYPE OF BILAYER REARRANGEMENT

The bilayer rearrangement hypothesis discussed above is very general and, as mentioned earlier, several variations can be considered. One particular type of rearrangement that could occur is shown in Figure 1 and is discussed briefly below. Specifically it is possible that proteins enter chloroplasts via a special type of vesicle traffic that has some similarities to receptor-mediated endocytosis. It is possible to envision this mechanism for translocation into chloroplasts because transport across two membranes is required. This hypothetical mechanism would allow proteins to be translocated across two membranes without actually traversing or passing through either one. This vesicle mechanism is analogous to several other situations in intracellular protein traffic where the translocation of proteins from one location to another occurs via vesicle movement. Examples include movement of proteins from the lumen of the endoplasmic reticulum to the golgi, from one stack of the golgi to another, from the golgi to the external medium, or even from the external medium to the lysosome or vacuole, as occurs in receptor-mediated endocytosis.

The particular scheme shown in Figure 1 utilizes the hypothesis that membrane fusion events proceed via an intermediate involving a hexagonal H_{II} lipid organization (Verkleij *et al.*, 1984). In this respect, it is relevant that chloroplastic membranes are rich in nonbilayer lipids (see above). It is also relevant that transit peptides and presequences are membrane-active agents. Other membrane-active peptides are known to cause membrane fusion events similar to those necessary for vesicle formation and dissolution. The vesicle is postulated to fuse quickly with the inner membrane because of the close proximity of the two membranes. Indeed, it is possible that fusion of the vesicle with the inner membrane begins before formation of the vesicle from the outer membrane is complete.

The model shown in Figure 1 is consistent with all the current information concerning protein translocation into chloroplasts. The experimental observations discussed above in terms of the general lipid rearrangement hypothesis can also be considered in terms of the more specific hypothesis shown in Figure 1. Only one of these will be mentioned here, i.e. the energy requirements of translocation. It has been demonstrated that energy in the form of ATP hydrolysis is needed for vesicle formation during translocation from the ER to the golgi and from one golgi stack to another (Balch *et al.*, 1986). Although the function of ATP hydrolysis has not been elucidated in these systems, it may be that ATP serves a similar function during translocation into chloroplasts. One attractive hypothesis is the reversible phosphorylation of envelope membrane proteins by protein kinases.

The discussion presented in this manuscript emphasizes the point that many important questions regarding the mechanism of protein translocation into chloroplasts remain to be answered. The lipid rearrangement hypothesis in its general form as well as the specific hypothesis presented in Figure 1 are highly speculative and will almost certainly need to be modified as our knowledge about the import process increases. However, they can serve as useful alternatives for interpreting experimental results and can serve as models for designing future experiments.

REFERENCES

Balch WE, Elliott MM, Keller DS (1986). ATP-coupled transport of vesicular stomatitis virus G protein between the endoplasmic reticulum and the golgi. J Biol Chem 261:14681-14689.

Batenburg AM, Demel RA, Verkleij AJ, de Kruijff B (1988). Penetration of the signal sequence of Escherichia coli phoE protein into phospholipid model membranes leads to lipid-specific changes in signal peptide structure and alterations of lipid organization. Biochem 27:5678-5685.

Batenburgh AM, van Esch JH, Leunissen-Bijvelt J, Verkleij AJ, de Kruijff B (1987). Interaction of melitin with negatively charged phospholipids: consequences for lipid organization. FEBS Lett 223:148-154.

Carde JP, Joyard Jacques, Douce Roland (1982). Electron microscopic studies of envelope membranes from spinach plastids. Biol Cell 44:315-324.

Cline K (1986). Import of proteins into chloroplasts. Membrane integration of a thylakoid precursor protein reconstituted in chloroplast lysates. J Biol Chem 261:14804-14810.

Cremers FFM, Voorhout WF, van der Krift TP, Leunissen-Bijvelt JJM, Verkleij AJ (1988). Visualization of contact sites between outer and inner envelope membranes in isolated chloroplasts. Biochim Biophys Acta 933:334-340.

Daum G (1985). Lipids of mitochondria. Biochim Biophys Acta 822:1-42.

Davis NG, Model P (1985). An artificial anchor domain: hydrophobicity suffices to stop transfer. Cell 41:607-614.

de Kruijff B, Cullis P, Verkleij A, Hope M, Van Echteld C, Taraschi T (1984). Lipid polymorphism and membrane function. In Martinosi A (ed): "Enzymes of Biological Membranes", New York: Plenum Press, pp131-204.

della-Cioppa G, Bauer SC, Klein BK, Shah DM, Fraley RT, Kishore GM (1986). Translocation of the precursor of 5-enolpyruvylshikimate-3-phosphate synthase into chloroplasts of higher plants in vitro. Proc Natl Acad Sci USA 83:6873-6877.

della-Cioppa G, Kishore GM (1988). Import of a precursor protein into chloroplasts is inhibited by the herbicide glyphosate. EMBO J 7:1299-1305.

Douce R, Block MA, Dorne A-J, Joyard J (1983). The plastid envelope membranes: their structure, composition, and role in chloroplast biogenesis. Subcell Biochem 10:1-84.

Douglas MG, McCammon MT, Vassarotti A (1986). Targeting proteins into mitochondria. Microbiol Rev 50:166-178.

Eilers M, Schatz G (1988). Protein unfolding and the energetics of protein translocation across biological membranes. Cell 52:481-483.

Engelman DM, Steitz TA (1981). The spontaneous insertion of proteins into and across membranes: the helical hairpin hypothesis. Cell 23:411-422.

Eytan G, Almary T (1983). Melittin-induced fusion of acidic liposomes. FEBS Lett 156:29-32.

Furuya S, Okada M, Ito A, Aoyagi H, Kanmera T, Kato T, Sagara Y, Horiuchi T (1987). Synthetic partial extension peptides of P-450(SCC) and adrenodoxin precursors: effects on the import of mitochondrial enzyme precursors. J. Biochem 102:821-832.

Hinz G, Flugge UI (1988). Phosphorylation of a 51-kDa envelope membrane polypeptide involved in protein translocation into chloroplasts. Eur J Biochem 175:649-659.

Ito A, Ogishima T, Ou W, Omura T, Aoyagi H, Lee S, Mihara H, Izumiya N (1985). Effects of synthetic model peptides resembling the extension peptides of mitochondrial enzyme precursors on import of the precursors into mitochondria. J Biochem 98:1571-1582.

Kaiser ET, Kezdy FJ (1987). Peptides with affinity for membranes. Ann Rev Biophys Biophys Chem 16:561-581.

Karlin-Neumann GA, Kohorn BD, Thornber JP and Tobin EM (1985). A chlorophyll *a/b* protein encoded by a gene containing an intron with characteristics of a transposable element. J Mol Appl Genet 3:45-61.

Lubben TH, Bansberg J, Keegstra K (1987). Stop-transfer regions do not halt translocation of proteins into chloroplasts. Science 238:1112-1114.

Lubben TH, Theg SM, Keegstra K (1988). Transport of proteins into chloroplasts. Photosynth Res 17:173-194.

Perara E, Rothman RE, Lingappa VR (1986). Uncoupling translocation from translation: implications for transport of proteins across membranes. Science 232:348-352.

Pfanner N, Hartl F-U, Neupert W (1988). Import of proteins into mitochondria: a multi-step process. Eur J Biochem 175:205-212.

Reitveld A, de Kruijff B (1986). Phospholipids as a possible instrument for translocation of nascent proteins across biological membranes. Bioscience Reports 6:775-782.

Roise D, Horvath SJ, Tomich JM, Richards JH, Schatz G (1986). A chemically synthesized pre-sequence of an imported mitochondrial protein can form an amphiphilic helix and perturb natural and artificial phospholipid bilayers. EMBO J 5:1327-1334.

Roise D, Schatz G (1988). Mitochondrial presequences. J Biol Chem 263:4509-4511.

Roise D, Theiler F, Horvath SJ, Tomich JM, Richards JH, Allison DS, Schatz G (1988). Amphiphilicity is essential for mitochondrial presequence function. EMBO J 7:649-653.

Schmidt GW, Mishkind ML (1986). The transport of proteins into chloroplasts. Ann Rev Biochem 55:879-912.

Singer SJ, Maher PA, Yaffe MP (1987). On the translocation of proteins across membranes. Proc Natl Acad Sci USA 84:1015-1019.

Skerjanc IS, Shore GC, Silvius JR (1987). The interaction of a synthetic mitochondrial signal peptide with lipid membranes is independent of transbilayer potential. EMBO J 6:3117-3123.

Soll J (1988). Purification and characterization of a chloroplast outer-envelope-bound, ATP-dependent protein kinase. Plant Physiol 87:898-903.

Soll J, Fischer I, Keegstra K (1988). A GTP-dependent protein kinase is localized in the outer envelope membrane of pea chloroplasts. Planta, in press.

Sprague SG, Staehelin LA (1984). Effects of reconstitution method on the structural organization of isolated chloroplast membrane lipids. Biochim Biophys Acta 777:306-322.

Tamm LK (1986). Incorporation of a synthetic mitochondrial signal peptide into charged and uncharged phospholipid monolayers. Biochem 25: 7470-7476.

Verkleij AJ, Leunnissen-Bijvelt J, de Kruijff B, Hope M, Cullis PR (1984). Non-bilayer structures in membrane fusion. In Evered O, Whelan J (eds): "Cell Fusion", London: Pitman Publishing, pp 45-59.

Verner K, Schatz G (1988). Protein translocation across membranes. Science 241:1307-1313.

von Heijne G (1988). Transcending the impenetrable: how proteins come to terms with membranes. Biochim Biophys Acta 947:307-333.

Walter P, Lingappa VP (1986). Mechanism of protein translocation across the endoplasmic reticulum membrane. Ann Rev Cell Biol 2:499-516.

Wickner WT, Lodish HF (1985). Multiple mechanisms of protein insertion into and across membranes. Science 230:400-407.

Photosynthesis, pages 359–361

REGULATION IN PHOTOSYNTHESIS

Ulrich Heber

Institute of Botany and Pharmaceutical Botany, University of Würzburg, D-87 Würzburg, Germany

Whereas the roots of modern photosynthesis research extend far into the past, knowledge about molecular details of the photosynthetic process is of recent origin. Mapping of photosynthetic carbon metabolism by Calvin was essentially complete, but not yet undisputed in the early 60's. Only a few years before, it has been recognized that chloroplast membranes, which were named thylakoids by Menke, could provide what Arnon called assimilatory or reducing power. They could phosphorylate ADP and reduce NADP which, in turn, are needed for carbon reduction. Hill proposed how two photosystems could cooperate to oxidize water and reduce NADP, and French recognized that the chloroplast system engaged in harvesting light and facilitating its conversion into biologically useful energy is much more complex than indicated by the chemical nature of its individual pigment components.

Although much further progress has been made during the last decades, insight into the molecular mechanisms of photosynthesis is still far from being complete. Much effort has recently gone into the study of detail. However, it is difficult to deduce the functioning of a clock after taking it apart and looking at its individual components. It must be admitted that listening to ticking is also insufficient. Obviously, the study of interaction is important. As complex a process as photosynthesis must be subject to regulation. In 1969, Bassham and Krause identified regulatory steps in carbon metabolism. This work has led to the discovery of the thioredoxin system which activates key stroma enzymes in the light and links their activity status to the availability of electrons in the thylakoid system. Although it may be viewed as providing only coarse control of carbon flux, it is essential for reconciling conflicting interests of the chloroplast thylakoid system and stroma enzymes. High rates of electron transport are difficult to maintain when the electron acceptor NADP is unavailable and phosphorylation is hampered by a lack of ADP. The concentrations of both solutes should be expected to be low when high levels of ATP and NADPH are needed to make carbon reduction proceed rapidly. Enzyme activation by the thioredoxin system and additional control of enzyme activity by substrates and effectors facilitate rapid carbon reduction at low ratios of ATP to ADP and

NADPH to NADP. This leaves NADP available as an electron acceptor even at high light intensities thereby preventing the over-reduction of the electron transport chain, a condition known to expose chloroplasts to photoinactivation.

Surprisingly, NADP availability permits effective charge separation in the reaction center of photosystem I even when leaves exposed to high intensity irradiation close their stomata under water stress restricting CO_2 access. Under these conditions, photosystem II activity is regulated down and NADPH and ATP consumption through a slow photorespiratory CO_2 turnover is sufficient to maintain a redox gradient between photosystem II and photosystem I which appears to set the stage for the effective conversion of light energy into heat. Whereas at low light intensities light energy is converted into redox energy so rapidly that losses of energy by fluorescence or radiationless decay are insignificant, at high light intensities dissipation of quantum energy via heat is so effective that it decreases fluorescence emission. Under conditions of excessive irradiation, the coordinated and harmless conversion of light energy into heat directly within the photochemical system is essential to avert damage to the photosynthetic apparatus. There is the question where dissipative processes are predominant, in the reaction centers or the antenna complexes. On the grounds of a close correlation between the rate of radiationless energy dissipation in the antennae and the level of the xanthophyll zeaxanthin formed by violaxanthin de-epoxidation, a functional relationship between zeaxanthin availability and energy dissipation seems possible. There is much current interest in the role of the xanthophyll cycle in protecting the photosynthetic apparatus against the destructive effects of excessive light.

In order not only to survive but actually to thrive, some plants need more and others less light. Those that actually prefer shade may suffer under high intensity irradiation, and the reverse is true for others. Shade plants need to be particularly adaptive. They cannot be wasteful with resources and need to restrict respiration. Whenever the sun penetrates a shading canopy, they need to make the best use of its light without suffering damage. As the sun moves, and as shading leaves rustle, exposure times to light intensities which permit accelerated photosynthesis vary. They may be very short. This requires fast light activation of stroma enzymes and, in high light, the formation of acceptor pools which are large enough to permit significant carbon gain by carboxylation even after the sun has gone.

Attempts to understand photosynthesis and its regulation can be justified on various grounds. Man's survival depends on the use of energy in several forms. His basic need, food, is derived from plants. As the human population increases, the demand for food likewise increases. There are many problems to solve. One of the questions that needs to be answered is whether we can produce plants of higher photosynthetic capacity than plants currently in agricultural use. In the area of molecular biology, we have tools at our disposal to alter plants which extend far beyond those used in conventional plant breeding. It needs no mention that the latter has been impressively successful. To direct and not to waste efforts, it is necessary to know how metabolism is regulated and where major control points reside. Only at these points does interference promise major alterations in plant responses. In a metabolic chain, all components are essential for its functioning, and, in consequence, all must take part in control. However, at least since George Orwell, we know that although all are equal in a democracy, some are more equal than others. Research in metabolic regulation can provide molecular biologists with information on where major control steps in photosynthesis are, what their relative significance is and how significance changes as flux conditions change. We know that light limits photosynthesis when its intensity is low, but not when it is high. For CO_2, the reverse situation exists. However, as all components of a system take part in control, we may wish to alter enzymic components of the photosynthetic apparatus instead of changing light or CO_2 in order to obtain changes in carbon flux. Kacser and Burns have provided an experimental approach and a theoretical framework to answer questions of control quantitatively. It rests on the observation of changes in flux, when changes in one of the components of a metabolic system are made. Only a few attempts have been made so far in photosynthesis to use their approach which promises to extend our knowledge of regulation. It may eventually be found useful in practical applications.

Photosynthesis, pages 363–374

REGULATION OF EXCITATION ENERGY DISTRIBUTION BY STATE TRANSITIONS

John Biggins, Doug Bruce and Pamela B. Gibbs

Section of Biochemistry, Brown University Providence, Rhode Island 02912 (J.B. and P.B.G), and Department of Biology, Brock University, St. Catherines, Ontario L2S 3A1, Canada (D.B.)

INTRODUCTION

The regulation of excitation energy distribution between the two photosystems in photosynthesis was first demonstrated in the green alga *Chlorella pyrenoidosa* (Bonaventura and Myers, 1969) and the red alga *Porphyridium cruentum* (Murata, 1969). The photoregulatory process, now referred to as the state transition, has since been observed in a wide variety of green algae, cyanobacteria and higher plants. Bonaventura and Myers (1969) also showed that changes in the rate of oxygen evolution accompanied the fluorescence changes, and they suggested that control of the distribution of energy serves to correct imbalance in turnover of the two photosystems to ensure efficient linear electron transport. Murata (1969) suggested that the mechanism of excitation energy redistribution could involve structural changes in the photosynthetic apparatus that might modify the distances between and/or the mutual orientation of pigments (see Fork and Satoh, 1986; Williams and Allen, 1987; and Biggins and Bruce, 1988 for reviews).

Experimentally the state transition can be visualized at room temperature by monitoring the effect of light preferentially absorbed by PSI on the fluorescence yield of PSII Chl *a*. Alternatively, the effect of preillumination of cells using light absorbed by either PSI or PSII on the fast fluorescence induction, or low temperature fluorescence emission spectra have been used. The latter reveals dramatic changes in the yields of fluorescence originating from the Chl *a* of PSI and PSII, confirming that a partitioning of excitation energy occurs between the two photosystems in response to preillumination wavelengths.

MODELS FOR THE STATE TRANSITION

The formulation of a satisfactory model for the state transition requires that three aspects be addressed :

1. A mechanism for sensing the imbalance of the relative activities of PSI and PSII to serve as a trigger for the transition,
2. A physical mechanism for alteration of the pathway of excitation energy transfer from the antenna to PSI and PSII, and
3. A biochemical mechanism for inducing changes within the photosynthetic apparatus to control the pathway of energy transfer.

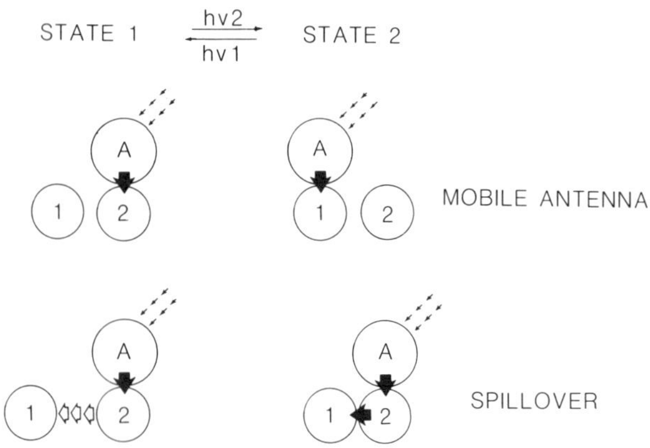

Figure 1. Models for the state transition. The solid arrows signify tight energy coupling. A decrease in spillover in state 1 is depicted by the broken arrow between PSII and PSI. A = antenna.

Two radically different models have been proposed to account for the state transition. As shown schematically above in Figure 1, the most obvious way of redistributing excitation energy between the two photosystems is to relocate a proportion of the antenna population between PSI and PSII. In the mobile antenna model, relocation of the antenna directly changes the absorbance cross section and, therefore, photochemical turnover of each photosystem. This model predicts a change in direction of the fraction of the excitation energy absorbed by the relocated antenna. Changes in the kinetics of excitation energy transfer may accompany changes in the energy transfer pathway.

The second model, referred to as the spillover model, historically originated from a careful analysis of enhancement by Myers and Graham (1963) shortly after formulation of the Z-scheme for electron transport.

They concluded that quanta delivered to PSII in excess might be transferred directly to PSI via "spillover" to maintain high quantum efficiency in the photosynthetic unit. In the current view of the spillover model for the state transition it is assumed that the antenna remains physically associated with PSII and maintains efficient energy coupling in both light states. The relative amounts of excitation energy reaching PSII and PSI are controlled by change in the rate of excitation energy transfer from PSII antenna Chl *a* to PSI. An increase in spillover increases the amount of excitation reaching PSI at the expense of PSII leading to state 2. Conversely, a decrease in spillover results in the conversion to state 1. In the spillover model it is also assumed that the probability of excitation energy transfer from PSII to PSI is controlled by either the distance between the two photosystems in the thylakoid or changes in orientation of the pigments or both. Such a conformational change required could be very minor when one considers that the distance dependence for the excitation transfer rate is $1/R^6$ for weak intermolecular coupling (Kenkre *et al*., 1974). Predictions from this model with respect to excitation energy transfer contrast with the mobile antenna model in that only changes in rate constants of excitation transfer are expected.

It should be stressed that *both* the spillover and the mobile antenna models will result in complementary changes in absorbance cross section for PSI and PSII. The two models differ in that the cross section change in the mobile antenna model is an exclusive function of absorption by the antenna, whereas the spillover model predicts that cross section changes should also include a contribution of absorption by PSII Chl *a in addition* to that of the antenna. Contrary to numerous implications in the literature we emphasize that differences between the two models is not one of a cross section change versus spillover *per se*.

HIGHER PLANTS AND GREEN ALGAE

The thylakoids of higher plants and green algae are morphologically complex and are arranged into stacked and unstacked regions. It is believed that the membranes are heterogeneous and PSII and PSI are distributed laterally between the stacked and unstacked regions respectively (Staehelin, 1986). The antenna complex in these organisms is the light harvesting Chl *a/b* protein (LHC) which is an intrinsic membrane component associated primarily with PSII in the stacked regions. This heterogeneous localization of PSI and PSII within the thylakoid morphology has necessitated consideration of the mobile antenna model in these organisms. Mechanistic details have been proposed based upon insights gained from the effect of cations on thylakoid structure and excitation energy distribution, and postulates concerning the regulatory role of the reversible protein phosphorylation of the LHC .

Initiation of the transition to state 2 is thought to be triggered by the reduction of the intersystem pool of PQ by illumination conditions that favor PSII over PSI. The PQ reduction results in kinase activation leading to the phosphorylation of threonine residues on a surface-exposed segment of the LHC located in the stacked regions of the thylakoids. The resulting change in surface charge density promotes lateral migration of the phosphorylated LHC to the unstacked regions of the thylakoids resulting in the uncoupling of antennae serving PSII and connection to PSI. Corresponding changes in absorbance cross section are then expected for PSI and PSII. The transition to state 1 is reversed by kinase deactivation, dephosphorylation of the phospho-LHC and migration of the LHC back to the stacked region followed by coupling to PSII.

Although this proposal has wide acceptance (Williams and Allen, 1987) many issues remain to be clarified such as the biochemistry of the kinase activation and localization of the kinase and phosphatase. Also several fundamental physical and structural details are now controversial (Wendler and Holzwarth, 1987, Bhalla *et al.*, 1988).

ORGANISMS CONTAINING PHYCOBILINS

The cyanobacteria, red algae, and cryptophytes all utilize phycobilins as accessory photosynthetic pigments. In the case of cyanobacteria and red algae, the phycobiliproteins are organized into antenna complexes called phycobilisomes, which are localized on the stromal surface of the thylakoids (Gantt, 1981). There is considerable evidence that the phycobilisomes are closely associated with PSII. For example, tight energetic coupling occurs in vesicle preparations bearing phycobilisomes (Katoh and Gantt, 1979, Dilworth and Gantt, 1981), and the distribution of phycobilisomes on the thylakoid surface correspond to that of PSII within the membrane (Morschel and Muhlethaler, 1983). Even highly purified preparations of PSII contain core polypeptides from the phycobilisome (Bowes *et al.*, 1983).

It has been established that the thylakoids of cyanobacteria and red algae are not arranged into stacked and unstacked regions, and there is no precedence for consideration of any gross lateral heterogeneity, or partitioning of large regions of the thylakoids into domains of PSII or PSI as appears to be the case in higher plants (Staehelin, 1986). The cryptophytes contrast with the cyanobacteria and red algae in that their thylakoids show some limited membrane appression and their phycobiliproteins are densely packed in the lumen of the thylakoid (Gantt *et al.*, 1971). The cryptophytes also contain an intramembrane light-harvesting complex containing Chl *a, c* and carotenoid, a feature also common to the chromophyte algae.

It now appears that the mechanism of the state transition in red algae and cyanobacteria is different from that in the higher plants. Although some aspects concerning the biochemistry remain controversial, there is strong evidence that the spillover model satisfactorily accounts for the most critical data. The most incisive experiments have relied on an analysis of the fluorescence from the various Chl *a* forms based upon insights developed by Butler and coworkers (Butler, 1978), and the use of picosecond time-resolved spectroscopy (see Biggins and Bruce, 1988 for review).

The extent of excitation energy transfer from PSII to PSI can be estimated by measuring the variable fluorescence in PSI Chl *a* emission at 77K (Ley and Butler, 1976). Although there are exceptions, in the majority of cases PSI has no intrinsic variable fluorescence and the yield of PSI emission is independent of its redox state. The variable fluorescence observed in these cases is assumed to have arisen by energy transfer from PSII because the kinetics are identical to those of the PSII emission (Butler and Kitajima, 1975; Ley and Butler, 1976; Butler, 1978).

In a detailed study of the millisecond rise kinetics of fluorescence from PSI and PSII at 77K in red algae, Ley and Butler (1980) showed that the rate of spillover increased in state 2. Further evidence favoring the spillover model has been obtained more directly by Bruce *et al.* (1985) by measuring the picosecond time-resolved fluorescence emission at 77K in cyanobacteria and red algae in state 1 and state 2. No differences were observed in the decay kinetics of the phycobilin species and the largest change observed was for that originating from PSII Chl *a*. A larger initial decay lifetime in state 1 indicated that much of the PSII Chl *a* was uncoupled from any further energy transfer to PSI Chl *a*. These data show that the energy coupling between the phycobilisome and PSII is maintained in both states and are consistent with a decrease in spillover in state 1. The observations were subsequently confirmed by determination of the corresponding fluorescence rise times using instrumentation with streak camera detection (Bruce *et al.*, 1986b).

The involvement of a conformational change in the mechanism of the state transition has been indicated by a variety of experimental approaches. Biggins (1983) and Bruce and Biggins (1985) showed that red algae and cyanobacteria can be chemically cross-linked in the light states using glutaraldehyde, and that after fixation the cells could not be reconverted. These findings implicate the existence of a conformational change occurring in the state transition that involves functional protein. Fork and coworkers (Fork *et al.*, 1979; Fork and Satoh, 1983) showed that the state transition would not proceed below the lipid phase transition in the thermophilic cyanobacterium *Synechococcus lividus*. These data strongly suggest the requirement for membrane fluidity and, therefore, the possible mobility of components within the thylakoid bilayer.

The possible involvement of conformational changes occurring during the state transition was studied directly using polarized optical techniques (Bruce and Biggins, 1985). Changes in the linear dichroism of oriented cyanobacteria were observed and it was concluded that the entire phycobilisome-PSII assembly may change orientation with respect to the plane of the thylakoid.

The role of protein phosphorylation in the state transition of cyanobacteria and red algae remains controversial. Although ca 12 phosphorylated polypeptides were observed in *P. cruentum* and *A. nidulans*, no changes were detected between the phosphorylated polypeptide profiles of cells in state 1 and state 2 (Biggins *et al.*, 1984b). These findings were confirmed and extended by Schuster *et al.* (1984) and Kirschner and Senger (1986) who were also unable to detect any light-dependent protein phosphorylation. On the other hand, Allen and coworkers (Allen *et al.*, 1985; Sanders *et al.*, 1986) claim that the transition to state 2 in *Synechococcus* 6301 is promoted by the phosphorylation of polypeptides on the phycobilisome and PSII in state 1. In a reinvestigation of the *in vitro* studies of Allen *et al.* (1985), Biggins and Bruce (1987) reported that:

1. The observed fluorescence changes were not indicative of a state transition,
2. The fluorescence changes were independent of protein kinase activity, and
3. The reported protein kinase activity was not light dependent.

Recently Sanders and Allen (1988) have now agreed with observations 1) and 2) and retracted their earlier claims. However, they have reiterated the necessity for light for the thylakoid protein kinase activity and extended their study of the divalent cation requirement. They now report that Zn^{2+} as well as Ca^{2+} can substitute for Mg^{2+}. Biggins and Bruce (1987) found that the cyanobacterial thylakoid kinase was inactive in the presence of Ca^{2+}.

Allen and coworkers (Allen *et al.*, 1985; Sanders *et al.*, 1986) have proposed a mobile antenna model for the state transition in cyanobacteria and red algae as a special case of a general formulation in photosynthesis for the control of excitation energy transfer by reversible protein phosphorylation (Allen and Holmes, 1986). However, in a recent report (Mullineaux and Allen, 1988), they have now modified their model from that of phycobilisome mobility on the membrane surface to an uncoupling of a fraction of PSII from the phycobilisome followed by movement of the detached PSII. The change in distance between the detached PSII and PSI then increases spillover from the *uncoupled* PSII to PSI. This revised model does not allow for the transfer of excitation energy from the phycobilisome antenna to PSI in either state 1 or state 2, or spillover from PSII that remains coupled to the phycobilisome Both of their proposals are inconsistent with the low temperature millisecond rise kinetics for PSI and PSII fluorescence (Ley and Butler, 1980), the picosecond time-resolved fluorescence decays (Bruce *et al.*, 1985), and the ultrafast fluorescence rise data (Bruce *et al.*, 1986). Recently Brimble and Bruce (personal communication) have demonstrated both a wavelength-dependent and a dark/ DCMU-light state transition in a mutant cyanobacterium lacking the antenna phycobilisome. This important observation unequivocally establishes that the regulation of excitation energy distribution in these organisms is controlled by events occurring within the thylakoid.

As a working hypothesis Biggins *et al.* (1984b) proposed that changes in spillover in cyanobacteria and red algae are promoted by alterations in charge distribution and electrostatic interactions between the two photosystems. This proposal was based primarily upon the requirement for coupled PSI cyclic transport for the transition to state 1 (Satoh and Fork, 1983; Biggins *et al.*, 1984a) which was later shown to occur in ca 15-20 photochemical turnovers (Biggins and Bruce, 1985). Our proposal also stipulates that the ion gradients be localized to some extent (Dilley *et al.*, 1987) because ion gradients must certainly be maintained by cells in state 2.

Biggins and coworkers have reported on the pathways of excitation energy transfer and wavelength-dependent changes in energy distribution in several cryptophytes. Snyder and Biggins (1987a, b) found that the mechanism of energy redistribution in *Cryptomonas ovata* was very similar to the state transition in cyanobacteria and red algae. However, they observed that the fluorescence changes did not correlate with the predicted changes in rates of electron transport. Bruce *et al.* (1986a, 1987) showed that although the picosecond fluorescence lifetimes of phycobilin, Chl c_2 and Chl *a* were dependent on preillumination wavelength, the emissions from PSI and PSII were independent. Therefore, both Bruce *et al.* (1987) and Snyder and Biggins (1987a, 1987b) have tentatively concluded that preillumination uncoupled the transfer of excitation energy from phycobilins to both PSII and PSI, in a process that may be a mechanism for photoprotection.

CHROMOPHYTA

Although there has been intensive investigation of the state transitions in higher plants, green algae, red algae, cyanobacteria and cryptophytes, the vast majority of microalgae active in oceanic primary production are species that contain membrane-bound Chl-*a*-*c*-carotenoid antennae for light harvesting (Larkum and Barrett, 1985). These species are taxonomically diverse and include the *Phaeophyta*, *Chrysophyta*, *Bacillariophyta*, and *Haptophyta*. The dinoflagellates are related but contain a soluble pigment-protein complex containing carotenoid (peridinin) that is functional in light harvesting (Boczar *et al.*, 1980). The chromophyte algae are unique in that their thylakoids are invariably arranged in bands of three, and a girdle lamella is present (Gibbs, 1970). It is not clear whether this morphological arrangement can be considered to be equivalent to the appression of thylakoids leading to the stacked regions of higher plant chloroplasts, or whether there is reason to suspect an analogous heterogeneity in the lateral distribution of PSI and PSII.

The functional role of the Chl *a*-*c*-carotenoid antenna in the transfer of excitation energy to reaction center Chl *a* has been established (Haxo and Blinks, 1950). With regard to state transitions, Owens (1986) investigated energy redistribution in the diatom *Phaeodactylum tricornatum* and only observed intensity-dependent changes. He suggested that energy redistribution might be regulated in this species by adjusting the level of energy transfer to PSII.

We have recently surveyed several species of chromophyte algae for their ability to respond to changes in wavelength. We now report that *Ochromonas danica* (Chrysophyta) undergoes a light state transition as judged by room temperature fluorescence transients and by analysis of the Chl *a* fluorescence emission at 77K following the preillumination of cells at room temperature (Gibbs and Biggins, 1988). Analysis of the conditions necessary to convert the cells to state 1 revealed a simple photochemical dose dependence as was observed previously for red algae (Biggins and Bruce, 1985). However, the kinetics of the room temperature transition more closely resembled those found in green algae. Elucidation of their mechanism will require studies of the kinetics of PSI and PSII fluoresence emission at 77K, ultrafast time-resolved fluorescence spectroscopy, and analysis of the possible role of protein phosphorylation.

CONCLUDING REMARKS

Although we have indicated that it is presently premature to conclude that the cryptophytes and certain species of chromophyte algae undergo conventional state transitions we suggest that the regulation of excitation energy distribution is most likely a general phenomenon in oxygenic photosynthesis. It is clear that the spillover model satisfactorily accounts for the physical mechanism in the organisms that have simple thylakoid arrangements, whereas the more advanced organisms with a heterogeneous thylakoid organization most probably utilize a mobile antenna mechanism. It remains to be established whether either of these two general models apply to the chromophyte algae which, morphologically, represent a transition in thylakoid complexity between these two extremes.

It is widely claimed that the regulation of excitation energy distribution in photosynthesis is a metabolic control process necessary to deal with chromatic fluctuations in natural water columns and terrestial leaf canopies. These claims are presently speculative because no published reports are available on the environmental significance of state transitions either from field experiments or controlled laboratory simulations. In the cyanobacteria and red algae there has been a good correlation between the state transition and change in rate of electron transport (Fork and Satoh, 1983; Biggins, 1983). However, the relevance of the fluorescence changes *in vivo*, and those induced by protein phosphorylation in higher plants, is less convincing. Clearly the field is open for further investigation into the mechanism of the state transitions and their significance in algal cell physiology and whole plant photosynthesis.

ACKNOWLEDGEMENTS

The authors' research was sponsored by the NSF, the USDA-CRGO, and the NSEC of Canada.

REFERENCES

Allen JF, Holmes NG (1986). A general model for regulation of photosynthetic function by protein phosphorylation. FEBS Lett 202:175-181.

Allen JF, Sanders CE, Holmes NG (1985). Correlation of membrane protein phosphorylation with excitation energy distribution in the cyanobacterium *Synechococcus* 6301. FEBS Lett 193:271-275.

Bhalla P, Sieburth LE, Bennett J (1988). Evidence for migration of two thylakoid phosphoproteins from exposed to appressed lamellae in isolated intact spinach chloroplasts. Eur J Biochem, in press.

Biggins J (1983). Mechanism of the light state transition in photosynthesis. 1. Analysis of the kinetics of cytochrome f oxidation in state 1 and state 2 in the red alga *Porphyridium cruentum.* Biochim Biophys Acta 724:111-117.

Biggins J, Bruce D (1985). Mechanism of the light state transition in photosynthesis.lll.Kinetics of the state transition in *Porphyridium cruentum.* Biochim Biophys Acta 806:230-236.

Biggins J, Bruce D (1987). The relationships between protein kinase activity and chlorophyll *a* fluorescence changes in thylakoids from the cyanobacterium *Synechococcus* 6301. In Biggins J (ed): "Progress in Photosynthesis Research," Vol 2, Dordrecht, Martinus Nijhoff Publishers pp 773-776.

Biggins J, Bruce D (1988). Regulation of excitation energy transfer in organisms containing phycobilins. Photosynth Res, in press.

Biggins J, Campbell CL, Creswell LL, Wood EA (1984a). Mechanism of the light state transition in *Porphyridium cruentum*. In Sybesma C (ed): "Advances in Photosynthesis Research," Vol 2, The Hague/Boston/ Lancaster, Martinus Nijhoff/Dr W Junk Publishers pp 303-306.

Biggins J, Campbell CL, Bruce D (1984b). Mechanism of the light state transition in photosynthesis. II. Analysis of phosphorylated polypeptides in the red alga *Porphyridium cruentum*. Biochim Biophys Acta 767:138-144.

Boczar BA, Prezelin BB, Markwell JP, Thornber JP (1980). A chlorophyll *c*-containing pigment-protein complex from the marine dinoflagellate, *Glenodinium sp.* FEBS Lett 120:243-247.

Bonaventura C, Myers J (1969). Fluorescence and oxygen evolution from *Chlorella pyrenoidosa*. Biochim Biophys Acta 189:366-383.

Bowes JM Stewart AC, Bendall DS (1983). Purification of photosystem II particles from *Phormidium laminosum* using the detergent dodecyl-β-D-maltoside. Properties of the purified complex. Biochim Biophys Acta 725:210-219.

Bruce D, Biggins J (1985). Mechanism of the light state transition in photosynthesis.V.77K linear dichroism of *Anacystis nidulans* in state 1 and state 2. Biochim Biophys Acta 810:295-301.

Bruce D, Biggins J, Steiner T, Thewalt M (1985). Mechanism of the state transition in photosynthesis.IV. Picosecond fluorescence spectroscopy of *Anacystis nidulans* and *Porphyridium cruentum* in state 1 and state 2 at 77K. Biochim Biophys Acta 806:237-246.

Bruce D, Biggins J, Steiner T, Thewalt M (1986a). Excitation energy transfer in the cryptophytes. Fluorescence excita-tion spectra and picosecond time-resolved emission spectra of intact algae at 77K. Photochem Photobiol 44:519-525.

Bruce D, Hanzlik CA, Hancock LE, Biggins J, Knox RS (1986b). Energy distribution in the photochemical apparatus of *Porphyridium cruentum*: picosecond fluorescence spectroscopy of cells in state 1 and state 2 at 77K. Photosynth Res 10:283-290.

Bruce D, Biggins J, Charbonneau S, Thewalt M (1987). Excitation energy transfer in *Cryptomonas ovata*. Preillumination dependent changes in 77K picosecond time resolved fluorescence emission spectra. In Biggins J (ed): "Progress in Photosynthesis Research" Vol 2, Dordrecht: Martinus Nijhoff Publishers pp 777-780.

Butler WL (1978). Energy distribution in the photochemical apparatus of photosynthesis. Ann Rev Plant Physiol 29:345-378.

Butler WL, Kitajima M (1975). Energy transfer between photosystem II and photosystem I in chloroplasts. Biochim Biophys Acta 396:72-85.

Dilley RA,Theg SM, Beard WA (1987). Membrane-proton interactions in chloroplast bioenergetics: localized proton domains. Ann Rev Plant Physiol 38:347-389.

Dilworth MF, Gantt E (1981). Phycobilisome-thylakoid topography on photosynthetically active vesicles of *Porphyridium cruentum*. Plant Physiol 67:608-612.

Fork DC, Satoh K (1983). State 1-state 2 transitions in the thermophilic blue-green alga (cyanobacterium) *Synechococcus lividus*. Photochem Photobiol 37:421-427.

Fork DC, Satoh K (1986). The control by state transitions of the distribution of excitation energy in photosynthesis. Ann Rev Plant Physiol 37:335-361.

Fork DC, Murata N, Sato N (1979). Effect of growth temperature on the lipid and fatty acid composition, and the dependence of temperature on light-induced redox reactions of cytochrome f and of light energy distribution in the thermophilic blue-green alga *Synechococcus lividus*. Plant Physiol 63:524-530.

Gantt E (1981). Phycobilisomes. Ann Rev Plant Physiol 32: 327-347.

Gantt E, Edwards MR, Provasoli L (1971). Chloroplast stucture of the Cryptophyceae. Evidence for phycobili-proteins within intrathylakoidal spaces. J Cell Biol 48:280-290.

Gibbs SP (1970). The comparative ultrastructure of the algal chloroplast. Ann Acad Sci (New York) 175:454-473.

Gibbs PB, Biggins J (1988). Regulation of the distribution of excitation energy in *Ochromonas danica*, an organism containing a chlorophyll-a/c/carotenoid light harvesting antenna. Photosynth Res, in press.

Haxo FT, Blinks LR (1950). Photosynthetic action spectra of marine algae. J Gen Physiol 33:389-422.

Katoh T, Gantt,E (1979). Photosynthetic vesicles with bound phycobilisomes from *Anabaena variablis*. Biochim Biophys Acta 546:383-393.

Kenkre VM, Knox RS (1974). Theory of fast and slow excitation transfer rates. Phys Rev Lett 33:803-806.

Kirschner J, Senger H (1986). Thylakoid protein phosphorylation in the red algae *Porphyridium cruentum*. In Akoyunoglou G, Senger H (eds): "Regulation of chloroplast differentiation," Plant Biology Vol 2, New York, AR Liss Inc Publishers pp 339-344.

Larkum AWD, Barrett J (1985) Light harvesting processes in algae. Adv Bot Res 10:1-210.

Ley AC, Butler WL (1976) Efficiency of energy transfer from photosystem 11 to photosystem 1 in *Porphyridium cruentum*. Proc Natl Acad Sci USA 73:3957-3960.

Ley AC, Butler WL (1980). Energy distribution in the photochemical apparatus of *Porphyridium cruentum* in state 1 and state 11. Biochem Biophys Acta 592:349-363.

Morschel E, Muhlethaler K (1983). On the linkage of exoplasmic freeze-fracture particles to phycobilisomes. Planta 158:451-457.

Mullineaux CW, Allen JF (1988). Fluorescence induction transients indicate dissociation of Photosystem 11 from the phycobilisome during the State-2 transition in the cyanobacterium *Synechococcus* 6301. Biochim Biophys Acta 934:96-107.

Murata N (1969a). Control of excitation transfer in photosynthesis. 1. Light-induced change of chlorophyll *a* fluorescence in *Porphyridium cruentum*. Biochim Biophys Acta 172:242-251.

Myers J, Graham J-R (1963). Enhancement in *Chlorella*. Plant Physiol 38:105-116.

Owens TG (1986). Light-harvesting function in the diatom *Phaeodactylum tricornatum*. 11.Distribution of excitation energy between the two photosystems. Plant Physiol 80:739-746.

Sanders CE, Holmes NG, Allen JF (1986). Membrane protein phosphorylation in the cyanobacterium *Synechococcus* 6301. Biochem Soc Trans 14:66-67.

Sanders CE, Allen JF (1988). Effects of divalent cations on the 77K fluorescence emission and on membrane protein phosphorylation in isolated thylakoids of the cyanobacterium *Synechococcus* 6301. Biochim Biophys Acta 934:87-95.

Satoh K, Fork DC (1983). The relationship between state 11 to state 1 transitions and cyclic electron flow around photosystem 1. Photosynth Res 4:245-256.

Schuster G, Owens GC, Cohen Y, Ohad I (1984). Thylakoid polypeptide composition and light-independent phosphorylation of the chlorophyll *a,b*-protein in *Prochloron*, a prokaryote exhibiting oxygenic photosynthesis. Biochim Biophys Acta 767:596-605.

Snyder UK, Biggins J (1987a). Excitation energy distribution in *Cryptomonas ovata*. In Biggins J (ed): "Progress in Photosynthesis Research," Vol 2, Dordrecht: Martinus Nijhoff Publishers pp 781-784.

Snyder UK, Biggins J (1987b). Excitation-energy distribution in the cryptomonad alga *Cryptomonas ovata*. Biochim Biophys Acta 892:48-55.

Staehelin LA (1986). Chloroplast structure and supramolecular organization of photosynthetic membranes. In Staehelin LA, Arntzen CJ (eds): "Ency Plant Physiol," New Series, Vol 19, Berlin, Heidelberg, New York and Tokyo, Springer-Verlag Publishers pp 1-84.

Wendler J, Holzwarth AR (1987). State transitions in the green alga *Scenedesmus obliquus* probed by time-resolved fluorescence spectroscopy and global data analysis. Biophys J 52:717-728.

Williams WP, Allen JF (1987). State 1/state 2 changes in higher plants and algae. Photosynth Res 13:19-45.

Photosynthesis, pages 375–391

LIGHT STRESS AND PHOTOPROTECTION RELATED TO THE CAROTENOID ZEAXANTHIN IN HIGHER PLANTS

Barbara Demmig-Adams[1], Klaus Winter, Almuth Krüger, Franz-Christian Czygan

Institut für Botanik und Pharmazeutische Biologie der Universität Würzburg, Mittlerer Dallenbergweg 64, 8700 Würzburg, FRG

INTRODUCTION

Light is the driving force for photosynthesis. However, when more light is absorbed by a leaf than can be utilized in photosynthesis, excitation energy can accumulate and ultimately destroy the photosynthetic apparatus. Various mechanisms such as leaf movements or increased reflection have been recognized which serve to avoid the absorption of an excess of light. Plants also possess a mechanism at the molecular level which can remove an excess of absorbed light by dissipating excess excitation energy in a harmless way directly within the photochemical system, as will be discussed in this paper.

RELATIONSHIP BETWEEN ENERGY DISSIPATION IN THE ANTENNA CHLOROPHYLL AND THE LEVEL OF THE CAROTENOID ZEAXANTHIN

When leaves were exposed to an excess of light, during a transition from low to high light or from normal air to CO_2-free air with 2% O_2 at a constant low PFD, a pronounced increase in the rate of radiationless energy dissipation in the antenna chlorophyll (K_D) was observed (Fig. 1). K_D was calculated from the quenching of chlorophyll *a* fluorescence (see Kitajima and Butler, 1975) determined in the light (Schreiber *et al.*, 1986). This fluorescence analysis suggests that an alternative quencher of fluorescence is formed in the antenna chlorophyll which competes with the reaction centers for excitation energy when light becomes excessive. A similar conclusion was also reached in a study using low temperature (77K) fluorescence which characterized a sustained type of fluorescence quenching (Demmig and Björkman, 1987).

[1]Permanent address: Department of Environmental, Population, and Organismic Biology, University of Colorado, Boulder, CO 80309-0334, USA

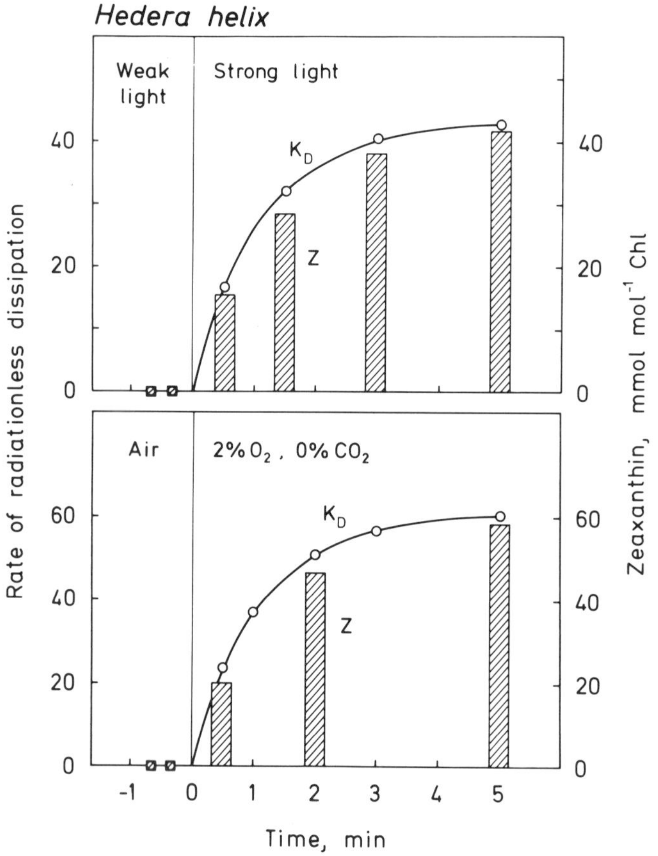

Figure 1. Effect of an excess of light on the rate of energy dissipation in the antenna chlorophyll (K_D) and the zeaxanthin content in shade leaves of *Hedera helix*. Leaves were either kept in normal air and the PFD was increased from 50 to 500 μmol photons m^{-2} s^{-1} (upper part) or were kept at a constant PFD of 100 μmol m^{-2} s^{-1} throughout the experiment and the composition of the gas stream was changed from normal air to CO_2-free N_2 with 2% O_2 (lower part). K_D was calculated from F_M determined in saturation pulses in the light after $F_M = K_F/(K_F + K_D)$ (Kitajima and Butler, 1975) with $K_F = 1$.

In the context of photoprotection a class of substances which comes to mind are the carotenoids, since chlorophyll in carotenoid-deficient mutants becomes bleached at very low light levels under the oxygen partial pressures of normal air (Krinsky, 1971). Pigment analyses performed by us revealed that the increase in K_D induced by either condition of excessive light was accompanied by an increase in the level of a particular carotenoid, the xanthophyll zeaxanthin (Fig. 1). It has long been known that light-dependent conversions of three xanthophylls take place in the thylakoid membrane in the so-called xanthophyll cycle (Hager, 1980; Yamamoto, 1979). The role of these reactions, however, had remained unknown.

Zeaxanthin would be an ideal candidate as an alternative consumer of excitation energy, since it is formed exclusively when light is excessive (Fig. 2) (Hager, 1980; Yamamoto, 1979). It is crucial that the protective substance should not be present and active under limiting light as it competes for excitation energy and therefore decreases the efficiency of

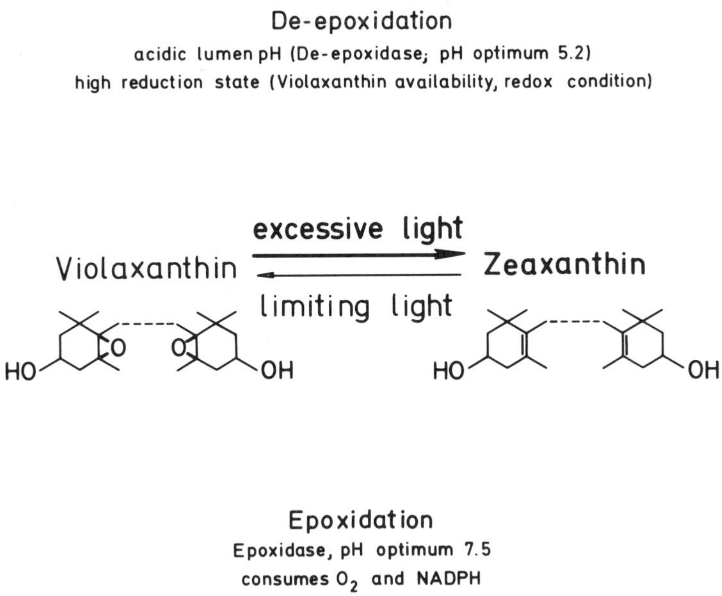

Figure 2. Schematic summary of the control mechanisms involved in the regulation of the equilibrium between zeaxanthin and violaxanthin in the xanthophyll cycle (compare Hager, 1980; Yamamoto, 1979) as well as of the difference in the structure of the di-epoxide violaxanthin and the de-epoxidized form zeaxanthin.

photosynthetic energy conversion. The de-epoxidase which catalyzes the formation of zeaxanthin is active only when the proton concentration on the inner side of the thylakoid membrane is high (Hager, 1980). Only when more photons are absorbed, and thus more protons accumulated, than are used in photosynthesis is the proton concentration high enough to allow zeaxanthin accumulation. In addition, the availability of the precursor violaxanthin is controlled by the redox state of "some" electron carrier of the electron transport chain (Yamamoto, 1979), which is also a measure of the balance between the absorption and consumption of photons in photosynthesis. Upon return to conditions of limiting light, the reconversion of zeaxanthin to violaxanthin by the O_2- and NADPH-dependent epoxidase (pH optimum 7.5) is favored.

RELATIONSHIP BETWEEN THE LIMIT FOR ZEAXANTHIN FORMATION AND PHOTOINHIBITORY INACTIVATION OF PHOTOCHEMISTRY

Evidence for photoinhibitory inactivation of the PS II reaction center has only been observed once the capacity for zeaxanthin formation has been exceeded. When a leaf kept in CO_2-free air (with 2% O_2) was exposed to increasing levels of PFD, and therefore increasing levels of excessive light, the zeaxanthin content increased and the violaxanthin content decreased in a similar fashion (Fig. 3). As long as the zeaxanthin level was rising, F_O fluorescence decreased. A decrease in F_O indicates increased radiationless energy dissipation in the antenna chlorophyll. As soon as the pool of the precursor violaxanthin was exhausted, F_O fluorescence began to rise. This increase in F_O fluorescence indicates an inactivation of the reaction centers of PS II (compare Demmig and Björkman, 1987; Kitajima and Butler, 1975) which occurs at the very point where the capacity for further zeaxanthin formation has been exceeded. This is also the point at which photochemical efficiency of PS II, F_V/F_M, showed a sustained decrease.

CHARACTERISTICS OF THE SUSTAINED DECREASE IN PHOTOCHEMICAL ACTIVITY INDUCED BY COMBINED LONG-TERM WATER AND LIGHT STRESS IN *Nerium oleander*

In nature, extreme levels of excessive light do not normally occur within such a short period of time as that depicted in the previous experiment but rather develop gradually. When a plant like *Nerium oleander* is exposed to water stress, photosynthetic activity decreases. With decreasing rates of photosynthesis, the existing light level, even when maintained constant, becomes increasingly excessive. As shown in Figure 4, both the rate of energy dissipation and the zeaxanthin content rose in *N. oleander* leaves as light became increasingly excessive over days following the cessation of watering. At the same time, a sustained decrease in photochemical efficiency of PS II was observed which could be largely accounted for by an increase in the rate of energy dissipation in the antenna chlorophyll. The dissipation process which was induced in

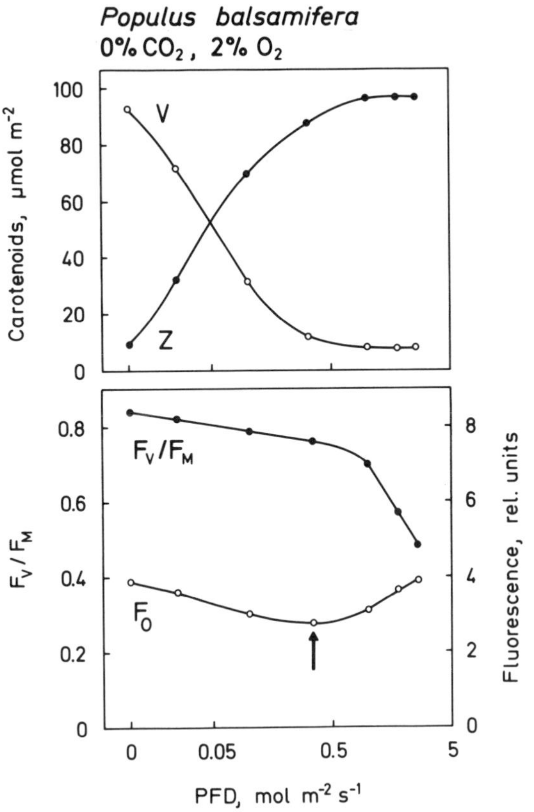

Figure 3. Effect of increasing PFD on the levels of violaxanthin (V) and zeaxanthin (Z) (upper part), and on F_O fluorescence and PSII photochemical efficiency, F_V/F_M, (lower part) in leaves of *Populus balsamifera* kept in CO_2-free air with 2% O_2. Fluorescence characteristics and pigment levels were determined in leaf segments which were predarkened for 5 min subsequent to removal from the leaves after 30 min at each PFD. (Data from Demmig *et al.*, 1987.)

N. oleander by a combination of long-term water stress and high light has identical characteristics to that of the dissipation process which can be induced in short-term treatments with an excess of light, for example, in soybean leaves (Fig. 5). However, these two responses exhibit strikingly different kinetics of recovery. Fluorescence quenching which was induced within a period of minutes also relaxed within minutes (Demmig-Adams *et al.*, 1989b). However the rate of K_D which increased during the combined high light and water stress treatment in *N. oleander* over a period of 12 days, exhibited almost no decrease overnight during a 12 h dark period (Demmig *et al.*, 1988). Since the correlation with zeaxanthin exists in both cases, we have proposed that even the rather sustained decrease in photochemical efficiency in *N. oleander* during water stress is a reflection of the same energy dissipation process (Demmig *et al.*, 1988), whereas previously such sustained decreases in photochemical efficiency had been considered to reflect damage to components of the photosynthetic membranes (see Powles, 1984).

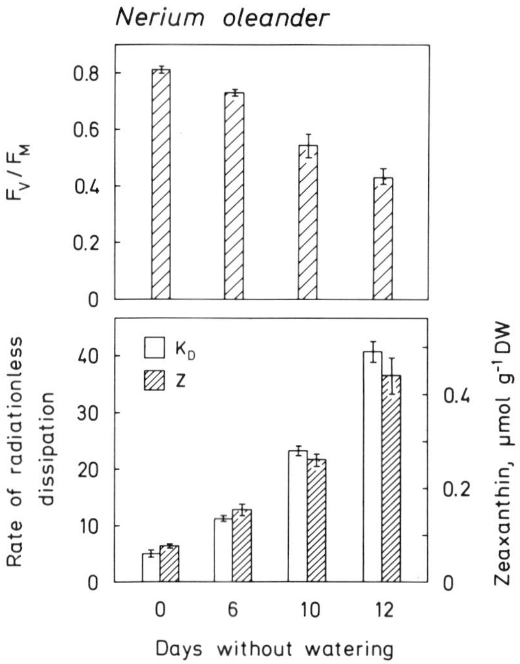

Figure 4. Effect of a combination of high light (12 h photoperiod, 1200 μmol m^{-2} s^{-1}) and water stress over 12 d on PSII photochemical efficiency, F_V/F_M (upper part), and on the rate of energy dissipation in the antenna chlorophyll (K_D) and the zeaxanthin content (lower part) of leaves of *Nerium oleander* determined at the end of the dark period. (Data from Demmig *et al.*, 1988.)

LIGHT RESPONSE OF CO_2 ASSIMILATION, ENERGY DISSIPATION, AND ZEAXANTHIN CONTENT IN PHOTOSYNTHESIZING LEAVES

In order to investigate further the properties of the dissipation process that is induced in short-term treatments with increasing levels of excessive light, the light response of CO_2 assimilation, the excess of light (I/A), radiationless energy dissipation (K_D), and zeaxanthin content (Z) were studied (Fig. 6). The quenching of fluorescence at closed PS II reaction centers in the light was used to calculate K_D. As long as photosynthesis was light-limited, the zeaxanthin content was zero. When CO_2 assimilation ceased to increase, K_D and the zeaxanthin content began to rise in a quantitatively similar fashion. The quenching of F_O fluorescence determined at each light level was exactly to the extent which would be expected from an increase in K_D, except for a small offset which developed between 0 and 400 μmol photons m^{-2} s^{-1} (Demmig-Adams *et al.*, 1989a). The quenching of F_M fluorescence observed at these low light levels was not accompanied by a decrease but by an increase in F_O and occurred while the zeaxanthin content was still zero. Part of this type of fluorescence quenching, but possibly not all, is related to a state shift favoring excitation of PS I (Demmig-Adams B, WW Adams III, U Schreiber, and K Winter, unpublished data). The

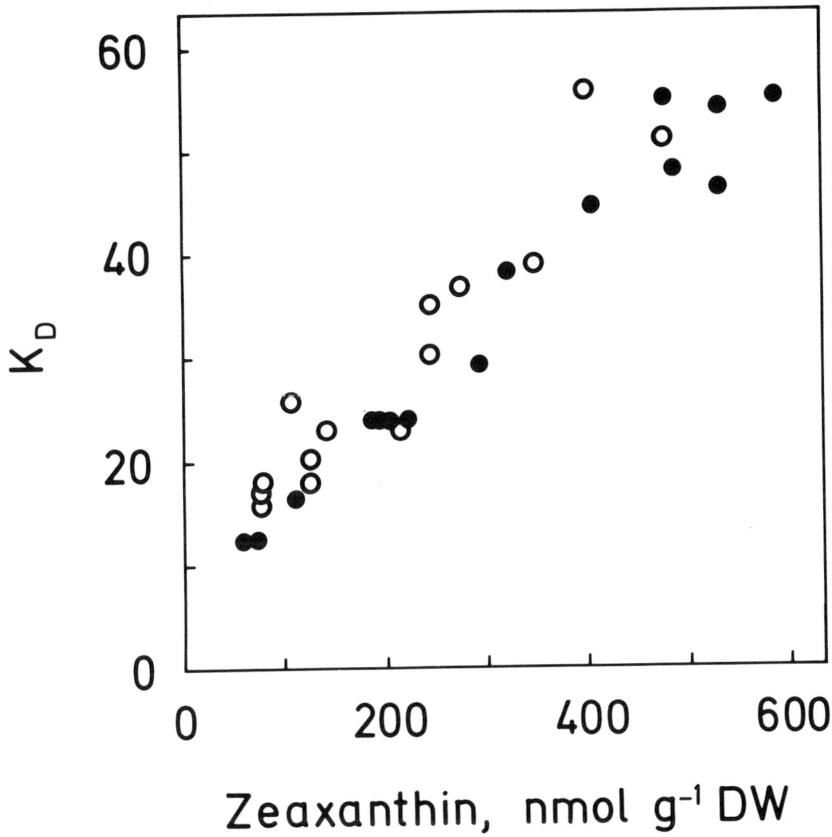

Figure 5. Relationship between the zeaxanthin content and the rate of energy dissipation in the antenna chlorophyll (K_D) determined at the end of the dark period in leaves of *Nerium oleander* during 12 d of high light (12 h photoperiod, 1200 μmol m^{-2} s^{-1}) and water stress (open circles; data from Demmig *et al.*, 1988), and the same parameters determined in the light in leaves of *Glycine max* exposed to a constant PFD of 200 μmol m^{-2} s^{-1} in 2% O_2, 0% CO_2 for different periods of time (0-30 min) (closed circles; data from Demmig-Adams *et al.*, 1989b).

predominant energy dissipation process over the range of PFDs considered here is the increase in K_D which occurs at light saturation of CO_2 assimilation and which is quantitatively related to the zeaxanthin content. Thus under these conditions, antenna quenching is responsible for the majority of the (rapidly relaxing) fluorescence quenching observed in the light which is currently described as "high energy state quenching" (Schreiber *et al.*, 1986). At even higher light levels, when the zeaxanthin content can no longer increase, an additional quench develops which does

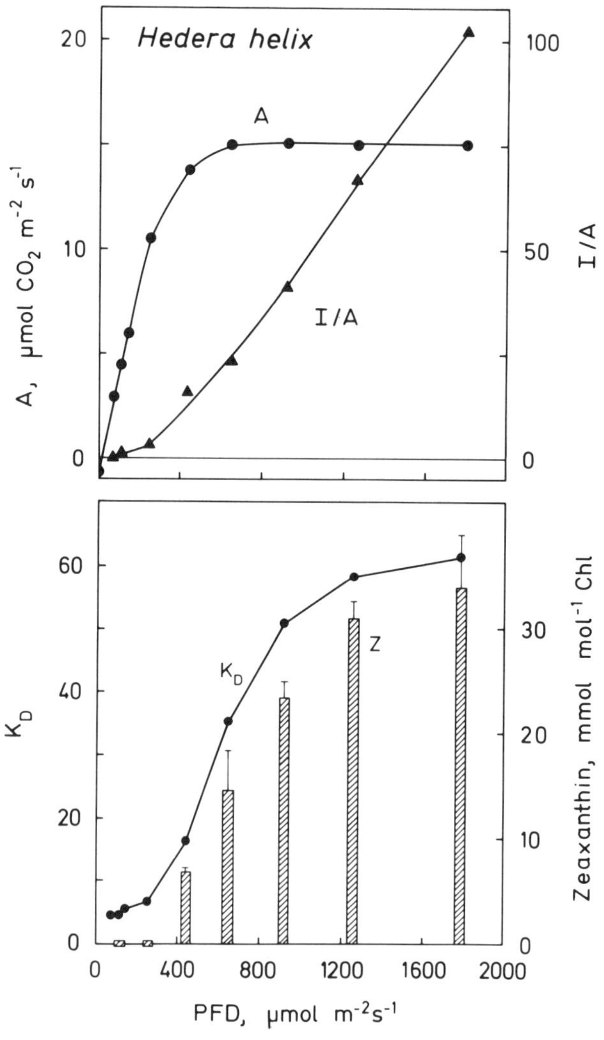

Figure 6. Light response of net CO_2 assimilation rate and the ratio PFD/assimilation rate (upper part), and the rate of energy dissipation in the antenna chlorophyll (K_D) determined from fluorescence measured in saturation pulses in the light, and the zeaxanthin content (lower part) in shade leaves of *Hedera helix*. The value of 12.5 from nonphotoinhibited control leaves was subtracted from all values of K_D. (Data from Demmig-Adams *et al.*, 1989a.)

not have the properties of an increase in K_D and which also shows a relationship with the photon yield of O_2 evolution different from that of the dissipation process in the antenna chlorophyll (Demmig-Adams B, WW Adams III, U Schreiber, and K Winter, unpublished data).

RELATIONSHIP BETWEEN ZEAXANTHIN AND A RAPID TYPE OF FLUORESCENCE QUENCHING

Fluorescence quenching induced by short-term exposure of leaves to light levels at which CO_2 assimilation was light saturated (compare Fig. 6) was rather rapidly reversible upon darkening of the leaves (Demmig and Winter, 1988), and could also be induced quite rapidly, such that it was questionable whether the kinetics of fluorescence quenching described above actually matched the kinetics of the enzymatic conversion of violaxanthin to zeaxanthin in the xanthophyll cycle. Figure 7 shows the induction and relaxation kinetics of fluorescence quenching in greater detail. In a leaf which, prior to the treatment, did not contain any

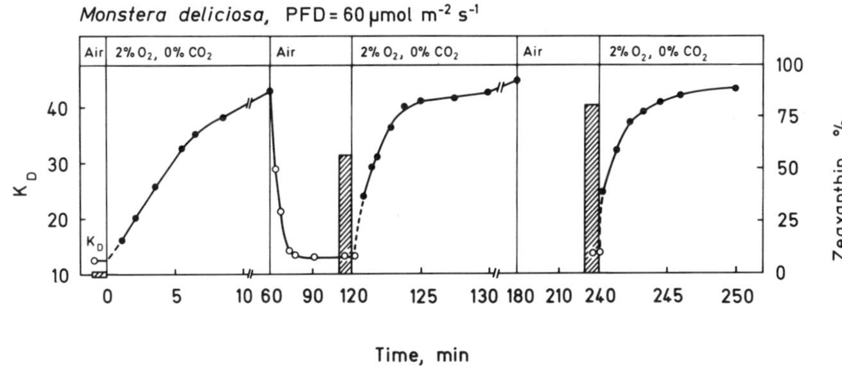

Figure 7. Time course of changes in the rate of energy dissipation in the antenna chlorophyll (K_D) determined from fluorescence measured in saturation pulses in the light in a leaf of *Monstera deliciosa*, exposed to a constant PFD of 60 μmol m^{-2} s^{-1} and subjected to successive treatments with 2% O_2, 0% CO_2. Also shown are zeaxanthin contents at the end of each 60 min period in air at 60 μmol m^{-2} s^{-1} which are given in % of the steady state content of zeaxanthin in 2% O_2, 0% CO_2 at 60 μmol m^{-2} s^{-1} determined in a comparable leaf. (Data from Demmig-Adams *et al.*, 1989b.)

zeaxanthin, the rate K_D increased over a period of 5-10 min upon exposure to conditions of excessive light which is consistent with the kinetics of the enzymatic formation of zeaxanthin in the xanthophyll cycle. Upon return to control conditions where light was nonexcessive, fluorescence quenching was fully reversible within 60 minutes but the zeaxanthin content did not return to zero. When light was made excessive for a second time, fluorescence quenching set in much more rapidly, and in the third cycle reached a considerable extent within a few seconds which is clearly too short a period of time for the enzymatic conversion of violaxanthin to zeaxanthin. However, since the leaf had retained a very high background level of zeaxanthin, it seems possible that even this extremely rapid type of fluorescence quenching is related to the presence of zeaxanthin. Furthermore, fluorescence quenching and zeaxanthin formation were found to be greatly retarded and to be closely matched in a leaf which did not initially contain any zeaxanthin during an exposure to high light at chilling temperatures, whereas in a leaf which contained zeaxanthin some fluorescence quenching developed more rapidly and without changes in the bulk content of zeaxanthin during an identical treatment (Demmig-Adams *et al.*, 1989c). Thus, this particular dissipation process has never been observed by us in the absence of zeaxanthin. The rapid development of fluorescence quenching in leaves which contain zeaxanthin would be expected if the action of zeaxanthin was controlled not only biochemically, by the enzymatic formation of zeaxanthin, but also by some additional, probably biophysical mechanism. Under excessive light the distance between chlorophyll and zeaxanthin molecules in the antennae, for example, might decrease during the conformational change which is known to accompany energization of the thylakoid membrane (Heber, 1969).

POSSIBLE MECHANISMS FOR THE INTERACTION BETWEEN ZEAXANTHIN AND CHLOROPHYLL

From our work, it is quite clear that zeaxanthin reacts with some excited state of chlorophyll and dissipates its energy. However, the exact nature of the interaction between chlorophyll and zeaxanthin is not known. Excitation of chlorophyll first results in the singlet excited state ($^1Chl^*$) after which the triplet excited state ($^3Chl^*$) may be formed by intersystem crossing. The reaction of this latter state with oxygen, giving rise to toxic singlet oxygen, could be prevented by triplet-triplet energy transfer from $^3Chl^*$ to zeaxanthin (Fig. 8). This process would provide protection to the photochemical system. However, it could not explain the quenching of fluorescence which originates from the singlet excited state of chlorophyll. Therefore zeaxanthin either interacts directly with $^1Chl^*$, or, alternatively, zeaxanthin may dissipate the energy from $^3Chl^*$ while another unidentified quencher of $^1Chl^*$ is formed simultaneously in stoichiometric amounts.

Since we have never observed an increase in K_D in the absence of zeaxanthin, it is likely that zeaxanthin itself can act as a fluorescence quencher, i.e. quencher of $^1Chl^*$. In addition, only the quenching of $^1Chl^*$, the species used for photochemistry, would require strict control of the kind provided by the xanthophyll cycle. Quenching of fluorescence in flash excitation experiments has been attributed to carotenoids in the triplet excited state (Mathis *et al.*, 1979) (compare legend of Figure 8). Under the conditions we used, however, the concentration of triplet excited states of carotenoid appear to be too low for this kind of

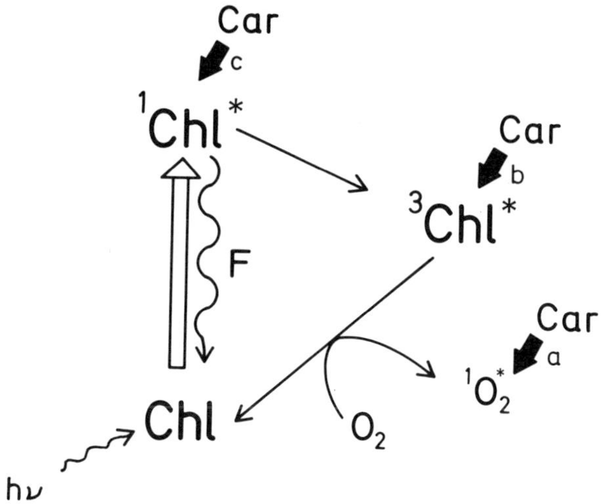

Figure 8. Schematic diagram of possible sites of general carotenoid action in photoprotection.
(a) Quenching of toxic singlet oxygen
(b) Triplet-triplet energy transfer:
$^3Chl^* + Car \rightarrow Chl + {}^3Car^* \rightarrow Chl + Car + heat$
(c) direct interaction with singlet excited chlorophyll of triplet excited carotenoids:
$^1Chl^* + {}^3Car^* \rightarrow Chl + Car + heat$
or carotenoid in the ground state (via rapidly reversible charge transfer):
$^1Chl^* + Car \rightarrow (Chl^- \, Car^+) \rightarrow Chl + Car$

interaction (P Mathis, pers. com.). A possible mechanism for a direct interaction between singlet excited chlorophyll and zeaxanthin is a rapidly reversible charge transfer. Such reactions involving carotenoid radicals occur readily *in vitro*, and have been reported for chlorophyll/carotenoid mixtures (Lafferty *et al.*, 1978).

ZEAXANTHIN AND PHOTOPROTECTION AT CHILLING TEMPERATURES

Since chilling temperatures greatly retard zeaxanthin formation and fluorescence quenching (Demmig-Adams *et al.*, 1989c), we investigated the effect of light stress at chilling temperatures on photochemistry in zeaxanthin-free leaves and also in leaves with a high zeaxanthin content. Prior to the chilling treatment these latter leaves were kept in CO_2-free air with 2% O_2 in low light, a treatment which led to the accumulation of zeaxanthin (Fig. 9). In the zeaxanthin-free leaves, PS II photochemical activity, F_V/F_M, and F_M showed a decrease during the chilling treatment

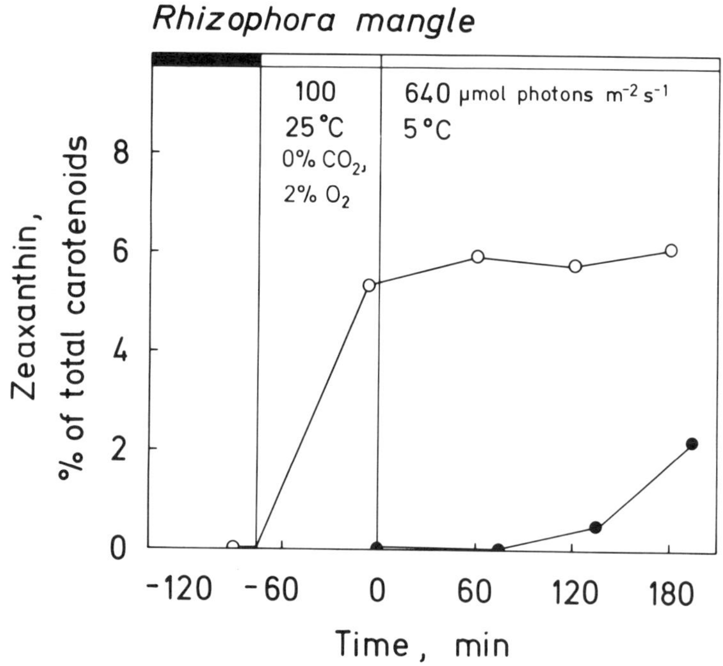

Figure 9. Time course of changes in the zeaxanthin content in leaves of the mangrove *Rhizophora mangle* which were predarkened (closed symbols) or preilluminated with 100 μmol photons m^{-2} s^{-1} at 25°C in 2% O_2, 0% CO_2 (open symbols) and then exposed to a PFD of 640 μmol m^{-2} s^{-1} in normal air at leaf temperatures of 5°C for 180 min. Samples were kept in the dark at room temperature for 5 min before fluorescence measurements were taken or samples for pigment analysis were frozen. (Data from Demmig-Adams *et al*., 1989c)

at a high PFD which did not recover (Fig. 10). The decrease in F_V/F_M and F_M in the pretreated leaf containing zeaxanthin, however, was rather rapidly and fully reversible. Thus, the leaves containing zeaxanthin were protected against this photoinhibitory inactivation of photochemistry by a chilling treatment at high PFD.

CHANGES IN CAROTENOID COMPOSITION IN LEAVES OF *Arbutus unedo* GROWING IN PORTUGAL

Leaves of the sclerophyllous shrub *Arbutus unedo* show a "midday depression" of water loss and CO_2 uptake under the natural conditions of the Portuguese summer, characterized by high light, high temperatures,

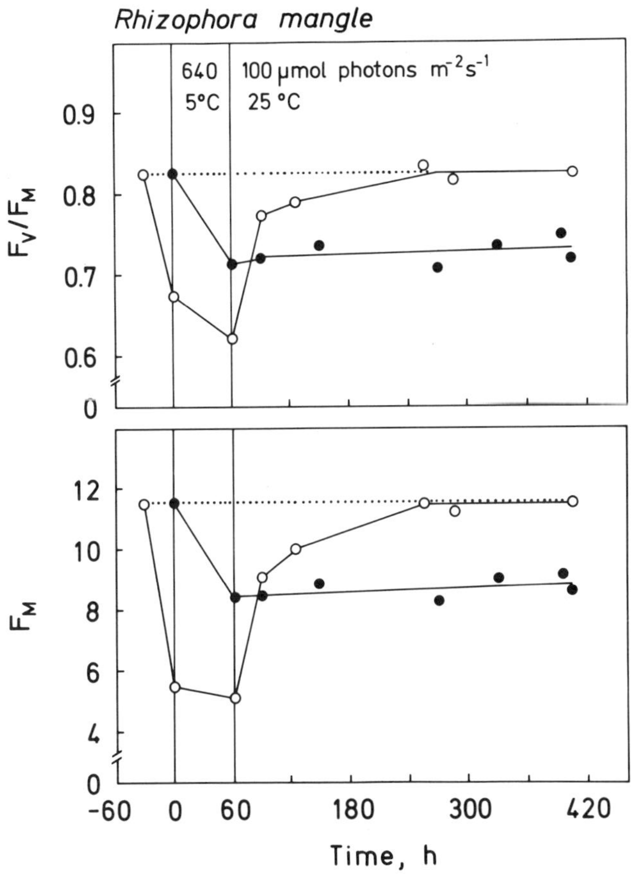

Figure 10. Time course of changes in PSII photochemical efficiency, F_V/F_M, (upper part) and maximum fluorescence, F_M, (lower part) in a predarkened leaf without zeaxanthin (closed symbols) and a preilluminated leaf containing zeaxanthin (open symbols) of *Rhizophora mangle*, both exposed to a PFD of 640 μmol m^{-2} s^{-1} in normal air at leaf temperatures of 5°C for 60 min. Preillumination was as described in the legend of Figure 9. Subsequent to the high light treatment both leaves were returned to 100 μmol photons m^{-2} s^{-1} and 25°C leaf temperature. (Data from Demmig-Adams *et al.*, 1989c.)

and aridity (Tenhunen *et al.*, 1980). We were able to demonstrate that this "midday depression" of gas exchange was paralleled by pronounced decreases in photochemical activity (Demmig-Adams *et al.*, 1989d). Whenever leaves were exposed to direct full sunlight, and leaf temperatures and irradiance incident on the leaves showed peak values, we observed a marked and reversible overall fluorescence quenching which was paralleled by a strong and largely reversible increase in the zeaxanthin level (Table I). For *A. unedo* growing in the field, energy dissipation in the antenna chlorophyll is thus greatly increased when leaves receive full sunlight. The increased activity of this process, however, was not the only effect on the photochemical system. In

Table I. Changes in PS II photochemical efficiency at room (approx. 298K) and low (77K) temperature and in the levels of zeaxanthin (Z), antheraxanthin (A), and violaxanthin (V), the sum of these three xanthophylls, and β-carotene (β-car) in sun and shade leaves of *Arbutus unedo* sampled prior to sunrise and during peak exposure to direct sunlight on 17 August 1987 in Portugal. Shade leaves were facing east-southeast and received 1-2 h of direct sunlight during midmorning. Sun leaves were facing southeast (SE), south (S), or southwest (SW). Leaf samples were kept in the dark at room temperature for 2 min before measurements were taken or samples were frozen. (Data from Demmig-Adams *et al.*, 1989d.)

Exposure and orientation	Photochemical efficiency of PS II		Pigment content, mmol carotenoid mol^{-1} Chl			
Time of day	F_V/F_M,298K	F_V/F_M,77K	Z	V	Z+A+V	β-car
Shade						
07.00	0.82	0.83	27	48	84	178
11.00	0.59	0.77	79	6	95	163
Sun, SE						
07.05	0.75	0.79	29	32	67	163
11.35	0.47	0.68	125	10	162	99
Sun, S						
07.12	0.75	0.80	36	37	88	173
13.05	0.46	0.67	130	5	151	137
Sun, SW						
07.15	0.76	0.78	42	50	101	173
16.15	0.41	0.62	122	1	144	130

addition to the overall quenching of room temperature and low temperature fluorescence indicative of an increase in K_D, there was, secondly, evidence for a state shift favoring excitation of photosystem I. A third effect was a specific further decrease of room temperature fluorescence (Table I) and of O_2 evolution with no effect on low temperature fluorescence, which would be expected from an inactivation of the water-splitting enzyme (Demmig-Adams *et al.*, 1989d; compare also Schreiber and Neubauer, 1987; Neubauer and Schreiber, 1987). The combination of the above modifications of the photochemical system has the potential to impose a true limitation for photosynthesis.

A. unedo showed another interesting feature regarding its carotenoid metabolism. The increase in zeaxanthin during peak irradiance was accompanied not only by a decrease in violaxanthin but also by a decrease in β-carotene (Table I). A reversible conversion of β-carotene to zeaxanthin has also been observed in some other species, both in strong light and at a constant low light level in CO_2-free air with 2% O_2 (Demmig *et al.*, 1987). This reversible conversion of β-carotene to zeaxanthin is therefore not induced by strong light *per se*, but rather by the imbalance between the absorption and consumption of photons in photosynthesis.

CONCLUSIONS

The energy dissipation process in the antenna chlorophyll which is induced by an excess of light is likely to be mediated by the xanthophyll zeaxanthin. This dissipation process sets in extremely rapidly in leaves which contain a background pool of zeaxanthin, suggesting a biophysical level of control. Energy dissipation in the antennae is responsible for fluorescence quenching under a large variety of conditions.

The interaction between zeaxanthin and chlorophyll represents a mechanism on the molecular level which protects the photosynthetic apparatus against destruction by excessive light. The mechanism is probably important for survival and productivity of plants in sunny and, for example, arid habitats.

ACKNOWLEDGEMENTS

This work was supported by grants from the Deutsche Forschungsgemeinschaft. I would like to thank my husband William W. Adams III for reading the manuscript and for many helpful discussions.

REFERENCES

Demmig B, Björkman O (1987). Comparison of the effect of excessive light on chlorophyll fluorescence (77K) and photon yield of O_2 evolution in leaves of higher plants. Planta 171:171-184.

Demmig B, Winter K, Krüger A, Czygan F-C (1987). Photoinhibition and zeaxanthin formation in intact leaves. A possible role of the xanthophyll cycle in the dissipation of excess light energy. Plant Physiol 84:218-224.

Demmig B, Winter K (1988). Characterisation of three components of non-photochemical fluorescence quenching and their response to photoinhibition. Aust J Plant Physiol 15:163-177.

Demmig B, Winter K, Krüger A, Czygan F-C (1988). Zeaxanthin and the heat dissipation of excess light energy in *Nerium oleander* exposed to a combination of high light and water stress. Plant Physiol 87:17-24.

Demmig-Adams B, K Winter, A Krüger, F-C Czygan (1989a). Light response of CO_2 assimilation, dissipation of excess excitation energy, and zeaxanthin content of sun and shade leaves. Plant Physiol, in press.

Demmig-Adams B, K Winter, A Krüger, F-C Czygan (1989b). Zeaxanthin and the induction and relaxation kinetics of the dissipation of excess excitation energy in leaves in 2% O_2, 0% CO_2. Plant Physiol, in press.

Demmig-Adams B, K Winter, A Krüger, F-C Czygan (1989c). Zeaxanthin synthesis, energy dissipation, and photoprotection of photosystem II at chilling temperatures. Plant Physiol, in press.

Demmig-Adams B, WW Adams III, K Winter, A Meyer, U Schreiber, JS Pereira, A Krüger, F-C Czygan, OL Lange (1989d). Photochemical efficiency of PSII, photon yield of O_2 evolution, photosynthetic capacity, and carotenoid composition during the "midday depression" of net CO_2 uptake in *Arbutus unedo* growing in Portugal. Planta, in press.

Hager A (1980). The reversible, light-induced conversions of xanthophylls in the chloroplast. In FC Czygan (ed): "Pigments in Plants," Fischer, Stuttgart, pp 57-79.

Heber U (1969). Conformational changes of chloroplasts induced by illumination of leaves *in vivo*. Biochim Biophys Acta 180:302-319.

Kitajima M, Butler WL (1975). Quenching of chlorophyll fluorescence and primary photochemistry in chloroplasts by dibromothymoquinone. Biochim Biophys Acta 376:105-115.

Krinsky NI (1971). Functions. In O Isler, H Gutmann, U Solms, (eds): "Carotenoids," Birkhäuser, Basel, pp 669-716.

Lafferty J, Truscott TG, Land EJ (1978). Electron transfer reactions involving chlorophyll *a* and *b* and carotenoids. J Chem Soc Faraday 75:2869-2872.

Mathis P, Butler WL, Satoh K (1979). Carotenoid triplet state and chlorophyll fluorescence quenching in chloroplasts and subchloroplast particles. Photochem Photobiol 30:603-614.

Neubauer C, Schreiber U (1987). The polyphasic rise of chlorophyll fluorescence upon onset of strong continuous illumination: I. Saturation characteristics and partial control by the photosystem II acceptor side. Z Naturforsch 42c:1246-1254.

Powles SB (1984). Photoinhibition of photosynthesis induced by visible light. Annu Rev Plant Physiol 35:15-44.

Schreiber U, Schliwa U, Bilger W (1986). Continuous recording of photochemical and non-photochemical chlorophyll fluorescence quenching with a new type of modulation fluorometer. Photosyn Res 10:51-62.

Schreiber C, Neubauer U (1987). The polyphasic rise of chlorophyll fluorescence upon onset of strong continuous illumination: II. Partial control by the photosystem donor side and possible ways of interpretation. Z Naturforsch 42c:1255-1264.

Tenhunen JD, Lange OL, Braun M, Meyer A, Lösch R, Pereira JS (1980). Midday stomatal closure in *Arbutus unedo* leaves in a natural macchia and under simulated habitat conditions in an environmental chamber. Oecologia 47:365-367.

Yamamoto HY (1979). Biochemistry of the violaxanthin cycle in higher plants. Pure Appl Chem 51:639-648.

Photosynthesis, pages 393–406

INTERACTIONS BETWEEN ELECTRON TRANSPORT AND CARBON ASSIMILATION: REGULATION OF LIGHT-HARVESTING AND PHOTOCHEMISTRY

Peter Horton

Department of Biochemistry and Research Institute for Photosynthesis, University of Sheffield, Western Bank, Sheffield, S10 2TN, UK

INTRODUCTION

Concepts of control of phosphorylating electron transfer chains are based upon feedback effects of the high energy state on the rate of electron transfer. The experimental foundation of this concept is the restriction of electron transfer when ATP/ADP ratios are high and the observed stimulation upon supplying ADP or in the presence of uncouplers. This was first observed in respiratory systems and the existence of the phenomenon is rationalised in terms of metabolic regulation, i.e. respiratory substrates are conserved when the ATP supply is adequate. However, a thermodynamic control must exist at some point since there is an energetic limit on the ability of the electron transfer system to pump protons against an increasing proton motive force. In mitochondria the proton motive force and the phosphorylation potential can be close to equilibrium. In photosynthetic systems it is also possible to demonstrate a similar photosynthetic control; isolated thylakoids readily show uncoupler- and ADP-stimulated electron transport. A quantitative approach indicated a sensitivity coefficient of almost 1 for the oxidation of plastoquinone by the cytochrome *b/f* complex in coupled thylakoids under high irradiance (Heber *et al.*, 1988). However, observations of intact chloroplasts have consistently shown a lack of equilibrium between ΔpH and the phosphorylation potential (Giersch *et al.*, 1980) and recent work has indicated relatively oxidised levels on either side of the PQ oxidation site under high light (Weis *et al.*, 1987). The high rate of ATP production in high light is associated with a large ΔpH, yet electron transport proceeds at rates equivalent to "uncoupling," indicating the absence of photosynthetic control.

The photosynthetic electron transport system *in vivo* is an integral part of the metabolic processes of the plant cell (for review, see Horton,

1985). Moreover, unlike respiratory substrates, light cannot be "stored" and there is actually no *physiological* reason for photosynthetic control involving ΔpH suppression of electron transport in the classical sense (although, of course the thermodynamic restriction mentioned above must still apply). Indeed, it could be argued that build-up of electrons on the acceptor side of PSII is a situation to be avoided in high light since this is most likely a photoinhibitory condition (Kyle *et al.*, 1985).

It is necessary therefore to answer two questions (Figure 1): Is control of electron transport necessary *in vivo* and what mechanisms may be operative? Two reasons may be invoked to suggest that the thylakoid membrane is not a passive interface between light and carbon metabolism. Firstly, in low light, efficient light utilisation needs to be ensured, and secondly, in high light, excess excitation needs to be dissipated.

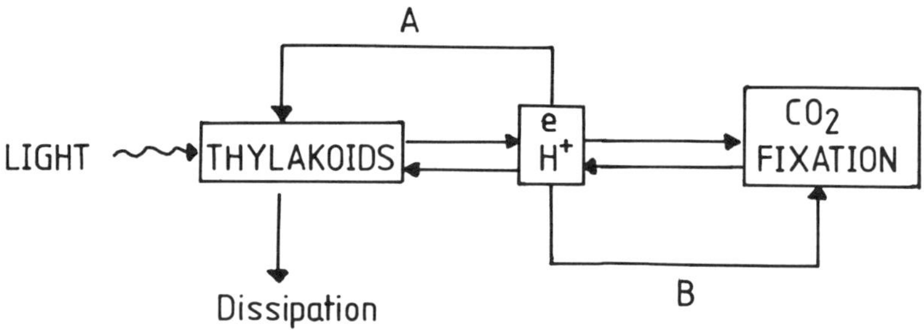

Figure 1. Possibilities for regulation of electron transport in photosynthesis. Passive control of thylakoid function via the turnover of the ATP and NADPH pools is indicated by the reversible horizontal arrows, excess light leaking as dissipation. *Active* control is indicated by A (regulation of light harvesting and electron transfer by the redox state and degree of energisation of the thylakoid membranes to ensure both the correct balance between PSII and PSI excitation and the harmless dissipation of excess) and B (modulation of Ferredoxin-NADP reductase, ATP synthase and enzymes of the reductive pentose phosphate pathway by changes in redox state and stromal pH).

REGULATION IN LOW LIGHT

The attainment of maximum quantum yield in low light requires that PSII and PSI are excited at "appropriate" rates. For linear electron transport, this means equal rates of excitation, but, *in vivo*, the relative rates of excitation of PSII and PSI need to be adjusted to ensure optimum redox poise and the provision of the ratios of ATP:NADPH demanded by metabolism (Horton, 1985). This adjustment has to be a dynamic one because the spectral quality of irradiance is variable with time and the required ATP:NADPH depends upon the particular metabolic pattern. Reversible phosphorylation of a 'mobile' population of LHCII has been identified as the molecular mechanism that controls the adjustment of the relative absorption cross sections of PSII and PSI (Horton, 1983). There is clear evidence that the level of LHCII phosphorylation responds to changes, not only in spectral quality of light (Horton and Black, 1980) but also in the relative demand for ATP:NADPH (Fernyhough *et al.*, 1983; 1984). It was suggested, therefore, that thylakoid protein phosphorylation is the mechanism by which the redox state of the electron transfer system is controlled and the correct ATP:NADPH ratio supplied either via the "Q" cycle or PSI-driven cyclic electron flow (Horton, 1985). The existence of the state transitions (state 1 = maximum transfer to PSII, state 2 = maximum transfer to PSI) *in vivo* are well-documented under special conditions of low intensity, narrow-band irradiance designed to excite PSII or PSI specifically. Recently, using both protoplasts (Horton and Hague, 1988) and leaves (Quick and Stitt, personal communication) it has been possible to demonstrate state transitions during 'normal' levels of irradiance. This advance has come as a result of progress in the analysis of the quenching of chlorophyll fluorescence.

During photosynthesis the yield of chlorophyll fluorescence from PSII is found to be lower than that observed in the dark-adapted state. Several processes contribute to this quenching of fluorescence, each of which is indicative of regulation of the electron transport and light-harvesting systems to balance irradiance and photosynthetic assimilation. Fluorescence quenched photochemically (qQ) is indicative of dissipation of excitation energy through electron transport, mainly (but *not* exclusively) to reduce glycerate 3-phosphate.

Fluorescence lost non-photochemically (qNP) occurs as a result of increases in thermal decay or through energy transfer to non-fluorescent pigments (e.g. PSI). Using saturating light pulses, a technique developed by Bradbury and Baker (1981), Quick and Horton (1984b), and Schreiber *et al.* (1986), qQ and qNP can be distinguished. qNP is due to at least three processes: qE (ΔpH-dependent quenching), qT (lowering of fluorescence due to state transitions), and qI (quenching associated with photoinhibitory treatment). In an attempt to resolve these components without use of inhibitors, the kinetics of relaxation of quenching have been studied (Horton and Hague, 1988). Three phases of relaxation were tentatively assigned to qE, qT and qI respectively.

At present, there is controversy regarding each of these three non-photochemical processes. For example it has been suggested that qE results from ΔpH-induced conversion of PSII reaction centres from an active fluorescent state to an inactive quenched state (Weis and Berry,

1987). Conversely, the quenching of Fo associated with qE (Horton and Hague, 1988), suggested that qE involves alteration in the antenna, a proposal supported by the recent work which indicates that zeaxanthin is required for qE formation (Demmig *et al.*, this volume). Similarly, the evidence for associating the middle (qT) phase with a state-transition is still not conclusive, despite the effects of NaF. Finally, it is unclear as to whether the slow qI component represents reversal of photoinhibition of the PSII reaction centres or the formation of zeaxanthin in the antenna (Demmig *et al.*, 1987).

The relaxation of fluorescence quenching can be measured upon darkening by continuing to illuminate with a low intensity measuring beam, and giving light saturation pulses at appropriate intervals to record Fm. Figure 2 shows relaxation curves for Fm recorded in this way for a period of 25 min following illumination. Two phases of relaxation were observed, whilst experiments on a longer time-scale revealed a third phase with a $t_{1/2}$ = 30 min which is essentially irreversible on the timescale of Figure 2. This "irreversible" phase, qI, was found to comprise approx. 20% quenching. The rapidly relaxing phase ($t_{1/2}$= 30 s) is assigned to qE, whilst the slow phase ($t_{1/2}$= 8 min) has kinetics appropriate for reversal of a protein phosphorylation-dependent state transition (qT). Evidence to support this view came from the observations that NaF, a protein phosphatase inhibitor blocks relaxation and that far-red light stimulates both the rate and amplitude of relaxation and only in the absence of NaF (Figure 2). Furthermore, low intensity 660nm light (80μE $m^{-2}s^{-1}$) generates only the slowly relaxing phase of quenching, an effect which is eliminated by simultaneous illumination with 720nm light (Hague and Horton, unpublished). The amplitude of the qT phase was maximum only at very low light intensity (Horton and Hague, 1988). The presence of qT in low light provides strong support for its role in enhancing photosynthetic efficiency, whilst its decline at higher light is consistent with the control by ΔpH (Fernyhough *et al.*, 1984). Indeed, the level of LHCII phosphorylation in isolated chloroplasts was found to increase when ΔpH was decreased following reduction of light intensity (Table I).

Experimental evidence supports the existence of an antiparallel relationship between qE and qT, with qE acting as the signal that indicates the level of energisation. Recent data shows that ΔpH alone is not sufficient to suppress protein phosphorylation, the development of qE being necessary (Oxborough *et al.*, 1987). This may be highly significant since qE is a process involved in *dissipation* of excitation energy, and demarking a shift in emphasis of thylakoid regulation from utilisation to dissipation.

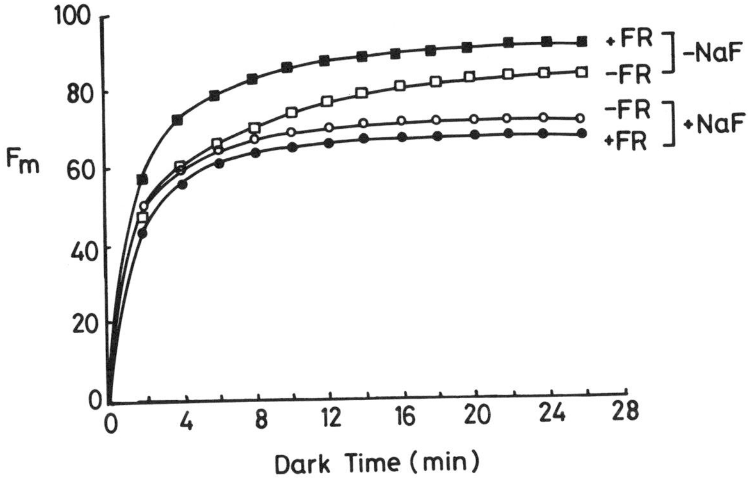

Figure 2. Relaxation of Fm after illumination of isolated protoplasts measured by pulsing every 2 mins. 50 mM NaF (○, ●) was added 30s prior to darkening (○, □) or turning on far-red light (L, ■).

Table I. Phosphorylation of LHCII and ΔpH (measured by 9-amino-acridine fluorescence) in maize mesophyll chloroplasts in the presence of pyruvate following successive changes in light intensity.

Conditions	ΔpH (arb. units)	LHC phosphorylation (cpm)
5 min dark	0	164
+ 7.5 min high light	0.28	1141
+ 7.5 min low light	0.19	3633
+ 7.5 min high light	0.25	2155

The close relationship between qT and photosynthesis is also shown by the data in Figure 3. The re-establishment of the induction period for O_2 evolution strongly correlates with the relaxation kinetics of qT. Similarly, the amount of O_2 evolved during each of the saturating pulses used to determine the relaxation of quenching also decays with the same kinetics as qT. Consistent with this rationale is the fact that NaF prevents the re-establishment of induction and re-illumination in the presence of NaF induces multiple oscillations in the rate of O_2 evolution. Whilst induction and oscillatory behaviour are complex phenomena involving a multitude of processes, these observations point to an important role for the

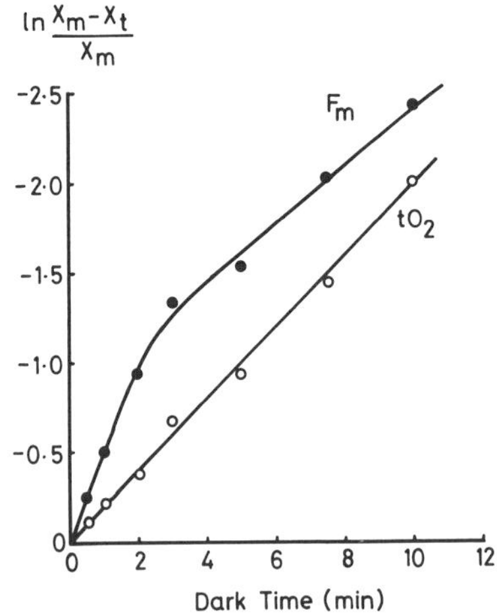

Figure 3. Semilogarithmic plots of the relaxation of Fm (•) and the length of the induction period for O_2 evolution in protoplasts. Xm and Xt refer to the fully dark-adapted value and that observed t min after darkening.

establishment of the ATP/NADPH stoichiometry and its control by protein phosphorylation (Quick and Horton, 1984a; Horton, 1985).

REGULATION IN HIGH LIGHT

At light levels approaching saturation and beyond, another reason arises for regulation of thylakoids. Excess light is generally acknowledged to be deleterious. Not only can pigments become photooxidised but the PSII reaction centre is sensitive to light-induced inactivation (photoinhibition), most likely as a result of damage to the D1 polypeptide upon over-oxidation of the donor side of PSII or over-reduction of the acceptor pool. Also, the requirement of avoiding over-reduction and maintaining optimum redox poise applies as much in high light as in low light. Hence, processes that prevent over-reduction are beneficial and these are of two kinds: a) alternative electron transfer pathways that keep QA oxidised and prevent passive (mass action) redox feedback to PSII due to limitation by carbon assimilation. b) dissipation of excitation energy by non-radiative decay in the antennae pigments or in the reaction centre.

The quantitative analysis of the quenching of chlorophyll fluorescence provides a means of assessing these processes (Weis and Berry, 1987; Horton and Hague, 1988; Oxborough and Horton, 1988; Horton *et al.*,

1988; Krause and Laasch, 1987; Krause *et al.*, 1988). In the first instance the complementary nature of qQ and qNP means that non-radiative dissipation is increased as photochemistry becomes light saturated (Horton, 1987). It is significant that qQ does not decrease as much as expected from the measured quantum yield of electron transport. Thus at high light there is a decline in intrinsic yield of PSII (Φp), the quantum yield of electron transport (Φs) corrected for the fraction of excitation reaching open centres (qQ). The observed correlation between qNP and Φp suggests that energisation converted PSII into a photochemically inactive quenched state (Weis and Berry, 1987). A non-linear relationship was also obtained between Φs and qQ in isolated chloroplasts using methyl viologen as an electron acceptor (Figure 4), exactly as observed in protoplasts and leaves.

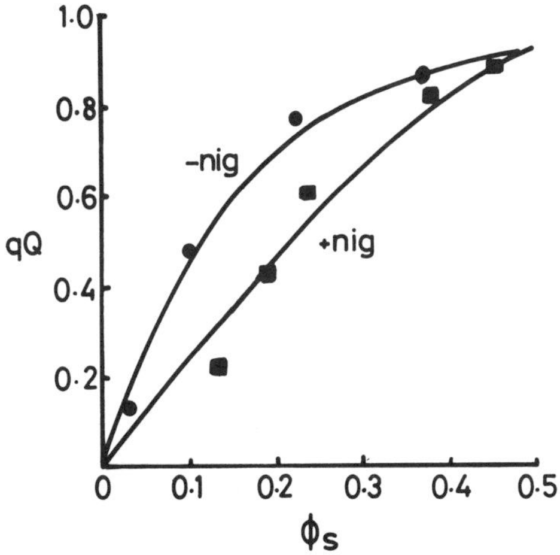

Figure 4. Quantum yield (Φs) and qQ in isolated thylakoids illuminated with light intensities between 50 and 3000 μE $m^{-2}s^{-1}$ in the presence of methyl viologen. Control (•) and in the presence of 1μM nigericin (□).

The value for Φp declined by over 50% in high light compared to low light, again similar to *in vivo* changes. The decline in Φp was still observed in the presence of antimycin A when qE was very neglegible (Figure 5). There was a strong correlation between the increase in ΔpH and the decrease in Φp. Addition of an uncoupler eliminated the lowering of Φp in high light. The conclusion would be that the observed decline in Φp in isolated thylakoids is due to ΔpH and not qE. In vivo, both qE and ΔpH will act to reduce Φp. These feedback effects mean that excitation density in the antenna of PSII will be reduced when irradiance is in excess of the capacity of metabolism, ΔpH acting as the regulatory signal. It should be pointed out that these feed-back effects may reflect aspects of the proton-motive force different from that driving photophosphorylation or controlling the rate of oxidation of

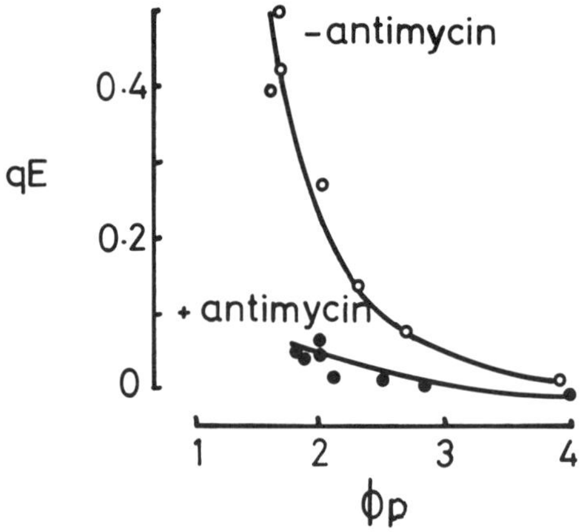

Figure 5. Relationship between Φp and qE in isolated pea chloroplasts illuminated with light intensities between 0 and 1080 μE $m^{-2}s^{-1}$. Control (I) and +3μM anitmycin A (•)

plastoquinol. For example, ΔpH as measured by 9-aminoacridine fluorescence can be eliminated by dibuccaine without significant effect upon qE (Krause *et al*., 1988). Localised proton domains may well be involved in driving photophosphorylation whilst feedback control responds to the bulk-phase ΔpH. As a corollary, it is clear that feedback control of the input is an essential part of the optimisation of photosynthesis; ΔpH has to be high enough to drive photophosphorylation yet not *restrict* electron transport, ATP/ADP and NADPH/NADP ratios need to be high enough to drive carbon assimilation yet not inhibit electron transport or photophosphorylation.

It is suggested that the ΔpH can induce a cycle around PSII and thus cause the observed decrease in Φp. One possibility is that it is the lumen pH, which is increasingly acidic in high light, that is modifying the activity of the donor side of PSII and so allowing a PSII cycle. Experiments with isolated PSII particles are consistent with this proposal (Figure 6). Lowering the pH below 6.5 induced a decrease in both Φs and Φp. Therefore as the environment of PSII became more acidic, Φp is lowered by some dissipative mechanism. Although the Fv/Fm ratio was also found to decline sharply at low pH, the decrease in Fv/Fm did not start to occur until around pH 5.3 and declined very rapidly to negative values below approximately pH 4.1. It is important to stress that the relationship between Fv/Fm and pH was markedly different from the relationship between Φp and pH; Φp fell when the pH was decreased below 6.5 and at pH 5.3 it had fallen by 50%, the Fv/Fm remaining at a value over 0.7.

These data also indicate that qE quenching could also be due to acidification of the thylakoid lumen. A requirement for a larger ΔpH to

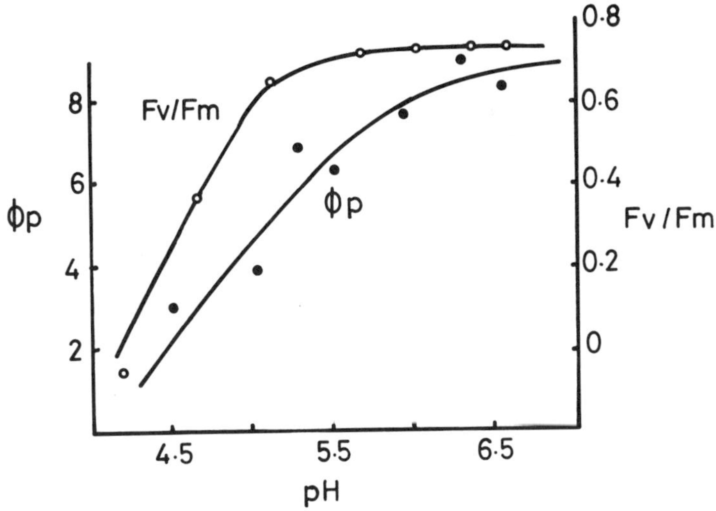

Figure 6. Effect of pH on Φp and Fv/Fm in isolated PSII particles. Φp was measured by O_2 evolution in the presence of 2,6-dichlorobenzoquinone and ferricyanide. Fv/Fm was recorded in the presence of ferricyanide.

cause qE formation than to decrease Φp is predicted from the data presented here and is supported by experiments on thylakoids (Oxborough and Horton, 1988). Estimates of ΔpH in high light are in the range 2.5 - 3.0 with a stromal pH of 7.5 - 8.0, suggesting that the lumen pH is within the range of 4.5 to 5.5. Thus the 50-60% decrease in Φp recorded *in vivo* is entirely consistent with the data in Figure 6. The data also imply that the PSII cycle and qE will act in concert to dissipate excitation both non-photochemically and photochemically in response to light-induced ΔpH formation. The effect of both phenomena would be to prevent either over-reduction or over-oxidation of PSII.

Evidence has been obtained for PSII cycling in high light (Falkowski *et al.*, 1988), upon donor side limitation on adding ADRY reagents (Heber *et al.*, 1979), and upon ATP addition (Schreiber and Reinits, 1986). The data described here are consistent with these reports and suggests that PSII cycling is an important mechanism for energy dissipation in high light. The pathway of this cycle is presently unknown but because qQ as determined by light-pulsing and DCMU-addition are always very close (Quick and Horton, 1984b) the electron must be flowing through the QB site. The change in redox potential of cytochrome *b*-559 between 6 and 5 and its photoxidation at low pH (Horton and Cramer, 1976) suggest that this component is involved in the PSII cycle. It is also of note that because the proposed cycling is measured in terms of the qQ parameter, its capacity would seem to be saturated at intensities about 2-3 times saturating for photosynthesis (when observed qQ values approach zero).

CONCLUSIONS

The reversible decline in the efficiency of light-harvesting and photochemistry are normal events during the diurnal cycle of plants. The thylakoid membrane has three lines of defense when exposed to high light. Photochemical dissipation *via* a PSII cycle, qE, and a sustained quenching in the antenna through zeaxanthin formation. Thylakoid function is under the control of as yet imperfectly understood combinations of redox state and energy state; all three of these processes are in some way related to ΔpH and redox state. Both qE and qI are associated with a decline in qT (Oxborough *et al.*, 1987, Demmig *et al.*, 1987). Protein phosphorylation effects PSII electron transfer directly (Horton and Lee, 1984) and also potentiates a PSII cycle (Horton and Lee, 1983).

Recent work suggests that zeaxanthin formation is necessary for qI *and* qE (Demmig *et al.*, this volume). All of these observations, together with the extreme diversity in the fluorescence changes in different species (Demmig *et al.*, 1987; Horton *et al.*, 1988) point to a complex regulatory system acting at both the PSII reaction centre and the LHCII. Previous papers have discussed PSII (Oxborough and Horton, 1988, Horton *et al.*, 1988). Here, a model is proposed that attempts to relate qE, qT and qI to changes in the LHCII. It is known that zeaxanthin is a component of LHCII (Peter and Thornber, 1988) and that LHC is the protein substrate involved in the phosphorylation-dependent state transition (Black *et al.*, 1984). Eight states of LHCI can be described, depending on protonation, phosphorylation and the presence of zeaxanthin. The model in Figure 7 describes five of these, protonation being restricted to the zeaxanthin-LHC (LHC-Z). The equilibria presented immediately predict the antagonism between qE and qT. Moreover, by suggesting that the state of LHCII affects its ability to act as substrate for the kinase/phosphatase or de-epoxidase/epoxidase enzymes, it is possible to explain other observations. e.g. if LHC is the preferred kinase substrate, then formation of LHC-Z (in sustained high light) would decrease LHC-P content, accounting for the decline in qT as qI increases. Furthermore, the model predicts two forms of phosphorylated LHCII, one (LHC-P) that could transfer excitation to PSI and another (LHC-PZ) that is entirely dissipative; this could provide an explanation for the inconsistent reports regarding the effect of phosphorylation on the rate of excitation of PSI. Though speculative, this model suggests some new experiments and extends the established rationale of the dynamic nature of thylakoid

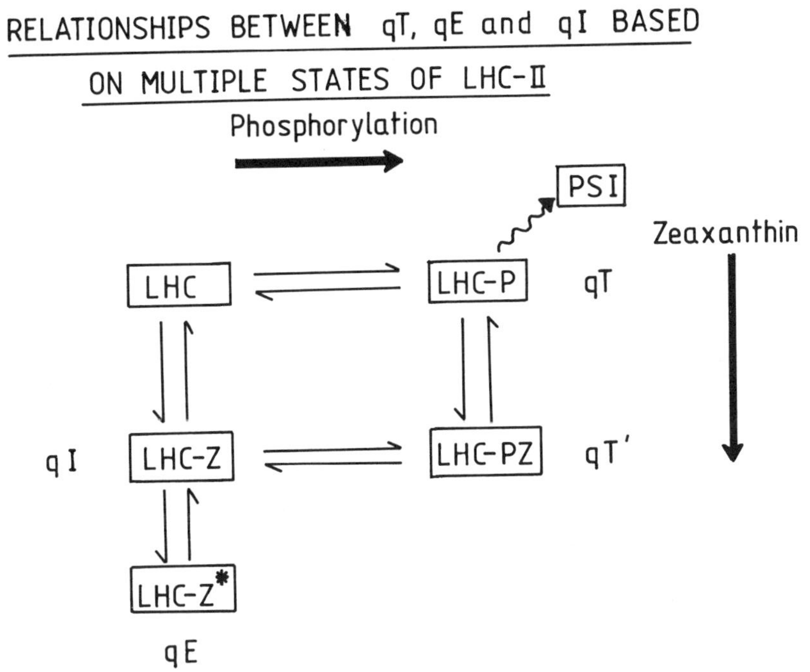

Figure 7. Model to explain the relationships between different processes of non-photochemical quenching based upon multiple forms of LHCII, which can be phosphorylated or dephosphorylated (horizontal transitions) or subject to violoxanthin de-epoxidation or zeaxanthin epoxidation (vertical transition). ΔpH drives the alteration in conformation that results in qE.

organisation that was based on the changes brought about by phosphorylation. It suggests that the fate of excitation energy absorbed by LHCII is dependent on its phosphorylation state, its protonation state and the presence of zeaxanthin.

REFERENCES

Black MT, Foyer CH and Horton P (1984). An investigation into the ATP requirement for phosphorylation of thylakoid proteins and for the ATP-induced decrease in yield of chlorophyll fluorescence in mature and developing chloroplasts. Biochim Biophys Acta 767:557-562.

Bradbury M and Baker NR (1981). Analysis of the phase of the *in vivo* chlorophyll fluorescence induction curve. Biochim Biophys Acta 635:542-551.

Demmig B, Winter K, Kruger A and Czygan F-C (1987). Photoinhibition and zeaxanthin formation in intact leaves. Plant Physiol 84:218-224.

Demmig B, Cleland RE and Björkman O (1987). Photoinhibition, 77K chlorophyll fluorescence quenching and phosphorylation of the light-harvesting chlorophyll-protein complex of photosystem II in soybean leaves. Planta 172:378-385.

Falkowski PG, Kolber Z and Fujita Y (1988). Effect of redox state on the dynamics of photosystem II during steady state photosynthesis in eukaryotic algae. Biochim Biophys Acta 933:432-443.

Fernyhough P, Foyer CH and Horton P (1983). Influence of metabolic state on the level of phosphorylation of the light-harvesting chlorophyll protein complex in chloroplasts isolated from maize mesophyll cells. Biochim Biophys Acta 725:155-161.

Fernyhough P, Foyer CH and Horton P (1984). Increase in level of thylakoid protein phosphorylation in maize mesophyll chloroplasts by decrease in the transthylakoid pH gradient. FEBS Lett 176:133-138.

Giersch C, Heber U, Kobayashi Y, Inoue Y, Shibata K and Heldt H (1980). Energy charge, phosphorylation potential and proton motive force in chloroplasts. Biochim Biophys Acta 590:59-73.

Heber U, Kirk MR and Boardman NK (1979). Photoreactions of cytochrome *b*-559 and cyclic electron flow in photosystem II of intact chloroplasts. Biochim Biophys Acta 546:292-306.

Heber U, Neimanis S and Dietz K-J (1988). Fractional control of photosynthesis by the QB protein, the cytochrome *f/b* complex and other components of the photosynthetic apparatus. Planta 173:267-274.

Horton P (1983). Control of electron transport by the thylakoid protein kinase. FEBS Lett 152:47-52.

Horton P (1985). Interactions between electron transport and carbon assimilation. In Barber J and Baker NR (eds): "Photosynthetic Mechanisms and the Environment," Amsterdam: Elsevier, pp 135-187.

Horton P (1987). Interplay between environmental and metabolic factors in the regulation of electron transport in higher plants. In Biggins J. (ed): "Progress in Photosynthesis Research Vol II," Dordrecht: Martinus-Nijhoff, pp 681-688.

Horton P and Black MT (1980). Activation of adenosine 5' triphosphate-induced quenching of chlorophyll fluorescence by reduced plastoquinone. FEBS Letters 119:141-144.

Horton P and Cramer WA (1976). Stimulation of photosystem I induced oxidation of chloroplast cytochrome *b*-559 by pre-illumination and by low pH. Biochim Biophys Acta 430:122-134.

Horton P and Hague A (1988). Studies on the induction of chlorophyll fluorescence in barley protoplasts IV. Resolution of non-photochemical quenching. Biochim Biophys Acta 932:107-115.

Horton P and Lee P (1983). Stimulation of a cyclic electron transfer pathway around photosystem II by phosphorylation of chloroplast thylakoid proteins. FEBS Lett 164:81-84.

Horton P and Lee P (1984). Phosphorylation of chloroplast thylakoids decreases the maximum capacity of photosystem II electron transfer. Biochim Biophys Acta 767:563-567.

Horton P, Oxborough K, Rees D and Scholes JD (1988). Regulation of the photochemical efficiency of photosystem II; consequences for the light response of field photosynthesis. Plant Physiol and Biochem 26:453-460.

Krause GH and Laasch H (1987) Energy-dependent chlorophyll fluorescence quenching in chloroplasts correlated with quantum yield of photosynthesis. Z Naturforsch 42c:581-584.

Krause GH, Laasch H and Weis E (1988) Regulation of thermal dissipation of absorbed light energy in chloroplasts indicated by energy-dependent fluorescence quenching. Plant Physiol Biochem 26:445-452.

Kyle DJ, Ohad I and Arntzen CJ (1985). Membrane protein damage and repair; selective loss of quinone protein function in chloroplast membranes. Proc Natl Acad Sci USA 81:4070-4074.

Oxborough K and Horton P (1987). Characterisation of the effects of antimycin A upon high energy state quenching of chlorophyll fluorescence (qE) in spinach and pea chloroplasts. Photosyn Res 12:119-128.

Oxborough K and Horton P (1988). A study of the regulation and function of energy-dependent quenching in pea chloroplasts. Biochim Biophys Acta 934:135-143.

Oxborough K, Lee P and Horton P, (1987). Regulation of thylakoid protein phosphorylation by high energy state quenching. FEBS Lett 221:211-214.

Peter GF and Thornber JP (1988) The antennna components of photosystem II with emphasis on the major pigment-protein, LHC IIb. In Scheer and Schneider (eds) "Photosynthetic Light-Harvesting Systems," Berlin New York: Walter de Gruter & Co, pp 175-186.

Quick WP and Horton P (1984a). Studies on the induction of chlorophyll fluorescence in barley protoplasts I. Factors affecting the observation of oscillations in the yield of chlorophyll fluoresence and the rate of oxygen evolution. Proc R Soc Lond B 220:361-370.

Quick WP and Horton P (1984b). Studies on the induction of chlorophyll fluorescence in barley protoplasts II. Resolution of fluorescence quenching by redox state and the transthylakoid pH gradient. Proc R Soc Lond B 220:371-382.

Schreiber U and Rienits KG (1986). ATP-induced photochemical quenching of variable chlorophyll fluorescence. FEBS Lett 211:94-104.

Schreiber U, Schliwa U and Bilger B (1986). Continuous recording of photochemical and non-photochemical chlorophyll fluorescence quenching with a new type of modulation fluorometer. Photosyn Res 10:51-62.

Weis E and Berry J (1987). Quantum efficiency of photosystem II in relation of energy dependent quenching of chlorophyll fluorescence. Biochim Biophys Acta 894:198-208.

Weis E, Ball T and Berry J (1987). Photosynthetic control of electron transport in leaves of *Phaseolus vulgaris*: evidence for regulation of photosystem 2 by the proton gradient. In Biggins J (ed): "Progress in Photosynthesis Research Vol II," Dordrecht: Martinus-Nijhoff, pp 553-556.

Photosynthesis, pages 407–424

REGULATION OF PHOTOSYNTHETIC CO_2 ASSIMILATION DURING SUNFLECKS.

Robert W. Pearcy

Department of Botany, University of California, Davis, CA 95616

INTRODUCTION

Leaves at the top of a canopy may experience long periods of high light that approximate steady-state conditions. However, within canopies or in the understory where rapid fluctuations of sun and shade often predominate, most assimilation may occur under transiently changing light conditions. The environmental and physiological controls on photosynthesis during transient light changes are not necessarily the same as those that determine photosynthetic performance under steady-state conditions. There has been much progress in understanding the controls on photosynthetic performance under steady state conditions. By comparison, relatively little is known about the environmental and physiological controls on utilization of dynamic light environments.

Essentially all of the components of the photosynthetic apparatus of a leaf respond either directly or indirectly to a change in light, but on very different time scales. Charge separation and electron transport occur instantaneously (on a physiologists time scale) leading to rapid increases in the supply of ATP and NADPH. However, increases in metabolites to concentrations capable of supporting maximal photosynthetic rates requires longer times. Finally, slow changes in activation state of enzymes and stomatal opening will lead to a slow increase in photosynthesis towards a final steady-state level. To complicate matters further, metabolite concentrations also exhibit slow changes caused by the light activation of the enzymes that regulate the flux through the pools. Since light fluctuations can occur on a time scale much faster than stomatal opening or enzyme activation, the response to any given light change will in part be a function of the previous light environment.

In this chapter, I will review work designed to examine the determinants of photosynthetic performance under dynamic light regimes. Much of this work has focused on a tropical forest understory species, *Alocasia macorrrhiza* (L.) G. Don, which may depend strongly on utilization of sunflecks to maintain net carbon gain and growth in its deeply shaded natural habitat.

DYNAMICS OF NATURAL LIGHT REGIMES

Within canopies of both forests and crops, natural light regimes are far from constant. In forest understories, the very low diffuse light is punctuated by sunflecks lasting from a few seconds to minutes and reaching photon flux densities (PFD) as much as 20-fold greater than the shade light. The duration is a function of the gap size in the canopy as well as movement of the canopy in the wind. Penumbral effects caused by the finite (1/2°) angular diameter of the solar disc result in a reduction of the maximum PFD of small sunflecks under tall canopies (Miller and Norman, 1971). Overstory canopies generally are clumped so that when viewed from the understory there are areas with more gaps while others have few gaps. As a result, sunflecks tend to occur in clusters separated by periods when there are few or none. Cloudiness is also important in determining the frequency of sunflecks. In climates where clouds build in the afternoon, for example, sunflecks will be restricted to the morning.

Studies in a variety of forests show that sunflecks make an important contribution to the light available for photosynthesis of understory plants. Measurements with a variety of sensors and techniques have shown that on mostly clear days between 35 and 70% of the daily photon flux is contributed by sunflecks (Björkman and Ludlow, 1972; Pearcy, 1983, 1987; Chazdon and Fetcher, 1984). In tropical forests, approximately the same proportion of total daily carbon gain by understory leaves can be attributed to sunflecks (Björkman *et al.*, 1972, Pearcy 1987, Pearcy and Calkin, 1983). Little is known about the role of sunfleck utilization in the carbon gain of leaves in crop canopies.

THE PHENOMENOLOGY OF SUNFLECK USE

Natural sunfleck regimes are too complicated to allow for an easy sorting of the various factors that may be limiting assimilation. Within a gas exchange chamber, parts of the leaf may be receiving a bright sunfleck while other parts may be below light saturation while the system measures the average response, which in this case is not very meaningful. Thus, most of the studies designed to examine the factors regulating carbon gain during sunflecks are laboratory-based using lamp and shutter systems that can simulate sunflecks with respect to PFD and timing, but avoid the problems of light gradients and lack of reproducibility. I have designated these simulated sunflecks, as lightflecks in order to distinguish them from their natural counterparts. Another problem encountered in studies of sunfleck utilization is that in order to analyze the dynamics of assimilation, very fast responding systems are needed to cope with the rapid changes. Although capabilities are changing rapidly, the computer and data acquisition systems required for this task are still not readily adapted to field work. The speed of response of gas exchange systems is still a major limitation to the study of fast transients since leaf chambers and infrared gas analyzers hold significant volumes of gas that cause lags in the response of the system. Some CO_2 analyzers have time constants of 1 to 2 s but this is still slower than some of the components of the response of assimilation rate to a light change.

Nevertheless, progress has been made in sorting out both the significance of sunfleck utilization to assimilation of understory plants and the internal and external controls that regulate their utilization. The effect of many of these controls are illustrated in Figs. 1 and 2. After a long period in low light, the maximum assimilation rate achieved in successive lightflecks increases because of photosynthetic induction (Fig. 1). The processes regulating induction continue in the low light period between lightflecks so that assimilation in each succeeding lightfleck increases. Thus, a lightfleck, in effect, primes the leaf so that is better able to utilize subsequent lightflecks. After a lag of 2 to 3 min, stomatal conductance increases in parallel with induction. The rate of induction is as fast with 1 min lightfleck/2 min low light as it is with continuous light (Chazdon and Pearcy, 1986a). However, with shorter lightflecks separated by 2 min of low light, as in Fig. 1, the rate of induction is slower. Induction loss following the transfer back to low light is much slower than induction gain in continuous light.

The dynamic responses to individual lightflecks at various times during the induction sequence are shown in Fig. 2. Only a low assimilation rate is achieved during the first lightfleck but CO_2 fixation continues at an elevated rate for nearly 1 min after the lightfleck. This postillumination CO_2 fixation is also evident for lightflecks given to a fully induced leaf or to one which has partially lost induction. The postillumination CO_2 burst commonly observed for leaves of C_3 plants when light is decreased is not observed following short lightflecks because of a long time constant for photorespiratory metabolism (Kirschbaum, personal communication), and the necessity that metabolites of the glycollate pathway build up to high levels before the rate of CO_2 release becomes high enough to observe. After 1-min long lightflecks given to fully induced leaves, a clear postillumination burst is present (data not shown).

The capability for postillumination CO_2 fixation results in shade plants being able to utilize short lightflecks with high efficiency. A lightfleck utilization efficiency (Pearcy *et al.*, 1985; Chazdon and Pearcy,

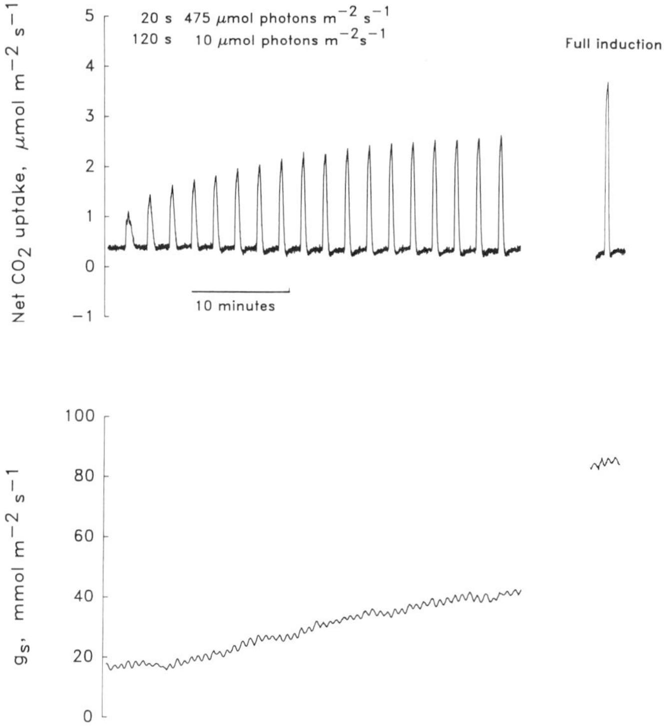

Figure 1. Response of net CO_2 uptake and stomatal conductance (g_s) of a shade-grown *Alocasia macrorrhiza* leaf to a series of 20-second lightflecks separated by 2 minutes of low light. The leaf had been in low light for 4 hours prior to the first lightfleck. The response at full induction was obtained by exposing leaf to high light for 45 minutes then lowering it for 2 minutes prior to the lightfleck.

1986b) can be calculated by comparing the integrated carbon gain due to a lightfleck to the amount that would have occurred at the same light intensity and duration under steady-state conditions. This efficiency allows for a comparison of assimilation due to lightflecks at different leaf induction states and for lightflecks of different duration and PFD. Calculated lightfleck utilization efficiencies for 5 s lightflecks given to fully induced leaves are in the range of 150% and may be greater than 200% for 2-s lightflecks, Efficiencies decrease with increasing lightfleck length in part because the postillumination CO_2 fixation becomes a smaller fraction of the total. However, other factors must be involved since efficiencies in fully induced leaves may fall below 100 percent. It is unlikely that the appearance of a significant photorespiratory burst in

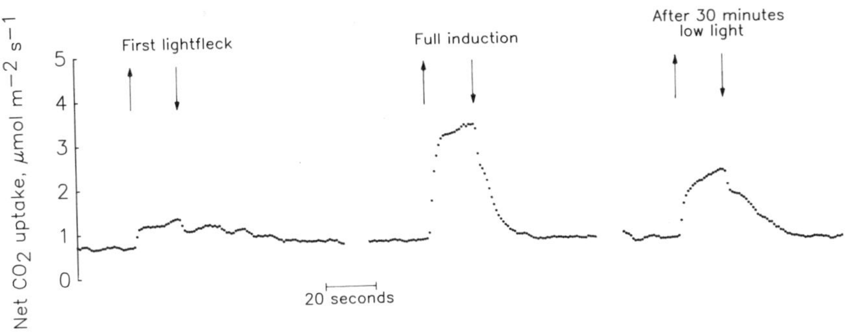

Figure 2. Dynamics of the response of photosynthetic CO_2 uptake of *Alocasia macrorrhiza* to 20-second lightflecks given to an uninduced leaf (first lightfleck), after full induction, or following a return to low light for 60 minutes.

long lightflecks and its absence in short lightflecks directly contributes to the reduced efficiencies for the former since oxygenation of RuBP should occur in both. However, it is possible that intercellular CO_2 pressures may remain higher during short as compared to long lightflecks because of the use of inorganic carbon pools and that this contributes to the increased efficiencies by reducing oxygenation rates. Significant pools of inorganic carbon may exist in mesophyll cells (Oja *et al.*, 1986).

This largely phenomenological account of the response of assimilation rate to lightflecks underscores the complex controls on their utilization by understory plants. There are multiple factors each with different time constants and involving different regulatory mechanisms involved in the overall response. Below, I consider these factors and their role in regulating the response of gas exchange to lightflecks, starting with the slowest and working towards those with the fastest responses to light changes.

REGULATION BY STOMATA

From Fig. 1, above, it can be seen that the increasing assimilation during the sequence of lightflecks corresponds to an increase in stomatal conductance. The role of stomata in the induction response of photosynthesis has been debated almost from the beginning of research on this phenomena. Algae and other intact organisms lacking stomata, and higher plant protoplasts and chloroplasts exhibit an induction requirement (Edwards and Walker, 1984). In many examples, induction is considerably faster for systems without stomata, but Dromgoole (1987) has reported induction responses for brown algae that are as long or longer than those commonly observed for leaves. More typically, however, induction in chloroplast systems and algae is complete within 2 to 4 min (Edwards and Walker, 1984). By contrast, leaves take 10 to 60 min to reach maximum assimilation rate. A role for the stomata is further suggested by the concurrent increases in stomatal conductance and assimilation rate. Indeed, the assimilation rate one minute after a light increase is highly dependent on the initial stomatal conductance (Kirschbaum and Pearcy, 1988a).

In principle, the separation of stomatal from biochemical limitations is achieved through calculation of intercellular CO_2 pressures (p_i) and determination of the response of assimilation to variations in p_i (an A/p_i curve). If stomatal opening accounted for all of the induction response, then changes in assimilation and p_i should all fall along a single A/p_i curve equal to that determined under steady-state conditions. It is frequently observed that the large change in assimilation rate during induction is associated with only a small change in p_i (Usuda and Edwards, 1984; Pearcy *et al.*, 1985). Consequently, it could be concluded that the origin of the induction response is primarily biochemical since initial values of assimilation versus p_i under these circumstances fall considerably below the steady-state values achieved after induction is complete.

Such a simple interpretation needs to be viewed with caution, since both a cuticular conductance and the possibility of a "patchy" stomatal behavior lead to uncertainty in the calculation of p_i during induction. A small cuticular conductance to water vapor (g_c) can lead to a significant overestimation of p_i, and therefore of biochemical limitations, when stomatal conductance (g_s) is also small as it is early in induction. Direct measurement of g_c is not possible for a stomata-bearing leaf surface, but Kirschbaum and Pearcy (1988a) have indirectly estimated it to be approximately 3 mmol m^{-2} s^{-1} for *Alocasia*. Subtracting the cuticular transpiration causes a greater variation in the calculated p_i during induction and consequently leads to the conclusion of a greater role for the stomata, especially when the initial conductances at the beginning of induction are low (Fig. 3). However, biochemical limitations to gas exchange are still evident during the first 10 min of induction but after that further increases in assimilation are explained by increases in stomatal conductance.

The non-uniform behavior of stomata resulting in a mosaic of open and closed patches (Terashima *et al.*, 1988; Downton *et al.*, 1988) also

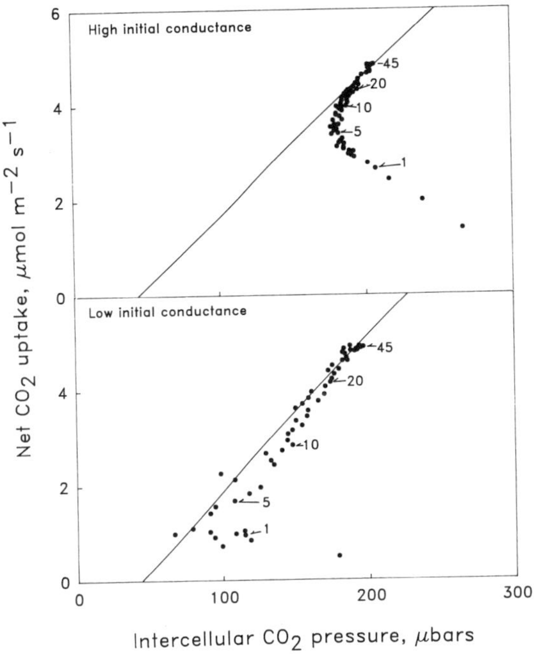

Figure 3. Relationship between photosynthetic CO_2 uptake rate and intercellular CO_2 pressure during induction for leaves with an initially high (32 mmol m^{-2} s^{-1}) or an initially low (7 mmol m^{-2} s^{-1}) stomatal conductance at the beginning of induction. The intercellular CO_2 pressures were calculated assuming a cuticular conductance of 3.2 mmol m^{-2} s^{-1}. The lines are calculated relationships between CO_2 uptake rate and intercellular CO_2 pressure using the modified model of Farquhar *et al.* (1979; Kirschbaum and Farquhar, 1982) that was fit to the points obtained late in induction. The numbers in the figures give the times in minutes since the increase in light.

leads to a significant overestimation of p_i in standard gas exchange calculations. Patchy stomatal patterns occur in leaves treated with ABA and possibly play an important role in the gas exchange responses to stress (Downton *et al*, 1988). In 2% O_2, quenching of variable fluorescence is a function of p_i (Schreiber and Bilger, 1987) which can be viewed using video imaging (Omasa *et al.*, 1987). Using this technique, Daley, Berry, Ball, and Raschke (personal communication) have observed patchy stomatal behavior in both induction and with ABA treatment in cocklebur (*Xanthium strumarium*). However, for *Alocasia*, no patches were apparent during induction and they were much less distinct than in cocklebur with ABA treatment. Thus, the biochemical limitation evident in the induction response of *Alocasia* is probably not a consequence of errors in p_i estimates. Stomatal patchiness may be more of a problem for estimation of p_i in heterobaric leaves such as *Xanthium* with bundle sheath extensions that effectively isolate individual interveinal areas (Terashima, *et al.*, 1988) than in homobaric leaves like those of *Alocasia*.

Because of hydromechanical constraints, stomata cannot open sufficiently fast to have much impact on assimilation during sunflecks shorter than a few minutes' duration. Stomatal opening in response to a light increase often exhibits a lag of several minutes followed by a slow increase requiring 10 to 30 minutes to reach a new steady state. However, opening occurring during one lightfleck could function to improve carbon gain during subsequent lightflecks. For *Alocasia*, the maximum g_s achieved in response to lightflecks was not reached until 20 min after the lightfleck (Fig. 4). Moreover, nearly 90 minutes were required for g_s to return to the original level in low light. The dynamics of the response to shorter lightflecks were similar except the maximum g_s achieved depended on the lightfleck length. Very similar dynamic stomatal responses have been observed for blue light pulses in bean and Commelina leaves (Iino *et al.*, 1985; Zeiger *et al.*, 1985).

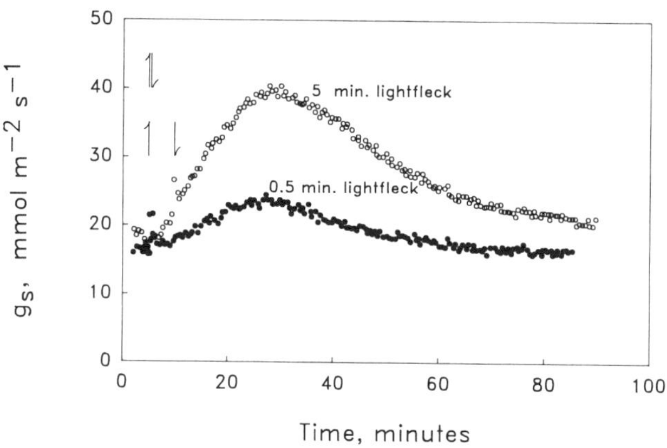

Figure 4. Response of stomatal conductance of a shade-grown *Alocasia macrorrhiza* leaf to a 0.5- and a 5-minute lightfleck. The arrows show the times of light increases and decreases.

The underlying mechanisms governing the dynamics of stomatal opening and closing are poorly understood. The blue light response of guard cells may involve a regulation of electrogenic ion pumping (Assmann *et al.*, 1985). The time course of electrogenic ion pumping is faster but otherwise similar to the stomatal responses observed here. Guard-cell protoplast NADP-malic dehydrogenase has been shown to be light-activated (Gotow *et al.*, 1985). Thus, several metabolic steps occurring on short time scales may be involved. A dynamic model involving an initial fast-changing component step with a shorter time constant for the increase than the decrease, coupled to two slower components accurately simulated the responses to single lightflecks (Kirschbaum *et al.*, 1988). The asymmetry of the initial biochemical component was important in determining the magnitude of the response and the time required to reach maximum conductance. This step may correspond to a light-activation and deactivation of some enzyme or an interconversion of some photoreceptor. The second and third components may correspond to increases and decreases in osmotica and the resultant water movement, respectively. These components had relatively long time constants that were identical for increases and decreases. The model deviated from the measured conductance under flashing light or when conductance was high because of an apparent dependence of the time constants on the value of conductance itself. Further refinements could be achieved by making the time constants some function of conductance or some other factor correlated with conductance, but at present there is no information on the appropriate functions or their mechanistic basis.

REGULATION BY LIGHT ACTIVATION OF RUBISCO

Gas exchange measurements discussed above suggest that during the first 10 min of induction biochemical limitations to photosynthesis predominate and are gradually removed as induction proceeds. Five Calvin cycle enzymes exhibit light activation (Edwards and Walker, 1984). At the biochemical level, the capacity of a leaf for CO_2 uptake depends on the *in vivo* activity of the carboxylating enzyme, ribulose 1,5-bisphosphate carboxylase (Rubisco). This activity is in turn a function of the total amount of enzyme present, the regulatory mechanisms governing its activity, and the concentration of CO_2 and RuBP at the enzyme. Light activation of either Rubisco or of enzymes involved in RuBP regeneration, which would influence the supply of RuBP, could therefore account for the biochemical limitations present early in induction. However, the enzymes involved in RuBP regeneration appear to be light activated within 1 min of a transition from dark to light and therefore would not appear to be candidates to account for the much slower increase in biochemical capacity for assimilation occurring over the first 10 min of induction. The light regulation of Rubisco, on the other hand, has a time course more similar to induction of CO_2 uptake. This enzyme exhibits complex light regulation involving activation by magnesium and CO_2 (Miziorko and Lorimer, 1983), a tight-binding inhibitor, carboxyarabinitol 1-phosphate (CA-1-P) that is synthesized in the dark or low-light and degraded in high light (Seemann *et al.*, 1985), RuBP binding to non-catalytic sites, and an activating enzyme, Rubisco activase (Salvucci *et al.*, 1986). There are species differences in the relative importance of each of these mechanisms and some seem to lack CA-1-P (Kobza and Seemann, 1988).

The relationship between light regulation of Rubisco and the induction response of *Alocasia* was examined by Seemann *et al.* (1988). The initial k_{cat} of Rubisco, which should reflect the *in vivo* capacity of the enzyme, increased 3.5 fold to a maximum value 10 minutes after the light increase (Fig. 5B). The decrease in initial k_{cat} following a decrease in light was much slower and was not complete after 45 min (Fig 5D). For both the light increase and decrease there was a good correspondence between the dynamics of leaf gas exchange (Fig. 5A and C) and the dynamics of Rubisco light regulation (Fig. 5B and D). In *Alocasia*, light regulation of Rubisco was mediated jointly by synthesis and degradation of CA-1-P and by Mg^{2+} and CO_2 activation (Seemann *et al.*, 1988).

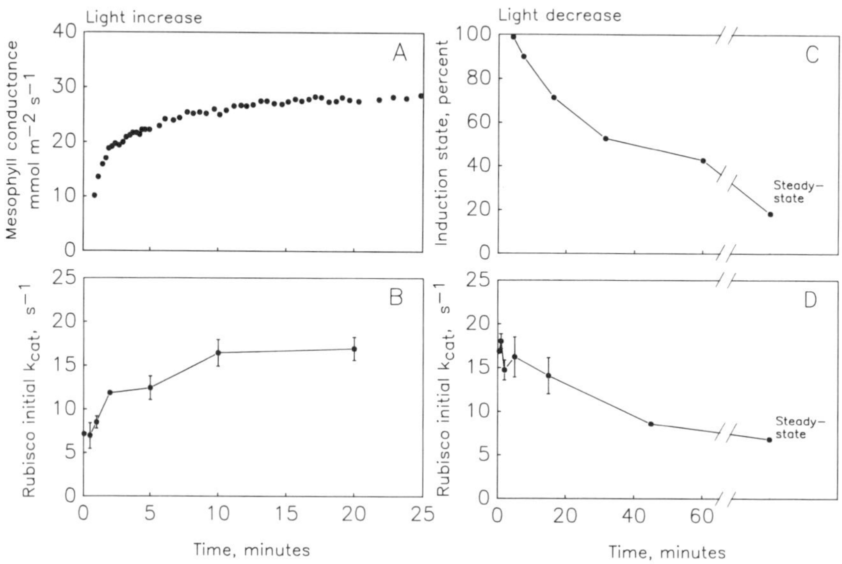

Figure 5. The time course of mesophyll conductance (A), induction state (C) and the initial K_{cat} of Rubisco (B, D) for *Alocasia macrorrhiza* leaves following a light increase from 10 to 500 μmol photons m^{-2} s^{-1} (A,B) or a light decrease from 500 to 10 μmol photons m^{-2} s^{-1} (C, D). Induction state was measured as the photosynthetic rate measured 1 min after a light increase expressed relative the fully-induced photosynthetic rate (Chazdon and Pearcy, 1986b). The data for Rubisco are from Seemann *et al.* (1988).

REGULATION BY A FAST ACTIVATING COMPONENT

In the first minute or so following the increase in light, the biochemical capacity for CO_2 fixation as derived from gas exchange analysis is lower than expected from the *in vivo* Rubisco activity. Thus, another limitation that may disappear in the first few minutes of induction may be present in *Alocasia* leaves. Resolution of transients in this time scale is difficult using gas exchange techniques because of the slow gas-exchange system responses. To overcome this limitation as much as possible, Kirschbaum and Pearcy (1988b) utilized a fast-response analyzer having small cell volumes (LI-COR Instruments model 6250) and a small-volume, single-sided leaf chamber. The chamber, which was clamped on to the abaxial surface of the leaf and took advantage of the fact that *Alocasia* leaves are hypostomatous, was only 4 mm deep and enclosed 22 cm^2 leaf area in a volume of 4 ml. With this system the response to a step change in CO_2 pressure in the chamber was complete in less than 2 s.

Measurements during the first phases of induction showed that when the light increased, assimilation exhibited an immediate increase followed by a second, slower increase over the next 60 to 90 s to a plateau. Thereafter, assimilation increased again because of the Rubisco light regulation and stomatal opening discussed above. The CO_2 dependence of the increase from 2 to 90 s was consistent with an increase in RuBP regeneration capacity but not with activation of Rubisco. After full induction, transfer back to low light resulted in a loss of the activation of this fast component with a $t_{1/2}$ of about 300 s. In low light, a fairly high activation state of this component was maintained since there was no limitation on the immediate increase in assimilation up to about 50 % of the rate achieved at 90 s. The activation state of this component, however, was quite sensitive to the PFD in the low light period, and was lower for darkened leaves as compared to those given 10 $\mu mol\ m^{-2}\ s^{-1}$.

The biochemical basis of this fast activating component is uncertain. Light activation of enzymes involved in RuBP regeneration is generally complete in 1-2 min and requires only low PFD for full activity. Both the timing and PFD dependence are therefore consistent with activation of one or more of these enzymes being involved. However, autocatalytic buildup of metabolite pools cannot be ruled out. Kobza and Edwards (1987) concluded on the basis of extractable enzyme activities and metabolite pool size changes during induction, that pool sizes are limiting during the fast induction following 5 min of dark in wheat leaves. Other factors, including a light-activation requirement for coupling-factor CF_1 (Strotmann and Bickel-Sandkötter, 1984) also could play a role. Further studies at the biochemical level will be required to determine the roles of these factors in the fast induction response.

REGULATION OF POSTILLUMINATION CO_2 ASSIMILATION

Shade plants can make more efficient use of short lightflecks than steady-state light of equal photon flux density because of a continued uptake of CO_2 for a few seconds after the lightfleck. This postillumination CO_2 fixation was first reported by McAlister (1939) for wheat leaves and has now been demonstrated for leaves of many higher plant species (Laisk *et al.* 1984), algae (Miyachi 1979) and higher plant chloroplasts (Miyachi and Hogetsu, 1970). McAlister concluded that the postillumination CO_2 fixation occurred because of a buildup of a "CO_2 combining intermediate" and Laisk *et al.* (1984) used the term "assimilatory power" to describe the sum of all energetically rich compounds that cause the postillumination CO_2 fixation. More recently, the term "assimilatory charge" was adopted by Laisk *et al* (1987), which implies more appropriate units.

In order for postillumination CO_2 fixation to contribute significantly to photosynthesis during lightflecks, the pools of metabolites comprising the assimilatory charge of the leaf must build up rapidly. Evidence for a rapid buildup of assimilatory charge can be obtained from simultaneous measurements of CO_2 and O_2 exchange during lightflecks (Fig 6). Under a steady state condition, the rate of each component step in assimilation, including electron transport, photophosphorylation, RuBP regeneration, and carboxylation can be only as fast as the slowest step. However, during a light increase, the rates of the component steps can become uncoupled from each other if pools of intermediates are available to accept electrons and energy and therefore function in effect as a "capacitance," storing the assimilatory charge. Since O_2 evolution occurs very early in the photosynthetic mechanism whereas CO_2 fixation is much later, O_2 evolution should be transiently higher than the CO_2 uptake rate early in a lightfleck whereas after the lightfleck, O_2 evolution rate should decrease immediately but CO_2 uptake should continue for a short time. This transient displacement of O_2 and CO_2 exchange during a 20-s lightfleck can be seen in Fig. 6. Integrated totals for O_2 evolution and CO_2 exchange were approximately equal. The initial peak of O_2 evolution only occurred when PFD was high. At either 160 μmol photons m^{-2} s^{-1}, which was at the transition from light limitation to saturation, or 56 μmol photons m^{-2} s^{-1}, no peak was evident (Kirschbaum and Pearcy, 1988c). Thus, if light is limiting for electron transport rate, metabolite pool buildup should be reduced or at least delayed, since most if not all available metabolites go to support assimilation during the lightfleck proper. Consequently, significant postillumination CO_2 fixation following short lightflecks should only occur when the PFD during the lightfleck is above saturating values, which has been shown in measurements of lightfleck utilization efficiency (Pearcy *et al.*, 1987).

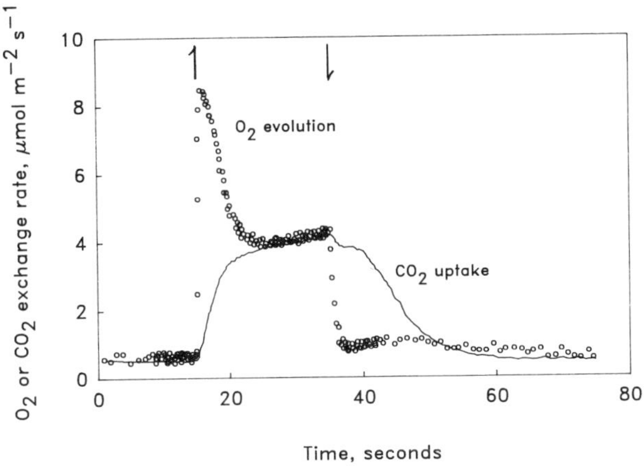

Figure 6. Time course of O_2 evolution and CO_2 uptake for an *Alocasia macrorrhiza* leaf during a 20-second lightfleck. The measurements were made at 25°C, 350 μbar CO_2, and 1850 μbar O_2 pressures. From Kirschbaum and Pearcy (1988).

Potentially, any pool intermediate between the O_2 evolution step and CO_2 fixation could contribute to the assimilatory charge of the leaf. This would include, electron transport intermediates, ATP and $NADPH_2$, the proton gradient across the thylakoid membrane plus reduced Calvin cycle intermediates. The plastoquinone pool is large relative to other electron carriers (Avron, 1981) but is unlikely to contribute to assimilatory charge during lightflecks since its use after the lightfleck would require photosystem 1 light. Any preferential redirection of energy to PSI after the lightfleck would show up as reduced O_2 evolution after the lightfleck, which was not observed. Pools of NADP and ATP in chloroplasts are small relative to the flux rate through them and thus contribute little to assimilatory charge. Consequently, the pools comprising assimilatory charge must be reduced Calvin cycle intermediates. Use of intermediates other than RuBP would require postillumination ATP synthesis, which depends on the pH gradient established and the buffering capacity of the thylakoids (Avron, 1981).

Metabolite pool size measurements reported by Sharkey *et al.* (1986) for leaves rapidly killed just before, during, and after lightflecks provide insight into the identity of the major pools contributing to the postillumination CO_2 fixation of *Alocasia* leaves (Table I). RuPB and triose phosphate pools increased during a 5-s lightfleck with most of the increase occurring in the latter, while PGA pools decreased. After 1 min in low light the pools of RuBP and triose phosphate had largely returned

Table I. Photosynthetic-metabolite pool-size changes in shade-grown leaves of *Alocasia macrorrhiza* at the beginning, end, and 1 min after 5-s lightflecks, and the potential postillumination CO_2 fixation that could result from these changes.

	Triose phosphates	RuBP	PGA
		(μmol m^{-2})	
5-s lightfleck[1]			
start	7.0 ± 1.0[2]	1.4 ± 0.4	23.3 ± 6.1
end	23.3 ± 3.1	3.7 ± 1.5	10.0 ± 2.7
1 min after lightfleck	6.0 ± 1.1	1.7 ± 0.7	16.2 ± 0.7
Total change in post-illumination period	16.3	2.0	6.2
potential postillumination CO_2 fixation	9.8	2.0	

[1] The 5-s lightflecks were 500 μmol photons m^{-2} s^{-1} imposed on a background of 10 μmol photons m^{-2} s^{-1}. The data are from Sharkey *et al.* (1986).
[2] mean ± 1 SE for 4 to 6 determinations.

to levels equal to those measured before the lightfleck. If all of the RuBP and triose phosphate pool size changes were a consequence of consumption of these metabolites in support of postillumination CO_2 fixation, then decreases of 1 mol of RuBP and triose phosphate should result in 1 and 3/5 mol CO_2 fixed, respectively. Use of the triose phosphate pool would have required 9 μmol m^{-2} of ATP, a value that seems reasonable in light of measurements by Hangartner and Good (1982) The steady-state photosynthetic rate of these leaves was 2.6 μmol m^{-2} s^{-1}. Given this steady-state rate and the observed pool-size changes, the efficiency of lightfleck utilization should have been 190%. Of course, other metabolites may contribute and some triose phosphate may be diverted elsewhere. Moreover, RuBP oxygenation and subsequent photorespiration would reduce this efficiency. Indeed the smaller increase in PGA suggests that not all of the assimilatory charge contributed to postillumination CO_2 fixation. Overall, the agreement with the 140 to 160% maximums observed for 5-s lightflecks seems quite good.

CONCLUDING REMARKS

The regulation of photosynthetic use of sunflecks is a consequence of both the induction requirement of the photosynthetic apparatus and the capacity of the leaf for postillumination CO_2 fixation. The importance of each will depend on the frequency and duration of the sunflecks. When they are longer than a few seconds and infrequent so that induction does not carry over, the limitations imposed by the induction requirement will predominate. However, with frequent short-duration sunflecks, such as would occur under canopies in windy conditions, postillumination CO_2 fixation may be an important contributor to photosynthetic carbon gain in understory habitats. Under real forest canopies where there are often periods of frequent sunflecks separated by periods with few or none, both processes are likely to play an important role.

Species differences in the capacity to utilize lightflecks have been demonstrated (Chazdon and Pearcy, 1986b) but at present we have little knowledge of how or why species differ in this regard. The regulatory processes identified so far as important in the response to sunflecks are common to all plants, whether or not they depend in a significant manner on the utilization of sunflecks. These mechanisms may function to maintain the exchange ratio between resources (i.e. water and CO_2 through the stomata) and at the biochemical level to hold metabolite concentrations within bounds that maximize the steady-state efficiency of photosynthesis or enhance the ability to respond to a change in environmental conditions (Woodrow and Berry, 1988). Why the changes are relatively slow and whether or not a more rapid response would be beneficial under the transient light conditions in canopies are unanswered questions. For example, would faster stomatal responses improve the utilization of dynamic light regimes or would additional energetic costs outweigh any benefits? Direct experimental examination of these possibilities is difficult but mechanistically based dynamic models of photosynthesis could prove useful for examining alternatives. Such a model is now under development.

ACKNOWLEDGEMENTS

This research was supported by grant 85-CRCR-1620 from the USDA Competitive Grants Program and grant BSR 86-00065 from the National Science Foundation. I thank Dr. Bill Pfitsch for his valuable comments on the manuscript.

REFERENCES

Assmann SA, Simoncini L, Schroder J (1985). Blue light activates electrogenic ion pumping in guard cell protoplasts of *Vicia faba.* Nature 318:285-287.

Avron M (1981). Photosynthetic electron transport and photophosphorylation. In Hatch MD, Boardman NK (eds): "The Biochemistry of Plants, Vol 4," New York: Academic Press, pp 163-191.

Bartholomew GA, Vleck D, Vleck C (1981). Instantaneous measurements of oxygen consumption during pre-flight cooling in spingid and saturniid moths. J Exp Biol 90:17-32.

Björkman O, Ludlow M (1972). Characterization of the light climate of a Queensland rainforest. Carnegie Inst Wash Year Book 71:85-94.

Björkman O, Ludlow M, Morrow P (1972). Photosynthetic performance of two rainforest species in their habitat and analysis of their gas exchange. Carnegie Inst Wash Year Book 71:94-102.

Chazdon RL, Fetcher N (1984). Photosynthetic light environments on a lowland tropical forest in Costa Rica. J Ecol 72:553-564.

Chazdon RL, Pearcy RW (1986a). Photosynthetic responses to light variation in rain forest species. I. Induction under constant and fluctuating light conditions. Oecologia 69:517-523.

Chazdon RL, Pearcy. R. W. (1986b). Photosynthetic responses to light variation in rain forest species. II. Carbon gain and light utilization during lightflecks. Oecologia 69:524-531.

Downton WJS, Loveys BJ, Grant WJR (1988). Stomatal closure fully accounts for the inhibition of photosynthesis by abscisic acid. New Phytol 108:263-266.

Dromgoole FI (1987). Photosynthesis of marine algae in fluctuating light. Funct Ecol 1:377-386.

Edwards G, Walker D (1983). "C_3, C_4: Mechanisms, and Cellular and Environmental Regulation of Photosynthesis." Berkeley: University of California Press, pp 542

Farquhar GD, Von Caemmerer S, Berry JA (1980). A biochemical model for photosynthetic CO_2 assimilation in leaves of C_3 species. Planta 119:78-90.

Gotow K, Tanaka K, Kondo N, Kobayashi K, Kyono K (1985). Light activation of NADP-malate dehydrogenase in guard cell protoplasts from *Vicia faba* L. Plant Physiol 79:829-832.

Hangarter RP, Good NE (1982). Energy thresholds for ATP synthesis in chloroplasts. Biochim Biophys Acta 681:397-404.

Iino M, Ogawa T, Zeiger E (1985). Kinetic properties of the blue-light response of stomata. Proc Natl Acad Sci 82:8019-8023.

Kirschbaum MUF, Farquhar GD (1984). Temperature dependence of whole-leaf photosynthesis in *Eucalyptus pauciflora* Sieb. ex Spreng. Aust J Plant Physiol 11:519-538.

Kirschbaum MUF, Gross LJ, Pearcy WW (1988). Observed and modelled stomatal responses to dynamic light environments in the shade plant *Alocasia macrorrhiza*. Plant Cell Environ 11:111-121.

Kirschbaum MUF, Pearcy RW (1988a). Gas exchange analysis of the relative importance of stomatal and biochemical factors in photosynthetic induction in *Alocasia macrorrhiza*. Plant Physiol 86:782-785.

Kirschbaum MUF, Pearcy RW (1988b). Gas exchange analysis of the fast phase of photosynthetic induction in *Alocasia macrorrhiza*. Plant Physiol 87:818-821.

Kirschbaum MUF, Pearcy RW (1988c). Concurrent measurements of O_2 and CO_2 exchange during lightflecks in *Alocasia macrorrhiza* (L.). G. Don. Planta 174:527-533.

Kobza J, Edwards GE (1987). The photosynthetic induction response in wheat leaves: net CO_2 uptake, enzyme activation, and leaf metabolites. Planta 171:549-559.

Kobza J, Seemann JR (1987). Mechanisms for the light regulation of ribulose-1,5-bisphosphate carboxylase activity and photosynthesis in intact leaves. Proc Nat Acad Sci USA 85:3815-3819.

Laisk A, Kiirats O, Eichemann H, Oja V (1987). Gas exchange studies of carboxylation kinetics in intact leaves. In Biggins J (ed). "Progress in Photosynthesis Research Vol IV," Dordrecht: Martinus Nijhoff, pp 245-252.

Laisk A, Kirats O, Oja V (1984). Assimilatory power (postillumination CO_2 uptake in leaves. Measurement, environmental dependences and kinetic properties. Plant Physiol 76: 723-729.

Miller E, Norman JM (1971). A sunfleck theory for plant canopies. II. Penumbra effect: intensity distributions along sunfleck segments. Agron J Agronomy J 63:739-748.

Miyachi S (1979). Light-enhanced dark CO_2 fixation. In M Gibbs and E Latzko (eds): "Encyclopedia of Plant Physiology Photosynthesis II Vol 6," Berlin: Springer Verlag pp 68-76.

Miyachi S, Hogetsu D (1970). Light-enhanced carbon dioxide fixation in isolated chloroplasts. Plant Cell Physiol 11:927-936.

Mizioriko HM, Lorimer GH (1983). Ribulose-1,5-bisphosphate carboxylase-oxygenase. Annu Rev Biochem 52:507-535.

Oja V, Laisk A, Heber U (1986). Light-induced alkalinization of the chloroplast stroma in vivo as estimated from the CO_2 capacity of intact sunflower leaves. Biochim Biophys Acta 849:355-365.

Omasa K, Shimazaki K-I., Aiga I, Larcher W, Onoe M (1987). Image analysis of chlorophyll fluorescence transients for diagnosing the photosynthetic system of attached leaves. Plant Physiol 84:748-752.

Pearcy RW (1983). The light environment and growth of C_3 and C_4 tree species in the understory of a Hawaiian forest. Oecologia 58:19-25.

Pearcy RW (1987). Photosynthetic gas exchange responses of Australian tropical forest trees in canopy, gap and understory microenvironments. Functional Ecology 1:169-178.

Pearcy RW, Calkin H (1983). Carbon dioxide exchange of C_3 and C_4 tree species in the understory of a Hawaiian forest. Oecologia 58:26-32.

Pearcy RW, Chazdon RL, Kirschbaun MUF (1987). Photosynthetic utilization of lightflecks by tropical forest plants. Progress in Photosynthesis Research Vol IV, pp 257-260.

Pearcy RW, Osteryoung K, Calkin H (1985). Photosynthetic responses of Hawaiian tree species to dynamic light environments. The time course of CO_2 uptake and carbon gain during sunflecks. Plant Physiol 79:896-902.

Salvucci ME, Portis AE, Ogren WL (1986). Light and CO_2 response of Ribulose-1,5-bisphosphate carboxylase/oxygenase activation in *Arabidopsis* leaves. Plant Physiol 80:655-659.

Schreiber U, Bilger W (1987). Rapid assessment of stress effects on plant leaves by chlorophyll fluorescence measurements. In Tenhunen JD, Cantarino FM, Lange OL, Oechel WC (eds): "Plant Response to Stress: A functional Analysis of Mediterranean Ecosystems," Berlin: Springer-Verlag, pp 27-51.

Seemann JR, Berry JA, Freas SM, Krump MA (1986). Regulation of ribulose bisphosphate carboxylase activity *in vivo* by a light-modulated inhibitor of catalysis. Proc Natl Acad Sci 82:8024-8028.

Seemann JR, Kirschbaum MUF, Sharkey TD, Pearcy RW (1988). Regulation of ribulose 1,5-bisphosphate carboxylase activity in *Alocasia macrorrhiza* in response to step changes in irradiance. Plant Physiol, in press.

Sharkey TD, Seemann JR, Pearcy RW (1986). Contribution of metabolites of photosynthesis to post illumination CO_2 assimilation in response to lightflecks. Plant Physiol 82:1063-1068.

Strotmann H, Bickel-Sandkötter S (1984). Structure, function and regulation of chloroplast ATPase. Annu Rev Plant Physiol 35:97-120.

Terashima I, Wong S-C, Osmond CB, Farquhar GD (1988). Characterization of non-uniform photosynthesis induced by abcisic acid in leaves having different mesophyll anatomies. Plant Cell Physiol 29:385-394.

Usuda H, Edwards GE (1984). Is photosynthesis during the induction period in maize limited by the availability of intercellular carbon dioxide. Plant Science Letters 37:41-45.

Woodrow IE, Berry JA (1988). Enzymatic regulation of photosynthetic CO_2 fixation in C_3 plants. Annu Rev Plant Physiol 39:533-594.

Zeiger E, Iino M, Ogawa T (1985). The blue light responses of stomata: pulse kinetics and some mechanistic implications. Photochem Photobiol 42:759-763.

Photosynthesis, pages 425–436

FUNCTION OF THE FERREDOXIN/THIOREDOXIN SYSTEM IN REDUCTIVE ACTIVATION OF TARGET ENZYMES OF ISOLATED INTACT CHLOROPLASTS

N.A. Crawford, M. Droux, N.S. Kosower and B.B. Buchanan

Division of Molecular Plant Biology, Hilgard Hall, University of California, Berkeley, CA 94720

INTRODUCTION

The ferredoxin/thioredoxin system, composed of ferredoxin, ferredoxin/thioredoxin reductase (FTR), and thioredoxin, has been shown to link light to the regulation of enzyme activity in numerous reconstituted assays (Buchanan, 1980; Cséke and Buchanan, 1986; Edwards *et al.*, 1985; Jacquot, 1984; Leegood *et al.*, 1985). In this system, thioredoxins, small proteins with an active site consisting of two half-cystine residues which undergo reversible reduction and oxidation, are reduced in the light by ferredoxin via a catalytically active thiol group on FTR. In the case of biosynthetic enzymes--e.g., fructose-1,6-bisphosphatase (FBPase) and NADP-malate dehydrogenase (NADP-MDH) --each subunit contains a regulatory disulfide group that is reduced to the sulfhydryl level by reduced thioredoxin, thereby leading to activation of the enzyme (DeCottignies *et al.*, 1988; Marcus *et al.*, 1988). Recent results with a reconstituted thylakoid assay containing homogeneous protein components provide evidence that the ferredoxin/thioredoxin system serves as a regulatory thiol chain that links light to target enzymes via the noncyclic electron transport chain (Droux *et al.*, 1987a, 1987b). Specific inhibition of light activation of NADP-MDH by an antibody against FTR suggests the ferredoxin/thioredoxin system is also active in lysed chloroplasts (Droux *et al.*, 1984).

Aside from findings with lysed chloroplasts and reconstituted assays, the only physiological evidence for the ferredoxin/thioredoxin systems rests on experiments showing that thioredoxin *m* is more reduced in the light than in the dark in isolated intact chloroplasts (Scheibe, 1981) and cyanobacterial cells (Darling *et al.*, 1986). If the current model of the ferredoxin/thioredoxin system is correct, one should see in addition to the thioredoxin *m* change in such studies a light-dependent reduction of the catalytically active FTR disulfide, as well as interacting thiol groups of thioredoxin *f* (Tsugita *et al.*, 1983) and the target enzymes (DeCottignies *et al.*, 1988; Marcus *et al.*, 1988).

To test this point, we have developed a technique to monitor thiol changes in components of the ferredoxin/thioredoxin system in isolated intact chloroplasts maintained under aerobic conditions. The method is based on use of monobromobimane (mBBr)--a membrane-permeable thiol-labeling reagent that reacts rapidly with available sulfhydryl groups to yield stable fluorescent products (Kosower and Kosower, 1987). Proteins of interest are immunoprecipitated and thiols are quantified by combining sodium dodecylsulfate polyacrylamide gel electrophoresis (SDS-PAGE) and fluorescence techniques. The results provide evidence that the ferredoxin/thioredoxin system functions in enzyme regulation in intact chloroplasts by way of the light-induced thiol changes identified in earlier reconstitution experiments.

EXPERIMENTAL PROCEDURES

The materials and methods used in this investigation will be described in full elsewhere (Crawford, Droux, Kosower, and Buchanan, Arch Biochem Biophys, in press).

RESULTS AND DISCUSSION

According to the procedure developed, each protein is isolated from broken chloroplasts by precipitation with a polyclonal antibody (Fig. 1). Important to the success of the procedure is the use of relatively large amounts of affinity purified antibody (50 μg for each experiment) and the inclusion of Triton X-100 and SDS to eliminate nonspecific contamination by ribulose-1,5-bisphosphate carboxylase/oxygenase (rubisco) and other proteins during precipitation. Results obtained with members of the ferredoxin/thioredoxin system containing a catalytically active thiol group are summarized below.

Thioredoxin *f*.

After immunoprecipitation and analysis on SDS-PAGE, thioredoxin *f* was found to be more highly fluorescent from light- than dark-treated chloroplasts (Fig. 2, lanes D and L of Treatments 1 and 2). That this increase in labeling was due to net reduction rather than conformational change (the most recent evidence shows that thioredoxin *f* contains 3 cyst(e)ine groups; P. Schürmann personal communication) is supported by the retention of the light effect in these two treatments--i.e., Fluorescence Treatment 1 (without SDS) and Treatment 2 (with SDS). In

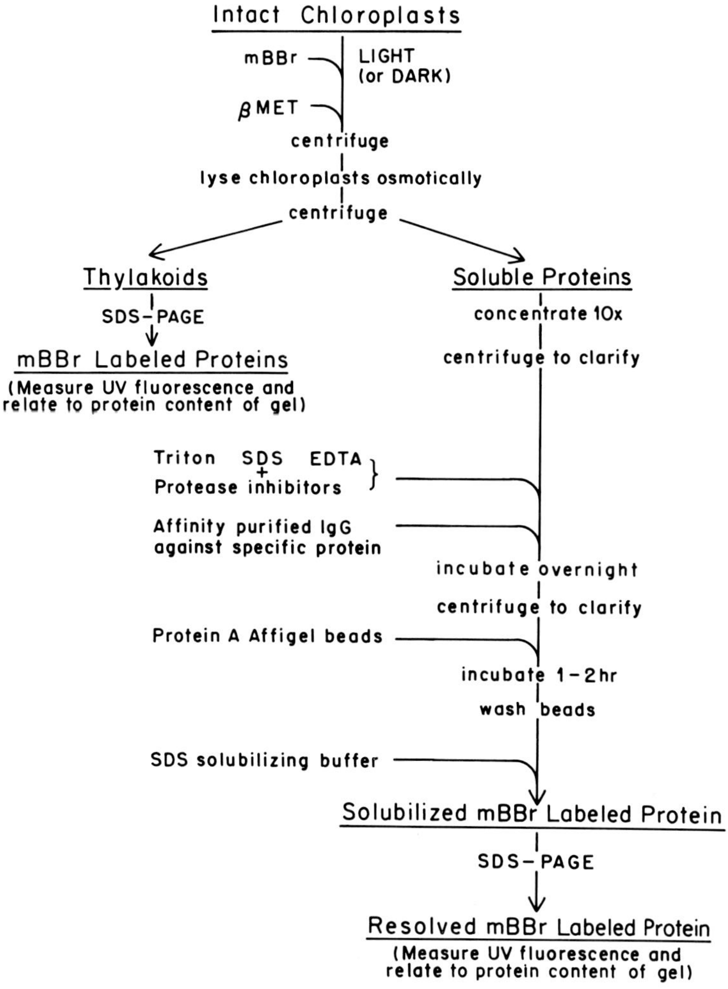

Figure 1. Flow chart of the protocol for labeling chloroplast proteins with mBBr in the light or dark and resolving them on SDS-PAGE.

the case of Treatment 2, the immunoprecipitated thioredoxin *f* sample, which had previously been labeled with mBBr in the light or dark as in Treatment 1, was denatured with SDS in the absence of reductant (2-mercaptoethanol) to expose potential sulfhydryl groups not available without denaturation and was then again treated with mBBr. This modification of the procedure resulted in only a slight increase in the

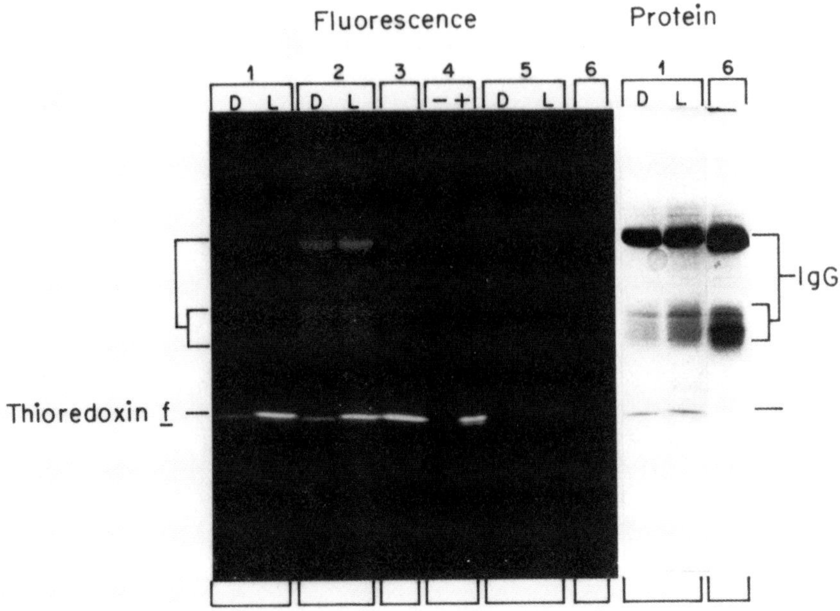

Figure 2. Demonstration of the photoreduction of thioredoxin *f* in intact chloroplasts. The numbering of the fluorescence treatments correspond to the stained protein. Other details are given in the text.

fluorescence of the thioredoxin *f* band from chloroplasts kept in the dark and the band from illuminated chloroplasts was still more highly labeled (Fig. 2, lane D and L of Treatment 2). That the protein precipitated from chloroplasts in Fig. 2 was reduced thioredoxin *f* was confirmed by (*i*) its reduction by DTT in lysed chloroplasts (Treatment 3); (*ii*) comparison to pure authentic thioredoxin *f* labeled with mBBr in the presence and absence of DTT (Treatment 4); (*iii*) dilution by added unlabeled purified thioredoxin *f* (Treatment 5); and (*iv*) absence from a control in which preimmune IgG replaced thioredoxin *f* antibody (Treatment 6).

Based on intensity of fluorescence determined by scanning photograph negatives, thioredoxin *f* was fully oxidized in the dark and almost fully reduced in the light. The low level of incorporation of mBBr in oxidized thioredoxin *f*, either in pure form or in dark adapted chloroplasts (see Fig. 2), is attributed to the above-mentioned "third" cysteine residue that is not a part of the active site (cys-gly-pro-cys). When corrected for values obtained for samples in which mBBr label was incorporated in the dark, thioredoxin *f* was found to be 95% reduced in the light (Table I).

Thioredoxin *m*.

When intact chloroplasts were incubated and treated with mBBr in the light, the fluorescence of thioredoxin *m*, like thioredoxin *f*, greatly increased relative to dark controls (data not shown). Thioredoxin *m* was recovered as its two isoforms described by Maeda *et al.* (1986). The authenticity of the recovered product as thioredoxin *m* was confirmed by as described above for thioredoxin *f*.

Scheibe (1981) found in the first light/dark transition that thioredoxin *m* was 62-77% reduced in the light and 8-30% reduced in the dark. Our values with mBBr support this conclusion but suggest a somewhat lower level of reduced thioredoxin *m* in the dark (4%) and a higher level in the light (75 and 87%) dependent on the isoform measured (Table I). These levels of photoreduction of thioredoxin *m* are sufficient to activate NADP-MDH according to the studies of Rebeille and Hatch who have reported that very high reduced-to-oxidized thioredoxin ratios are required for activation of the enzyme (Rebeille and Hatch, 1986a, 1986b).

A noteworthy point made in our study concerns the entry of DTT into chloroplasts. When DTT was added to dark-adapted chloroplasts that were labeled with mBBr prior to lysis, the incorporation of fluorescent label in thioredoxin *m* was only about 10% that observed for controls lysed prior to DTT addition (data not shown). Thus, according to our results, DTT enters plastids, but thioredoxin reduction under these conditions is clearly limited. Similar findings were made with thioredoxin *f* and FTR. Such low levels of reduced thioredoxins generated by DTT under these conditions are seemingly sufficient to enhance carbon dioxide assimilation as was observed in model studies carried out in the dark with isolated intact chloroplasts (Werdan *et al.*, 1975).

Ferredoxin-thioredoxin reductase.

Like the f and m thioredoxins, FTR was found to be photoreduced in intact chloroplasts. That the observed fluorescence increase was due to net reduction, rather than conformational change, was confirmed by retention of the light effect in samples treated with SDS as described above for thioredoxin *f*.

Table I: Extent of reduction of components of the ferredoxin/ thioredoxin system in light- or dark-treated intact chloroplasts.

Protein	% Reduction		
	Dark	Light	DTT control
Thioredoxin *f*	0	95 ± 9.4	100
Thioredoxin *m*			
13 kDa	4 ± 3	87 ± 4	101 ± 1
11 kDa	4 ± 5	75 ± 6	97 ± 5
FTR	3 ± 0.2	38 ± 5	31 ± 2
NADP-MDH	2 ± 1	33 ± 5	74 ± 22

Reduction of FTR occurred on the catalytically active disulfide group that was identified earlier on the 13 kDa) subunit of the enzyme (Droux *et al.*, 1987a,b). The extent of reduction of FTR was 38% in the light and 3% in the dark (Table I). The finding that the extent of reduction in the light was relatively low compared to that of the thioredoxins (see above) is in agreement with earlier evidence obtained with a reconstituted system. The results from both lines of study suggest that the redox potential and turnover of the FTR disulfide group are sufficient to promote the higher level of reduction (about 90%) of the thioredoxins *f* and *m* required for activation of their respective target enzymes, NADP-MDH (Droux *et al.*, 1987b) and FBPase (Droux *et al.*, 1987a). In the future, it will be of interest to measure the redox potential of the FTR disulfide and relate that potential to the value reported for the active disulfide of thioredoxin *m* (E'o = -330 mV) (Rebeille and Hatch, 1986a).

Target Enzymes.

The above results are consistent with the conclusion, drawn from enzyme studies, that the ferredoxin/thioredoxin system functions as a regulatory thiol chain, linking noncyclic electron transport to enzyme regulation. Here, photoreduced ferredoxin serves as reductant for the catalytically active disulfide group of FTR, which, in turn, transfers the reducing equivalents to thioredoxins f and m. The terminal member of the chain is a target enzyme, e.g., NADP-MDH or FBPase, which contain a regulatory disulfide site that, when reduced, leads to activation of catalysis (Droux *et al.*, 1987a, 1987b). To date, a photoreduction of such a target enzyme has not been reported for an intact chloroplast system. We, therefore, addressed this problem using the procedure described above.

Fig. 3 illustrates that light increased the incorporation of mBBr label into both NADP-MDH and FBPase. In the case of NADP-MDH, the light-induced increase in fluorescence corresponds to a 33% reduction when compared to a DTT-reduced purified standard for the enzyme (Table I). This increase was the result of a net photoreduction. When subjected to the SDS treatment developed for thioredoxin *f*, the NADP-MDH subunit from light treated chloroplasts showed a 23% greater fluorescence than its dark counterpart. Expectedly, the fluorescence background was increased markedly by this SDS treatment, as previously unaccessible sulfhydryl groups became available to react with mBBr. This above value correlates well with the value of 33% calculated from the data of Fig. 3. We estimate these values for extent of reduction to be minimal because of difficulty in precipitating reduced enzyme, i.e., reduced NADP-MDH precipitated only about 50% as effectively as the oxidized enzyme.

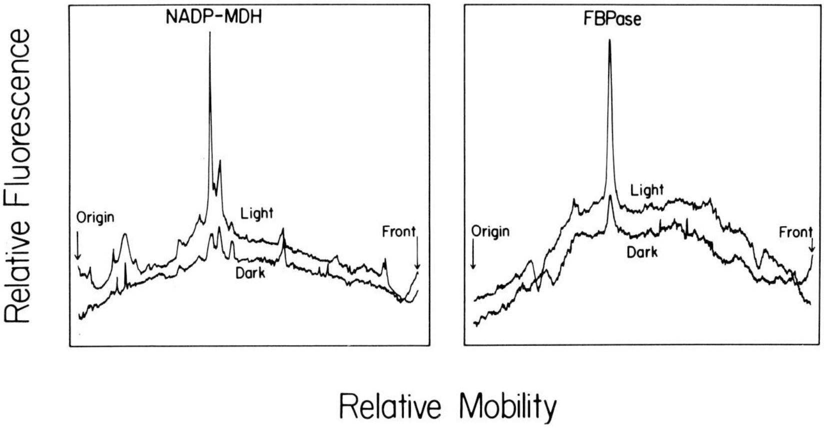

Figure 3. Effect of light on the incorporation of mBBr into NADP-MDH and FBPase of intact chloroplasts.

Because of the close proximity of the heavy chain of gamma globulin to the FBPase subunit, it was not possible to quantitate the light induced thiol changes in FBPase with the mBBr technique. While scanning of fluorescence was not a problem, the Coomassie blue stained FBPase could not be resolved from an IgG band. If one makes the assumption that the oxidized and reduced forms of the FBPase precipitate equally and that the 1 mM DTT control for the enzyme is fully reduced, the FBPase was oxidized in the dark and was 25% reduced under our conditions in the light.

Time Course of Thioredoxin Reduction and Enzyme Activation.

The development of an mBBr technique applicable to intact chloroplasts makes it possible to determine the kinetics of reduction of thioredoxins *f* and *m* relative to the activation of target enzymes. In this connection, we consistently found that reduction of FTR and of thioredoxins *f* and *m* preceded activation of target enzymes. In the experiment shown in Fig. 4, the components of the ferredoxin/thioredoxin system were rapidly reduced in the light and the levels of reduced thioredoxin *f* and *m* reached steady state after 2 min. Activation of the target enzymes, NADP-MDH and FBPase, occurred more slowly and in this experiment reached steady state after 10 min. When the light was turned off, the activity of NADP-MDH and FBPase decreased less rapidly than oxidation of the thioredoxin and FTR. It should be noted that while the rate of light-mediated enzyme activation varied from one experiment to another (full activation was achieved in the 3 to 10 min timeframe), the reduction of FTR and thioredoxins *f* and *m* consistently preceded activation of the target enzymes. Similarly, the oxidation of the regulatory thiol proteins occurred prior to deactivation of the enzymes in the dark. The results thus show a correlation between the reduction of thioredoxins and the activation of NADP-MDH and FBPase that is in accord with previous evidence indicating a central role for the ferredoxin/thioredoxin system in the light activation of these enzymes (Buchanan, 1980; Cséke and Buchanan, 1986; Edwards *et al.*, 1985; Jacquot, 1984; Leegood *et al.*, 1985).

In recent interesting work on the structure of light activated chloroplast enzymes, the regulatory reduction sites have been identified for NADP-MDH (DeCottignies *et al.*, 1988), FBPase (Marcus *et al.*, 1988), and PRK (Porter *et al.*, 1988). In the case of NADP-MDH and FBPase, the regulatory site consists of a redox active group containing two cyst(e)ines separated by 4 or 5 amino acids that, by analogy with these enzymes from other sources, are not part of the active site. The results of Fig. 5 provide experimental evidence on this point. When added to illuminated intact chloroplasts just prior to turning off the light,

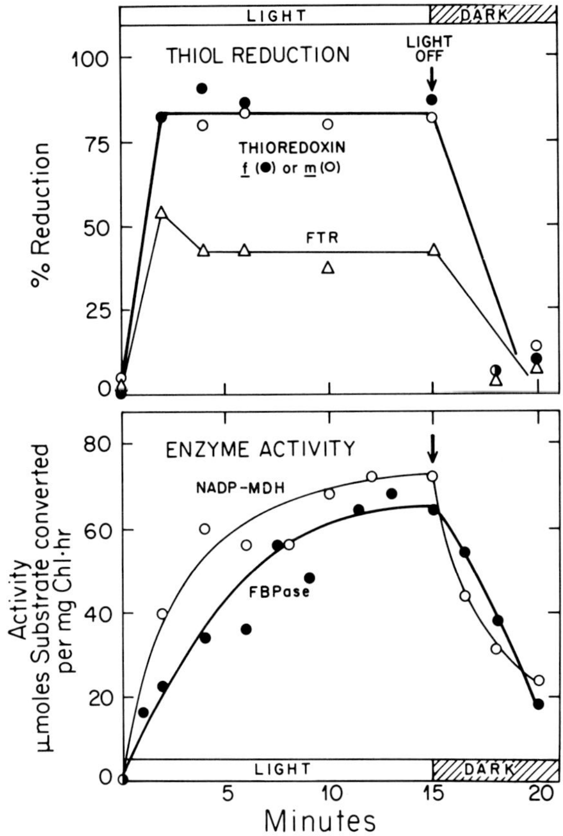

Figure 4. Time course of the light-dependent reduction of components of the ferredoxin/thioredoxin system and concurrent activation of target enzymes in intact chloroplasts.

mBBr maintained the activated state of NADP-MDH and FBPase in the dark (Fig. 5, right panel). This result supports the conclusion that the regulatory dithiol groups of both enzymes are separate from the active site. A similar conclusion was earlier reached for FBPase on the basis of kinetic data (Soulie *et al.*, 1985).

A different situation exists for PRK. Here, the two regulatory cyst(e)ines are separated by 39 amino acids (Porter *et al.*, 1988) and one of the cyst(e)ines has been concluded to be a part of the active site (Omnaas *et al.*, 1985). The present results support this conclusion. In contrast to NADP-MDH and FBPase, mBBr rapidly inactivated PRK

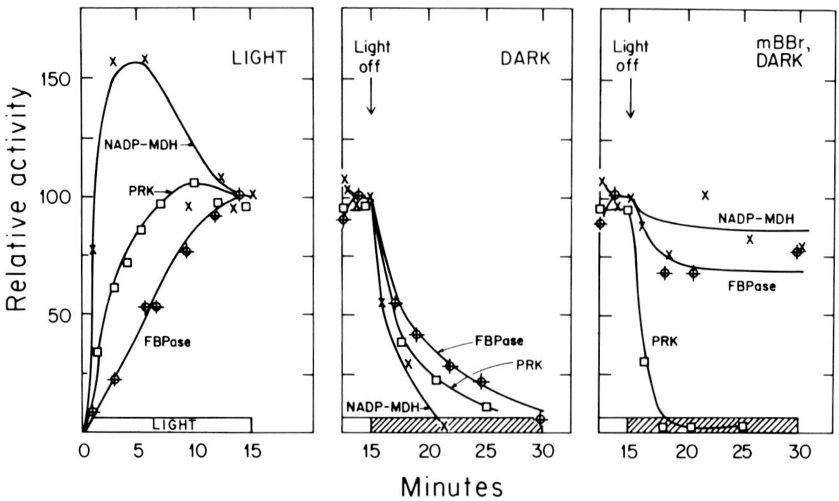

Figure 5. Effect of mBBr on the light-induced activity of NADP-MDH and FBPase target enzymes after change from light to dark conditions.

upon transfer to dark conditions (Fig. 5, right panel). Furthermore, PRK was also rapidly inactivated when mBBr was added to chloroplasts kept in the light. Results similar to those of Fig. 5 were obtained with NEM for PRK as well as FBPase and NADP-MDH (data not shown).

CONCLUSIONS

The present results provide evidence that the ferredoxin/thioredoxin system functions in the light mediated activation of enzymes of isolated intact chloroplasts in the manner established with reconstituted enzyme assays. The kinetics of reduction of catalytically active thiols of the system relative to the kinetics of activation of target enzymes are in accord with the reaction sequence, Light → Chlorophyll → Ferredoxin → FTR → Thioredoxin *f* or *m* → Target Enzyme. The extent of light-dependent reduction of each photoactive thiol group is also commensurate with this conclusion.

In addition to documenting the function of the ferredoxin-thioredoxin system as a regulatory thiol chain, the current work supports recent conclusions on the regulatory reduction sites of chloroplast FBPase, NADP-MDH, and phosphoribulokinase. Our data also give new information on the relation between the regulatory and active sites for each of these thioredoxin-linked enzymes.

ACKNOWLEDGEMENTS

This work was supported by a grant from the National Science Foundation.

REFERENCES

Buchanan, BB (1980). Role of light in the regulation of chloroplast enzymes. Annu Rev Plant Physiol 31:341-374.

Cséke C, Buchanan BB (1986). Regulation of the formation and utilization of photosynthate in leaves. Biochim Biophys Acta 853:43-63.

Darling AJ, Rowell P, Steward WDP (1986). Light-induced transitions in the redox state of thioredoxin in the N2-fixing cyanobacterium, *Anabaena cylindrica*. Biochim Biophys Acta 850:116-120.

DeCottignies P, Schmitter JM, Miginiac-Maslow M, Le Marechal P, Jacquot JP, Gadal P (1988). Primary structure of the light-dependent regulatory site of corn NADP-malate dehydrogenase. J Biol Chem, 263: 11780-11785.

Droux M, Crawford NA, Buchanan BB (1987a). Mechanism of thioredoxin-linked activation of chloroplast fructose-1,6-bisphosphatase. Comptes-Rendus, Acad Sci Paris 305:335-341.

Droux M, Jacquot JP, Miginiac-Maslow M, Gadal P, Huet, JC, Crawford NA, Yee BC, Buchanan BB (1987b). Ferredoxin-thioredoxin reductase: a catalytically active dithiol group links photoreduced ferredoxin to thioredoxin functional in photosynthetic enzyme regulation. Arch Biochem Biophys 252:426-439.

Droux M, Jacquot JP, Suzuki A, Gadal P (1984). Ferredoxin-thioredoxin reductase: purification and substrate requirements. In Sybesma C, (ed): "Advances in Photosynthesis Research," Vol 3, Nijhoff/Junk, Dordrecht, The Netherlands, pp 533-536.

Edwards GE, Nakamoto H, Burnell JN, Hatch MD (1985). Pyruvate, Pi dikinase and NADP-malate dehydrogenase in C4 photosynthesis: properties and mechanisms of light/dark regulation. Annu Rev Plant Physiol 36:255-286.

Jacquot JP (1984). Post-translational modifications of proteins in higher plant chloroplasts: enzyme regulation by thiol-disulfide interchange. Physiol Vég 22:487-507.

Kosower NS, Kosower E (1987). Thiol labeling with bromobimanes. Methods Enzymol 143:76-84.

Leegood RC, Walker DA, Foyer C (1985). Regulation of the Benson-Calvin cycle. In Barker J, Baker NR (eds): "Topics in Photosynthesis," Vol. 6, Elsevier/North Holland, Amsterdam, pp 189-258.

Maeda K., Tsugita A, Dalzoppo D, Vilbois F, Schürmann P (1986). Further characterization and amino acid sequence of *m*-type thioredoxins from spinach chloroplasts. Eur J Biochem 154:197-203.

Marcus F, Moberly L, Latshaw SP (1988). Comparative amino acid sequence of fructose-1,6-bisphosphatases. Identification of a region unique to the light-regulated enzyme. Proc Natl Acad Sci USA, 85: 5379-5383.
Omnaas J, Porter MA, Hartman FC (1985). Evidence for a reactive cysteine at the nucleotide binding site of spinach ribulose-5-phosphate kinase. Arch Biochem Biophys 236:646-653.
Porter MA, Stringer DC, Hartman FC (1988). Characterization of the regulatory thioredoxin site of phosphoribulokinase. J Biol Chem 263:123-129.
Rebeille F, Hatch MD (1986a). Regulation of NADP-malate dehydrogenase in C4 plants: effect of varying NADPH to NADP ratios and thioredoxin redox state on enzyme activity in a reconstituted system. Arch Biochem Biophys 249:164-170.
Rebeille F, Hatch MD (1986b). Regulation of NADP-malate dehydrogenase in C4 plants: relationship among enzyme activity, NADPH to NADP ratio and thioredoxin redox states in intact maize mesophyll chloroplasts. Arch Biochem Biophys 249:171-179.
Reeves SG, Hall DO (1980). Higher plant chloroplasts and grana: general preparative procedures (excluding high carbon dioxide fixation ability chloroplasts). Methods Enzymol 69:85-101.
Scheibe R (1981). Thioredoxin *m* in pea chloroplasts: concentration and redox state under light and dark conditions. FEBS Lett 133:301-304.
Soulie JM, Buc J, Riviere M, Ricard J (1985). Equilibrium binding of thioredoxin *fβ* to chloroplastic fructose bisphosphatase. Evidence for a thioredoxin site distinct from the active site. Eur J Biochem 152:565-568.
Tsugita A, Maeda, K., Schürmann P (1983). Spinach chloroplast thioredoxins in evolutionary drift. Biochem Biophys Res Commun 115:1-7.
Werdan K, Heldt HW, Milovancev M (1975). The role of pH in the regulation of carbon fixation in the chloroplast stroma. Studies on CO_2 in the light and dark. Biochim Biophys Acta 396:276-292.

Photosynthesis, pages 437–455

CALCIUM REGULATION OF LOCALIZED TO DELOCALIZED PROTON GRADIENT SWITCHING IN THYLAKOIDS: THE 8 kDa CF_0 SUBUNIT IS PART OF THE Ca^{+2} GATING STRUCTURE[1]

Gisela G. Chiang and Richard A. Dilley

Department of Biological Sciences, Purdue University, West Lafayette, Indiana 47907

INTRODUCTION

Energy transduction of oxidation-reduction potential energy into the driving force for ATP formation in chloroplast thylakoids, bacteria, and mitochondria almost certainly has an intermediary energy transfer step utilizing proton electrochemical potential gradients (Ferguson, 1985; Dilley *et al.*, 1987). Important research goals in understanding these bioenergetics processes now focus on the mechanisms whereby protons, released by the redox reactions, move to and through the CF_0-CF_1 energy coupling complex and couple the H^+ flux to ATP formation. We seem a long way from answering the elusive questions of how protons cross the membrane in the CF_0 structure or how they interact with the CF_1 protein subunits in the catalytic steps of ATP formation, although progress is being made on both fronts (Cain *et al.*, 1986; Boyer, 1987). Another important question that has recently seen considerable progress toward resolution, is the ongoing controversy about whether the energy coupling proton gradients are solely bulk-to-bulk phase (delocalized) or in some instances membrane domain-localized.

Work from several laboratories indicates that localized proton gradients can, in some instances, be coupled to ATP formation (Ort *et al.*, 1976; Sigalat *et al.*, 1985; Horner and Moudrianakis, 1983; Beard and Dilley, 1986; Pick *et al.*, 1987), while in other circumstances delocalized gradients are involved (Ort *et al.*, 1976; Sigalat *et al.*, 1985; Beard and Dilley, 1986; Pick *et al.*, 1987; Gräber, 1982; Vinkler *et al.*, 1980; Davenport and McCarty, 1980). Results from this laboratory led us to the hypothesis that switching from localized to delocalized H^+ fluxes is a

[1]This work is dedicated to Dr. Stacy French on the occasion of his 80th birthday. Supported in part by grants from the National Science Foundation and the U.S. Department of Agriculture.

reversible event (Beard and Dilley, 1988) and that Ca^{+2} ions are involved in some type of gating action that regulates the distribution of protons between localized and delocalized energy coupling gradients (Chiang and Dilley, 1987). Fig. 1 shows a "cartoon" depicting the hypothesized Ca^{+2} regulated H^+ gate between domain-localized H^+ gradients and the lumen. Factors which favor localized proton gradient coupling are the maintenance of high levels of divalent cations, particularly Ca^{+2}, and having thylakoids with high coupling efficiency (Chiang and Dilley, 1987). Delocalized coupling is induced by storing thylakoids in high levels of KCl or NaCl (Sigalat *et al.*, 1985; Beard and Dilley, 1988), displacing Ca^{+2} ions by lipid soluble Ca^{+2} chelators (Chiang and Dilley, 1987), or by activating basal electron flow so as to accumulate protons in membrane domains without the phosphorylating system utilizing the protons in energy-linked H^+ efflux (Beard and Dilley, 1988; Chiang and Dilley, 1987; Beard *et al.*, 1988). The model shown in Fig. 1 is not, of course, a final, exact portrayal of what goes on in membrane-proton interactions, but rather a working hypothesis around which to design experiments.

An aspect which seems basically correct, even if not well understood in detail, is that the thylakoid membrane has sequestered domains with two classes of proton buffering groups. There is an array of near 30-40 nmol (mg Chl)$^{-1}$ of anomalously low pK_a amine groups ($pK_a \approx 7.2$-7.8, see Dilley *et al.*, 1987, for further details) and a much larger buffering array, perhaps 200 nmol (mg Chl)$^{-1}$, of carboxyl groups which have pK_a values in the range of pH 5 (Beard and Dilley, 1988; Beard *et al.*, 1988). Data support the involvement of both buffering arrays in interactions with protons in localized domains while the protons are building up the gradient needed to energize ATP formation (Beard and Dilley, 1988; Beard *et al.*, 1988; Theg *et al.*, 1988). Therefore, an important conceptual scheme, embodied in the Fig. 1 model, that we use in thinking about proton gradients involved in energy coupling, is that protons are first released by the redox reactions into membrane localized domains, from which they either traverse directly out through the CF_0-CF_1 in the energy coupled H^+ flux, or they build up sufficient acidity in the domains to protonate the putative gate carboxyl groups and displace the Ca^{+2}, allowing protons to equilibrate freely with the lumen.

Having obtained evidence for a Ca^{+2} gating function in the reversible switching between the two types of H^+ gradients, we are continuing to test the hypothesis further with two approaches: one is to find ways of perturbing the putative Ca^{+2} gating function with Ca^{+2} antagonists and other membrane effectors; the other approach involves identifying the membrane component(s) which may be interacting with Ca^{+2} in the apparent gating function. This report concerns the initial findings which show that Ca^{+2} antagonists of the phenothiazine group -- which interact

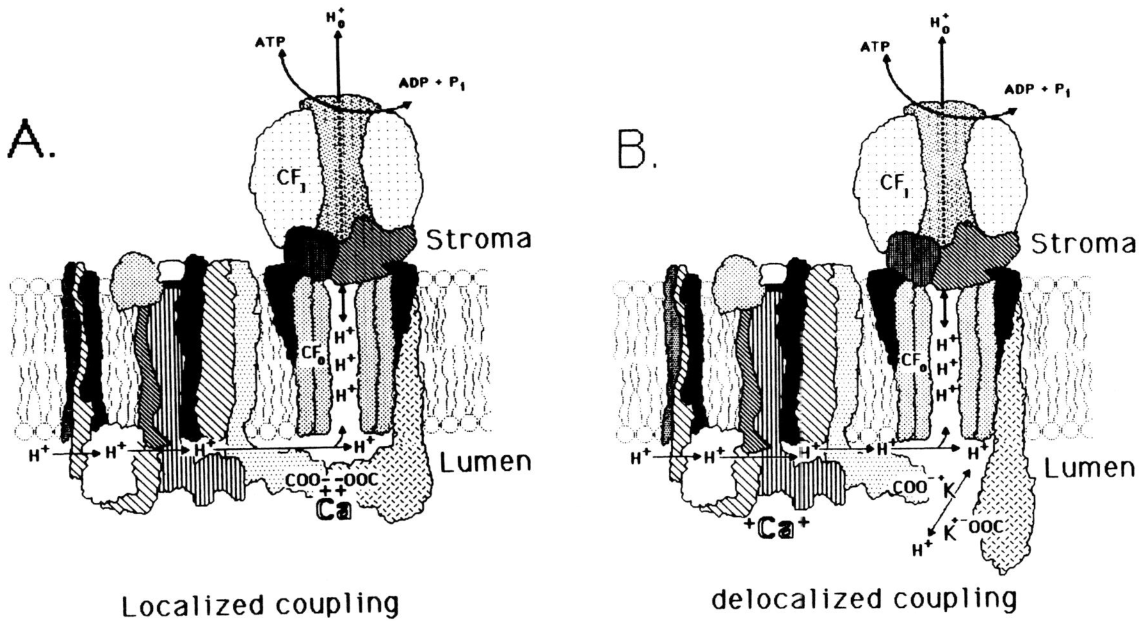

Figure 1. A model for a possible gating of proton fluxes between localized (A) or delocalized (B) energy-coupling gradients. (A) A portion of a thylakoid membrane is shown depicting, in a generalized and as yet speculative way, several intrinsic membrane proteins participating to form a localized proton diffusion domain from the proton-releasing reactions in H_2O oxidation and plastoquinol oxidation into the CF_O channel.

with Ca^{+2}-binding calmodulin proteins -- apparently interact with the thylakoid proton flux gating mechanism. Other experiments indicate that certain thylakoid proteins (one of them a CF_0 component) interact with Ca^{+2} in the proton gating function.

The approach we used for the present studies depends on the interesting fact that the well-known calmodulin antagonists, chlorpromazine (CPZ) and trifluoperazine (TF) [both of the phenothiazine group (Levin and Weiss, 1979)] bind tightly to thylakoid membranes when Ca^{+2} is abundant and localized proton gradients are favored. The phenothiazine drugs bind tightly to calmodulins only when Ca^{+2} is bound to the Ca^{+2}-specific binding sites (Levin and Weiss, 1979; Prozialeck *et al.*, 1981). Moreover, chlorpromazine can be photoactivated by UV light and when bound to the tight binding site of calmodulin, covalent bond formation can occur leading to inactivation of calmodulin and, if [^{3}H]CPZ is used, radioactively tagging the calmodulin (Prozialeck *et al.*, 1981). We have used this property of CPZ and have isolated three thylakoid polypeptides whose photoaffinity labeling is Ca^{+2}-dependent.

The related question was also studied, whether CPZ bound at the Ca^{+2}-occupied sites has any effects on H^+ flux between membrane-localized domains and the lumen, or on ATP formation. We have recently developed assays that can distinguish between proton accumulation in membrane phase domains and accumulation in the thylakoid lumen (Beard *et al.*, 1988). The assay depends on the fact that permeable buffers such as pyridine increase the buffer capacity of the lumen but not the membrane domains. The operation of the H^+ gating function to cause delocalized H^+ gradients can thus be detected (Beard and Dilley, 1988; Beard *et al.*, 1988) as predictable effects of pyridine buffering action on lumenal H^+ accumulation and/or pyridine effects on the development of a sufficiently acidic delocalized lumen pH to energize ATP formation (when valinomycin and K^+ keep the electric field suppressed). So far, the data suggest that when Ca^{+2} is not displaced, thus keeping the Ca^{+2}-controlled H^+ gate closed, H^+ accumulation occurs within the membrane phase and allows ATP formation to be driven with no pyridine effects; hence we call that localized gradient coupling (Beard and Dilley, 1988; Beard *et al.*, 1988). A corresponding set of effects of switching the H^+ flux into the lumen or maintaining localized H^+ gradients in driving ATP formation is detected in two types of post-illumination ATP formation (Beard *et al.*, 1988). When CPZ is tightly bound to the Ca^{+2}-occupied sites, we detect a partial blockage of proton flux connection between the membrane domains and the lumen. If we have properly interpreted the results, they give added credence to the concept of a gated H^+ conducting channel connecting membrane domains and the bulk aqueous lumen.

METHODS

Spinach or pea plants, grown in growth chambers or the greenhouse, respectively, were used to prepare thylakoids according to Beard and Dilley, 1986). Low salt thylakoid storage media was 200 mM sucrose, 5 mM Hepes pH 7.5, 2 mM $MgCl_2$ and 0.5 mg/ml BSA. The high salt storage media was similar except that 100 mM KCl replaced the sucrose.

Phosphorylation activity driven by single-turnover flashes was assayed by the luciferin-luciferase method as described by Beard and Dilley (1986). Proton accumulation was assayed with a sensitive recording pH meter to determine the total extent of H^+ uptake in 5 Hz flash trains or in steady state illumination for either coupled or basal conditions. Hexokinase and glucose were added to recycle any ATP formed in the coupled case, permitting measurement of thylakoid H^+ uptake under phosphorylating conditions.

[^{3}H]Chlorpromazine (NEN Company) was photoactivated in the presence of thylakoids with a Mineralite UV source to induce covalent attachment of CPZ to thylakoid protein binding sites (Prozialeck *et al.*, 1981). After delipidation and SDS solubilization of the membranes, thylakoid proteins were run on SDS-PAGE gels, and labeled bands detected by slicing the gels and counting radioactivity in a scintillation counter (Laszlo *et al.*, 1984). The 8 kDa CF_0 subunit was purified by butanol extraction and ether precipitation followed by SDS-PAGE (Nelson *et al.*, 1977). [^{14}C]Dicyclohexylcarbodiimide (DCCD) was incorporated into the 8 kDa CF_0 protein according to Nelson *et al.* (1977). [^{3}H] chlorpromazine photoaffinity labeling was carried out on some samples prior to the [^{14}C] DCCD incorporation step.

RESULTS AND DISCUSSION

After finding evidence that Ca^{+2} ions play a specific role in regulating what we hypothesized is a gating function of H^+ fluxes between localized and delocalized phases (Chiang and Dilley, 1987), we began a survey of known pharmacologically active Ca^{+2} antagonists to test their effects on the thylakoid system. Two classes of agents were tested initially, those that are Ca^{+2} channel blockers, such as verapamil (Galizzi *et al.*, 1986) and calmodulin antagonists such as chlorpromazine (Levin and Weiss, 1979). The Ca^{+2} channel blockers had no detectable effects on any parameters we tested. However, chlorpromazine (CPZ) and trifluoperazine (TF), another phenathiazine-class calmodulin-directed Ca^{+2} antagonist, had very intriguing and selective results, to be discussed next.

Chlorpromazine is known from previous work (Good *et al.*, 1966) to inhibit ATP formation in the 5-20mM range, so we avoided high concentrations in final assay medium by giving the drug at up to 10 μM concentration only during the thylakoid storage stage (a 100-150 fold dilution occurs going into the assay medium). Using that protocol, CPZ or TF did not inhibit ATP formation in the assay medium, measured either as the ATP yield per flash or by the number of flashes required to reach the energization threshold (Table I). The interesting effect of CPZ or TF was to block pyridine-dependent postillumination ATP formation, denoted as PIP^-, measured in the traditional way where ADP and Pi are added after the light stage (after proton accumulation occurs in basal electron flow conditions). The right-hand column of Table I shows that PIP^- ATP yield was increased by the CPZ pretreatment in the absence of pyridine from near 23 to about 32 nmol ATP (mg Chl)$^{-1}$ but that no further increase in PIP^- ATP yield occurred when pyridine was added. The postillumination ATP yield occurring after a train of flashes during which ATP formation occurred (termed PIP^+, see legend of Table I) is always much less than the PIP^- ATP yield -- normally being about 4-7 nmol ATP (mg Chl)$^{-1}$ -- and this parameter was not affected by CPZ pretreatment. As shown earlier (Beard and Dilley, 1986), low salt-stored thylakoids have localized H^+ gradient characteristics (no effect of permeable buffers such as pyridine) based on two criteria; either the number of flashes (amount of protons) required for attaining the threshold energization for ATP formation, or for the PIP^+ mode of postillumination phosphorylation. However, when the flashes are given without ADP and Pi, greater proton accumulation occurs initially in the membrane domains, which we postulate causes displacement of the Ca^{+2} ions at the gating structure, allowing massive H^+ accumulation in the lumen and measurable pyridine buffering effects (Beard *et al.*, 1988). In that situation, even with low salt-stored thylakoids, pyridine enhances the PIP^- ATP yield as expected for delocalized gradient coupling (Beard *et al.*, 1988).

The CPZ effects on PIP^+ and PIP^- ATP yields were studied in more detail, including comparing low salt- and high salt-stored thylakoids. Table II, lines 2 and 3, show that the usual pyridine-dependent PIP^- increase in ATP yield of about 10 nmol ATP (mg Chl)$^{-1}$ (line 1 of Table II) was blocked by the CPZ or TF. The PIP^+ ATP yields were not affected by CPZ or TF. However, as in Table I, the CPZ or TF caused an increase in PIP^- yield in the absence of pyridine, in low salt-stored thylakoids, a remarkable effect that may be a key to understanding what is happening at the molecular level. We could conclude that the CPZ treatment allows the lumenal buffering groups to contribute to H^+ accumulation and subsequent PIP^- ATP formation. However, in that case, pyridine present in the lumen should, but does not, contribute to the H^+ buffering and to PIP^- ATP yields, as it does in the control (line 1,

Table I. Effect of Chlorpromazine on Phosphorylation Parameters at pH 7.5 using Low Salt-Stored Thylakoids.

Treatments	Pyridine	Energization lag Number of flashes	Yield/flash nmol ATP/mg chl	nmol ATP/mg chl PIP^+	PIP^-
low salt	-	45/51	0.52	4.0	23.3
	+	49/58	0.44	4.0	29.7
low salt, plus 5 μM CPZ	-	47/55	0.52	4.3	31.8
	+	50/61	0.50	4.7	28.3

Thylakoids were washed and resuspended in a low salt media containing 2 mM $MgCl_2$, 0.5 mg/ml BSA and 200 mM Sucrose or in the same media with 5 μM Chlorpromazine (CPZ). Flashes (100) were delivered at 5 Hz to thylakoids containing 15 μg Chl suspended at 10°C in 800 ml reaction mixture consisting of 50 mM Tricine-KOH pH 7.5, 10 mM sorbitol, 3 mM $MgCl_2$, 1 mM KH_2PO_4, 5 mM DTT, 0.1 mM methylviologen, 400 mM valinomycin, 5 mM diadenosine pentaphosphate and 10 μl luciferin-luciferase prepared from the LKB ATP assay kit. 0.1 mM ADP and 1 mM KH_2PO_4 were included in the above reaction mixture before flash excitation to ascertain the energization lag, yield/flash and PIP^+. The energization lag was determined by the number of flashes required before the first detectable rise in luminescence (number given before the slash) and by back extrapolation of the steady state rise in luminescence to the base-line (number after the slash). The standard deviation for the onset lag was not more than ± 2, for the yield/flash ± 0.08, for PIP^+ ± 0.9, and for PIP^- ± 2.0. PIP^+, post-illumination phosphorylation occurring after a flash train having ADP, Pi present during the flash excitation. PIP^-, post-illumination phosphorylation in the traditional mode, with ADP, Pi added after the last flash.

Table II). Alternatively, if CPZ bound at the Ca^{+2}-protein adduct site tends to block H^+ ions in the membrane domains from displacing the Ca^{+2} at the gating site, then one would expect more H^+ accumulation in the membrane domains, especially under basal conditions (no ADP) wherein protons are not being effluxed in the energy-linked H^+ efflux event. We showed in previous work that the membrane phase buffering groups can contribute protons to the PIP^- ATP yield, and argued that buffers such as pyridine cannot occupy sufficient space in the localized

Table II. Effect of Salts and Phenoziathazines on Postillumination Phosphorylation ATP Formation at pH 8.0.

Treatments		PIP$^+$ ATP yield nmol ATP/mg chl -Pyr	+Pyr	PIP$^-$ ATP yield nmol ATP/mg chl -Pyr	+Pyr
1.	low salt (control)	2.8 ± 0.5	3.0 ± 0.2	19.6 ± 0.4	29.8 ± 0.6
2.	low salt 4 mM CPZ	3.3 ± 0.1	3.4 ± 0.4	29.7 ± 2.6	26.2 ± 2.0
3.	low salt 5 mM TF	3.3 ± 0.4	3.0 ± 0.8	33.3 ± 3.1	29.1 ± 4.1
4.	high salt (control)	3.6 ± 0.1	5.3 ± 0.4	22.5 ± 1.6	32.4 ± 2.4
5.	high salt 4 mM CPZ	3.6 ± 0.4	5.4 ± 0.4	21.4 ± 1.1	31.7 ± 1.9

Thylakoids were washed and resuspended in either a low salt media described in Table I or in a high salt media in which 30 mM sucrose and 100 mM KCl replaced 200 mM sucrose. Either chlorpromazine (CPZ) or trifluoperazine (TF) was added to the above resuspension mixtures as treatments during thylakoid storage, prior to dilution into the assay media. PIP$^+$ and PIP$^-$ ATP yields were determined as described in Table I. In PIP$^-$ experiments 100 flashes were delivered at 5 Hz.

domains to increase the proton content of the domains significantly (Beard *et al.*, 1988). Therefore, if CPZ blocks domain protons from going through the putative gate into the lumen, their greater accumulation in the domains should cause more PIP$^-$ ATP yield and there should be no pyridine effect on the PIP$^-$ ATP yield. Both of those results were observed. This interpretation requires further testing, and such experiments are in progress.

With high salt-stored thylakoids CPZ had no effect on the PIP^- or PIP^+ ATP yields with or without pyridine (Table II, lines 4 and 5). We interpret this result as owing to CPZ not binding to the hypothesized Ca^{+2}-protein gate structure, which is suggested to be in an open configuration in the high salt-stored thylakoid case. As an aside, but an important one, this can be taken as evidence that CPZ at these concentrations does not cause uncoupling, an important point already suggested by the increased PIP^- ATP yield in the CPZ-treated low salt-stored thylakoids and shown by a different analysis below. Moreover, the lack of CPZ effects in the high salt-stored samples is consistent with the notion that the phenothiazine Ca^{+2} antagonists do not bind to the Ca^{+2}-free form of the target proteins involved in certain Ca^{+2}-mediated intracellular signaling events [calmodulin, for instance (Levin and Weiss, 1979; Prozialeck *et al.*, 1981)].

One interpretation of the above results -- not the only one, but at present we favor it -- is that CPZ binds to the Ca^{+2}-occupied form of the putative H^+ flux gate structure in the low salt-stored thylakoids, and in so doing, blocks the H^+ fluxes into the lumen from the domains and/or blocks H^+ fluxes originating in the lumen from going out through the CF_0-CF_1 complex. In high salt-stored thylakoids, earlier data have suggested that Ca^{+2} ions are displaced from the gating structure by the K^+ ions (Chiang and Dilley, 1987), so, based on the calmodulin-Ca^{+2} antagonist model, the CPZ or TF would not bind tightly; hence there would be no detectable effects of CPZ or TF. It is known that phenathiazine Ca^{+2} antagonists bind tightly only to the Ca^{+2}-occupied calmodulins and when so bound, block the normal functions of the Ca^{+2}-calmodulin complex (Levin and Weiss, 1979). We have previously shown that low salt-stored thylakoids -- which maintain localized proton gradients with minimal proton flux into the lumen under phosphorylating conditions -- can be induced to accumulate large amounts of protons in the lumen by withholding ADP and Pi (Beard *et al.*, 1988). Without the H^+ efflux coupled to ATP formation, the excessive H^+ accumulation in the localized domains apparently protonates the carboxyl groups holding the Ca^{+2} in the closed gate configuration, causing displacement of the Ca^{+2} and opening the H^+ flux pathway into the lumen. That interpretation was supported by the observation that lumenally located pyridine caused increased postillumination phosphorylation of the traditional type (ADP and Pi added only in the dark, the PIP^- mode) in low salt-stored thylakoids, yet those thylakoids showed no pyridine-dependent postillumination ATP yield when ADP and Pi were present during the flash train (Beard and Dilley, 1988; Beard *et al.*, 1988). Direct measurement of lumenally-located pyridine effects on H^+ uptake in the coupled and basal conditions also support that interpretation (Beard *et al.*, 1988).

Therefore, a way of testing whether CPZ blocks the H^+ flux from the membrane domains through the gate into the lumen is to measure H^+ uptake into the lumen with and without CPZ pretreatment. Lumenal H^+ uptake is distinguished from membrane domain uptake by the former showing significantly increased uptake when pyridine is present in the lumen to buffer protons arriving there, allowing more H^+ uptake (Nelson *et al.*, 1971). Such an experiment is shown in Table III, and it is clear that 5 μM CPZ given to low salt-stored thylakoids, reduced by about 60-70%, the pyridine-dependent extra H^+ uptake driven by 100 flashes at 5 Hz. CPZ given to high salt-stored thylakoids did not inhibit the pyridine-dependent extra H^+ uptake (Table III). From past experiments (Chiang and Dilley, 1987) we expect the high salt storage conditions to displace Ca^{+2} from the H^+ flux gate. If, as seems to be the case, this system interacts with CPZ similarly to the calmodulins, then it is expected that when Ca^{+2} is displaced from the gate structure the CPZ will not bind tightly. This could account for the lack of CPZ effect on the pyridine-dependent increase in lumenal H^+ uptake in the high salt-stored thylakoids.

Could the CPZ inhibition of pyridine-dependent H^+ uptake be caused by some type of increased H^+ efflux in the CPZ-treated thylakoids? If so, an increased H^+ efflux rate constant would be detected. Our measurements of the k_d for H^+ efflux showed no increase in the CPZ-treated low salt-stored thylakoids. In both control and CPZ treated (5μM CPZ present during the storage, but not in the ΔH^+ assay, just as for the phosphorylation studies) the k_d was $0.037s^{-1}$ (data not shown). This point is also supported by the observation that CPZ treatment did not decrease the efficiency of phosphorylation measured by the length of the onset lag for ATP formation or ATP yield per flash (Table I), as would have been expected if the CPZ pre-treatment caused an increase in H^+ leak fluxes.

The functional studies are consistent with the hypothesis that CPZ tightly binds to the Ca^{+2}-occupied form of a gate or switch that regulates H^+ flux from membrane domains into the lumen. Experiments are presently in progress to test whether CPZ bound to the Ca^{+2}-regulated gating structure can block proton flux from the lumen out through the CF_0-CF_1. One way to do this is to test for CPZ effects on the acid-base jump ATP formation driven by succinic acid-derived protons in the lumen.

IDENTIFYING THE Ca^{+2} GATING PROTEINS

Chlorpromazine can be used directly as a photoaffinity probe for calmodulin proteins due to its being activated by UV light to a reactive species which can covalently bind to the calmodulin (Prozialeck *et al.*,

Table III. Effect of Chlorpromazine on Flash Induced Proton Uptake in the Absence and Presence of Pyridine at pH 7.5 (100 flashes delivered at 5 Hz)

Treatment		nmol H^+ (mg Chl)$^{-1}$ -Pyr	+Pyr	Difference +Pyr-(-)Pyr
Low salt		211 ± 4	338 ± 15	127
Low salt	+ 5 μM CPZ	226 ± 19	258 ± 9	32
Low salt	+ 1 mM $CaCl_2$	191 ± 11	303 ± 18	112
Low salt	+ 5 μM CPZ + 1 mM $CaCl_2$	183 ± 11	227 ± 3	44
High salt		192 ± 15	320 ± 14	128
High salt	+ 5 μM CPZ	187 ± 13	310 ± 10	123

Thylakoids were washed and resuspended in either a low salt or high salt media described in previous Tables. $CaCl_2$ and/or Chlorpromazine was/were added as indicated. 100 flashes were delivered at 5 Hz to thylakoids diluted in 2 ml (20 μg/ml) of reaction mixture consisting of 50 mM sorbitol, 0.3 mM Tricine pH 7.5, .15 mM KCl, 3 mM $MgCl_2$, 0.1 mM MV and 400 nM Valinomycin in the presence and absence of 5 mM pyridine. The reaction solution, thermostatted at 10°C, was adjusted to pH 7.5 prior to each assay and continuous recordings of pH obtained using a strip chart recorder connected to a Corning Model 12 pH meter.

1981). It is known that the photoaffinity labeling is quite specific for the Ca^{+2}-calmodulin complex, apparently because the reactive CPZ covalently bonds only to proteins it is tightly bound to prior to photoactivation; ie., Ca^{+2}-free calmodulin, BSA or hemoglobin did not show much binding of CPZ (Prozialeck *et al.*, 1981). We employed the photoaffinity labeling protocol with [^{3}H]CPZ added to thylakoids, followed by SDS-PAGE separation of thylakoid proteins. By comparing the label incorporation with low salt- and high salt-stored thylakoids, we found that three low molecular weight polypeptides are significantly more labeled in the low salt-stored sample -- the sample believed to have Ca^{+2} bound to the H^+ gate structure -- than to high salt-stored thylakoids. Figure 2 shows a representative result indicating the three major bands having significant differential labeling at 8, 11-12, and 13 kDa. The 8 kDa band runs close to the M_r of the DCCD-sensitive CF_0 subunit (Nelson *et al.*, 1977), so we

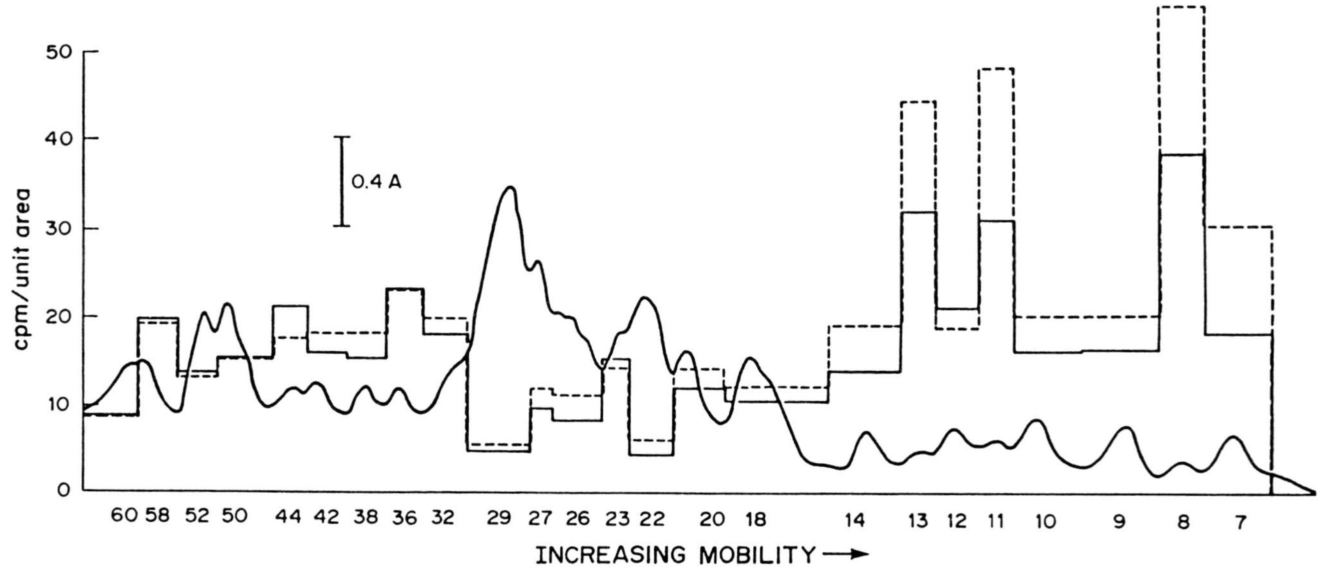

Figure 2. Coomassie stain band pattern (—) of an SDS-PAGE gel of low salt-stored thylakoids, after photoaffinity labeling with [^{3}H]chlorpromazine. Thylakoids from low or high salt-stored conditions were suspended in either of those media at 25 g Chl ml^{-1} with 5 M [^{3}H]chlorpromazine. Exposure to a Mineralit UV source was for 20 min at 4°C. Gels were run on the SDS solubilized thylakoid proteins with the standard techniques. Radioactivity of gel slices, expressed as CPM per unit area of stained band; — data from high salt-stored thylakoids, - - - - low salt-stored sample. Count data are averages from these gel lanes. The experiment was repeated once on a different day with different thylakoids, giving good agreement.

isolated that protein from [^{3}H]CPZ-labeled thylakoids, using the butanol extraction, ether precipitation technique (Nelson *et al.*, 1977). Figure 3 shows labeling associated with a prominent band in the 8 kDa region of the SDS-PAGE gel. Significant protein and ^{3}H counts occurred also in the 6-7 kDa region, caused either by a second protein present in the preparation, or to an altered form of the DCCD-sensitive 8 kDa protein.

To test these alternatives, we used the diagnostic [^{14}C]DCCD labeling which is common to a subunit in F_0 complexes of all energy transducing membranes (Nelson *et al.*, 1977). No other thylakoid polypeptides in the 6-12 kDa range are known to bind covalently DCCD with short term labeling with <100 μM DCCD. Thylakoids were first given UV photoactivation in the presence of [^{3}H]CPZ followed by 50 μM [^{14}C]DCCD. The whole membrane SDS-PAGE gel patterns (Fig. 4) from that treatment showed the most significant labeling for both isotopes in the 7-9 kDa region, suggesting that the 8 kDa CF_0 subunit is a common target site for both DCCD and CPZ labeling. Double labeled thylakoids were carried through the butanol-ether purification steps and both ^{14}C and ^{3}H labels were concentrated in the same bands (data not shown). The protein was somewhat smeared out in such gels, perhaps owing to effects on protein mobility resulting from labeling the small polypeptide with two ligands. These effects must be studied further, but it seems very likely that the 8 kDa CF_0 subunit is one of the thylakoid proteins that binds CPZ tightly and thus it is a candidate for the Ca^{+2} binding site involved in the proton flux gating function recently identified.

Whether either of the other two [^{3}H]CPZ labeled polypeptides are also CF_0 subunits remains to be tested. There are known CF_0 subunits in the 13-15 kDa range (Pick and Racker, 1979).

SPECULATIONS ON PHYSIOLOGICAL RELEVANCE

Functional criteria (H^+ fluxes from the membrane into the thylakoid lumen and the attendant postillumination ATP formation from lumenal proton sources) and structural identification, support the hypothesis that one (or more) of the CF_0 proton channel subunits is involved in a Ca^{+2}-regulated gate that controls H^+ fluxes between localized domains and the thylakoid lumen. As discussed earlier (Dilley *et al.*, 1987; Beard and Dilley, 1986, 1988), our working hypothesis is that thylakoids can utilize either localized or delocalized proton gradients for driving ATP formation. Data from different experimental approaches support the possibility of localized coupling (see Introduction), and few workers ever doubted that delocalized gradients *can* drive ATP formation. The important question now is to understand what regulates the switching between the two modes, and the conditions that dictate when localized or

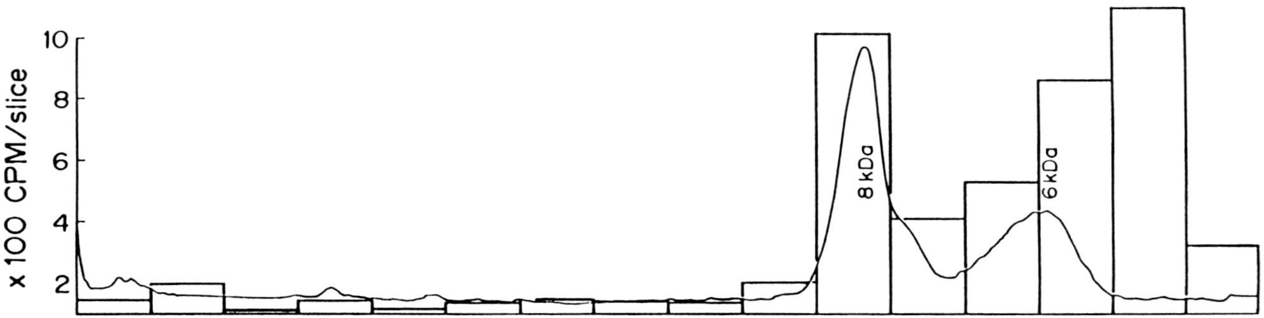

Figure 3. Coomassie stained SDS-PAGE gel of an 8 KDa CF_O preparation from photoaffinity labeled thylakoids. Thylakoids (33 μg Chl ml^{-1}) were photoaffinity labeled with 5 μM [^{3}H]chlorpromazine for 20 min in UV light, followed by butanol extraction and ether precipitation to obtain the 8 KDa CF_O protein. The main band is at 8 KDa. Radioactivity was counted in 5 mM gel slices and is given by the bar graph.

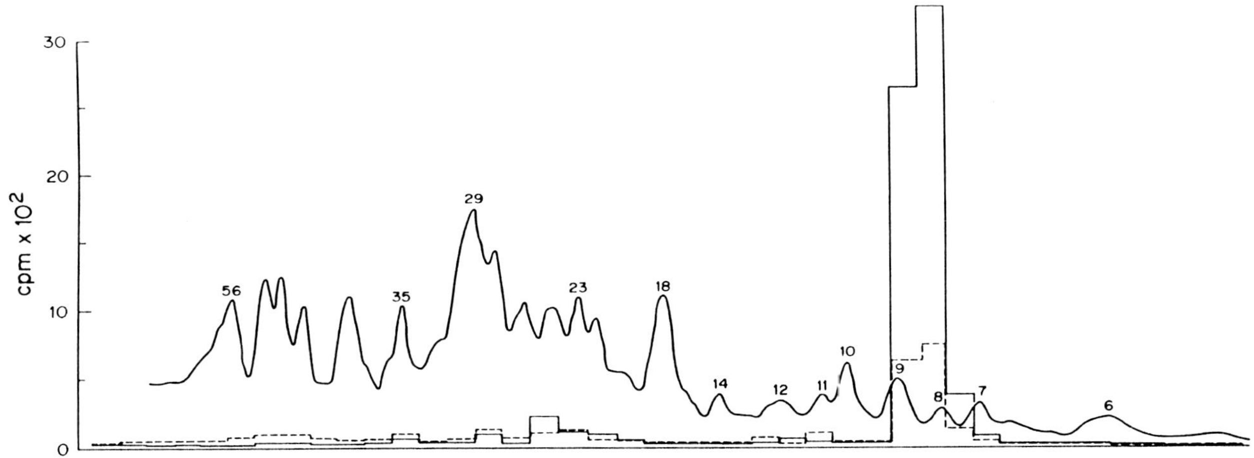

Figure 4. Optical density profile and dual labeling pattern for [^{14}C]DCCD and [^{3}H]chlorpromazine labeled low salt-stored whole thylakoid proteins. Low salt-stored chloroplast membranes (8 mg chlorophyll) were irradiated for 1 hour at 3°C with 254 nm UV light in the presence of 6 μM (22.3 Ci/mmol) [^{3}H]chlorpromazine (- - -) in the modification media containing 0.2 M sucrose, 5 mM Hepes, pH 7.5, 2 mM $MgCl_2$ and 0.5 mM $CaCl_2$. The chloroplasts were then washed once and resuspended in the same modification media as above with 400 nmol (54.8 mCi/mmol) [^{14}C]DCCD (—). After 1 hour incubation at 25°C the chloroplasts were washed and resuspended in 1 ml cold water. Samples of whole thylakoid proteins were delipidated by washing once in 95% acetone prior to loading approximately 150 μg protein/lane on 15% SDS-polyacrylamide slab gels. Gel lanes were sliced in 2 mm slices, digested by 5 h incubation at 80°C in 0.6 mo 3% (v/v) H_2O_2 and counted after addition of 10 ml tritosol scintillation cocktail.

delocalized proton gradients are called into play. Our working hypothesis, diagrammed in Fig. 1, as to how the H^+ gate may function can be stated in the following terms: (a) Under "normal," dark conditions there is sufficient Ca^{+2} ions in thylakoid membranes to keep the putative H^+ gate carboxyl groups in the Ca^{+2}-bound or gate closed form. Protons in the lumen are postulated to not have direct access to the CF_0-CF_1 complex in the closed gate configuration. (b) As redox turnovers are activated protons are initially released from the PSII water-oxidizing system and the plastoquinol oxidation into membrane-localized domains containing ≈ 30 nmol (mg Chl)$^{-1}$ amine buffers and ≈ 150-200 nmol (mg Chl)$^{-1}$ carboxyl groups (Beard *et al.*, 1988). In the resting state the domain carboxyls probably occur as internal salt bridges to amines or as -COO^-...Mg^{++} or K^+ salts. The domain amine groups are usually protonated. Protons accumulating in the domains are suggested to be in direct communication with the CF_0-CF_1 complex, based on experiments correlating the energization onset of ATP formation with the status of the domain buffer group protonation and on the Ca^{+2}-occupied or Ca^{+2}-displaced form of the H^+ gating structure (Dilley and Schreiber, 1984; Theg *et al.*, 1988). (c) When ADP and Pi are present, the development of the proton gradient in localized domains to the thermodynamic threshold (≈ 2.3 ΔpH units in the absence of an electric field contribution and under our usual ADP, Pi and ATP concentrations) can initiate ATP formation before the local pH is acidic enough to protonate the gate carboxyl groups and displace the Ca^{+2}. Preliminary evidence indicates that localized coupling can be maintained in steady state conditions, but this is still being tested (Ewy, R., Theg, S., and Dilley, R., work in progress). The model we have proposed is consistent with that, provided efficient coupling to ATP formation is possible, thus utilizing the proton gradient fast enough to keep the local pH above an as-yet-unspecified value. Moreover, other work, using steady state conditions, has been interpreted as consistent with localized coupling (Pick *et al.*, 1987; Opanasenko *et al.*, 1985). Our present view of the hypothesis is that when ADP or Pi are absent, or if coupling efficiency is low for some other reason, then the pH in the localized domains can become acidic enough to protonate the carboxyl groups proposed to be associated with Ca^{+2} binding to the putative H^+ gate structure. At that point the gate would open, allowing the localized proton gradient to equilibrate freely with the lumen and delocalized coupling behavior would be observed.

What physiological advantage would such dual coupling pathways have over a straightforward, solely delocalized model? One possibility, discussed in some detail by Dilley *et al.* (1987), is that by generally using localized proton gradients, the continuous (in the light) acidification of the lumen is avoided, thus minimizing the potential threat that H^+-K^+ or H^+-Na^+ exchanges pose as stimulators of net salt uptake and swelling. As an aside, given the high Pi levels in chloroplasts, and the fact that Pi is permeant enough ($t_{\frac{1}{2}}$ ≈ 30 s, Ort *et al.*, 1976) to accumulate in the lumen, it could be expected that Na^+ or KH_2PO_4 accumulation might occur if solely bulk phase H^+ gradients always occurred. On the other hand, thylakoids may have anion exchanges to regulate lumen Pi content. While the swelling response is not constant, even for apparently similar conditions (see above ref.), thylakoids can sometimes swell greatly due to net NaCl, KCl uptake, driven ultimately by acidification of the lumen. The variability in the swelling response suggests variations which are not understood in H^+-K^+ or H^+-Na^+ exchange activity. This is an area of osmoregulation that should be studied in the near term. We suggest that the capacity to do localized proton gradient coupling may be part of the osmoregulation strategy used by plants to avoid serious problems in organelle volume changes.

A related aspect which merits study is the effect on increased Mg^{++} and Ca^{+2} efflux to the stroma which could occur as protons flow into the lumen and neutralize $-COO^-$ groups. Increased divalent cation exchange into the stroma could have important regulatory effects on metabolic events in the stroma. The lumen has roughly 2-3 times more carboxyl group buffering than the membrane domains. An interesting question is whether there is any tendency for either of those carboxylate arrays to be neutralized by mono- or divalent cations in the resting state.

Another obvious "benefit" derived from localized gradient coupling is that the threshold energization level for driving ATP formation can be reached more quickly when filling localized domains compared to acidifying the larger volume and greater buffer capacity in the lumen. For example, at 18°C we commonly observe a two- to three-fold longer onset lag when delocalized gradients are established compared to localized coupling conditions (30-40 single turnover flashes compared to 10-15 with no added pyridine). Lag lengths of 30-40 single turnover flashes are equivalent to about 90-120 nmol H^+ (mg Chl)$^{-1}$. At moderate to low light intensities [giving rates of 100-200 mmol H^+ (mg Chl hr)$^{-1}$] this could mean 2-4 sec of illumination to reach the threshold H^+ uptake. Whether this is a physiologically important efficiency factor remains to be determined. But the possibility exists that localized proton gradient coupling may provide a more rapidly responding energy coupling system, a factor that could be physiologically useful in certain situations where leaves are regularly exposed to rapid fluctuations in light intensity (as in the "light fleck" phenomenon in forest environments).

REFERENCES

Beard WA, Dilley RA (1986). A shift in chloroplast energy coupling by KCl from localized to bulk phase delocalized proton gradients. FEBS Lett 201:57-62.

Beard WA, Dilley RA (1988). ATP formation onset lag and post-illumination phosphorylation initiated with single-turnover flashes. III. Characterization of the ATP formation onset lag and post-illumination phosphorylation for thylakoids exhibiting localized or bulk-phase delocalized energy coupling. J Bioenerg Biomemb 20:129-154.

Beard WA, Chiang G, Dilley RA (1988). ATP formation onset lag and post-illumination phosphorylation initiated with single-turnover flashes. II. Two modes of post-illumination phosphorylation driven by either delocalized or localized proton gradient coupling. J Bioenerg Biomemb 20:107-128.

Boyer PD (1987). The unusual enzymology of ATP synthase. Biochemistry 26:8503-8507.

Cain BD, Simoni RD (1986). Impaired proton conductivity resulting from mutations in the a subunit of F_1F_0 ATPase in *E. coli*. J Biol Chem 261:10043-10050.

Chiang G, Dilley RA (1987). Evidence for Ca^{2+}-gated proton fluxes in chloroplast thylakoid membranes: Ca^{2+} controls a localized to delocalized proton gradient switch. Biochemistry 26:4911-4916.

Davenport JW, McCarty RF (1980). The onset of photophosphorylation correlates with the rise in transmembrane electrochemical proton gradients. Biochim Biophys Acta 589:353-357.

Dilley RA, Theg SM, Beard WA (1987). Membrane-proton interactions in chloroplast bioenergetics. Annu Rev Plant Physiol 38:347-389.

Ferguson SJ (1985). Fully delocalized chemiosmotic or localized proton flow pathways in energy coupling? A scrutiny of experimental evidence. Biochim Biophys Acta 811:47-95.

Galizzi J-P, Borsotto M, Barhanin J, Fosset M, Lazdunski M (1986). Characterization and photoaffinity labeling of receptor sites for the Ca^{2+} channel inhibitors d-cis-diltiazen, (%)-bepridil, desmethoxyverapamil, and (+)-PN 200-110 in skeletal muscle transverse tubule membranes. J Biol Chem 261:1393-1397.

Gräber P (1982). Phosphorylation in chloroplasts: ATP synthesis driven by $\Delta\psi$ and by ΔpH of artificial or light-generated origin. Curr Topics Memb and Trans 16:215-245.

Good NE, Izawa S, Hind G (1966). Uncoupling and energy transfer inhibition in photophosphorylation. Curr Topics in Bioenerg 1:75-112.

Horner RD, Moudrianakis EN (1983). The effect of permeant buffers on initial ATP synthesis by chloroplasts using rapid mix-quench techniques. J Biol Chem 258:11643-11647.

Jagendorf AT, Uribe E (1966). ATP formation caused by acid-base transition of spinach chloroplasts. Proc Natl Acad Sci USA 55:170-177.

Klee CB, Crouch TH, Richman PG (1980). Calmodulin. Annu Rev Biochem 49:489-515.

Laszlo JA, Baker GM, Dilley RA (1984). Chloroplast thylakoid proteins having buried amine buffering groups. Biochim Biophys Acta 764:160-169.

Levin RM, Weiss B (1979). Selective binding of antipsychotics and psychoactive agents to the calcium-dependent activator of cyclic nucleotide phosphodiesterase. J Pharmacol Exp Ther 208:454-459.

Nelson N, Nelson H, Naim Y, Neumann J (1971). Effect of pyridine on the light-induced pH rise and post-illumination ATP synthesis in chloroplasts. Arch Biochem Biophys 145:263-267.

Nelson N, Eytan E, Notsani BE, Sigrist H, Sigrist-Nelson K, Gitler C (1977) Isolation of a chloroplast DCCD-binding proteolipid active in proton translocation. Proc Natl Acad Sci (US) 74:2375-2378.

Opanasenko VK, Red'ko TP, Kuz'mina VP, Yaguzhinsky LE (1985). The effect of gramicidin on ATP synthesis in pea chloroplasts: two modes of phosphorylation. FEBS Lett 187:257-269.

Ort DR, Dilley RA, Good, NE (1976). Photophosphorylation as a function of illumination time. II. Effects of permeant buffers. Biochim Biophys Acta 449:108-124.

Pick U, Racker E (1979). Purification and reconstitution of the DCCD-sensitive ATPase complex from spinach chloroplasts. J Biol Chem 254:2793-2799.

Pick U, Weiss M, Rottenberg H (1987). Anomalous uncoupling of photophosphorylation by palmitic acid and by gramicidin D. Biochemistry 26:8295-3802.

Prozialeck WC, Cimino M, Weiss B (1981). Photoaffinity labeling of calmodulin by phenothiazine antipsychotics. Mol Pharmacol 19:264-269.

Sigalat C, Haraux F, de Kouchkovsky F, Hung SPN, de Kouchkovsky Y (1985). Adjustable microchemiosmotic character of the proton gradient generated by systems I and II for photosynthetic phosphorylation in thylakoids. Biochim Biophys Acta 809:403-413.

Theg SM, Chiang G, Dilley RA (1988). Protons in the thylakoid membrane-sequestered domains can directly pass through the coupling factor during ATP synthesis in flashing light. J Biol Chem 263:673-681.

Uribe E, Jagendorf AT (1967). On the location of organic acids in acid-induced ATP synthesis. Plant Physiol 42:697-705.

Vinkler C, Avron M, Boyer PD (1980). Effects of permeant buffers on the initial time course of photophosphorylation and postillumination phosphorylation. J Biol Chem 255:2263-2266.

Photosynthesis, pages 457–473

BIOCHEMICAL STUDIES OF PHOTOSYNTHESIS: FROM CO_2 TO SUCROSE

Richard C. Leegood

Research Institute for Photosynthesis and Department of Plant Sciences, University of Sheffield, Sheffield, S10 2TN, U.K.

Regulation of photosynthetic carbon dioxide assimilation to sucrose is a complex process in which phosphate plays a key role (Walker, 1976). The product of chloroplast photosynthesis is triose phosphate and this esterified phosphate must be recycled to provide Pi to allow continuation of ATP synthesis in the chloroplast (Walker and Robinson, 1978). Pi therefore provides a feedback link from carbon assimilation to electron transport. Sucrose or organic acid synthesis in the cytosol releases Pi which then must re-enter the chloroplast *via* the phosphate translocator (Heber and Heldt, 1981) for continued triose phosphate export (Fig. 1). However, the rate of Pi recycling which occurs in the stroma as a result of starch synthesis is insufficient in C_3 plants to support maximal rates of photosynthesis (Walker, 1976). For this reason, Pi will limit chloroplast photosynthesis at suboptimal concentrations and in leaves must be present in the cytosol at a suitable concentration. The relative amounts of Pi within the chloroplast and cytoplasm will influence whether triose-P can be exported from the stroma *via* the phosphate translocator and Pi is thus an important factor which determines the cellular partitioning of carbon between starch and sucrose synthesis.

Orthophosphate also plays a direct role in the modulation of the activity of certain enzymes, both within the chloroplast [e.g. FBPase and SBPase (Charles and Halliwell, 1980; Woodrow *et al.*, 1983), RuBP carboxylase (Heldt *et al.*, 1978), NADP-glyceraldehyde-P dehydrogenase (Wolosiuk and Buchanan, 1978), Ru5P kinase (Anderson, 1973; Gardemann *et al.*, 1983)] and in the cytoplasm [e.g. sucrose-P synthase (Harbron *et al.*, 1981; Doehlert and Huber, 1983, 1984; Stitt *et al.*, 1988) and FBPase (Harbron *et al.*, 1981; Stitt *et al.*, 1985)].

RELATIONSHIPS BETWEEN PI STATUS AND CONTENTS OF PHOSPHORYLATED INTERMEDIATES

The work of Stitt has shown that FBPase in the cytosol acts as a valve for triose-P utilization in sucrose synthesis, such that a threshold

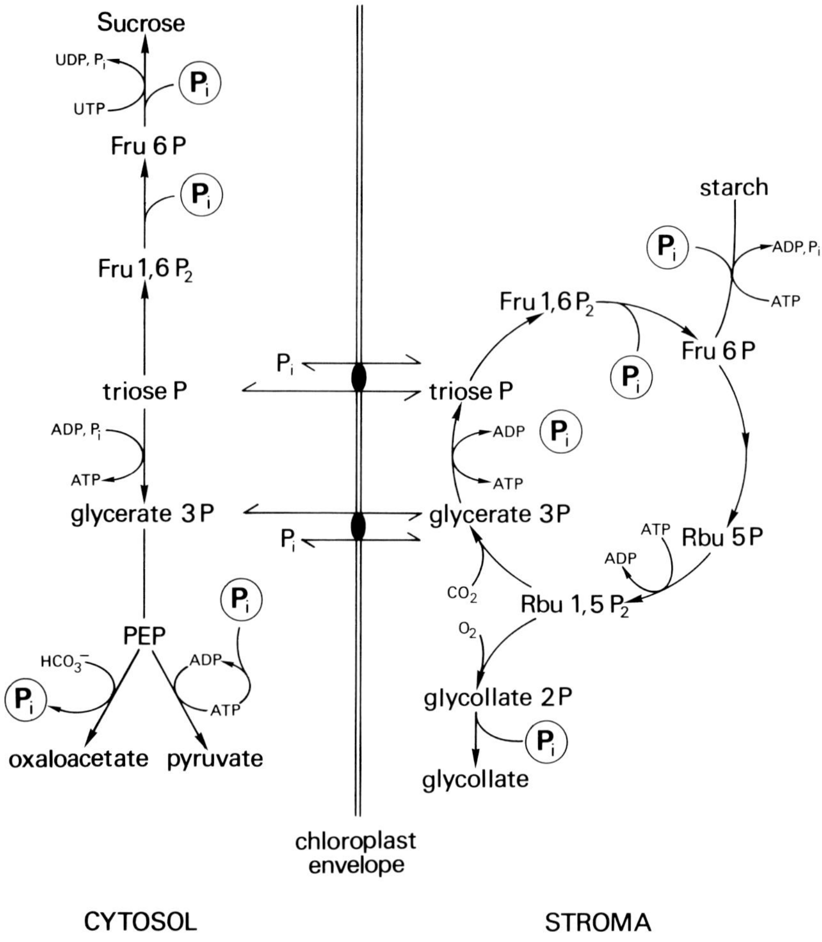

Figure 1. Simplified representation of points at which Pi may be recyled during photosynthesis.

level must be built up before withdrawal of triose-P from the Calvin cycle can occur (Heldt and Stitt, 1987). However, if increases in the rate of sucrose synthesis were driven entirely by increases in substrates, then large amounts of phosphorylated intermediates would be required for high rates of sucrose synthesis. The consequence would be depletion of Pi to levels where it might start to limit sucrose synthesis by preventing transport of triose-P across the chloroplast envelope. In practice this trade-off between amounts of intermediates and the amount of Pi is avoided by changes in the substrate affinity of the enzymes FBPase and sucrose-P synthase such that cytosolic fluxes can be altered over a substantial range of rates of CO_2 fixation without appreciable changes in the contents of phosphorylated intermediates. For example, the stability of the pools of hexose-P and UDP-glucose (Gerhardt *et al.*, 1987; Heldt and Stitt, 1987) suggests light modulation of sucrose-P synthase. A similar situation obtains in the Calvin cycle, where light-induced increases in the affinities of FBPase and SBPase for their substrates occur (Charles and Halliwell, 1980; Woodrow *et al.*, 1983). In the cytosol, regulation of the conversion of triose phosphate to F6P is achieved by modulation of the affinity of the cytosolic FBPase for its substrate by fructose 2,6-bisP (Cseke *et al.*, 1982) which is a competitive inhibitor and which also increases the sensitivity of the enzyme to inhibition by AMP and Pi. The amount of fructose 2,6-bisP is governed by the relative concentrations of PGA, DHAP, F6P, and Pi through the differential action of these metabolites on the activation of the enzymes fructose 6-phosphate 2-kinase and fructose 2,6-bisP phosphatase (Stitt *et al.*, 1987). If, for example, sucrose accumulates in the cytosol an increased level of F6P simultaneously activates of kinase and inactivates the phosphatase, leading to an increase in fructose 2,6-bisP. This action serves to decrease export of triose phosphate from the chloroplast.

Sucrose-P synthase is subject to light modulation which involves a change in its sensitivity to inhibition by Pi (Stitt *et al.*, 1988). It is also stimulated by G6P (Doehlert and Huber, 1983,1984) such that the G6P/Pi ratio provides a means of adjusting the activity of this enzyme to the supply of carbon from the chloroplasts.

REGULATION OF THE METABOLICALLY ACTIVE POOL OF PI

As far as Pi metabolism is concerned we can regard the leaf cell as having two major compartments, the metabolic pool of the cytoplasm and chloroplasts, and the essentially non-metabolic pool of the vacuole. In the long-term, the Pi pool in the cytoplasm remains relatively constant. For example, when sycamore cells are transferred to either phosphate-deficient or phosphate-rich media the Pi of the cytoplasm is maintained at the expense of the vacuolar Pi pool (Rebeille *et al.*, 1983). A similar situation obtains in barley leaves (Foyer and Spencer, 1986) and it is therefore highly unlikely that substantial overall changes in the Pi content of the cytoplasm occur in the long term. However, short-term changes in the cytoplasmic Pi concentration would not be appreciably ameliorated by transport of Pi from the vacuole, which is relatively slow compared with Pi transport across the plasmalemma and chloroplast envelope (Woodrow *et al.*, 1984, Fliege *et al.*, 1978; Bligny *et al.*, 1988).

SYMPTOMS OF PI DEFICIENCY IN VIVO

A good deal of our knowledge of the function of phosphate *in vivo* has come from the use of Pi-sequestering agents. Analogues of glucose such as D-mannose and 2-deoxy-D-glucose are phosphorylated by hexokinase, but many plants are unable subsequently to metabolise these phosphorylated compounds at appreciable rates (Herold and Lewis, 1977). Hence cytoplasmic Pi becomes sequestered into organic compounds. Recently, glycerol and choline have also been identified as phosphate-sequestering agents in sycamore cell suspension cultures (Bligny *et al.*, 1988) and glycerol is also effective in leaves (although choline is not) (Leegood *et al.*, 1988). Glycerol presumably enters metabolism through the action of glycerol kinase, as it leads to large increases in glycerol-3-P. In barley leaves, phosphate deficiency can be largely explained by sequestration of Pi into glycerol-3-P (Leegood *et al.*, 1988).

Growth in limiting Pi is an alternative way of manipulating the metabolically active pool of Pi (Dietz and Foyer, 1986; Brooks, 1986; Brooks *et al.*, 1988; Rao *et al.*, 1987a,b; Prinsley and Leegood 1986), although the symptoms are not always identical to those of plants fed with Pi-sequestering agents. In general, the symptoms of induced Pi deficiency are as follows:

(i) The inhibition of photosynthesis by feeding Pi-sequestering agents or growth in low Pi can be rapidly reversed by feeding Pi to leaves (Sivak and Walker, 1986; Dietz and Foyer, 1986; Leegood *et al.*, 1988).

(ii) Phosphate-sequestering agents reduce light-saturated rates of photosynthesis and light saturation of photosynthesis occurs at a lower irradiance, while there is no effect on the light dependence of carbon assimilation at lower, limiting, irradiance. Thus carbon metabolism is inhibited without any pronounced effects upon electron transport in limiting light (Walker and Osmond, 1986; Leegood *et al.*, 1988). Similar behaviour is observed after feeding Li^+, which inhibits the cytosolic FBPase and hence reduces the rate of Pi recycling by sucrose synthesis (Stitt and Schreiber, 1988). Plants grown in limiting Pi also show no decrease in quantum yield (Brooks, 1986; Abadia *et al.*, 1987).

(iii) During photosynthesis, the reduction of glycerate-3-P to triose P is maintained near equilibrium (Dietz and Heber, 1984), thus the ratio $[PGA].[ATP].[NAD(P)H].[H^+]/[triose\ P].[ADP].[NAD(P)^+].[Pi]$ remains close to the equilibrium constant ($9.8.10^{-6}$). The ratio PGA/triose-P allows estimation of the ratio $[ATP].[NAD(P)H]/[ADP].[NAD(P)^+].[Pi]$, the assimilatory force (Heber *et al.*, 1986). PGA/triose-P ratios are increased by Pi-sequestering agents, indicating a limitation by Pi on the generation of assimilatory force (Prinsley and Leegood, 1986; Leegood *et al.*, 1988) which arises by decreased availability of stromal Pi. Similar observations have been made in Pi-limited chloroplasts (Giersch and Robinson, 1987, Furbank *et al.*, 1987a,b).

(iv) Supply of Pi-sequestering agents leads to an increase in thylakoid energisation (Sivak and Walker, 1987), specifically of the rapidly-relaxing component of non-photochemical quenching of chlorophyll fluorescence (Leegood *et al.*, 1988), as would be predicted if the rate of photophosphorylation was limited by Pi, leading to an increase of ΔpH.

(v) Promotion of oscillatory behaviour and lowering of the irradiance threshold at which oscillations occur has been shown following feeding of Pi-sequestering agents to leaves (Walker and Sivak, 1985; Walker and Osmond, 1986; Sivak and Walker, 1986, 1987; Leegood *et al.*, 1988).

(vi) Pi-sequestering agents such as mannose favour the partitioning of fixed carbon into starch (Chen-She *et al.*, 1975). Starch synthesis is promoted when the ratio PGA/Pi in the stroma is high (Heldt *et al.*, 1977), thereby activating ADP-glucose pyrophosphorylase (Sanwal *et al.*, 1968). Thus when isolated chloroplasts are illuminated in a medium deficient in Pi, starch synthesis is favoured (Walker, 1976, Heldt *et al.*, 1977). Similarly, starch synthesis is promoted if plants are grown in limiting Pi (Dietz and Foyer, 1986; Rao *et al.*, 1987a).

(vii) The supply of Pi-sequestering agents or growth on low Pi decreases pools of phosphorylated intermediates (Prinsley and Leegood, 1986; Dietz and Foyer, 1986; Brooks, 1986; Rao *et al.*, 1987b; Sicher and Kremer, 1988).

(viii) Secondary effects may also occur when Pi-sequestering agents are supplied to leaves. Phosphate-sequestering agents have effects on stomata, either causing stomatal opening in the case of mannose (Harris *et al.*, 1983; Morison and Batten, 1986) or stomatal closure in the case of glycerol (Leegood *et al.*, 1988). The accumulation of large amounts of glycerol-3-P in barley has a pronounced secondary effect on photosynthesis by extending the period of the oscillations in the rate of photosynthesis (Fig. 2).

It has been proposed that the supply of Pi may be a crucial factor regulating oscillatory behaviour (Laisk and Walker, 1986, Stitt, 1986), and that a reactivation of sucrose synthesis may be needed before Pi can be recycled in the cytosol to reinitiate photosynthesis during oscillations (Stitt, 1987; Stitt and Schreiber, 1988). Glycerol-3-P is an effective competitive inhibitor of the phosphate translocator (Fliege *et al.*, 1978)

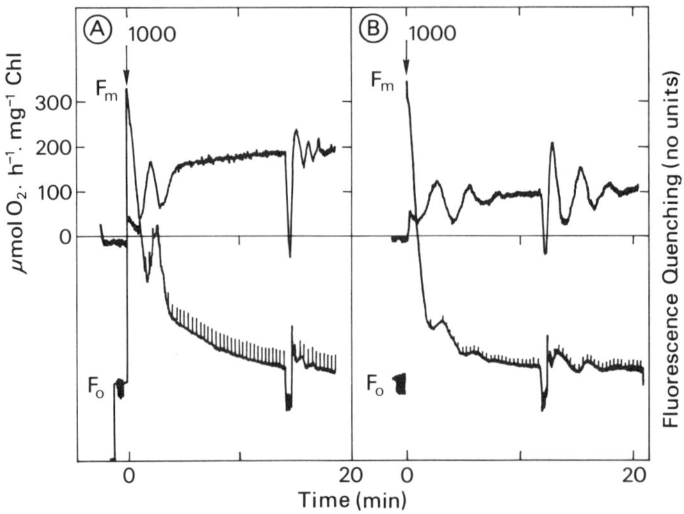

Figure 2. Influence of feeding glycerol (50mM) to barley leaves upon CO_2 -saturated rates of photosynthesis and changes in chlorophyll fluorescence. A, control; B, glycerol-treated. From Leegood *et al*., (1988).

and would therefore delay the supply of Pi from the cytosol (Leegood *et al*., 1988).

(ix) O_2 insensitivity. An observation frequently made with leaves illuminated at low temperatures or at higher than ambient concentrations of CO_2 is the lack of stimulation of photosynthesis by 2% O_2 (Jolliffe and Tregunna, 1968, 1973; Canvin, 1978; Cornic and Louason, 1980; McVetty and Canvin, 1981). This lack of stimulation by 2% O_2 is not due to the absence of photorespiration (Arrabaca *et al*., 1981; Harris *et al*., 1983; Sharkey, 1985). The key to the nature of this inhibition by low O_2 is the influence which the phosphate-sequestering agent, mannose, has on photosynthesis in spinach leaves, where treatment with mannose induces O_2 insensitivity (Fig. 3). The most feasible explanation for this anomalous effect of 2% O_2 is therefore that the system is impaired in respect of its ability to recycle Pi (Leegood, 1985; Sharkey, 1985), although it should be noted that leaves from plants grown in limiting Pi do not exhibit O_2 insensitivity (Brooks *et al*., 1988).

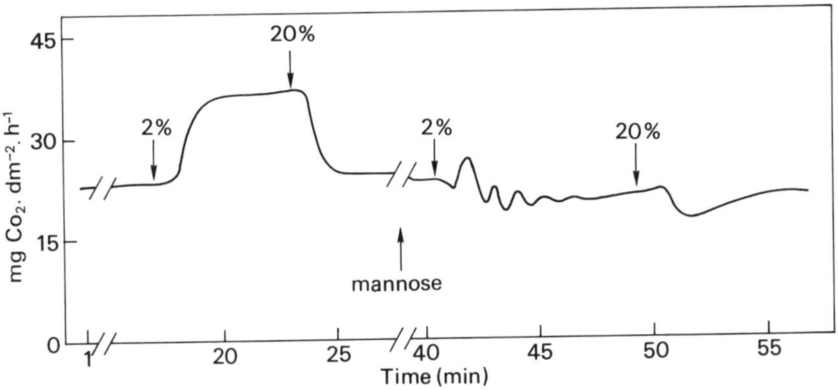

Figure 3. Influence of mannose on the O_2 sentivity of photosynthetic CO_2 uptake by spinach leaves. Redrawn from Harris *et al*., (1983).

REGULATION AND LIMITATION OF PHOTOSYNTHESIS BY PHOSPHATE STATUS

An important feature of most of the above symptoms of changes in phosphate status within leaves is that they do not distinguish between **regulation** by phosphate and **limitation** by phosphate. The regulation of metabolism is a systemic property and all individual steps will contribute to some degree to the control of flux. Every step has a control coefficient (Kacser and Burns, 1973) which is a discrete measure of the ability of that step to regulate the flux in a given circumstance. Many steps, of course, will have a minor influence on the flux, but only if the system is pushed to extremes will one step come to limit the process. Thus in C_3 plants the properties of RuBP carboxylase account for the kinetics of the increase in the rate of CO_2 assimilation with increasing CO_2 at low CO_2 concentrations (Farquhar *et al*., 1980). Even when this limitation by RuBP carboxylase activity occurs, not all of the regulation is invested in RuBP carboxylase. In other words, the occurrence of a limitation does not imply a control coefficient of 1.0 for the step which is limiting.

In this context, **regulation** by phosphate can be considered to comprise those phosphate-dependent changes which can be accommodated by the system without any overall changes in flux. Examples can be seen in the dampening of oscillatory behaviour by Pi (Fig. 4) and in the changes in metabolites that occur without any change in the the photosynthetic rate (Table I). Table I shows metabolite contents in air in control and phosphate-fed spinach leaves. The amounts of triose P, RuBP and hexose P, the ATP/ADP ratio and the assimilatory force were higher in phosphate-fed leaves, although the rate of CO_2 uptake was similar in both

Table I. Effect of feeding phosphate to spinach leaf discs in air on the assimilation rate (A), metabolite levels, and on the ratio of PGA to triose P. Data are means ± SE from four separate experiments. The leaf temperature was 11-12°C. The irradiance was 750 μmol quanta.m^{-2}.s^{-1} . From Leegood and Furbank (1986).

Metabolite (nmol.mg^{-1} Chl)	control	+ Pi
G6P	177 ± 37	247 ± 19
F6P	89 ± 15	115 ± 19
FBP	32 ± 2	43 ± 9
RuBP	116 ± 39	152 ± 18
PGA	267 ± 49	295 ± 33
triose P	44 ± 5	69 ± 6
ATP/ADP	1.34 ± 0.46	2.0 ± 0.48
PGA/triose P	6.3 ± 1.4	4.3 ± 0.2
Assimilatory force (M^{-1})	109 ± 17	146 ± 9
A (μmol.m^{-2}.s^{-1})	13.5 ± 1.6	13.4 ± 1.7

treatments. The important point here is that the regulatory importance of phosphate was likely to have been greater in control leaves, and the burden of regulation has shifted, but phosphate was not limiting photosynthesis. On the other hand there are clearly circumstances in which the phosphate status limits photosynthesis, as revealed by a stimulation of the flux by feeding Pi when the leaves are photosynthesising in low O_2 (Fig. 4).

This distinction between regulation and limitation must be borne in mind when we consider the use of symptoms such as oscillations, lack of O_2 sensitivity or the triose-P/PGA ratio as probes of phosphate status. The occurrence of oscillations and O_2 insensitivity and phosphate limitation (or feedback limitation, Sharkey, 1985) are therefore in no sense synonymous. O_2 insensitivity is not of necessity an indication of a Pi-regeneration limitation (Sage *et al.*, 1988), although it is certainly an indicator of the regulatory importance of regulation by Pi.

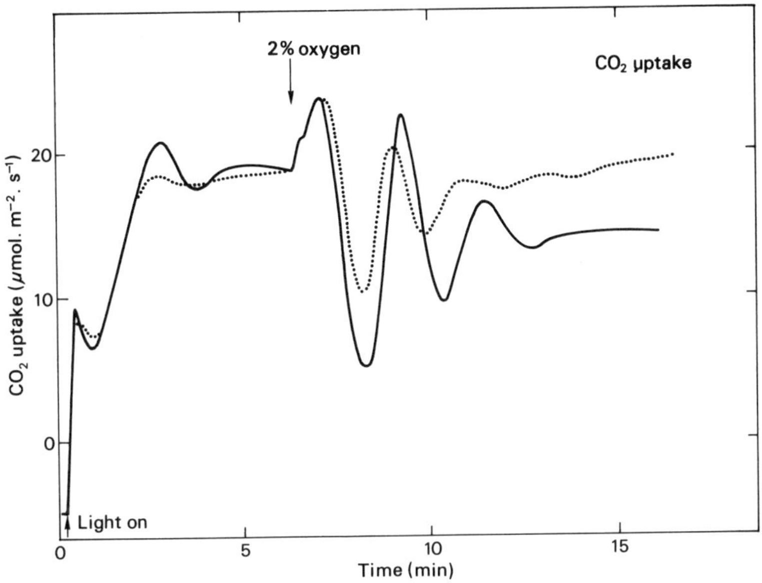

Figure 4. Influence of Pi-feeding (····) upon the rate of CO_2 uptake before and after the transition from 20% to 2% O_2. The concentration of CO_2 was 360 $\mu l.l^{-1}$ CO_2, the leaf temperature 11°C and the irradiance 1500 $\mu mol.m^{-2}.s^{-1}$. From Leegood and Furbank (1986).

MECHANISMS OF LIMITATION OF PHOTOSYNTHESIS BY PHOSPHATE DEFICIENCY IN VIVO

When leaves are photosynthesising in high CO_2 and high light the maximum capacity for triose P utilisation in the processes of starch and sucrose synthesis may be reached (Walker, 1976; Badger *et al.*, 1984; Sharkey, 1985; Walker and Sivak, 1985). This means that the rate of Pi recycling is limited. Perhaps the clearest example of this is to be found in an O_2-insensitive mutant of *Flaveria linearis* which is deficient in the cytosolic FBPase and hence has a diminished ability to make sucrose (Sharkey *et al.*, 1988). However in many situations, even in high CO_2 and high light, the total maximum capacity for carbohydrate synthesis cannot be the limitation on phosphate recycling because phosphate often stimulates the rate of CO_2 fixation dramatically (Sivak and Walker, 1986) (by up to 100% at low temperatures in 2% O_2; Leegood and Furbank 1986). This means that the concentration of Pi in the cytosol is also an important factor and suggests that, as with isolated chloroplasts, photosynthesis *in vivo* has a relatively sharp phosphate optimum (cytosolic), departure from which leads to a phosphate limitation on transport across the chloroplast envelope. In the context of inhibition of carbon assimilation by 2% O_2, it is worth noting that isolated chloroplasts have a higher phosphate optimum when oxygenation is suppressed (Usuda and Edwards, 1982).

The mechanism by which Pi limits electron transport has recently received attention. It has always been difficult to reconcile the rather low Km (Pi) of the ATPase measured *in vitro* with the rather high concentrations of Pi measured in the stroma (Furbank *et al.*, 1987b). However recent evidence suggsts both that the conditions of measurement are critical to estimation of the Km (Pi) (Quick and Mills, 1988) and that the *in vivo* Km (Pi) may be higher than previously supposed (Furbank *et al.*, 1987b). In addition it is clear from studies of isolated chloroplasts that the phenomenon of photosynthetic control is an extremely complex one, involving multiple sites of action (Furbank *et al.*, 1987a).

These considerations raise the further question as to why a switch to low O_2 and diminished photorespiration influences and/or exacerbates the phosphate status of leaves adversely. This is likely to be an extremely complex syndrome, involving direct effects on adenylate and pyridine nucleotide pools, as well as those mediated directly by changes in phosphate status. For example, production of glycollate-2-P, and hence release of phosphate within the chloroplast by glycollate-2-P phosphatase, would be greatly diminished in 2% O_2 (Fig. 1). Following a transition to 2% O_2, pools of photorespiratory intermediates built up in air would continue to be converted to glycerate and thence to PGA, which would impose a burden on ATP-generating reactions. On the other hand, NADH normally generated during glycine decarboxylation would be unavailable (e.g. for ATP synthesis). Whether or not these events result in a limitation by phosphate on CO_2 assimilation will depend upon whether phosphate released by starch synthesis within the stroma or by sucrose synthesis in the cytosol can be recycled rapidly enough to substitute for the shortfall in Pi and energisation.

PHYSIOLOGICAL RELEVANCE OF PERTURBATIONS IN PHOSPHATE STATUS

Recently, considerable interest has focussed on the circumstances in which phosphate can limit photosynthesis in the absence of Pi-sequestering agents and in conditions other than high CO_2 and high light. Such a situation is found at low temperatures. Evidence of the regulatory importance of changes in phosphate status at low temperatures is provided by a number of observations. At lower temperatures:

(i) Phosphorylated metabolites in illuminated leaves increase (Leegood 1985; Dietz and Heber, 1986).

(ii) The PGA/triose-P ratio decreases (Leegood and Furbank 1986; Sharkey *et al.*, 1986; Dietz and Heber 1986; Kobza and Edwards 1987).

(iii) There is an increase in oscillatory behaviour in both gas-exchange and chlorophyll fluorescence (Sivak and Walker, 1986; Leegood and Furbank, 1986) (Fig. 4).

(iv) Lack of O_2 sensitivity of photosynthetic carbon assimilation occurs (Cornic and Louason, 1980; Jolliffe and Tregunna, 1968,1973; McVetty and Canvin, 1981; Leegood and Furbank, 1986; Sharkey *et al.*, 1986).

(v) Increased sensitivity to Pi-sequestering agents such as glycerol occurs (Leegood *et al.*, 1988).

(vi) There is an increase in the Pi optimum for photosynthesis in isolated chloroplasts (Leegood and Walker 1983; Mächler *et al.*, 1984).

(vii) Lastly, it is apparent that abrupt transfer of leaves of many plants to a sufficiently low temperature results in a limitation of the rate of photosynthetic carbon assimilation by phosphate, as revealed by a stimulation of photosynthetic carbon assimilation by feeding Pi at lower temperatures (Labate and Leegood, 1988).

Acclimation of photosynthetic carbon assimilation to changes in temperature is readily observed in the long-term, e.g., by increased O_2 sensitivity of photosynthesis after growth at low temperatures (Cornic and Louason, 1980; Sage and Sharkey, 1987) (Fig. 5) and by increases in photosynthetic rates (Badger *et al.*, 1982). In the shorter term, exposures of even a few hours to low temperatures in the dark can lead to significantly increased rates of carbon assimilation and relief of the symptoms of phosphate deficiency in leaves of barley (Labate and Leegood, 1988). Recovery from phosphate limitation presumably depends upon the rate and extent of phosphate movement out of the vacuole in response to lowered temperatures.

In considering the relationship between Pi and contents of photosynthetic intermediates, we should note that at low temperatures it appears that the threshold content of metabolites required to drive sucrose synthesis may be increased because of a large temperature-dependent decrease in the affinity of FBPase for its substrate at low temperatures and a high Q_{10} for sucrose-P synthase (Stitt 1987), or even inactivation of FBPase (Weeden and Buchanan 1983). This raises the question of to what extent increases in the rate of sucrose synthesis at low temperatures are accommodated by increases in metabolite pools and/or by increases in Pi in the cytosol at low temperatures. The simplest explanation for the

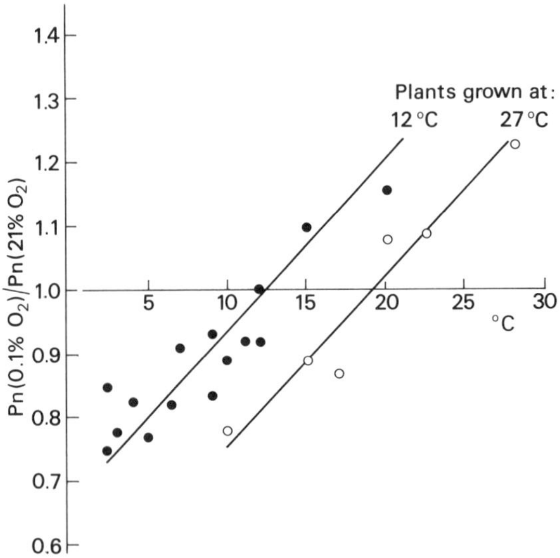

Figure 5. Acclimation to low temperature in respect of O_2 sensitivity in plants of white mustard grown at 12°C and at 27°C. Redrawn from Cornic and Louason (1980).

limitation by phosphate at low temperatures is that Pi is sub-optimal in the cytosol. An increase in the rate of sucrose synthesis could then come about by increasing the cytosolic pool of phosphate. However, phosphorylated metabolites increase at low temperatures (Leegood 1985; Dietz and Heber 1986) so that this phosphate deficiency could also manifest itself as a limitation of sucrose synthesis by low levels of metabolites. However, recent evidence shows that although phosphorylated intermediates increase in the dark at low temperatures, CO_2 fixation is still limited by Pi (Labate CA and Leegood RC, unpublished data).

REFERENCES

Abadia, J., Rao, I.M. and Terry, N. (1987) Thylakoid structure and function in relation to leaf phosphate status in sugar beet. In Biggins J (ed): "Progress in Photosynthesis Research." Dordrecht: Martinus Nijhoff, pp 755-758.

Anderson LE (1973). Regulation of pea leaf ribulose-5-phosphate kinase activity. Biochim Biophys Acta 321:484-488.

Arrabaca MC, Keys AJ, Whittingham CP (1981). Effect of temperature on photosynthetic and photorespiratory metabolism. In Akoyunoglou G (ed): "Photosynthesis" Vol IV Philadelphia: Balaban Int Sci Services, pp 463-470.

Badger MR, Björkman O, Armond PA (1982). An analysis of photosynthetic response and adaptation to temperature in higher plants: temperature acclimation in the desert evergreen *Nerium oleander* L. Plant Cell Environ 5:85-99.

Badger MR, Sharkey TD, von Caemmerer S (1984) The relationship between steady-state gas-exchange of bean leaves and the levels of carbon-reduction-cycle intermediates. Planta 160:305-313.

Bligny R, Roby C, Douce R (1988). Phosphorus-31 nuclear magnetic resonance studies in higher plant cells. In Pfeffer PE, Gerasimowicz WV (eds): "Nuclear Magnetic Resonance in Agriculture" Cleveland: CRC Press (in press).

Brooks A (1986). Phosphorus nutrition and photosynthesis of spinach leaves: effects on ribulose-1,5-bisphosphate carboxylase activation, quantum yield and contents of some Calvin cycle intermediates. Aust J Plant Physiol 13:669-688.

Brooks A, Woo KC, Wong SC (1988). Effects of phosphorus nutrition on the response of photosynthesis to CO_2 and O_2, activation of ribulose bisphosphate carboxylase and amounts of ribulose bisphosphate and 3-phosphoglycerate in spinach leaves. Photosynthesis Res 15:133-141.

Canvin DT (1978). Photorespiration and the effect of oxygen on photosynthesis. In Siegelman H, Hind G (eds): "Photosynthetic Carbon Assimilation." New York: Plenum Press, pp 61-76.

Cornic G, Louason G (1980). The effects of O_2 on net photosynthesis at low temperature (5 C). Plant Cell Environ 3:149-157.

Charles SA, Halliwell B (1980). Properties of freshly purified and thiol-treated spinach chloroplast fructose bisphosphatase. Biochem J 185:689-693.

Chen-She S-H, Lewis DH, Walker DA (1975). Stimulation of photosynthetic starch formation by sequestration of cytoplasmic orthophosphate. New Phytol 74:383-392. Cseke C, Weeden NF, Buchanan BB, Uyeda K (1982). A special fructose bis-phosphate functions as a regulatory metabolite in green leaves. Proc Natl Acad Sci USA 79:4322-4326.

Dietz K-J, Heber U (1984). Rate-limiting factors in leaf photosynthesis. I. Carbon fluxes in the Calvin cycle. Biochim Biophys Acta 767:432-443.

Dietz K-J, Foyer CH (1986). The relationship between phosphate status and photosynthesis in leaves. Reversibility of the effects of phosphate deficiency on photosynthesis. Planta 167:376-381.

Dietz K-J, Heber U (1986). Light and CO_2 limitation of photosynthesis and states of the reactions regenerating ribulose-1,5-bisphosphate or reducing 3-phosphoglycerate. Biochim Biophys Acta 848:392-401.

Doehlert DC, Huber SC (1983). Regulation of spinach leaf sucrose phosphate synthase by glucose-6-phosphate, inorganic phosphate, and ΔpH. Plant Physiol 73:989-994.

Doehlert DC, Huber SC (1984). Phosphate inhibition of spinach leaf sucrose phosphate synthase as affected by glucose-6-phosphate and phosphoglucoseisomerase. Plant Physiol 76:250-253.

Farquhar GD, von Caemmerer S, Berry JA (1980). A biochemical model of photosynthetic CO_2 assimilation in leaves of C species. Planta 149:78-90.

Fliege R, Flügge U-I, Werdan K, Heldt HW (1978). Specific transport of inorganic phosphate, 3-phosphoglycerate and triosephosphates across the inner membrane of the envelope in spinach chloroplasts. Biochim Biophys Acta 502:232-247.

Foyer CH, Spencer C (1986). The relationship between phosphate status and photosynthesis in leaves. Effects on intracellular Pi, photosynthesis and assimilate partitioning. Planta 167:369-375.

Furbank RT, Foyer CH, Walker DA (1987a). Interactions between ribulose-1,5-bisphosphate carboxylase and stromal metabolites. II. Corroboration of the role of this enzyme as a metabolite buffer. Biochim Biophys Acta 894:165-173.

Furbank RT, Foyer CH, Walker DA (1987b). Regulation of photosynthesis in isolated spinach chloroplasts during orthophosphate limitation. Biochim Biophys Acta 894:552-561.

Gardemann A, Stitt M, Heldt HW (1983). Control of CO_2 fixation. Regulation of spinach ribulose-5-phosphate kinase by stromal metabolite levels. Biochim Biophys Acta 722:51-60.

Gerhardt R, Stitt M, Heldt HW (1987). Subcellular metabolite levels in spinach leaves: regulation of sucrose synthesis during diurnal alterations in photosynthetic partitioning. Plant Physiol 83:399-407.

Giersch C, Robinson SP (1987). Regulation of photosynthetic carbon metabolism during phosphate limitation in isolated spinach chloroplasts. Photosynthesis Res 14:211-227.

Harbron S, Foyer CH, Walker DA (1981). The purification and properties of sucrose phosphate synthetase from spinach leaves: the involvement of this enzyme and fructose bisphosphatase in the regulation of sucrose biosynthesis. Arch Biochem Biophys 212:237-246.

Harris GC, Cheesbrough JK, Walker DA (1983). Effects of mannose on photosynthetic gas exchange in spinach leaf discs. Plant Physiol 71:108-111.

Heber U, Heldt HW (1981). The chloroplast envelope: structure, function and role in leaf metabolism. Annu Rev Plant Physiol 32:139-168.

Heber U, Neimanis S, Dietz K-J, Viil J (1986). Assimilatory power as a driving force in photosynthesis. Biochim Biophys Acta 852:144-155.

Heldt HW, Chon CJ, Maronde D, Herold A, Stankovic ZS, Walker DA, Kraminer A, Kirk MR, Heber U (1977). Role of orthophosphate and other factors in the regulation of starch formation in leaves and isolated chloroplasts. Plant Physiol 59:1146-1155.

Heldt HW, Chon CJ, Lorimer GH (1978). Phosphate requirement for the light activation of ribulose-1,5-bisphosphate carboxylase in intact spinach chloroplasts. FEBS Lett 92:234-240.

Heldt HW, Stitt M (1987). The regulation of sucrose synthesis in leaves. In Biggins J (ed): "Progress in Photosynthesis Research." Dordrecht: Martinus Nijhoff, pp 675-684.

Herold A, Lewis DH (1977). Mannose and green plants: occurrence, physiology and metabolism, and use as a tool to study the role of orthophosphate. New Phytol 79:1-40.

Jolliffe PA, Tregunna EB (1968). Effect of temperature, CO_2 concentration, and light intensity on oxygen inhibition of photosynthesis in wheat leaves. Plant Physiol 43:902-906.

Jolliffe PA, Tregunna EB (1973). Environmental regulation of the oxygen effect on apparent photosynthesis in wheat. Can J Bot 51:841-853.

Kacser H, Burns JA (1973). The control of flux. Symp Soc Exp Biol 27:65-104.

Kobza J, Edwards GE (1987). Influences of leaf temperature on photosynthetic carbon metabolism in wheat. Plant Physiol 83:69-74.

Labate CA, Leegood RC (1988). Limitation of photosynthesis by changes in temperature. Factors affecting the response of carbon-dioxide assimilation to temperature in barley leaves. Planta 173:519-527.

Laisk A, Walker DA (1986). Control of phosphate turnover as a rate-limiting factor and possible cause of oscillations in photosynthesis: a mathematical model. Proc R Soc Lond B 227:281-302.

Leegood RC (1985). Regulation of photosynthetic CO_2-pathway enzymes by light and other factors. Photosynthesis Res 6:247-259.

Leegood RC, Walker DA (1983). The role of transmembrane solute flux in regulation of CO_2 fixation in chloroplasts. Biochem Soc Trans 11:74-76.

Leegood RC, Furbank RT (1986). Stimulation of photosynthesis by 2% O_2 at low temperatures is restored by phosphate. Planta 168:84-93.

Leegood RC, Labate CA, Huber SC, Neuhaus HE, Stitt M (1988). Phosphate sequestration by glycerol and its effects on photosynthetic carbon assimilation by leaves. Planta: in press.

Mächler F, Schnyder H, Nösberger J (1984). Influence of inorganic phosphate on photosynthesis of wheat chloroplasts. I. Photosynthesis and assimilate export at 5°C and 25°C. J Exp Bot 153:481-487.

McVetty PBE, Canvin DT (1981). Inhibition of photosynthesis by low oxygen concentrations. Can J Bot 59:721-725.

Morison JIL, Batten GD (1986). Regulation of mesophyll photosynthesis in intact wheat leaves by cytoplasmic phosphate concentrations. Planta 168:200-206.

Prinsley RT, Leegood RC (1986). Factors affecting photosynthetic induction in spinach leaves. Biochim Biophys Acta 849:244-253.

Quick P, Mills JD (1988). Changes in the apparent affinity of CF_0- CF_1 for its substrate during photophosphorylation. Biochim Biophys Acta 932:232-239.

Rao IM, Abadia J, Terry N (1987a). Leaf phosphate status and its effects on photosynthetic carbon partitioning and export in sugar beet. In Biggins J (ed): "Progress in Photosynthesis Research." Dordrecht: Martinus Nijhoff, pp 751-754.

Rao IM, Abadia J, Terry N (1987b). The role of orthophosphate in the regulation of photosynthesis *in vivo.* In Biggins J (ed): "Progress in Photosynthesis Research." Dordrecht: Martinus Nijhoff, pp 325-328.

Rebeille F, Bligny R, Martin JB, Douce R (1983). Relationship between the cytoplasm and the vacuolar phosphate pool in *Acer pseudoplatanus* cells. Arch Biochem Biophys 225:143-148.

Sage RF, Sharkey TD (1987). The effect of temperature on the occurrence of O_2 and CO_2 insensitive photosynthesis in field grown plants. Plant Physiol 84:658-664.

Sage RF, Sharkey TD, Seemann JR (1988) The in-vivo response of the ribulose-1,5-bisphosphate carboxylase activation state and the pool sizes of photosynthetic metabolites to elevated CO_2 in *Phaseolus vulgaris* L. Planta 174:407-416.

Sanwal GG, Greenberg E, Hardie J, Cameron EC, Preiss J (1968). Regulation of starch biosynthesis in plant leaves. Activation and inhibition of ADP-glucose pyrophosphorylase. Plant Physiol 43:417-427.

Sharkey TD (1985). O_2-insensitive photosynthesis in C plants. Its occurrence and a possible explanation. Plant Physiol 78:71-75.

Sharkey TD, Stitt M, Gerhardt R, Heineke D, Raschke K, Heldt HW (1986). Limitation of photosynthesis by carbon metabolism. II. O_2 insensitive photosynthesis results from a limitation of triose phosphate utilization. Plant Physiol 81:1123-1129.

Sharkey TD, Kobza J, Seemann JR, Brown RH (1988) Reduced cytosolic fructose-1,6-bisphosphatase activity leads to loss of O_2 sensitivity in a *Flaveria linearis* mutant. Plant Physiol 86:667-671.

Sicher RC, Kremer DF (1988). Effects of phosphate deficiency on assimilate partitioning in barley seedlings. Plant Sci 57:9-17

Sivak MN, Walker DA (1986). Photosynthesis *in vivo* can be limited by phosphate supply. New Phytol 102:499-512.

Sivak MN, Walker DA (1987). Oscillations and other symptoms of limitation of *in vivo* photosynthesis by inadequate phosphate supply to the chloroplast. Plant Physiol Biochem 25:635-648.

Stitt M (1986). Limitation of photosynthesis by carbon metabolism. I. Evidence for excess electron transport in leaves carrying out photosynthesis in saturating light and CO_2. Plant Physiol 81:1115-1122.

Stitt M (1987). Limitation of photosynthesis by sucrose synthesis. In Biggins J (ed): "Progress in Photosynthesis Research." Dordrecht: Martinus Nijhoff, pp 685-692.

Stitt M, Herzog B, Heldt HW (1985). Control of sucrose synthesis by fructose-2,6-bisphosphate. V. Modulation of the spinach leaf cytosolic fructose-1,6-bisphosphate. Plant Physiol 79:590-598.

Stitt M, Gerhardt R, Wilke I, Heldt HW (1987). The contribution of fructose 2,6-bisphosphate to the regulation of sucrose synthesis during photosynthesis. Physiol Plant 69:377-386.

Stitt M, Schreiber U (1988). Interaction between sucrose synthesis and photosynthesis. III. Manipulation of the relation between sucrose synthesis, electron transport and the Calvin cycle. J Plant Physiol: in press.

Stitt M, Wilke I, Feil R, Heldt HW (1988). Coarse control of sucrose-phosphate synthase in leaves: Alterations of the kinetic properties in response to the rate of photosynthesis and the accumulation of sucrose. Planta 174:217-230.

Usuda H, Edwards GE (1982). Influence of varying CO_2 and orthophosphate concentrations on rates of photosynthesis, and synthesis of glycolate and dihydroxyacetone phosphate by wheat chloroplasts. Plant Physiol 69:469-473.

Walker DA (1976). Plastids and intracellular transport. In Stocking CR, Heber U (eds): "Encyclopedia of Plant Physiology" Vol. 3. Berlin: Springer Verlag, pp 85-136.

Walker DA, Robinson SP (1978). Chloroplast and cell. A contemporary view of photosynthetic carbon assimilation. Ber Deutsch Bot Ges 91:513-526.

Walker DA, Osmond CB (1986). Measurement of photosynthesis *in vivo* with a leaf disc electrode: correlations between light dependence of steady-state photosynthetic O_2 evolution and chlorophyll *a* fluorescence. Proc Roy Soc Lond B 227:267-280.

Walker DA, Sivak MN (1985). Can phosphate limit photosynthetic carbon assimilation *in vivo*? Physiol Veg 23:829-841.

Weeden NF, Buchanan BB (1983). Leaf cytosolic fructose-1,6-bisphosphatase. A potential target site in low temperature stress. Plant Physiol 72:259-261.

Wolosiuk RA, Buchanan BB (1978). Activation of chloroplast NADP-linked glyceraldehyde-3-phosphate dehydrogenase by the ferredoxin/thioredoxin system. Plant Physiol 61:669-671.

Woodrow IE, Murphy DJ, Walker DA (1983). Regulation of photosynthetic carbon metabolism. The effect of inorganic phosphate on stromal sedoheptulose-1,7-bisphosphatase. Eur J Biochem 132:121-123.

Woodrow IE, Ellis JR, Jellings A, Foyer CH (1984). Compartmentation and fluxes of inorganic phosphate in photosynthetic cells. Planta 161:525-530.

Photosynthesis, pages 475–501

LIMITATION TO CO_2 FIXATION BY PHOTOSYNTHETIC PROCESSES[1]

Ian E. Woodrow
Bureau of Rural Resources, Department of Primary Industries and Energy, Canberra, ACT 2600, Australia

INTRODUCTION.

Numerous studies have sought to define the way in which external variables such as water, nitrogen, CO_2 and temperature limit the rate of photosynthesis. But relatively little attention has been given to the limitation imposed by internal variables such as enzyme concentrations and kinetic constants. With the recent developments in molecular biology, it is the influence of these internal variables on metabolic fluxes that is of increasing interest to scientists wishing to alter the metabolism of plants in a quantitative fashion.

A major difficulty in determining how internal variables limit the rate of photosynthesis is that, unlike the supply of H_2O or CO_2, these variables cannot be altered readily. One approach to the problem is to use models of the photosynthetic system to predict the sensitivity of the various fluxes to

---

[1] **Abbreviations.** A, assimilation rate; bl_1, boundary layer diffusion of CO_2; bl_2, boundary layer diffusion of water vapour; bl_3, boundary layer transfer of convective heat; c, mole fraction CO_2 (subscripts: a, ambient; s, leaf surface; i, intercellular spaces); C_p, molar specific heat of air; D, molecular diffusivity (subscripts: h, sensible heat; w, water vapour; c, CO_2); d, characteristic dimension of leaf; E, evaporation rate; g_c, conductance to CO_2 (additional subscripts: s, stomata; b, boundary layer); g_w, conductance to water vapour (same additional subscripts to g_c); g_{bh}, boundary layer conductance to sensible heat; H, flux density of sensible heat; J, metabolic flux; K, Michaelis constant (additional subscripts: c, carboxylation; o, oxygenation); n, number of stomata per unit leaf area; P, independent variable with respect to which control coefficients are defined; PFD, photon flux density; Q, external effector (i.e. effector whose concentration can be manipulated as an "independent" variable); Rubisco, ribulose 1,5-bisphosphate carboxylase/oxygenase; R_c, the carboxylase reaction of rubisco; R_o, the oxygenase reaction of Rubisco; RE, radient emittance; S_1, stomatal diffusion of CO_2; S_2, stomatal diffusion of water vapour; T, temperature (subscripts: a, ambient; l, leaf; s, surface); u, windspeed; v, catalytic velocity of a reaction; V, maximum catalytic velocity (subscripts: c, carboxylation; o, oxygenation); w, mole fraction water vapour (subscripts as for c); ϵ, surface emissivity; σ, Stephan-Boltzmann constant (5.67×10^{-8}W m^{-2}K^{-4}); φ, radiant flux density.

changes in the magnitude of internal variables. These predictions can then be used to devise the best strategies for quantitatively altering the metabolic pathways of photosynthesis.

In this paper a model of the photosynthetic system of a single leaf (Woodrow *et al* 1986) is analysed. The model utilises both mechanistically based and empirically determined rate equations. There are few assumptions underlying the model, and the analysis yields expressions for the degree to which several internal variables limit the steady state fluxes of the system. The data are relevant to steady states occuring over the minutes to hours time range.

MATERIALS AND METHODS.

Most data used in this paper are derived from previous studies of soybeans by Woodrow *et al* (1986), Ball *et al* (1986) and Woodrow and Mott (unpublished results). Transpiration rates, CO_2 assimilation rates and leaf temperatures were measured with a compensating gas exchange system similar to that described by Field *et al* (1982). Gas exchange parameters were calculated according to Ball (1987). The boundary layer conductance to water vapour was 3.5 mol $m^{-2}s^{-1}$ distributed equally between the two surfaces of the leaf. Ambient CO_2 and O_2 were varied as described by Woodrow and Mott (1988).

DESCRIPTION OF THE PHOTOSYNTHETIC SYSTEM.

A model of the photosynthetic system of a single leaf (Fig. 1) is used in the analysis of flux limitation. The model consists of four resolved components: (1) the leaf boundary layer, (2) the stomata, (3) exchange of longer wavelength radiation between the leaf and the environment, and (4) ribulose 1,5-bisphosphate carboxylase/oxygenase (Rubisco). The leaf boundary layer mediates three transfer processes between the bulk atmosphere and the leaf surface: (1) CO_2 diffusion ($C_a \rightleftharpoons C_s$), (2) water vapour diffusion and associated latent heat transfer ($W_a \rightleftharpoons W_s$), and (3) heat transfer by forced and natural convection ($H_a \rightleftharpoons H_s$). The stomata mediate the diffusion of water vapour ($W_s \rightleftharpoons W_i$) (and associated latent heat transfer) and CO_2 ($C_s \rightleftharpoons C_i$) between the leaf surface and the intercellular spaces. Heat is also transferred in the form of longer wavelength radiation by a process that is independent of both the stomata and boundary layer and that depends upon the leaf and environmental temperatures. Rubisco catalyses two reactions: an oxygenase and a carboxylase reaction which involve the stromal O_2 and CO_2 pools, respectively.

The intercellular CO_2 pool (C_i), which is assumed to be in equilibrium

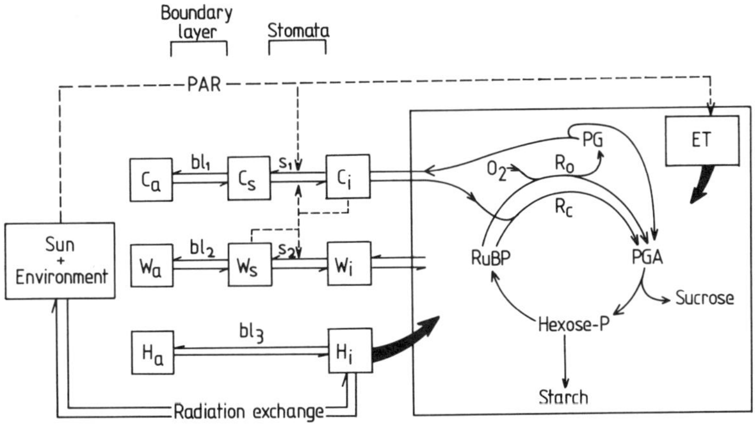

Figure 1. Scheme of the photosynthetic system of C_3 plants showing the four basic components examined in this paper: the leaf boundary layer, the stomata, the exchange of longer wavelength radiation, and the two reactions catalysed by ribulose 1,5-bisphosphate carboxylase/oxygenase. The boundary layer mediates three transfer processes between the bulk atmosphere and the leaf surface: CO_2 diffusion (bl_1), water vapour diffusion (bl_2), and heat transfer by convection (bl_3). The stomata mediate CO_2 (S_1) and water vapour (S_2) diffusion between the leaf surface and intercellular spaces. Stomatal conductance is assumed to be directly influenced by the internal CO_2 and surface water vapour concentrations, and by leaf temperature. The latter also affects the oxygenase and carboxylase reactions. It is assumed that the concentration of O_2 in the stroma is equal to that in the bulk air.

with the stromal CO_2 pool (i.e. there is no liquid phase resistance to CO_2 diffusion), is influenced by diffusion through the stomata and the rates of photorespiration, respiration and carboxylation. It is also assumed that (1) the variation in stomatal conductance (g_s) observed as w_s and c_i are changed is due to essentially direct effects of these pools on g_s, and (2) increases in the rate of transpiration do not change leaf water potential such that the kinetic characteristics of Rubisco are altered significantly. It should be emphasized that, although only four basic components are considered in the following analysis, no simplifying assumptions are made regarding the remainder of the photosynthetic system or other interacting structural and metabolic components.

A steady state is defined as occurring when the rate of CO_2 assimilation has approached a largely constant value which is maintained for a period of several minutes. Therefore, the degree to which components limit the steady state flux, as defined in this study, is relevant to photosynthetic rates in the short term (i.e. minutes to hours) rather than the long term (i.e. days to years). There is, however, no reason why the same analysis could not be applied to longer term fluxes, or even to non-steady state fluxes.

DEFINITION OF FLUX LIMITATION.

If the activities of each of the components (enzymes etc) of a metabolic system are perturbed sequentially and, after each perturbation, the flux recorded, then a wide variety of responses are to be expected. Some perturbations may effect a flux increase, others a decrease, and some may effect no change in flux at all. Therefore, when describing the importance of enzyme activities in determining the flux, it is quite inadequate to attempt to separate components into "rate limiting" and "non-rate limiting" groups (i.e. components whose activities have a large and small effect on the flux, respectively). Proper description of limitation to steady state fluxes requires quatification of the degree to which infinitesimal changes in parameters, such as enzyme activities, affect the various fluxes of the system.

In the ensuing analysis, the degree to which components of the photosynthetic system limit the flux is defined using the "control coefficient" (see Higgins, 1965; Kacser and Burns, 1973; Savageau, 1976; Crabtree and Newsholme, 1985). This coefficient is a dimensionless quantity which relates changes in the flux (J) to infinitesimal changes in an "independent" variable (P) such as the activity of an enzyme:

$$C_P^J = \frac{\partial J}{\partial P}\frac{P}{J} = \frac{\partial \ln J}{\partial \ln P}. \tag{1}$$

In subsequent equations a subscript symbolising the component of the system that the control coefficient refers to (eg. bl_1, S_1, R_c) will be used instead of an abbreviation of the P variable itself. The "independent" variables used in this study are given in the Appendix.

Individual control coefficients can be determined for each component with respect to any steady state flux and can have positive or negative values, or can be zero. The magnitude of the coefficient provides a measure of the "contribution" of the component to determining or limiting the flux. If a single component has a control coefficient of unity, then, in the vicinity of the steady state, the flux is directly proportional to the P variable for that component. A negative control coefficient indicates that an increase in the P variable effects a decrease in the flux.

Two other types of coefficients are used in calculating control coefficients. The first, the "combined response coefficient" (Kacser and Burns, 1973), describes the response of the flux (J) to an infinitesimal change in the concentration of an independently variable external effector (Q):

$$R_Q^J = \frac{\partial J}{\partial Q}\frac{Q}{J}. \tag{2}$$

In the present experiments, Q is the ambient mole fraction of CO_2 or O_2.

The second coefficient, the "elasticity coefficient" (Kacser and Burns, 1973), describes the effect of an infinitesimal change in the concentration of a substrate or effector (S) on the rate of an isolated reaction or process (v_i):

$$\mathcal{E}_S^i = \frac{\partial v_i}{\partial S}\frac{S}{v_i}. \tag{3}$$

In the case where the substrate or effector is an "independent" variable (i.e. an external effector), the elasticity coefficient is called a "special elasticity coefficient" (Kacser and Burns, 1973) and is defined as follows:

$$^{\kappa}\mathcal{E}_Q^i = \frac{\partial v_i}{\partial Q}\frac{Q}{v_i}, \tag{4}$$

where Q is the concentration of the external effector.

The elasticity coefficients used in this study are calculated from the rate equations for the various stomatal, boundary layer and biochemical processes (see Appendix). It must be emphasized that there is a fundamental difference between the control and elasticity coefficients. Whereas the control coefficient defines the response of the *whole* system to changes in the properties of a single component, the elasticity coefficient defines the response of the *isolated* component to changes in the concentration of a molecular species affecting that component (see Kacser and Burns, 1973 for a more detailed discussion of these coefficients).

CALCULATION OF CONTROL COEFFICIENTS.

Control coefficients can be determined using both direct and indirect methods. The most direct involve altering enzyme activity by either adding specific inhibitors or genetically changing enzyme concentrations or kinetic constants (see Kacser and Burns, 1973). One of the indirect methods involves measuring the response of the metabolic flux of interest to a change in the concentration of an external effector (i.e. an effector whose concentration can be manipulated as an "independent" variable) and comparing this response to the predicted influence of the effector on the "target" components (i.e. the components whose activities are directly influenced by the external effector). Knowing the change in flux and the change in activity of a single component is generally sufficient to calculate the control coefficient for that component. This latter strategy is employed in this study for deriving two equations used in calculating the control coefficients (see eqs. 7, 8 and 9).

The ambient CO_2 pool interacts with the leaf boundary layer only. A given rise in c_a can, therefore, be equated with a rise in the boundary layer conductance to CO_2 (g_{bc}) — i.e. certain changes in c_a and g_{bc} produce equivalent changes in the metabolic fluxes of the system. This forms the basis of an equation relating response, control and special elasticity coefficients (see appendix):

(5) $$R^J_{c_a} = C^J_{bl_1} \, {}^{\kappa}\mathcal{E}^{bl_1}_{c_a}.$$

The response coefficient ($R^J_{c_a}$) can be determined empirically from the slope of the flux versus c_a curve. The special elasticity coefficient (${}^{\kappa}\mathcal{E}^{bl_1}_{c_a}$) can be derived by differentiating the rate equation for CO_2 diffusion across the leaf boundary layer (see Appendix) and is given by:

(6) $${}^{\kappa}\mathcal{E}^{bl_1}_{c_a} = g_{bc}\, c_a / v_{bl_1}$$

where $v_{bl_1} = g_{bc}(c_a - c_s)$.

It should be noted that the equations describing CO_2 and H_2O diffusion are those of a binary transport system (see Appendix) — i.e. the flux is directly proportional to the concentration gradient. The model can be refined by using the "ternary transport equations" (see Jarman, 1974) which take into account the interactions between H_2O and CO_2.

The control coefficient for boundary layer conductance to CO_2 is, therefore, given by:

(7) $$C^J_{bl_1} = \frac{\partial J}{\partial g_{bc}} \frac{g_{bc}}{J} = \frac{\partial J}{\partial c_a} \frac{v_{bl_1}}{g_{bc} J}$$

where $\partial J / \partial c_a$ is the slope of the J versus c_a curve (n.b. w_a, T_a, $[CO_2]$ and the *PFD* must be held constant while c_a is varied around the normal steady state value). If one defines this control coefficient for the rate of photosynthetic CO_2 assimilation then, in the steady state, $J = A = v_{bl_1}$ and eq (7) simplifies to:

(8) $$C^A_{bl_1} = \frac{\partial A}{\partial c_a} \frac{1}{g_{bc}}$$

where $\partial A/\partial c_a$ is the slope of the assimilation versus c_a curve at the air temperature, *PFD* and levels of O_2, water vapour and CO_2 under consideration.

It is assumed in the model that the ambient O_2 concentration is equal to that in the chloroplast stroma. Changes in the ambient concentration of this gas can, therefore, be considered to interact directly with the carboxylase and oxygenase reactions of Rubisco. The relationship between the control coefficients for these two components and the response coefficient for O_2 is:

$$R^J_{O_2} = C^J_{R_c}\ {}^{\kappa}\mathcal{E}^c_{O_2} + C^J_{R_o}\ {}^{\kappa}\mathcal{E}^o_{O_2}. \tag{9}$$

Like $R^J_{c_a}$, $R^J_{O_2}$ can be quantified by holding the external variables at their steady state values and varying the O_2 concentration (see Woodrow and Mott, 1988). The special elasticity coefficients can be derived by differentiating the rate equations for the catalytic velocities of Rubisco:

$$v_c = \frac{V_c[CO_2]K_o}{K_cK_o + K_o[CO_2] + K_c[O_2]}; \tag{10}$$

$$v_o = \frac{V_o[O_2]K_c}{K_cK_o + K_o[CO_2] + K_c[O_2]}; \tag{11}$$

where K_o, K_c, V_o and V_c are the Michaelis constants for O_2 and CO_2 and the maximum catalytic velocities for oxygenation and carboxylation, respectively. Recent evidence suggests that changes in the level of RuBP may affect V_o and V_c only and not alter K_c and K_o. Therefore, the following elasticities may be quite independent of the RuBP concentration:

$${}^{\kappa}\mathcal{E}^c_{O_2} = \frac{-[O_2]K_c}{K_cK_o + K_oc_i + K_c[O_2]}; \tag{12}$$

$${}^{\kappa}\mathcal{E}^o_{O_2} = \frac{K_oc_i + K_cK_o}{K_cK_o + K_oc_i + K_c[O_2]}. \tag{13}$$

Values of 264 μbar for K_c and 395 mbar for K_o (Jordan and Ogren, 1984) were used in calculating the elasticities.

Over the range of conditions used in the present experiments, it is also assumed that: (1) changes in the activation state and kinetic properties of Rubisco due to alterations in light intensity and associated changes in the stromal levels of H^+, $Mg^{2+}$and other effectors do not greatly alter K_c and

K_o; and (2) in the vicinity of the normal stromal CO_2 and O_2 concentrations these effectors influence photosynthesis primarily by altering the catalytic properties of Rubisco.

Several other equations involving control coefficients can be formulated using the "connectivity property" (Kacser and Burns, 1973) which shows how local responses to initially local effects (given by the elasticity coefficients) give rise to systemic responses (given by the control coefficients). Connectivity equations with respect to each effector or intermediate "pool" can be written as follows:

$$\sum_{P=1}^{n} C_P^J \, \mathcal{E}_S^P = 0, \tag{14}$$

where S stands for the shared substrate or effector. It must be emphasized that the special elasticity coefficients (indicated by the superscript κ) do not enter into the connectivity equations.

Connectivity equations with respect to c_s, c_i, w_s and T_l can be written as follows (see appendix):

$$C_{bl_1}^J \, \mathcal{E}_{c_s}^{bl_1} + C_{S_1}^J \, \mathcal{E}_{c_s}^{S_1} = 0; \tag{15}$$

$$C_{S_1}^J \, \mathcal{E}_{c_i}^{S_1} + C_{S_2}^J \, \mathcal{E}_{c_i}^{S_2} + C_{R_c}^J \, \mathcal{E}_{c_i}^{R_c} + C_{R_o}^J \, \mathcal{E}_{c_i}^{R_o} = 0; \tag{16}$$

$$C_{bl_2}^J \, \mathcal{E}_{w_s}^{bl_2} + C_{S_1}^J \, \mathcal{E}_{w_s}^{S_1} + C_{S_2}^J \, \mathcal{E}_{w_s}^{S_2} = 0; \tag{17}$$

$$C_{bl_3}^J \, \mathcal{E}_{T_l}^{bl_3} + C_{S_1}^J \, \mathcal{E}_{T_l}^{S_1} + C_{S_2}^J \, \mathcal{E}_{T_l}^{S_2} + C_{R_c}^J \, \mathcal{E}_{T_l}^{R_c} + C_{R_o}^J \, \mathcal{E}_{T_l}^{R_o} + C_{RE}^J \, \mathcal{E}_{T_l}^{RE} = 0. \tag{18}$$

The elasticity coefficients can be derived by differentiating the various rate equations (see appendix) and are given by:

$$\mathcal{E}_{c_s}^{bl_1} = -g_{bc} \, c_s / v_{bl_1};$$
$$\mathcal{E}_{c_s}^{S_1} = g_{sc} \, c_s / v_{S_1};$$
$$\mathcal{E}_{c_i}^{S_1} = [(c_s - c_i)\frac{\partial g_{sc}}{\partial c_i} - g_{sc}] c_i / v_{S_1};$$
$$\mathcal{E}_{c_i}^{S_2} = (w_i - w_s)\frac{\partial g_{sw}}{\partial c_i} / v_{S_2};$$
$$\mathcal{E}_{c_i}^{R_c} = (K_o K_c + [O_2] K_c) / Y;$$

$$\mathcal{E}^{R_o}_{c_i} = -c_i K_o / Y;$$
$$\mathcal{E}^{bl_2}_{w_s} = g_{bw} w_s / v_{bl_2};$$
$$\mathcal{E}^{S_1}_{w_s} = w_s(c_s - c_i)\frac{\partial g_{sc}}{\partial w_s}/v_{S_1};$$
$$\mathcal{E}^{S_2}_{w_s} = [(w_i - w_s)\frac{\partial g_{sw}}{\partial w_s} - g_{sw}]w_s/v_{S_2};$$
$$\mathcal{E}^{bl_3}_{T_l} = T_l/(T_l - T_a);$$
$$\mathcal{E}^{S_1}_{T_l} = (c_s - c_i)T_l\frac{\partial g_{sc}}{\partial T_l}/v_{S_1};$$
$$\mathcal{E}^{S_2}_{T_l} = (w_i - w_s)T_l\frac{\partial g_{sw}}{\partial T_l}/v_{S_2};$$
$$\mathcal{E}^{R_o}_{T_l} = \frac{\partial V_c}{\partial T_l}\frac{T_l}{V_c} - \frac{(K_o + [O_2])T_l}{Y}\frac{\partial K_c}{\partial T_l} + (\frac{T_l}{K_o} - \frac{(K_c + c_i)T_l}{Y})\frac{\partial K_o}{\partial T_l};$$
$$\mathcal{E}^{R_c}_{T_l} = \frac{\partial V_o}{\partial T_l}\frac{T_l}{V_o} - \frac{(K_c + c_i)T_l}{Y}\frac{\partial K_o}{\partial T_l} + (\frac{T_l}{K_c} - \frac{(K_o + [O_2])T_l}{Y})\frac{\partial K_c}{\partial T_l};$$
$$\mathcal{E}^{RE}_{T_l} = -4,$$

where $Y = K_c K_o + K_o c_i + K_c[O_2]$ and $\partial x/\partial y$ is the slope of the x versus y curve at the steady state value of y. It is important to emphasize that, when making a plot of x versus y, all of the other parameters affecting x should be held constant. A clear distinction must be made between measuring systemic responses, such as $\frac{\partial A}{\partial c_a}$, and properties of individual processes, such as $\frac{\partial g_{sc}}{\partial c_i}$. In the case of $\frac{\partial A}{\partial c_a}$, all external variables other than c_a are held constant and the response of the system (i.e. A) to c_a measured. In the case of $\frac{\partial g_{sc}}{\partial c_i}$, however, a property of an individual process is being measured *in situ*. Therefore, the external variables are manipulated such that T_l, w_s and the *PFD* remain constant at their steady state values and only c_i changes. The fact that c_a, c_s, A and other parameters change does not matter because only stomatal conducatance to CO_2 is being examined. In contrast to this sort of *in situ* measurement, the slopes of the V_c, V_o, K_c and K_o versus temperature curves were calculated from results of *in vitro* experiments on Rubisco (Jordan and Ogren, 1984).

Two additional equations involving the control coefficients of elements of the photosynthetic system can be derived (see Appendix):

(19) $$C^J_{RE}\, H - C^J_{bl_3}\, \varphi = 0$$

(20) $$\frac{C^J_{bl_3}}{H} - \frac{C^J_{S_2}}{E} - \frac{C^J_{bl_2}}{E} = 0,$$

where E, H and φ are the evaporation rate, heat flux density and radiant flux density, respectively.

The preceeding equations (5, 9, 15–18, 19 and 20) can be solved simultaneously to yield values for the control coefficients. The solutions can be expressed in matrix form as follows:
(21)

$$\begin{bmatrix} C^J_{bl_1} \\ C^J_{bl_2} \\ C^J_{bl_3} \\ C^J_{S_1} \\ C^J_{S_2} \\ C^J_{R_c} \\ C^J_{R_o} \\ C^J_{RE} \end{bmatrix} = \begin{bmatrix} \kappa\varepsilon^{bl_1}_{ca} & 0 & 0 & 0 & 0 & 0 & 0 & 0 \\ \varepsilon^{bl_1}_{c_s} & 0 & 0 & \varepsilon^{S_1}_{c_s} & 0 & 0 & 0 & 0 \\ 0 & 0 & 0 & \varepsilon^{S_1}_{c_i} & \varepsilon^{S_2}_{c_i} & \varepsilon^{R_c}_{c_i} & \varepsilon^{R_o}_{c_i} & 0 \\ 0 & \varepsilon^{bl_2}_{w_s} & 0 & \varepsilon^{S_1}_{w_s} & \varepsilon^{S_2}_{w_s} & 0 & 0 & 0 \\ 0 & 0 & \varepsilon^{bl_3}_{T_l} & \varepsilon^{S_1}_{T_l} & \varepsilon^{S_2}_{T_l} & \varepsilon^{R_c}_{T_l} & \varepsilon^{R_o}_{T_l} & \varepsilon^{RE}_{T_l} \\ 0 & 0 & -\varphi & 0 & 0 & 0 & 0 & H \\ 0 & -1/E & 1/H & 0 & -1/E & 0 & 0 & 0 \\ 0 & 0 & 0 & 0 & 0 & \kappa\varepsilon^{R_c}_{O_2} & \kappa\varepsilon^{R_o}_{O_2} & 0 \end{bmatrix}^{-1} \begin{bmatrix} R^J_{c_a} \\ 0 \\ 0 \\ 0 \\ 0 \\ 0 \\ 0 \\ R^J_{O_2} \end{bmatrix}$$

The inversion of the 8 x 8 matrix can be performed using the program outlined in the Appendix.

The control coefficients, $C^J_{bl_1}, C^J_{bl_2}$ and $C^J_{bl_3}$, reflect the degree to which the "partial processes" of the boundary layer — conductance to CO_2, water vapour and heat, respectively — determine or limit the flux, J. In reality these three conductances do not change independently of each other and overall control coefficients for the boundary layer can be written with respect to parameters such as wind speed or the characteristic dimensions of the leaf. For example, the sensivity of the flux to the leaf dimension (d) can be defined as follows (see Appendix):

$$C^J_d = -\frac{1}{2}(C^J_{bl_1} + C^J_{bl_2} + C^J_{bl_3}). \tag{22}$$

Similar equations can be written for the two stomatal control coefficients. For example, the sensitivity of the flux to the number of stomata (n) is given by:

$$C^J_n = \frac{\partial J}{\partial n}\frac{n}{J} = C^J_{S_1}\frac{\partial g_{sc}}{\partial n}\frac{n}{g_{sc}} + C^J_{S_2}\frac{\partial g_{sw}}{\partial n}\frac{n}{g_{sw}}. \tag{23}$$

If one assumes that $n \propto g_{sc}$ and $n \propto g_{sw}$ (see Appendix) then

$$C^J_n = C^J_{S_1} + C^J_{S_2}. \tag{24}$$

When considering the rate of CO_2 assimilation or fluxes such as sucrose and starch synthesis and RuBP utilization, it is possible to make several assumptions that simplify the equations outlined above. Woodrow *et al* (1986) showed that the influence on CO_2 assimilation of boundary layer conductance to water vapour ($C^A_{bl_2}$) and heat ($C^A_{bl_3}$), stomatal conductance to water vapour ($C^A_{S_2}$) and emission of longer wavelength radiation (C^A_{RE}) is generally not very significant until the ambient temperature deviates significantly from that which is optimal for photosynthesis.

The calculation of control coefficients can, therefore, be simplified by assuming that $C^J_{bl_2}, C^J_{bl_3}, C^J_{RE}$ and $C^J_{S_2}$ are equal to zero. In other words, it is assumed that T_l and w_s are "independent" variables. Under these conditions four equations involving control and elasticity coefficients can be written:

(25) $$C^J_{bl_1}\ {}^{\kappa}\mathcal{E}^{bl_1}_{c_a} = R^J_{c_a};$$

(26) $$C^J_{bl_1}\ \mathcal{E}^{bl_1}_{c_s} + C^J_{S_1}\ \mathcal{E}^{S_1}_{c_s} = 0;$$

(27) $$C^J_{S_1}\ \mathcal{E}^{S_1}_{c_i} + C^J_{R_c}\ \mathcal{E}^{R_c}_{c_i} + C^J_{R_o}\ \mathcal{E}^{R_o}_{c_i} = 0;$$

(28) $$C^J_{R_c}\ {}^{\kappa}\mathcal{E}^{R_c}_{O_2} + C^J_{R_o}\ {}^{\kappa}\mathcal{E}^{R_o}_{O_2} = R^J_{O_2}.$$

The solutions to these equations can be written in matrix form:

(29) $$\begin{bmatrix} C^J_{bl_1} \\ C^J_{S_1} \\ C^J_{R_c} \\ C^J_{R_o} \end{bmatrix} = \begin{bmatrix} {}^{\kappa}\mathcal{E}^{bl_1}_{c_a} & 0 & 0 & 0 \\ \mathcal{E}^{bl_1}_{c_s} & \mathcal{E}^{S_1}_{c_s} & 0 & 0 \\ 0 & \mathcal{E}^{S_1}_{c_i} & \mathcal{E}^{R_c}_{c_i} & \mathcal{E}^{R_o}_{c_i} \\ 0 & 0 & {}^{\kappa}\mathcal{E}^{R_c}_{O_2} & {}^{\kappa}\mathcal{E}^{R_o}_{O_2} \end{bmatrix}^{-1} \begin{bmatrix} R^J_{c_a} \\ 0 \\ 0 \\ R^J_{O_2} \end{bmatrix}$$

and the program described in the Appendix can be used to invert the 4 x 4 matrix. Expressions for the control coefficients can also be derived by simultaneously solving eqs. (25–28):

(30) $$C^J_{bl_1} = R^J_{c_a} / {}^{\kappa}\mathcal{E}^{bl_1}_{c_a}:$$

(31) $$C^J_{S_1} = -C^J_{bl_1}\ \mathcal{E}^{bl_1}_{c_s} / \mathcal{E}^{S_1}_{c_s};$$

(32) $$C^J_{R_o} = \frac{R^J_{O_2}\ \mathcal{E}^{R_c}_{c_i} - C^J_{S_1}\ \mathcal{E}^{S_1}_{c_i}\ {}^{\kappa}\mathcal{E}^{R_c}_{O_2}}{{}^{\kappa}\mathcal{E}^{R_o}_{O_2}\ \mathcal{E}^{R_c}_{c_i} - \mathcal{E}^{R_o}_{c_i}\ {}^{\kappa}\mathcal{E}^{R_c}_{O_2}};$$

(33) $$C^J_{R_c} = \frac{R^J_{O_2} - C^J_{R_o}\ {}^{\kappa}\mathcal{E}^{R_o}_{O_2}}{{}^{\kappa}\mathcal{E}^{R_c}_{O_2}}.$$

It is possible to simplify the system further by treating c_i as an external "independent" variable and neglecting the influence of the stomatal and boundary layer components. This treatment, which was developed by Woodrow and Berry (unpublished work), gives an indication of the importance of the two Rubisco activities within the context of the photosynthetic biochemistry. Two equations involving response, control, and special elasticity coefficients can be written (see Appendix):

(34) $$R^J_{\mathrm{CO_2}} = C^J_{R_c}\ {}^{\kappa}\mathcal{E}^{R_c}_{\mathrm{CO_2}} + C^J_{R_o}\ {}^{\kappa}\mathcal{E}^{R_o}_{\mathrm{CO_2}};$$

(35) $$R^J_{\mathrm{O_2}} = C^J_{R_c}\ {}^{\kappa}\mathcal{E}^{R_c}_{\mathrm{O_2}} + C^J_{R_o}\ {}^{\kappa}\mathcal{E}^{R_o}_{\mathrm{O_2}}.$$

The simultaneous solution of eqs. (34) and (35) gives expressions for the control coefficients:

(36) $$C^J_{R_c} = \frac{{}^{\kappa}\mathcal{E}^{R_o}_{\mathrm{O_2}}\ R^J_{\mathrm{CO_2}} - {}^{\kappa}\mathcal{E}^{R_o}_{\mathrm{CO_2}}\ R^J_{\mathrm{O_2}}}{{}^{\kappa}\mathcal{E}^{R_o}_{\mathrm{O_2}}\ {}^{\kappa}\mathcal{E}^{R_c}_{\mathrm{CO_2}} - {}^{\kappa}\mathcal{E}^{R_o}_{\mathrm{CO_2}}\ {}^{\kappa}\mathcal{E}^{R_c}_{\mathrm{O_2}}};$$

(37) $$C^J_{R_o} = \frac{R^J_{\mathrm{CO_2}} - {}^{\kappa}\mathcal{E}^{R_c}_{\mathrm{CO_2}}\ C^J_{R_c}}{{}^{\kappa}\mathcal{E}^{R_o}_{\mathrm{CO_2}}}$$

where $R^J_{\mathrm{CO_2}} = \frac{\partial J}{\partial c_i} c_i / J$ and c_i is manipulated as an "independent" variable.

Kacser and Burns (1973) demonstrated that the sum of all of the control coefficients for a defined system and flux (J) is unity as long as the parameters with respect to which the control coefficients are defined are proportional to the velocities of their respective reactions and processes — i.e.

(38) $$\sum_{P=1}^{n} C^J_P = 1.$$

It is, therefore, possible to gain some indication of the importance of the rest of the components of the system by subtracting the control coefficients defined above from unity. For example, in the latter case where c_i is treated as an independent variable the total "limitation" imposed on the flux by all elements of the photosynthetic "biochemistry" other than Rubisco is given by:

(39) $$C^J_{\mathrm{rest}} = 1 - C^J_{R_c} - C^J_{R_o}.$$

Table 1. Basic data used in calculating the elasticity and control coefficients (eq 21) for soybean leaves at a *PFD* of 1000 μmol quanta $m^{-2}s^{-1}$.

Variable	Value	Variable	Value
c_a	312 μmol mol^{-1}	A	20.60^a μmol $m^{-2}s^{-1}$
c_s	304^a "	T_a	297.48 ^0K
c_i	247^a "	T_l	298.16 "
w_a	26.2 mmol mol^{-1}	$\partial g_{sc}/\partial c_i$	-1370^b mol $m^{-2}s^{-1}$
w_s	26.9^a "	$\partial g_{sw}/\partial c_i$	-2190^b "
w_i	31.1^a "	$\partial g_{sc}/\partial w_s$	13.40^b "
$[O_2]$	210 "	$\partial g_{sw}/\partial w_s$	21.45^b "
g_{sc}	0.38^a mol $m^{-2}s^{-1}$	$\partial g_{sc}/\partial T_l$	-0.049^b mol $m^{-2}s^{-1}$ $^0K^{-1}$
g_{sw}	0.60^a "	$\partial g_{sw}/\partial T_l$	-0.079^b "
g_{bc}	2.55^a "	$\partial V_c/\partial T_l$	0.103^c μmol mg^{-1} min^{-1} $^0K^{-1}$
g_{bw}	3.50^a "	$\partial V_o/\partial T_l$	0.035^c "
E	114.0^a W m^{-2}	$\partial K_c/\partial T_l$	31.25^c μbar $^0K^{-1}$
H	77.4^d "	$\partial K_o/\partial T_l$	9.62^c mbar $^0K^{-1}$
φ	425.7^d "		

[a]calculated using equations summarised by Ball (1987)
[b]determined empirically (see main text)
[c]determined from data of Jordan and Ogren (1984)
[d]see appendix and Campbell (1977)

RESULTS AND DISCUSSION.

The control coefficients for the various components and parameters of the photosynthetic system were calculated with respect to the rate of CO_2 assimilation using eq 21. In order to write this equation, the individual elasticity coefficient elements of the 8 x 8 matrix were determined for soybeans at a *PFD* of 1000 μmol quanta $m^{-2}s^{-1}$, an air temperature of 298.48°K and ambient mole fractions of CO_2, O_2 and water vapour of 312 μmol mol^{-1}, 210 mmol mol^{-1} and 26.2 mmol mol^{-1}, respectively (see Table 1), and are shown in the following:

$$\begin{bmatrix} C^A_{bl_1} \\ C^A_{bl_2} \\ C^A_{bl_3} \\ C^A_{S_1} \\ C^A_{S_2} \\ C^A_{R_c} \\ C^A_{R_o} \\ C^A_{RE} \end{bmatrix} = \begin{bmatrix} 39.0 & 0 & 0 & 0 & 0 & 0 & 0 & 0 \\ -38.0 & 0 & 0 & 5.33 & 0 & 0 & 0 & 0 \\ 0 & 0 & 0 & -5.46 & -.879 & .635 & -.365 & 0 \\ 0 & 36.22 & 0 & 0.998 & -5.61 & 0 & 0 & 0 \\ 0 & 0 & 438.5 & -40.4 & 97.1 & 0.117 & 18.76 & -4 \\ 0 & 0 & -425.7 & 0 & 0 & 0 & 0 & 77.4 \\ 0 & \frac{-8.77}{1000} & \frac{12.92}{1000} & 0 & \frac{-8.77}{1000} & 0 & 0 & 0 \\ 0 & 0 & 0 & 0 & 0 & -.198 & 0.802 & 0 \end{bmatrix}^{-1} \begin{bmatrix} 0.72 \\ 0 \\ 0 \\ 0 \\ 0 \\ 0 \\ 0 \\ -.40 \end{bmatrix}$$

A BASIC program, based upon that published by Phipps (1986) (see Appendix), was used to invert the 8x8 matrix and solve the above equation (see Table 2 for control coefficients). It should be emphasized that, in order to determine the control coefficients with respect to other fluxes of the photosynthetic system, one must alter the response coefficients (i.e. they must be calculated with respect to the flux of interest) and two of the equations involving control coefficients (eqs 19 and 20).

Table 2. Control coefficients for several components and parameters of the photosynthetic system.

Component	Symbol	Value*
Boundary layer		
— CO_2 diffusion	$C^A_{bl_1}$	0.018
— H_2O diffusion	$C^A_{bl_2}$	3.3×10^{-4}
— sensible heat transfer	$C^A_{bl_3}$	0.018
Leaf Characteristic		
Dimension	C^A_d	-0.018
Wind Speed	C^A_u	0.018
Stomata		
— CO_2 diffusion	$C^A_{S_1}$	0.132
— H_2O diffusion	$C^A_{S_2}$	0.026
— number	C^A_n	0.158
Radiant Emittance	C^A_{RE}	0.097
Rubisco		
— carboxylation	$C^A_{R_c}$	1.027
— oxygenation	$C^A_{R_o}$	-0.245

*The control coefficients were calculated with respect to the rate of CO_2 assimilation using the 8 x 8 matrix (eq 21) for soybeans at a *PFD* of 1000 μmol $m^{-2}s^{-1}$, an ambient temperature of 297.48°K and ambient mole fractions of CO_2, O_2 and water vapour of 312 μmol mol^{-1}, 210 mmol mol^{-1} and 26.2 mmol mol^{-1}, respectively.

The control coefficients shown in Table 2 provide a quantitative estimate of the degree to which the rate of photosynthesis was limited by various components under the specific conditions of the experiment. The largest coefficients belong to the Rubisco carboxylase and oxygenase components with values of 1.03 and -0.25, respectively. The magnitude of the carboxylase coefficient is in accord with the model of Farquhar *et al* (1980) which specifies that, at a largely rate-saturating *PFD*, the rate of photosynthesis is limited by Rubisco activity. The negative value associated with the oxygenase reaction indicates that an increase in the activity of this reaction would cause a decrease in the steady state rate of CO_2 fixation.

Woodrow and Mott (1988) showed that, as the *PFD* is lowered, both Rubisco coefficients decrease in magnitude and approach zero at a *PFD* of zero. They also analysed the control coefficients with respect to the rate of RuBP consumption (J_R) and found a sigmoid relationship between $C_{R_c}^{J_R}$ and *PFD* (Woodrow and Mott, unpublished work; see also Woodrow and Berry, 1988). It is noteworthy that $C_{R_c}^{J_R}$ is quite low up to a *PFD* of about 600 μmol quanta $m^{-2}s^{-1}$, despite [RuBP] being above the concentration of active sites at all but the lowest *PFD* values (Woodrow and Mott, unpublished results). This observation is discussed in detail by Woodrow and Berry (1988) who suggest that the notion of Rubisco being substrate-saturated at a concentration of RuBP slightly greater than that of active sites is incorrect. They estimate that Rubisco becomes significantly saturated (i.e. $> 80\%$), and therefore rate limiting, when [RuBP] (which does not include RuBP that is bound and unavailable for catalysis) approaches twice that of active sites and that the major mode of regulation of Rubisco activity is by changes in the enzyme activation state. The regulatory importance of the modulation of activation state is also discussed by Woodrow and Berry (1988).

The control coefficients for CO_2 diffusion through the boundary layer ($C_{bl_1}^A$) and stomata ($C_{S_1}^A$) are 0.018 and 0.132, respectively, and reflect roughly the magnitudes of the conductances of these components to CO_2. Both coefficients may vary considerably depending on the temperature (Woodrow *et al*, 1986), *PFD*, relative humidity (Woodrow, Ball and Berry, unpublished results), leaf characteristic dimension and wind speed (see Appendix and Campbell, 1977). For example, under the conditions used in the current experiments, Woodrow *et al* (1986) demonstrated that $C_{S_1}^A$ reaches a maximum value of about 0.3 at 36°C (Fig 2). If, however, the level of water vapour in the air were to become progressively lower, $C_{S_1}^A$ would approach unity and assimilation would be limited by the supply of CO_2.

The control coefficients for components involved in heat transfer ($C_{bl_2}^A$, $C_{bl_3}^A$, $C_{S_2}^A$ and C_{RE}^A) all have positive value because the temperature of the leaf is higher than both that of the surrounding air and that required for maximum photosynthesis (Woodrow *et al* 1986), and therefore any increase in the efficiency of heat (latent or sensible) loss from the leaf will result in an increase in the rate of photosynthesis. At more extreme temperatures, the role of these components in determining the rate of CO_2 assimilation becomes very important. Woodrow *et al* (1987) showed that $C_{bl_2}^A$ and $C_{bl_3}^A$ have large negative values under these conditions since any increase in boundary layer conductance results in an even greater deviation of leaf temperature from that which is optimal for assimilation. These authors concluded that, at extreme temperatures, the boundary layer is highly important in insulat-

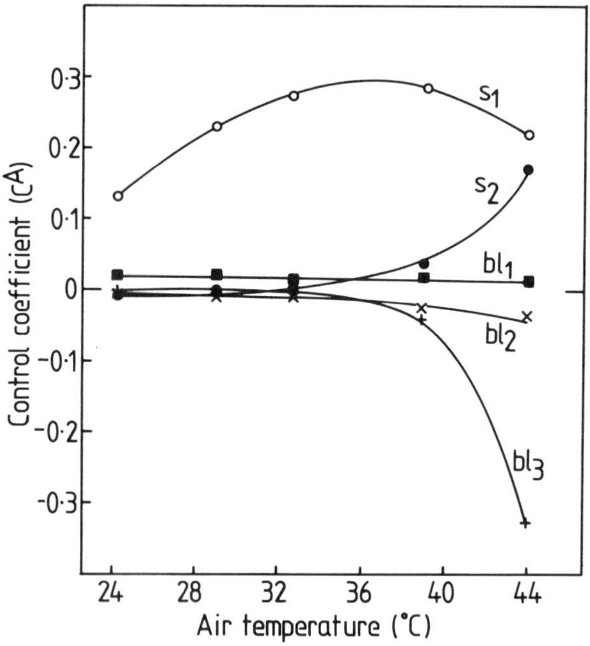

Figure 2. The relationship between the control coefficients for the individual stomatal and boundary layer components shown in Figure 1 and the ambient temperature. The coefficients were calculated using the 8 x 8 matrix (eq 21) by Woodrow *et al* (1986) and express the degree to which the various processes limit the rate of CO_2 assimilation (A) (see Appendix). The ambient CO_2, O_2 and water vapour mole fractions were 330 μmol mol^{-1}, 210 mmol mol^{-1} and 26.6 mmol mol^{-1}, respectively.

ing the leaf and, therefore, in maintaining high rates of photosynthesis.

It is interesting to note that the control coefficients for windspeed and leaf characteristic dimension are composed of components ($C^A_{bl_1}$, $C^A_{bl_2}$ and $C^A_{bl_3}$) that may have both positive and negative values. As noted above, there is a temperature at which $C^A_{bl_2}$ and $C^A_{bl_2}$ change sign. At this point, the increase in assimilation rate due to a rise in g_{bc} is balanced by the decrease due to the change in leaf temperature. At the temperature where $C^A_{bl_1} = -(C^A_{bl_2} + C^A_{bl_3})$, both C^A_d and C^A_u will be zero and the characteristic dimension and windspeed may be considered optimal for steady state photosynthesis under the prevailing conditions.

CONCLUSIONS.

The analysis presented in this paper provides (1) a means to describe quantitatively the degree to which components of the photosynthetic system limit steady state fluxes, and (2) a framework within which to assess the impact of changes to "independent" variables and, thereby, to devise strategies for altering the performance of the photosynthetic system. Considerable attention has been given to the role of the boundary layer, stomatal and Rubisco components in limiting the fluxes because accurate models of the kinetics of processes mediated by these components are available. But as research on the biochemistry of photosynthesis progresses, there should be little difficulty in incorporating other enzymic components into the model (see Woodrow, 1986).

This study and that of Woodrow and Mott (1988) indicate that the total limitation exerted by these additional enzymic components may be quite small compared to that exerted by Rubisco. Nevertheless, if the partial pressure of atmospheric CO_2 increases according to even the most conservative predictions and plants are not capable of adjusting protein pools to levels that are optimal for photosynthesis, the most limiting enzymic components of the photosynthetic system may be in the sucrose synthetic pathway and at the branch points in the PCR cycle (see Woodrow, 1986; Woodrow and Berry, 1988).

ACKNOWLEDGEMENTS.

I would like to acknowledge the not inconsiderable contributions of Joe Berry (Carnegie Institution, Stanford), Tim Ball (UC Davis) and Keith Mott (Utah State, Logan) to both the practical and theoretical aspects of this work. I am also grateful to Gordon Burch (Bureau of Rural Resources, Canberra), Graham Farquhar (ANU, Canberra) and Roger Gifford (CSIRO, Canberra) for their helpful comments.

REFERENCES.

1. Ball, J.T. (1987). Calculations related to gas exchange. In: Stomatal Function, ed. E. Zeiger, G.D. Farquhar, I.R. Cowan, pp. 445–76. Stanford: Stanford Univ. Press.
2. Ball, J.T., Woodrow, I.E., Berry, J.A. (1986). A model predicting stomatal conductance and its contribution to the control of photosynthesis under different environmental conditions. In: Progress in Photosynthesis Research, ed. J. Biggins, 4: 221–24. Dordrecht: Martinus Nijhoff.
3. Campbell, G.S. (1977) An Introduction to Environmental Biophysics New York, Springer-Verlag.
4. Crabtree, B., Newsholme, E.A. (1985). A quantitative approach to metabolic control. Curr. Top. Cell Regul. 25: 21–76.
5. Farquhar, G.D., von Caemmerer, S., Berry, J.A. (1980). A biochemical model of photosynthetic CO_2 fixation in leaves of C_3 species. Planta 149: 78–90.
6. Field, C., Berry, J.A., Mooney, H.A., (1982). A portable system for measuring carbon dioxide and water vapour exchange of leaves. Plant Cell and Environment, 5: 179–86.
7. Higgins, J. (1965). Dynamics and control in cellular reactions. In: Control of Energy Metabolism, ed. B. Chance, R.K. Estabrook, J.R. Williamson, pp. 13–46. New York:

Academic Press.
8. Jarman, P.D. (1974). The diffusion of carbon dioxide and water vapour through stomata. J. Exp. Bot. 25: 928–36.
9. Jordan, D.B., Ogren, W.L. (1984). A sensitive assay procedure for simultaneous determination of ribulose 1,5-bisphosphate carboxylase and oxygenase activities. Plant Physiol. 54: 237–45.
10. Kacser, H., Burns, J.A. (1973). The control of flux. Symp. Soc. Exp. Biol. 27: 65–104.
11. Phipps, J.E. Jr. (1986). The inversion of large matrices. BYTE pp. 181–188.
12. Savageau, M.A. (1976). Biochemical systems analysis. Menlo Park, Calif: Addison-Wesley.
13. Woodrow, I.E. (1986). Control of the rate of photosynthetic carbon dioxide fixation. Biochim. Biophys. Acta 851: 181–92.
14. Woodrow, I.E., Ball, J.T., Berry, J.A. (1986). A general expression for the control of the rate of photosynthetic CO_2 fixation by stomata, the boundary layer and radiation exchange. In: Progress in Photosynthesis Research, ed. J. Biggins, 4: 225-28. Dordrecht: Martinus Nijhoff.
15. Woodrow, I.E., Berry, J.A. (1988). Enzymatic regulation of photosynthetic CO_2 fixation in C_3 plants. Ann. Rev. Plant Physiol. Plant Mol. Biol. 39: 533–94.
16. Woodrow, I.E., Mott, K.A. (1988). A quantitative assessment of the degree to which RuBP carboxylase/oxygenase determines the rate of steady state photosynthesis during sun-shade acclimation in *Helianthus annuus*. Aust. J. Plant Physiol. 15:253–262.

Appendix

Definition of Control Coefficients. The control coefficients in this study are defined with respect to independent variables that are proportional to the velocities of the individual processes. The coefficients are as follows:
(a)Boundary Layer

(a1) $$C^J_{bl_1} = \frac{\partial J}{\partial g_{bc}} \frac{g_{bc}}{J},$$

(a2) $$C^J_{bl_2} = \frac{\partial J}{\partial g_{bw}} \frac{g_{bw}}{J},$$

(a3) $$C^J_{bl_3} = \frac{\partial J}{\partial g_{bh}} \frac{g_{bh}}{J},$$

where g_{bc}, g_{bw} and g_{bh} are the exchange conductances for CO_2, water vapour and heat, respectively.
(b) Stomata

(a4) $$C^J_{S_1} = \frac{\partial J}{\partial D_c} \frac{D_c}{J},$$

(a5) $$C^J_{S_2} = \frac{\partial J}{\partial D_w} \frac{D_w}{J},$$

where D_w and D_c are the molecular diffusivities for water vapour and CO_2, respectively. As, in reality, D_c and D_w are not independent, we can derive an overall stomatal control coefficient with respect to stomatal number per unit leaf area (n), for example. We can write that

$$\frac{\partial J}{\partial n} = \frac{\partial J}{\partial D_c}\frac{\partial D_c}{\partial n} + \frac{\partial J}{\partial D_w}\frac{\partial D_w}{\partial n}$$

which is equivalent to

$$\frac{\partial J}{\partial n}\frac{n}{J} = \frac{\partial J}{\partial D_c}\frac{D_c}{J}\frac{\partial D_c}{\partial n}\frac{n}{D_c} + \frac{\partial J}{\partial D_w}\frac{D_w}{J}\frac{\partial D_w}{\partial n}\frac{n}{D_w}$$

and

(a6) $$C_n^J = C_{S_1}^J + C_{S_2}^J$$

as $\frac{\partial D_c}{\partial n} n/D_c$ and $\frac{\partial D_w}{\partial n} n/D_w = 1$.

A similar "overall" control coefficient can be derived for the leaf boundary layer with respect to parameters such as the characteristic dimension of the leaf (d) or the wind velocity (u). We can write that

$$\frac{\partial J}{\partial d} = \frac{\partial J}{\partial g_{bc}}\frac{\partial g_{bc}}{\partial d} + \frac{\partial J}{\partial g_{bw}}\frac{\partial g_{bw}}{\partial d} + \frac{\partial J}{\partial g_{bh}}\frac{\partial g_{bh}}{\partial d}$$

which is equivalent to

(a7) $$C_d^J = C_{bl_1}^J \frac{\partial g_{bc}}{\partial d}\frac{d}{g_{bc}} + C_{bl_2}^J \frac{\partial g_{bw}}{\partial d}\frac{d}{g_{bw}} + C_{bl_3}^J \frac{\partial g_{bh}}{\partial d}\frac{d}{g_{bh}}.$$

Similarly, we can write that

(a8) $$C_u^J = C_{bl_1}^J \frac{\partial g_{bc}}{\partial u}\frac{u}{g_{bc}} + C_{bl_2}^J \frac{\partial g_{bw}}{\partial u}\frac{u}{g_{bw}} + C_{bl_3}^J \frac{\partial g_{bh}}{\partial u}\frac{u}{g_{bh}}.$$

In the case where forced convection dominates the heat and gas exchange processes, the conductances for a flat plate in laminar flow are approximated by

(a9) $$g_{bh} = D_H\, A_l\, Re^{\frac{1}{2}}\, Pr^{\frac{1}{3}} / V\, d$$

and

(a10) $$g_{bj} = D_j\, A_l\, Re^{\frac{1}{2}}\, Sc^{\frac{1}{3}} / V\, d$$

where Re (the Reynolds number)= ud/v, Pr (the Prandtl number)= v/D_H, Sc (the Schmidt number)= v/Dj, u is a velocity characteristic of the wind outside the leaf boundary layer, v the kinematic viscosity of air, d the characteristic dimension of the leaf, D_H the sensible heat diffusivity, D_j the molecular diffusivity for species j, A_l a constant which depends on leaf geometry and other factors, and V the molar volume.

Partial differentiation of these equations (a9 and a10) yields expressions for the partial differentials with respect to d and u in eqs. (a7) and (a8). The latter equations can now be written as:

(a11) $$C_d^J = -\frac{1}{2}(C_{bl_1}^J + C_{bl_2}^J + C_{bl_3}^J)$$

and

(a12) $$C_u^J = \frac{1}{2}(C_{bl_1}^J + C_{bl_2}^J + C_{bl_3}^J).$$

The elements of the matrix (i.e. $C_{bl_1}^J, C_{bl_2}^J$ and $C_{bl_3}^J$) can, therefore, be used to calculate the sensitivity of the various photosynthetic fluxes to changes in wind velocity and the characteristic dimension of the leaf.

(c) Radiant Emittance

(a13) $$C_{RE}^J = \frac{\partial J}{\partial \epsilon}\frac{\epsilon}{J}$$

where ϵ is the emissivity of the leaf surface.

(d) Rubisco

(a14) $$C_{R_c}^J = \frac{\partial J}{\partial k_c}\frac{k_c}{J},$$

(a15) $$C_{R_o}^J = \frac{\partial J}{\partial k_o}\frac{k_o}{J},$$

where k_c and k_o are the catalytic rate constants for carboxylation and oxygenation, respectively.

Rate Equations.

The model of the photosynthetic system (Fig 1) incorporates both general rate equations, which have been determined in studies of individual reactions and processes, and empirically determined rate data. The equations used in the analysis are:

(a) Boundary Layer

(a16) $$v_{bl_1} = g_{bc}(c_a - c_s),$$

(a17) $$v_{bl_2} = g_{bw}(w_s - w_a),$$

(a18) $$v_{bl_3} = C_p\, g_{bh}(\mathrm{T}_l - \mathrm{T}_a),$$

where g_{bc}, g_{bw} and g_{bh} are the exchange conductances for CO_2, water vapour and sensible heat, respectively, C_p the molar specific heat of air, and T_l, w_s, c_s and T_a, w_a, c_a are the temperature, water vapour mole fraction and CO_2 mole fraction at the leaf surface and in the bulk air, respectively.

(b) Stomata

(a19) $$v_{S_1} = g_{sc}(c_s - c_i),$$

(a20) $$v_{S_2} = g_{sw}(w_i - w_s),$$

where g_{sw} and g_{sc} are the exchange conductances for water vapour and CO_2, respectively.

(c) Radiant Emittance

(a21) $$\varphi = \epsilon\sigma \mathrm{K}^4$$

where ϵ is the emissivity of the surface, K is the Kelvin temperature, σ is the Stephan-Bolzmann constant, and φ is the emitted flux density.

(d) Rubisco

The rate equations for oxygenation and carboxylation are given in the main text.

Derivation of Equations.

The mathematical relationship between the combined response, special elasticity and control coefficients was derived by Kacser and Burns (1973). The basic principal underlying the derivation is that, in this study, certain increases in c_a and g_{bc} produce identical changes in v_{bl_1}. For small changes in c_a (δc_a),

(a22) $$ {}^{\kappa}\mathcal{E}^{bl_1}_{c_a} \frac{\delta c_a}{c_a} = \mathcal{E}^{bl_1}_{g_{bc}} \frac{\delta g_{bc}}{g_{bc}} $$

where

$$ {}^{\kappa}\mathcal{E}^{bl_1}_{c_a} = \frac{\partial v_{bl_1}}{\partial c_a} \frac{c_a}{v_{bl_1}} $$

and

$$ \mathcal{E}^{bl_1}_{g_{bc}} = \frac{\partial v_{bl_1}}{\partial g_{bc}} \frac{g_{bc}}{v_{bl_1}}. $$

The latter term, $\mathcal{E}^{bl_1}_{g_{bc}}$, equals unity because $v_{bl_1} \propto g_{bc}$ (see eq. a16).

In order to derive the relationship between $R^J_{c_a}$ and $C^J_{bl_1}$, consider the situation in which a small change in c_a is balanced by a change in g_{bc} such that there is no change in v_{bl_1} or in the flux, J. Under these conditions and in the limit as $\delta c_a \to 0$

$$ \delta J = 0 = \frac{\partial J}{\partial c_a} \delta c_a + \frac{\partial J}{\partial g_{bc}} \delta g_{bc} $$

which is equivalent to

(a23) $$ R^J_{c_a} \frac{\delta c_a}{c_a} + C^J_{bl_1} \frac{\delta g_{bc}}{g_{bc}} = 0. $$

Since

$$ \frac{\delta c_a}{c_a} = -\frac{\delta g_{bc}}{{}^{\kappa}\mathcal{E}^{bl_1}_{c_a}\, g_{bc}} $$

when c_a and g_{bc} changes balance each other, we can use this equation and eq. (a23) to write that

(a24) $$ R^J_{c_a} = C^J_{bl_1}\, {}^{\kappa}\mathcal{E}^{bl_1}_{c_a}. $$

The derivation of the "connectivity" equations (15–18) can be illustrated by considering a small change in c_s and the alterations in the P variables required to keep v_{bl_1} and v_{S_1}, and therefore the flux (J), constant. In the limit as $\delta c_s \to 0$

(a25) $$\mathcal{E}^{bl_1}_{g_{bc}} \frac{\delta g_{bc}}{g_{bc}} + \mathcal{E}^{bl_1}_{c_s} \frac{\delta c_s}{c_s} = 0$$

and

(a26) $$\mathcal{E}^{S_1}_{n} \frac{\delta n}{n} + \mathcal{E}^{S_1}_{c_s} \frac{\delta c_s}{c_s} = 0,$$

where n is the number of stomata per unit leaf area and

$$\mathcal{E}^{S_1}_{n} = \frac{\partial v_{S_1}}{\partial n} \frac{n}{v_{S_1}}.$$

Equations (a25) and (a26) give expressions for the fractional changes in g_{bc} and n which can be multiplied by the respective control coefficients to obtain the fractional changes in the flux (J). In the limit as $\delta c_s \to 0$

$$\delta J = 0 = \frac{\partial J}{\partial g_{bc}} \delta g_{bc} + \frac{\partial J}{\partial g_{sc}} \delta g_{sc}$$

which is equivalent to

(a27) $$C^J_{bl_1} \mathcal{E}^{bl_1}_{c_s} + C^J_{S_1} \mathcal{E}^{S_1}_{c_s} = 0$$

since both $\mathcal{E}^{bl_1}_{g_{bc}}$ and $\mathcal{E}^{S_1}_{n}$ equal unity ($g_{bc} \propto v_{bl_1}$ and $n \propto v_{S_1}$). The other connectivity equations can be derived in a similar manner.

The last two equations involving the control coefficients of components of the system (eqs. 19 and 20) can be derived as follows. Consider small changes in g_{bh} (boundary layer conductance to heat) and ϵ (the leaf emissivity to radiation) such that $\delta H = -\delta\varphi$ and there is no change in either T_l or the flux, J. As $\delta\varphi/\varphi = \delta\epsilon/\epsilon$ and $\delta H/H = \delta g_{bh}/g_{bh}$, we can write that

(a28) $$\frac{H\,\delta g_{bh}}{g_{bh}} = -\frac{\varphi\,\delta\epsilon}{\epsilon}.$$

And since there is no change in J, in the limit as $\delta\epsilon$ and $\delta g_{bh} \to 0$

$$\frac{\delta J}{J} = 0 = C^J_{RE} \frac{\delta\epsilon}{\epsilon} + C^J_{bl_3} \frac{\delta g_{bh}}{g_{bh}}$$

which, when solved simultaneously with eq. (a28), yields

(a29) $$C^J_{bl_3}/H - C^J_{RE}/\varphi = 0.$$

Consider also small changes in g_{bl_2}, g_{S_2} and g_{bh} such that $\delta v_{bl_2} = \delta v_{S_2} = -\delta H$. Under these conditions

(a30) $$v_{bl_2}\frac{\delta g_{bw}}{g_{bw}} = v_{S_2}\frac{\delta g_{sw}}{g_{sw}} = -H\frac{\delta g_{bh}}{g_{bh}}$$

And since there is no change in J, we can write that in the limit as $\delta g_{bw}, \delta g_{sw}$ and $\delta g_{bh} \to 0$

$$\frac{\delta J}{J} = 0 = C^J_{bl_2}\frac{\delta g_{bw}}{g_{bw}} + C^J_{S_2}\frac{\delta g_{sw}}{g_{sw}} + C^J_{bl_3}\frac{\delta g_{bh}}{g_{b}h}$$

which, when solved simultaneously with eq. (a30), yields

(a31) $$C^J_{bl_2}/E + C^J_{S_2}/E - C^J_{bl_3}/H = 0.$$

The equations for calculating the control coefficients for the two activities of Rubisco (eqs. 34 and 35) can be derived as follows. Consider changes in the catalytic rate constants for oxygenation (k_o) and carboxylation (k_c) that are equivalent — in terms of their impact on the catalytic velocities of oxygenation (v_o) and carboxylation (v_c) — to a given change in the level of CO_2 or O_2. Mathematically, these equalities can be expressed for small changes in $[O_2](\delta[O_2])$ as

(a32) $$^{\kappa}\varepsilon^{R_c}_{O_2}\frac{\delta[O_2]}{[O_2]} = \varepsilon^{R_c}_{k_c}\frac{\delta k_c}{k_c}$$

and

(a33) $$^{\kappa}\varepsilon^{R_o}_{O_2}\frac{\delta[O_2]}{[O_2]} = \varepsilon^{R_o}_{k_o}\frac{\delta k_o}{k_o},$$

and for small changes in $[CO_2]$ (δ $[CO_2]$) as

(a34) $$^{\kappa}\varepsilon^{R_c}_{CO_2}\frac{\delta[CO_2]}{[CO_2]} = \varepsilon^{R_c}_{k_c}\frac{\delta k_c}{k_c}$$

and

(a35) $$^{\kappa}\mathcal{E}^{R_o}_{CO_2}\frac{\delta[CO_2]}{[CO_2]} = \mathcal{E}^{R_o}_{k_o}\frac{\delta k_o}{k_o},$$

where $\mathcal{E}^{R_c}_{k_c} = \frac{\partial v_c}{\partial k_c}k_c/v_c$ and $\mathcal{E}^{R_o}_{k_o} = \frac{\partial v_o}{\partial k_o}k_o/v_o$ both of which are very close to unity.

In order to derive the relationship between response coefficients and control coefficients, consider the situation in which a small change in $[O_2]$ is balanced by changes in k_c and k_o such that there is no change in v_c and v_o or the flux (J). Under these conditions and in the limit as $\delta[CO_2], \delta k_c$ and $\delta k_o \rightarrow 0$

(a36) $$\delta J = 0 = \frac{\partial J}{\partial[O_2]}\delta[O_2] + \frac{\partial J}{\partial k_o}\delta k_o + \frac{\partial J}{\partial k_c}\delta k_c.$$

Solution of eqs. (a32), (a33) and (a36) yields

(a37) $$R^J_{O_2} = C^J_{R_c}\ {}^{\kappa}\mathcal{E}^{R_c}_{O_2} + C^J_{R_o}\ {}^{\kappa}\mathcal{E}^{R_o}_{O_2}.$$

A similar logic can be applied to a small change in $[CO_2]$ and balancing changes in k_o and k_c. In the limit as $\delta[CO_2], \delta k_c$ and $\delta k_o \rightarrow 0$

(a38) $$\delta J = 0 = \frac{\partial J}{\partial[CO_2]}\delta[CO_2] + \frac{\partial J}{\partial k_c}\delta k_c + \frac{\partial J}{\partial k_o}\delta k_o.$$

Solution of eqs. (a34), (a35) and (a38) yields

(a39) $$R^J_{CO_2} = C^J_{R_c}\ {}^{\kappa}\mathcal{E}^{R_c}_{CO_2} + C^J_{R_o}\ {}^{\kappa}\mathcal{E}^{R_o}_{CO_2}.$$

(4) Inversion of the Matrices.

The 8 x 8 and 4 x 4 matrices given in the main text can be inverted using the algorithums developed by Pan and Reif (1985). The BASIC program outlined below is based upon that published by Phipps (1986). In the output, the first and last elements of each row of the inverted matrix are multiplied by the first and last elements of the 1 x 8 or 1 x 4 matrix — i.e. by $R^J_{c_a}$ and $R^J_{O_2}$. For a detailed discussion of this program see Phipps (1986).

```
A Program for the Modified Pan-Reif Matrix-inversion Method
10 REM Real-entry matrix inversion (Pan-Reif method)
20 REM Program by Thomas E Phipps, Jr
30 REM Quartic modification of Newton's iteration
40 INPUT "RESPONSE COEFFICIENTS - Rca, RO2"; Rca, RO2
50 INPUT "MATRIX DIMENSION = "; N
60 REM D1 data in line 630
70 READ D1:  OPTION BASE 1
80 REM Can optionally insert DEFDBL A,B,E,X here
```

```
90 REM A is the matrix to invert, B is the approximate
     inverse of A, E is the error matrix for the inverse, and
     X is a temporary storage matrix as explained in
     the article.
100 DIM A(N,N),B(N,N),X(N,N),E(N,N)
110 FOR I=1 TO N: FOR J=1 TO N: PRINT I;J: INPUT A(I,J): NEXT:
     NEXT
120 INPUT "LARGEST ELEMENT IN MATRIX"; L
130 FOR I=1 TO N: FOR J=1 TO N: A(I,J)= A(I,J)/L: NEXT:NEXT
140 CLS:TIM$="00"
150 REM Eval.  t by Pan-Reif method
160 FOR 1=1 TO N:R0=0:S0=0:FOR J=1 TO N
170 R0=R0+ABS(A(I,J)):S0=S0+ABS(A(J,I)):NEXT J
180 X(1,I)=R0:E(1,I)=S0:NEXT I:T1=0:T2=0
190 FOR I=1 TO N:IF X(1,I)>T1 THEN T1=X(1,I)
200 IF E(1,I)>T2 THEN T2=E(1,I)
210 NEXT I:T=1/(T1*T2)
220 REM Eval.  initial B-matrix
230 FOR I=1 TO N:FOR J=1 TO N:B(I,J)=T*A(J,I):NEXT:NEXT :H=0
240 PRINT "ITER."TAB(10);"E(1,1)"TAB(28);"E(1,N)" TAB(46);
     "B(1,1)"TAB(64);"B(1,N)"
250 REM Eval.  error matrix E
260 FOR I=1 TO N:FOR K=1 TO N:Z=0:FOR J-1 TO N
270 Z=Z+B(I,J)*A(J,K):NEXT J:E(I,K)=-Z:NEXT K:NEXT I
280 FOR I=1 TO N:E(I,I)=1+E(I,I):NEXT I
290 BEEP:PRINT H;TAB(6);E(1,1);TAB(24);E(1,N);TAB(42);B(1,1);
     TAB(60);B(1,N)
300 IF H>50 THEN PRINT "STUCK!":BEEP:BEEP:GOTO 590
310 REM Test for escape from loop
320 FOR I=1 TO N:FOR J=1 TO N:IF ABS(E(I,J))> D1 THEN 340
330 NEXT :NEXT :GOTO 410
340 H=H+1
350 FOR I=1 TO N:
     X(1,I)=1+(1+(1+(1+E(I,I))*E(I,I))*E(I,I))*E(I,I):NEXT
360 FOR I=1 TO N:FOR J=1 TO N:E(I,J)=X(1,I)*E(I,J):NEXT
     :E(I,I)=1+E(I,I):NEXT
370 FOR I=1 TO N:FOR K=1 TO N:W=0:FOR J=1 TO N
380 W=W+E(I,J)*B(J,K):NEXT J:X(I,K)=W:NEXT K:NEXT I
390 FOR I=1 TO N:FOR J=1 TO N:B(I,J)=X(I,J):NEXT:NEXT
400 GOTO 250
410 REM output follows
420 T$=TIME$:BEEP:BEEP:BEEP:PRINT "Duration = ";T$
430 PRINT "No.  iter.=";H,"Matrix dim.="N,"Max.  error=";D1/L
440 INPUT "Another iteration (y,n)";C$:
    IF C$="Y" OR C$="y" THEN TIME$="00":GOTO 340
450 PRINT "The maximum error was calculated by multiplying
    A↑-1*A."
460 PRINT "As a check on the true error in the approximated
    inverse matrix"
470 PRINT"the maximum error in A*A↑-1 will now be determined"
480 ERASE X
490 DIM X(N,N)
500 FOR I=1 TO N:FOR J=1 TO N:FOR K=1 TO N
510 X(I,K)=X(I,K)+A(I,J)*B(J,K)
520 NEXT :NEXT :NEXT
530 FOR I=1 TO N:FOR J=1 TO N
```

```
540 TEST.VAL=ABS(X(I,J)):IF I=J THEN TEST.VAL=ABS(1-TEST.VAL)
550 IF TEST.VAL>MAX.ERROR THEN MAX.ERROR=TEST.VAL
560 NEXT :NEXT
570 PRINT "The max.  error found this way is "; MAX.ERROR
580 PRINT ""
590 PRINT "CONTROL COEFFICIENTS"
600 FOR I=1 TO N: PRINT (B(I,1)/L)*Rca + (B(I,N)/L)*RO2:NEXT
610 END
620 REM Error criterion D1 =
630 DATA 3E-6
```

In solving the 8 x 8 matrix shown in the "results" section, the following input parameters were used: `Rca=0.72; RO2=-0.4`; matrix dimension = 8; and then the elements of the matrix were entered sequentially along each row (I). The inversion process may take several minutes.

Index